# *Introduction to*
# *Wildland Fire*

*Introduction to*

# *Wildland Fire*

*Fire Management in the United States*

STEPHEN J. PYNE

History Department
University of Iowa
and
Fire Management Specialist, Seasonal
National Park Service



A Wiley-Interscience Publication

John Wiley & Sons New York Chichester Brisbane Toronto Singapore

***Library of Congress Cataloging in Publication Data:***

Pyne, Stephen J., 1949–
Introduction to wildland fire.

Includes index.
1. Wildfires. 2. Wildfires—Prevention and control. 3. Wildfires—United States. 4. Wildfires—United States—Prevention and control. I. Title.

SD421.P94 1984 363.3'7 83-17100
ISBN 0-471-09658-X

Printed in the United States of America

10 9 8 7 6 5 4

*For Sonja and Lydia*

# Preface

Within our solar system the Earth, and perhaps the Earth alone, is a fire planet. Only on Earth are combined the essential components of combustion. With lightning, it has a source of ignition; with atmospheric oxygen, an oxidizing agent; and with organic matter, a fuel. Jupiter has lightning, Mars has traces of free oxygen, and the moons of the outer planets have atmospheres rich in flammable hydrocarbons. But only the Earth contains all the essential constituents, the processes needed to mix them, and a suitable environment for their interaction. To complement its ignition source, the Earth also has a suppressant: water. The Earth can start fire, sustain fire, and suppress fire. The things that make Earth unique among the planets have made it hospitable to fire. And fire, in return, has had much to do with shaping the history of the planet.

The process of acquiring fire began with lightning. Not only did lightning make ignition possible, but it probably catalyzed the evolution of life. Life provided the other two essentials for combustion: atmospheric oxygen and fuel. As terrestrial life expanded, so did fire. With the appearance of the genus *Homo,* the geography and natural history of fire changed dramatically. Humans assumed control over the start, spread, and suppression of fire. Humans could manipulate fire in new ways and shape the fire environment to new effects. Fire was removed from areas where it had previously ranged, and it was introduced to landscapes that had not formerly known it. Thanks to its symbiosis with humans, fire has even spread to other planets. As it became enfolded into human society, fire changed its character, much as domestication has done for selected animals and plants. At the same time fire profoundly affected its human agents, much as it had affected terrestrial biota before. Even today combustion remains a fundamental property of planet Earth, fire persists as an evolutionary presence and an ecological process of great import, and control over fire continues to be one of the foundations of human culture.

This volume has two purposes and two audiences. It should serve as an introductory text for students of wildland fire and as a reference work for fire managers. It intends to show how fire is conceptualized and how, in the United States, it is managed. It is more of a how-to-see-it book than a how-to-do-it book. It is not an encyclopedia of data and techniques, nor a surrogate for the vocational training that most fire agencies mandate for their employees. It intends to complement the practical courses endorsed by the National Wildfire Coordinating Group, not to compete with them. It hopes, in brief, to identify, clarify, and consolidate the concepts of fire research and fire management and to describe the institutional environment that sustains these ideas.

Fire is a synthetic subject and fire management a synthetic art. Fire research must combine knowledge from many fields into a unique pattern. It intersects chemistry, in the form of combustion; physics, with fire behavior mechanisms; biology, with the interdependence of fire and ecosystems; meteorology, with the interaction of fire and atmosphere; culture, in the relationship between fire and human history. The list can be expanded. Even as fire, wildland fire shows unique characteristics compared to other forms of combustion. Because it integrates so many probabilistic processes and because not all of its important relationships are known, wildland fire remains a largely local and particularized phenomenon. There are, of course, general principles of combustion, fire behavior, fire weather, fire ecology, and fire practices by humans. But wildland fire—not physics, chemistry, or biology—must remain the object of attention. Fire uniquely integrates the many subjects it touches on. The point of intersection, moreover, tends to be local. Consequently, this volume tries to reconcile the general and the particular by first explaining concepts and then illustrating those principles with particular examples.

For fire management as a practice there is a second complication. Wildland fire was used before it was understood; it was acquired from nature, not invented in a laboratory; and the necessity to manage fire has dominated much of the thinking about it. Scientific research has followed practice, and practice and research are not fully integrated. There are, in a sense, two cultures: a high culture of researchers and a vernacular culture of practitioners. Each culture has its own concepts and language. Each describes the range of fire phenomena fully. Both cultures flourish, but the trend is to resolve more and more vernacular knowledge into rigorous scientific explanation and quantitative descriptions. In one sense, this is merely a process of translation. A hot fire is described as having a fireline intensity of 800 Btu/ft/sec(661.5 kcal/m/sec), for example. But in another sense, the process is one of transformation. By coupling fire to the high culture of scientific discipline, research acquires tools of greater analytical power and concepts of larger synthetic scope.

Both cultures, of course, have limits. But high culture has the power to progress. For all its flaws, the trend to express fire knowledge in terms of basic scientific concepts can only accelerate. The advent of modern electronic computers promises to reinforce this tendency by integrating larger amounts of data and applying that information to ever more specific sites and events. In this way research derives from fundamental concepts and processes—fire management from fire effects, fire effects from fire behavior, and fire behavior from combustion. The process is simultaneously reductionistic and synthetic. Progressively, too, such concepts interpenetrate and assimilate practical use. Field experience remains, and for the identifiable future must remain, the basis for practical management. Its working concepts, however, are being rapidly reformulated into the language of science. This volume follows that trend.

That the volume restricts its scope to fire management in the United States perhaps requires a word of explanation. This choice is not so severe a limitation as might seem at first blush. Wildland fire remains a particularized phenomenon, the more so when its cultural context is considered. By concentrating on the fuel types, fire climates, institutions, and fire history of the United States, it was possible to give the subject a significant degree of unity, both of style and content. Moreover, along with Canada, the United States was the first industrial nation with extensive forests to adopt fire control on a national scale. From the United States came the concept of systematic fire protection, the first programs of wholesale fire research, and an important exemplar that could be, and often was, disseminated to other countries. Excepting the tropics, the fire regimes of the United States encompass most of those typical of the world. The fire history of the United States includes most practices of fire use and control found throughout the world—from hunting and gathering societies, to agricultural economies, to the industrial order. Because of the universality of fundamental principles of combustion and fire behavior, because of the many examples of fire regimes apparent in American geography, and because of the panorama of fire practices evident in American history, this volume acquires more breadth than its title might indicate. In the fire history of the United States there is an abundant record of a country that successfully accommodated technology transfer, and in the present administration of fire agencies there is an important example of high technology applied to fire management. To see the problem of fire management worked out in all its complexity was worth a tradeoff—in this case, its restriction to a single country. It is hoped, nonetheless, that the volume will serve as an introduction to anyone, from any land, interested in the question of wildland fire.

As a state-of-the-art digest, this volume is intended to complement my earlier work, *Fire in America: A Cultural History of Wildland and Rural Fire. Fire in America* told its story from the point of view of fire and it brought that story up to the 1980s. But its vision was historical: it sought to harmonize fire history with cultural history. *Introduction to Wildland Fire* has a different perspective. Though it includes a synopsis of American fire history, neither its organization nor its vision is historical. Its structure builds instead upon a hierarchy of concepts from the physical chemistry of combustion to the art and science of fire management. Where it appeals to history, it does so either to examine the conceptual foundations of certain fire research and management activities, or to account for those aspects of fire management that can only be understood as a response to unique, historical needs. Enough is given to make the subject historically intelligible and to comment on the concept of fire as history, but this book is not history.

Science describes the natural universe: history—human history—builds moral universes. Because humans use fire, fire participates in that moral universe and cannot, ultimately, be divorced from it. In the final analysis the greatest questions about fire management may reside there. *Fire in America* was an attempt to describe how fire came to inhabit those moral universes and how, in America, it behaved. Some of fire's natural properties were delineated for understanding, but they were not the principal object of inquiry. *Introduction to Wildland Fire* reverses this proportion. It seeks to examine fire, and the human use of fire, within their natural worlds so far as this is possible. The two books are distinct, yet, I hope, complementary.

A few final words. To promote the readability of the text, citations have been kept to a minimum. Where a special reference is called for, one is given. But the important summary works are identified in the bibliographic essays that follow each chapter. These essays list the significant literature for the subject covered in the chapter, but they also single out particularly meritorious works for further reading. In general, these are the same sources I used in writing the chapter. Finally, I have hewn to a qualitative description of fire processes even where a quantitative description does exist. In some respects this decision complicates the earlier portions of the book. A mathematical formula can synopsize the interrelationships of many factors into a concise expression. To describe those same relations with words is laborious and often confusing. Yet it was felt that once introduced, mathematical descriptions would have to be followed throughout the work; that, because of their analytical and synthetic power, the elaboration of these expressions might absorb more and more of the book; that, although rendering some portions of the book more meaningful to some readers, it would make more of the book inaccessible to that general population of students and fire managers for whom the book was intended. The price of access is, in this case, a degree of wordiness and a certain clumsiness of logical development. By way of compromise, particularly in the chapter on combustion, I have retained many concepts that have acquired mathematical definition in the hopes that these concepts might furnish, for those who wish to pursue the subject in more depth, a point of departure. For better or worse, many of these expressions have entered the common discourse of fire management, and I believe that some introduction to them is preferable to none at all.

STEPHEN J. PYNE

*Iowa City, Iowa*
*and Rocky Mountain National Park*
*January 1984*

# Acknowledgments

Many people assisted me in the writing of this book. Special thanks goes to Craig C. Chandler, who gave the entire manuscript a critical review. When it became apparent that we were both writing texts on fire management for the same publisher, we agreed to exchange manuscripts to ensure that our contributions would be complementary. Obviously, Craig's views are not identical with mine, nor his preference for certain themes and particular kinds of data. That is why he, with the collaboration of other like-minded colleagues, wrote the book he did, and I, the book I have. But our correspondence has enlightened me, as it has improved the book. I am grateful. For the final views expressed, I am, of course, responsible.

I also wish to thank Mary Strottman, Elaine Melcher, and Mary Abbott Cannon who helped with portions of manuscript typing; the History Department of the University of Iowa, who made a computer terminal available for use as a word processor; and the Graduate College, who granted funds to use it. Gary Olson, provided by the department as a research assistant, helped with the coversion of English to SI units, with typing, and with other matters during the final stages of manuscript preparation—for all of which services I am grateful. John Kolp assisted on abstruse points of computer systems. James Pyne reviewed critically some early chapters. Thomas Mills read the section on fire economics, with helpful observations. And Lenny Dems more than once satisfied frantic requests for materials with an alacrity that should make him the envy of a Class I Service Chief.

And then there is Sonja, my wife, and Lydia, my daughter. Their love gave me reason to begin the book, and their patience, the opportunity to complete it. To both of them I offer my apologies for being so often away, and my gratitude for their many sacrifices.

S. J. P.

# *Contents*

Chapter Two

## Fire Behavior



Chapter Three

## Wildland Fuels

*Chapter Four*

## *Fire Weather*



*Part Two*

# *The Fire Regime*

*Chapter Five*

## *Fire and Life*

*Chapter Six*

## *Fire and Culture*



*Chapter Seven*

## *The Administration of Fire Regimes*

*Part Three*

# *Fire Management*

*Chapter Eight*

## *Programs for Fire Management*



*Chapter Nine*

## *Fire Suppression*

# *Abbreviations*

| | |
|---|---|
| AFM | Aviation and Fire Management |
| AFMO | Assistant Fire Management Officer |
| AID | Agency for International Development |
| AQCR | Air Quality Control Region |
| AQMA | Air Quality Maintenance Area |
| AS | Ammonium sulfate |
| BI | Burning index |
| BIA | Bureau of Indian Affairs |
| BIFC | Boise Interagency Fire Center |
| BLM | Bureau of Land Management |
| CCC | Civilian Conservation Corps |
| CDF | California Department of Forestry |
| CFP | Cooperative Fire Program |
| DAID | Delayed action ignition device |
| DAP | Diammonium phosphate |
| DOD | Department of Defense |
| DTA | Differential thermal analysis |
| ERC | Energy release component |
| FAO | Food and Agriculture Organization |
| FEES | Fire Economics Evaluation System |
| FEMA | Federal Emergency Management Agency |
| FFASR | Forest Fire and Atmospheric Sciences Research |
| FFF | Fire Fighting Fund |
| FIMS | Firescope Information Management System |
| FLI | Fire load index |
| FMA | Fire management area |
| FMC | Fuel moisture content |
| FMF | Fire management fund |
| FMO | Fire Management Officer |
| FMU | Fire management unit |
| FMZ | Fire management zone |
| FWIS | Forestry Weather Information System |
| FWS | Fish and Wildlife Service |
| GHQ | General headquarters |
| IAMS | Initial Attack Management System |
| IC | Ignition component |
| ICS | Incident command system |
| IR | (1) Interregional; (2) Infrared |
| LCC | Long-continuing current |
| LCPL | Least-cost-plus-loss |
| LFO | Large fire organization |
| LLC | Lifting level of condensation |
| LR | Lightning risk |
| MAB | Man and the Biosphere |
| MACS | Multiagency command system |
| MAFFS | Modular airborne fire fighting system |
| MCR | Man-caused risk |
| MEDC | Missoula Equipment Development Center |
| NAFC | North American Forestry Commission |
| NAS–NRC | National Academy of Sciences-National Research Council |
| NBS | National Bureau of Standards |
| NF | National forest |
| NFDRS | National Fire Danger Rating System |

| | |
|---|---|
| NFFL | Northern Forest Fire Laboratory |
| NFIL | National Fuel Inventory Library |
| NIFQS | National Interagency Fire Qualification System |
| NIIMS | National Interagency Incident Management System |
| NPS | National Park Service |
| NSF | National Science Foundation |
| NWCG | National Wildfire Coordinating Group |
| NARTC | National Advanced Resources Technology Center |
| NVC | Net value change |
| NWS | National Weather Service |
| OCC | Operations command center |
| OCD | Office of Civil Defense |
| OH | Overhead |
| OI | Occurrence index |
| PNF | Prescribed natural fire |
| PPAD | Preplanned area dispatch |
| RAWS | Remote automated weather station |
| REC | Roscommon Equipment Center |
| RFFL | Riverside Forest Fire Laboratory |
| SAF | Society of American Foresters |
| SC | Spread component |
| SCOPE | Science Committee for Problems of the Environment |
| SDEDC | San Dimas Equipment Development Center |
| SFFL | Southern Forest Fire Laboratory |
| SMDW | Strong mountain downslope wind |
| SRV | Snake River Valley |
| SWFFF | Southwest Forest Fire Fighters |
| TDR | Threat-determined response |
| TGA | Thermogravimetric Analysis |
| TL | Timelag |
| UNESCO | United Nations Educational, Scientific, and Cultural Organization |
| USFA | U.S. Fire Administration |
| WFCA | Western Forestry and Conservation Association |
| YUM | Yarding unmerchantable material |

*Introduction to*

# *Wildland Fire*

*Part One*

# The Fire Environment

*Chapter One*

# Wildland Combustion

A fire is an energy system that receives its driving power from combustion, the rapid transformation of stored chemical energy into kinetic energies of heat and motion. The combustion process is complex, but its physics and chemistry are wholly accessible to scientific analysis (Figure 1-1). As a physical event, combustion involves the creation, transfer, and absorption of heat. The distribution of heat is fundamental to any fire; heat is required to initiate combustion, and all the important chemical reactions are temperature-dependent. As a chemical event, combustion belongs with that general category of chemical reactions that, through oxidation and heat, both decompose and synthesize. There are oxidizing reactions that are not combustion—examples being the rusting of iron or the hardening of wax. Similarly, there are forms of chemical decomposition, by heat or the action of organisms, that are not combustion reactions but which act on similar substances and yield similar products. Combustion may or may not invove flame. If it does, the flaming zone creates a chemical environment of great significance. Many of the special properties of wildland fire as a chemical, physical, and biological process derive from the fact that most wildland fires do flame.

As a natural phenomenon, combustion thus obeys general principles of physics and chemistry; its fuels conform to general models of biological growth, interplay, and evolution; oxygen mixes with fire according to physical and meteorological processes. It is possible to speak of the physics, chemistry, biology, and even the culture of fire. For combustion proper, one can speak of its thermochemistry and its thermophysics. Although combustion is a universal process, it occurs under widely varying conditions. It assumes different forms, consumes different kinds of fuels, and gives off different sorts of products. Humans have learned to manipulate some of those conditions, and the understanding of fire is most thor-

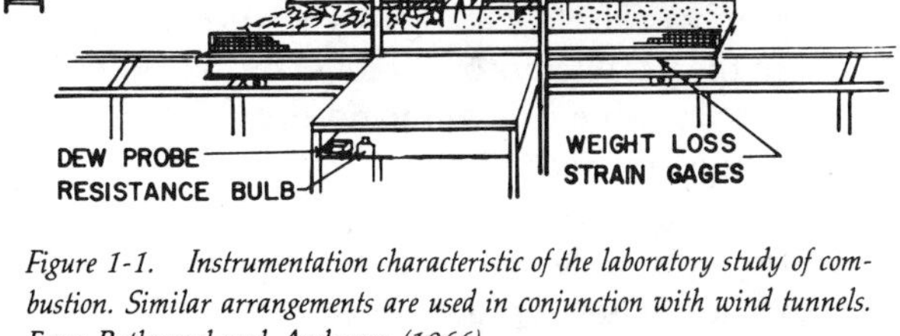

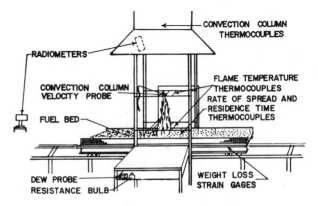


*Figure 1-1. Instrumentation characteristic of the laboratory study of combustion. Similar arrangements are used in conjunction with wind tunnels. From Rothermel and Anderson (1966).*

ough for those expressions of combustion over which control is most possible.

Simply put, between a Bunsen burner and a wildland conflagration there exists a continuum of fire types according to the degree of control humans exercise over them. Most industrial fires, such as furnaces, are *regulated* fires. Control exists over the ingredients of combustion, over the means by which they are brought together, and over the actual combustion chamber. By contrast, wildland fire is a *free-burning* fire. It burns natural, not refined fuels; it mixes evolved gases and oxygen by turbulent convection; its reaction zone is defined by arrays of natural fuels burning within topographic configurations and changing air masses. For quasi-free-burning fires, such as a campfire, some control over the fire is possible: some kinds of fuel can be added rather than other kinds, and the deliberate arrangement of fuels can shape a more suitable reaction zone. If the fire were situated in a fireplace, some control over convection (and thereby mixing) would be possible. Wildland fire is a free-burning fire that can propagate. It accepts natural fuels, mixes fuel and oxygen through coarse diffusion processes, and spreads. The reaction zone is not confined. For the fire to survive it must propagate into new fuels. How it spreads, and what shape it assumes, adds another dimension to the burning process: fire behavior.

Burning occurs within a combustion environment that contains the heat, oxygen, and fuel which must interact for combustion to take place. But the combustion environment occurs within a fire environment that, by directing the behavior of the fire as a whole, controls the shape and distribution of the combustion zones of the fire. The determinants of fire behavior are broad environmental considerations—wind, fuel complexes, topography. These parameters of fire behavior govern the parameters of combustion. To describe the reaction zone of a wildland fire, both sets of descriptors are needed.

Moreover, the fire environment introduces questions of scale into the burning process. At smaller scales, microprocesses of combustion tend to dominate the fire as a whole, guided by the chemistry of reactants and by the activities that set their reaction rates. At larger scales, such as those in the range of most wildland fires, the processes of burning are dominated by physical processes—the transfer of heat and the geometry of fuels and flame. The configuration of the fire as a whole, especially its flaming zone, creates the environment for combustion. For the largest of wildland fires, the physical processes merge with the meteorological. Diffusion processes, which guide the mixing of fuel and oxygen, rather than heat transfer processes may control the behavior of the fire as a whole.

## 1.1 PHASES OF COMBUSTION

Combustion occurs within a special environment —a reaction zone defined by the mixing of heat, fuel, and oxygen. Like any reaction, combustion shows a history: it begins, enjoys a period of sustained activity, and ends. These combustion phases are commonly known as ignition (initiation), propagation (spread), and extinction (extinguishment). With regulated, industrial fires the combustion environment—and the phases of combustion—can be manipulated. Gross natural fuels are refined into their most active ingredients before entering the reaction zone, oxygen replaces air, and carefully designed chambers provide the environment for injecting heat, fuel, and oxygen, for mixing them in proper proportions, and for ejecting their byproducts. With wildland fire, by contrast, almost none of the variables can be controlled, and the fire may be heterogeneous, displaying several combustion phases at any one time or at different times in the same place.

The complexity of wildland combustion derives from the complexity of its fuels, the complexity of the chemical processes that prepare and mix fuels with air, and the complexity of the environment in which these reactions occur: the immediate environment of the combustion zone and the larger fire environment of air, earth, and organisms which surround that combustion zone and interact with it. Wildland fuels are chemically complicated hydrocarbons existing in a solid state —complexes of organic matter evolved through natural selection and arranged into infinitely complicated patterns according to the dynamics of ecological systems. That they are solids means that a wildland fire must evolve by special processes the gases it requires for flaming combustion, and that these fuels consist of long, complex molecules means that wildland fires may have scores, if not hundreds, of elementary reactions within their combustion zone. Instead of a regulated, perhaps refined, oxydizing agent, there is an atmosphere created by geochemical evolution and organized into huge air masses of constantly changing position and characteristics, masses superimposed over a kaleidoscope of microclimates. Instead of an enclosed combustion chamber, wildland fires burn within gross topographic features.

The combustion environment is not confined within a chamber. Rather it is constantly refashioned by large features of the landscape and atmosphere according to principles that govern fire behavior. Fuel complexes, weather, topography, and the arrangement of other fires all make up the fire environment. Moreover, wildland fuels are not chemically uniform and wildland fires do not burn with uniform properties. Wildland fuels consist of different chemical constituents, each of which ignites at different temperatures and with different response times. Wildland fires, too, transfer heat by different means and at inconstant rates. The heat sources and heat sinks involved in wildland combustion, in brief, are variable.

The resulting fire environment shapes the reaction zone, and the reaction zone determines the actual environment of combustion. One consequence is that the flaming zone is not synonymous with the fire, flame shape is not identical with fire shape, and flame velocity is not the same as rate of fire spread. Another outcome is that a single fire may exhibit several reaction zones—each in a different phase of combustion and each responding to different parameters of fire behavior. Burning may take the form of flaming combustion or glowing combustion, or the two in combination. Glowing combustion results from oxidation on the surface of a solid fuel; flaming combustion results from the oxidation of gases. In the case of wildland fires its solid fuels must be prepared for burning through an assortment of thermal processes.

### *Preignition*

Burning begins with endothermic reactions that absorb energy and it ends with exothermic reactions that release energy. The endothermic reactions are known as preignition, the exothermic reactions as combustion, and the point of transition as ignition. *Preignition* is the mandatory process by which, through preheating and pyrolysis, the fuel is prepared to support combustion. The temperature of the solid fuel must be raised and, in the case of flaming combustion, suitable gases must be evolved out of it. The fuel temperature at which combustion proper begins is the *ignition temperature.* The response time needed to thermally prepare the fuel is the *delay time.* Both the ignition temperature and the delay time are variables. They depend on the physical and chemical properties of the fuel, on the character of the heat source, and on the type of combustion, flaming or glowing, that results. Consequently, the entire preignition process is itself variable, not constant.

*Preheating.* The chemistry of the fuel and its thermal properties will determine the form of combustion, especially its ability to flame. The processes of converting complex, inaccessible solids (solids often filled with adsorbed water) into usable fuels are known as dehydration and pyrolysis. *Dehydration* removes volatiles by the distillation of water and extractives. *Pyrolysis* breaks down the chemical structure of woody substances, such as cellulose; the newly liberated or reconstituted chemicals can then combust. The amount of energy required to bring fuels to ignition is the *heat of preignition,* which sums the *heat of dehydration* and the *heat of pyrolysis.* Because preheating takes both time and heat, ignition often proceeds in a sequence. It begins with easily ignited fuels that provide the heat to prepare and ignite more difficult fuels. For a given temperature, only certain fuels are available for combustion. As temperatures rise, more of the total fuel load becomes available fuel.

Preheating acts first on low-temperature volatiles that are present in many living organisms. Even a warm day is enough to evaporate some extractives. Continued preheating then operates on any adsorbed water within the fuel particle—its fuel moisture. *Fuel moisture content* (FMC) is highly variable, and different processes govern fuel moisture contents in living and dead fuels. Dead fuels show fuel moisture contents of 1–30% of ovendry weight, whereas live fuels will range from 80 to 200%. Regardless of the mechanism involved, the adsorbed water must be boiled off before the heating of the particle proper can begin (Figure 1-2). The fuel moisture may be prohibitively high. In this case all of the available heat is expended in evaporating water and none is applied to preparing the fuel upon which combustion depends. Such levels of fuel moisture are referred to as the *moisture of extinction* (Rothermel, 1972).

*Pyrolysis.* With water expelled, preheating advances into pyrolysis. This is a process (or set of processes) of thermal degradation, of chemical decomposition through the application of heat. Pyrolysis occurs in sequence, or rather along pathways of often competing sequences. Two general reaction pathways are recognized (Figure 1-3; Shafizadeh, 1968). One set of processes yields char and water, while the other decomposes to tar and volatiles. Both pathways compete for the same initial substances and for many of the same intermediary products, like levoglucosan. Both can proceed, to some extent, simultaneously. Altogether, the pyrolysis of wildland fuels yields combinations of volatiles, tars, carbonaceous char, and mineral ash. Volatiles, tars, and char contribute to combustion; mineral ash does not, and may even inhibit combustion. If pyrolysis occurs under controlled conditions without oxygen, it is known as *destructive distillation,* an industrial source of wood chemicals. If it occurs under wildland conditions, however, it leads to fire.

Which of these two pathways predominates depends on fuel properties and the temperature of the reaction zone. High temperatures favor the evolution of volatiles—flammable gases known

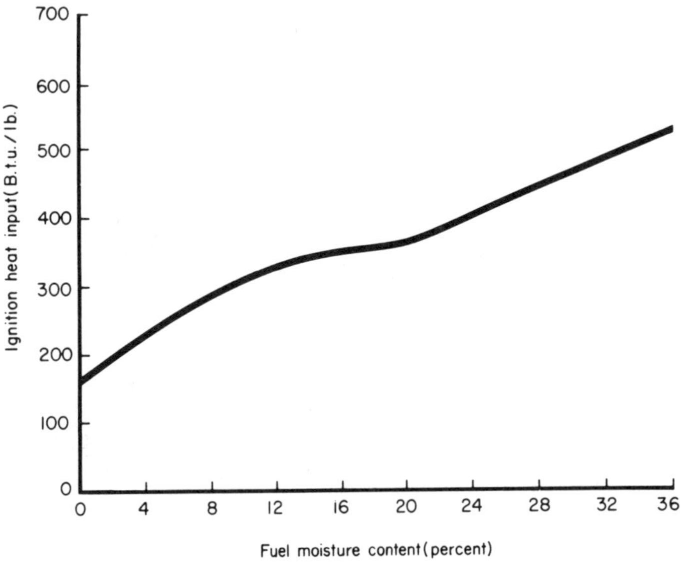


*Figure 1-2. Heat required to bring one pound of fuel to ignition as a function of fuel moisture content. From Anderson (1969).*

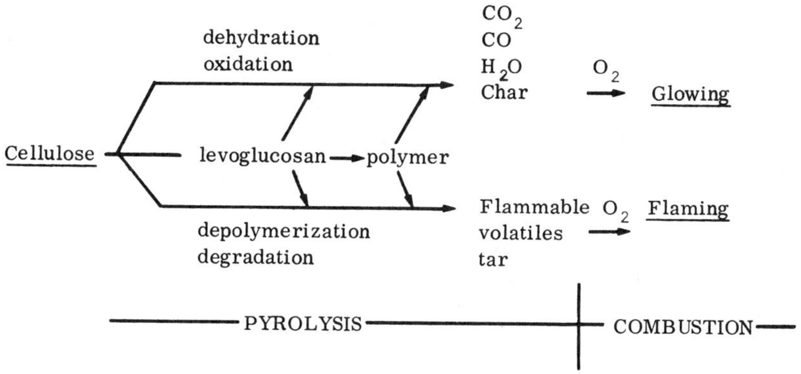


*Figure 1-3. Pathways for the pyrolysis of cellulose. From Philpot (1971).*

as *pyrolysates*—whereas low temperatures promote the production of tar and char. Flaming combustion with its high temperatures encourages the evolution of pyrolysates. But in wildland fuels the flame does not long remain at one site. The poor thermal conductivity of wood means that the pyrolysis of larger fuel particles or deeper fuelbeds will proceed too slowly to contribute to the flaming zone. The flaming front will move onward to fresh, more readily pyrolyzed fuels, while the residual fuels, now burning in a lower temperature environment, tend towards glowing combustion or char. Because char has ⅓ to ½ the thermal conductivity of wood, the processes of heat transfer will be slowed even further.

But the amount and rate of preheating does not itself account for pyrolysis products. The thermophysics and thermochemistry of fuel particles, and their organization into fuel arrays, also influence the process. In their heat contents, wildland fuels show a general uniformity, a product of a common evolutionary history (Susott et al., 1975). But a particle is not uniform in its chemistry, and a fuel array does not consist of a single type of particle uniformly distributed. Different portions of a woody fuel will preheat at different temperatures and to different effects. Extractives, such as lipids and terpenoid hydrocarbons will volatilize at low temperatures and yield high energy gases. Cellulose (the principal constituent of wood) shows thermal stability until particle temperatures of 480°F (250°C) are reached. At 620°F (325°C) it breaks down rapidly, evolving large quantities of flammable gases. Lignin resists thermal decomposition, leaving it more prone to char as a product and glowing combustion as a process. Some mineral constituents, like silica, have little effect on burning. Most, however, retard flaming combustion by promoting low temperature pyrolysis to tar and char. And of course there is fuel moisture to consider.

That wildland fuels contain easily distilled volatiles means that it is often possible to escalate pyrolysis and combustion through a graduated sequence of intensities, with low-temperature volatiles combusting to liberate the heat needed to pyrolyze high-temperature pyrolysates. To oversimplify the issue a little, this process helps to account for the paradox that living fuels can burn. The presence of high-energy, easily distilled volatiles in living fuels provides a sufficiently large heat source to overcome the otherwise prohibitively large heat sink created by the high fuel moisture content of living fuels.

Similarly, physical properties of fuels account for differential pyrolysis and combustion. Its surface-to-volume ratio describes the relative susceptibility of a particle mass to incident heat, and its *thermal conductivity,* or *diffusivity,* describes the particle's ability to transfer that heat from its surface to its interior. Woody fuels are notoriously poor conductors of heat, and larger particles may absorb more heat than they liberate during the passage of the flaming front. Larger particles may combust, but not flame. Flames require high-temperature pyrolysis, and most flames will occur in association with fine fuels. Although a certain degree of pyrolysis must precede the flaming zone, much will occur within the flaming zone since that is where the highest temperatures will be concentrated. The preheating of larger fuels will be controlled by conduction and, especially after passage of the flaming zone, combustion will be glowing. For fuel arrays the analogous properties are bulk density, porosity, and packing ratios—all of which describe the proportion of fine fuels to large, and the amount of exposure of individual surfaces to a heat source. Porous fuel arrays allow for more of the total fuel mass to be exposed to preheating and to burn simultaneously in the flaming zone.

*Measurements of Pyrolysis.* For a given fuel the heat of preignition will consist of the heat of pyrolysis, a constant, and the heat of dehydration, a variable that depends primarily on fuel moisture content. The reactions are endothermic. Their energy absorption can be recorded by differential thermal analysis (DTA) techniques, and their transformation of matter can be recorded by thermogravimetric (TG) means (Shafizadeh and DeGroot, 1976). The amount and rate of mass loss describe the evolution of flammable gases by pyrolysis. The amount of mass loss gives a measure of the total heat converted by combustion in all forms, glowing and flaming. The rate of mass loss, however, measures the efficiency of pyrolysis. High rates accompany high temperatures and indicate the rapid evolution of pyrolysates—this suggests flaming combustion. Low rates may simply represent the leaching away of volatiles or evolved gases, or the oxidation of char.

The efficiency of mass loss due to pyrolysis compared to the elapsed time available for pyrolysis gives the *reaction velocity.* High reaction velocities improve the likelihood of flaming combustion over glowing combustion; they suggest the rapid expulsion of pyrolysis products, water among them, that might otherwise smother oxidation (Figure 1-4). The character of combustion will thus depend on the character of pyrolysis. Equally, since the heat of combustion supplies the energy for preheating and pyrolysis, the properties of combustion will determine the properties of pyrolysis (Rothermel, 1972).

## Combustion

*Ignition.* Ignition marks the onset of combustion. It signals the predominance of exothermic reactions, whose net effect is to liberate energy, over endothermic reactions, whose net outcome is to absorb energy. In actuality, ignition is not a single process, but a continuum of processes; not the end of pyrolysis but an acceleration of it; not a singular event, completed once for a fire, but a continuous sequence of events, as individual particles within the fuel complex become separately involved. The heat released by ignition enhances

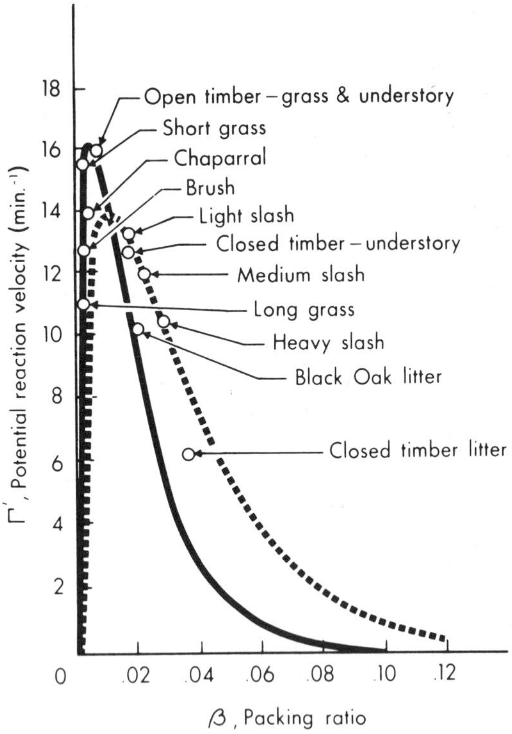


*Figure 1-4. Potential reaction velocity of typical wildland fuels. The two curves give extreme values for fuel surface-to-volume ratios: one for grass and brush, and the other for logging slash. From Rothermel (1972).*

pyrolysis, which reaches a peak within the flaming zone. Combustion appears to be uniform because the fuel particles brought to their ignition temperature are small relative to the size of the flaming zone and because the delay time is short compared to the general rate of spread. And ignition only seems to occur once because combustion can become self-sustaining. For a fire to spread, it requires that discrete new ignitions occur in unceasing succession. In establishing sustained combustion, the delay time is thus critical. If the delay time is short, the fire may spread rapidly; if too long, the fire may expire. In sum, the ignition process is fundamental to the overall rate of combustion, and the delay time is fundamental to the rate of ignition.

The speed and success of ignition depend on the properties of the fuel particle and the character of the source of heat applied to it (Figure 1-5). Certain physical and chemical properties of the particle influence its susceptibility to heating. A fuel particle with a high surface-area-to-volume ratio, for example, will heat more quickly than a larger-diameter particle, and a particle that is rich in easily volatilized extractives will show lower ignition temperatures than one that does not. So, also, ignition will vary with properties of the heat source—the mode of heat transfer, the intensity of heat transferred, and the duration of heating. The heat source may or may not be associated with flame. Fuel heating in the absence of flame leads to *spontaneous ignition;* fuel in the presence of flame leads to *pilot ignition.*

Consider, by way of example, the types of ignition that can result from a radiant heat source, one that transmits heat through radiation. The irradiated fuel particle may respond according to one of three states: nonignition, transient ignition, and persistent ignition. In the case of *nonignition,* the duration of heating is too brief or its intensity too low to bring the fuel element to its ignition temperature. In the case of *transient ignition,* combustion occurs but it will continue only as long as a certain heat intensity persists; if the irradiating heat is withdrawn, combustion will cease. The reason is that for every particle there

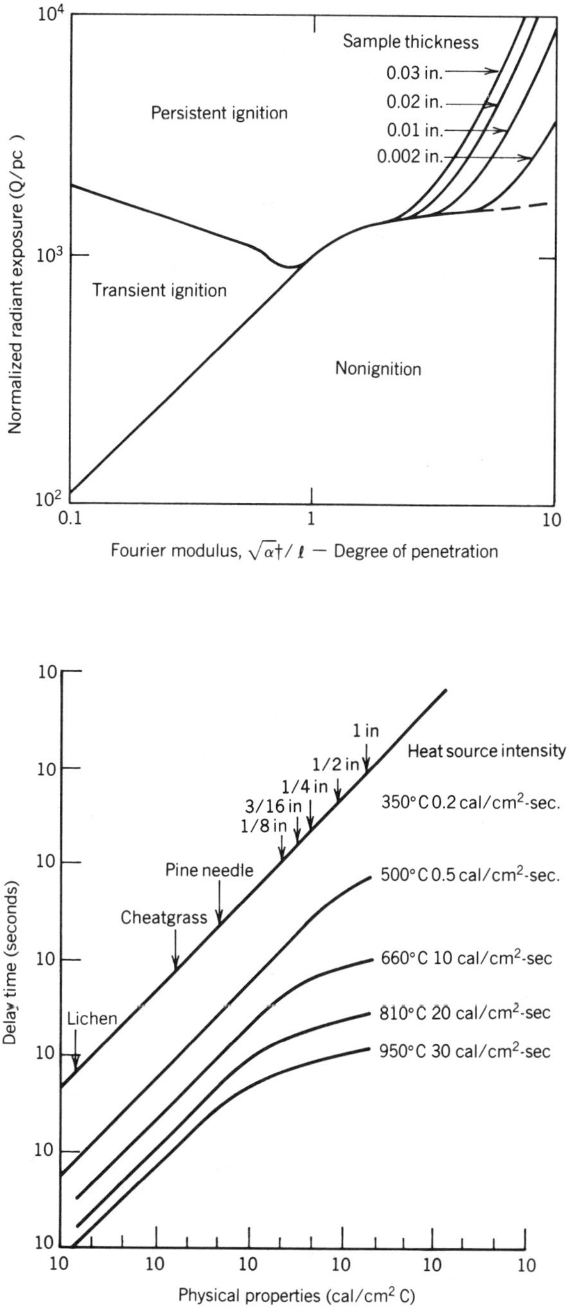


*Figure 1-5. Ignition. Upper diagram describes the generalized ignition behavior of cellulose according to the intensity of the radiant heat source and the conductivity of the fuel particle (expressed as a Fourier modulus). Lower diagram shows the relation of delay time to physical properties (sample density, specific heat, and thickness). From Anderson (1970).*

exists a certain threshold of heat intensity, its *critical irradiance,* below which ignition will not occur, no matter how long the heat is applied. In wildland fires, transient ignition is a common phenomenon. It often occurs in association with the flaming front, whose extreme temperatures char or scorch large fuels without leading to sustained combustion after the front has passed by. Similarly, breaking up a pile of logs (each member of which irradiates the others) often results in extinction of a fire. Finally, there is the case of *persistent ignition,* in which the range of heat intensities and durations is adequate for combustion to begin and endure (Anderson, 1970).

The mode of ignition and combustion will also affect the ignition temperature. Piloted ignition begins at lower temperatures than spontaneous ignition, and glowing combustion starts at lower temperatures than flaming combustion. The latter is an important fact in the production of firebrands and for the mixed burning characteristics of most wildland fires. Within the primary reaction zone, the normal mode of combustion is flame; outside that zone, glowing combustion is common.

*Flaming Combustion.* No part of a fire is more spectacular than its flaming front, and none is so significant for an understanding of fire and its management. Here the greatest rates of heat are released, the most intense combustion environment is fashioned, and most of the biochemical and ecological effects peculiar to wildland fire are realized. The flaming zone provides a ready index of fire behavior and effects. With flame length, there is a quantitative measure of resistance to control efforts and the lofting potential for firebrands; with flame height, a measure of lethal scorching of conifers and the likelihood of crowning; with flame depth, a measure of flame height and the probability of especially intense burning. Most systems for forecasting fire behavior, fire effects, and fire danger rating express the important attributes of the fire in terms of its flaming zone. Like ignitability, however, flammability is a relative concept, not a fixed constant.

The transition to flame requires both heat and gaseous fuels. If the fuel is initially gaseous, the combustion rate follows from the temperature of the reaction zone and the rates of air flow and mixing within it. But not all fuels enter the reaction zone as a gas; wildland fuels rarely do. If the fuel is liquid, then vaporization precedes flaming combustion. And if the fuel is solid, combustion may be partial in scope and heterogeneous in burning traits—with flames resulting only insofar as pyrolysis can evolve flammable gases from the solid and as these gases can be entrained within the reaction zone. Hence the zone of pyrolysis may be distinct from the zone of combustion, and flames may be correspondingly *attached* or *unattached* to the original fuel elements.

Flammability appears within limits set by temperature and mixing processes. When the temperature is low, the mixing of oxygen poor, and the evolution of a gas difficult, flame may be impossible or, if possible, the flames may be unattached, separated from the zone of vaporization and diffusion. Where the fuel is complex, such that different portions of it exist in different states of matter, combustion tends to be heterogeneous. The fire will feature both flaming combustion, in which gases can be evolved and ignited, and glowing combustion, in which oxidation occurs directly on the solid fuel. Pyrolysis, flaming combustion, glowing combustion—all may coexist within a common reaction zone, or they may exist in different portions of the fire or at different times during the larger history of the fire.

The description of flame in wildland fires requires two sets of different, but related parameters. One set pertains to flame within the combustion environment; the other within the fire environment (see Figure 1-6). Inside the reaction zone, *flame shape* describes the size of the zone; *flame velocity,* the rate of flame propagation through the fuel–oxygen mixture; and the *reaction intensity,* the overall rate of combustion. In regulated and confined fires, this reaction zone is identical to the fire; the combustion environment and the fire environment are the same. In wildland fires, they are not. The fire environment influences the distribu-

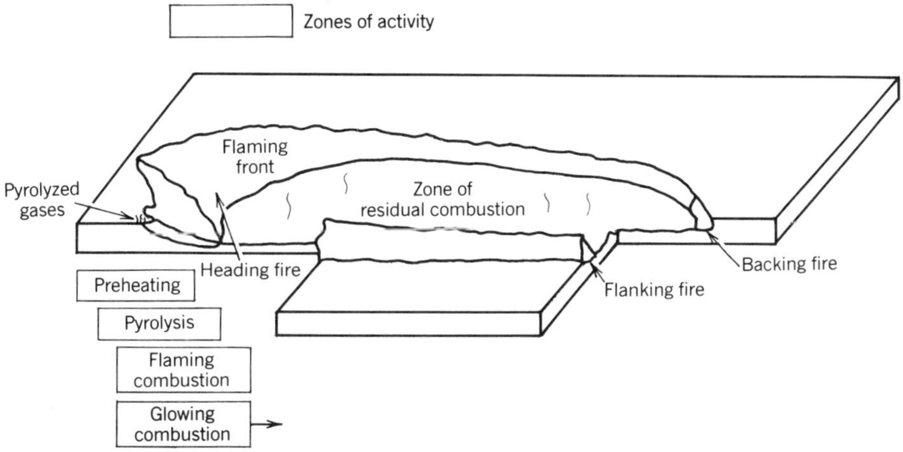


*Figure 1-6. The combustion environment and the fire environment. For a regulated fire, like a Bunsen burner, the combustion environment and the fire environment are identical. For a free-burning wildland fire, the two can be distinguished, though they remain keenly interrelated. In general, the gross properties of the fire environment govern fire behavior, manifest by the shape, intensity, and rate of spread of the flaming front. The properties of the flaming front, in turn, set the important parameters of the combustion environment. In fuelbeds composed wholly of fine fuels, the flaming front will embrace the entire zone of combustion. But in mixed fuelbeds typical of most wildland fires, combustion will diverge into two phases as the flaming front spreads outward. There will be a primarily flaming phase along the advancing perimeter, and a primarily glowing phase within that perimeter. Though the flaming front is the more spectacular phase, the zone of residual combustion may assume considerable dimensions, as in the picture (above), where snags and downed logs sustain a large but disorganized zone of burning. Photo courtesy of U.S. Forest Service.*

tion, shape, and intensity of the combustion environments. In addition to the flame descriptors common to all forms of flaming combustion, an additional set applies to flames in wildland fires in which the entire flaming zone moves. To flame shape must be added *fire shape*; to flame velocity, the *rate of spread* of the flaming front; to reaction intensity, the *fireline intensity,* the rate of energy release along the advancing front.

Consider, first, those universal properties that apply to all flames. Flame is a process that assumes a structure (Fristrom and Westenberg, 1969). The combustion reaction, and thereby flame shape, is dynamic; it is continually created as new fuel and oxygen enter it and as waste products leave. Flame shape thus reflects the distribution of heat, the temperature of the zone; of oxygen, the flow field that brings air into the reaction zone; of fuel, its state of matter and the size and shape of the gas-emitting area. The particular shape of a flame can be expressed in terms of its area and length.

But flame depends on the rates of flow, not merely their amount. *Flame velocity* integrates the rates at which fuel, heat, and oxygen enter the system. The flow rate of fuel controls the amount of potential energy available for combustion. Low heat production can result when there exists insufficient fuel for combustion—perhaps because gases take too long to evolve from solid fuels. Diffusion rates control combustion because they determine how much oxygen mixes with gaseous fuels. Temperature controls the rate of all chemical reactions, including all those associated with the combustion process. It determines how rapidly and in what quantities pyrolysis evolves gases, at what velocity diffusion by convective flow occurs, and how speedily oxygen and fuels can combine at the molecular level. Where fuels readily gasify, where temperature is high, and where diffusion is rapid, the rate of combustion is high and the flame velocity fast (Anderson, 1969).

In wildland fires flame shape is a far more complicated problem. The reason is that, unlike industrial fires, the reaction zone is not confined. The fire is larger than its flaming zone (or zones); the flaming zone moves as a unit; and its shape (fire shape) and the shape of the flames within it constantly change with alterations in the fire environment. Not only does flame propagate through a fuel–air mixture within the combustion zone, but the flaming perimeter of the whole fire moves. Fuel flow is not only upward but outward. Mixing is not a relatively simple case of air being drafted into the reaction site and diffused, but a complex aerodynamic regime that represents the interaction of ambient winds and fire-drafted air. Flame shape thus depends on those fire behavior processes that continually restructure the combustion zone; and flame velocity depends on the rate of fire spread.

Like all measures of velocity, these are relative —in this case to the fuel–air flow. Flame velocity and fire spread can both be expressed in terms of fuel flow rates; and flame shape and fire shape can be defined by the configuration of the gas-emitting areas. The flame of a Bunsen burner, which seems fixed, is actually in motion relative to the fuel–air mixture brought into the combustion zone; and the flaming front of a wildland fire, so apparently in motion, can be considered as fixed in its location while the surrounding landscape passes through it. Contemplating a flaming front in fixed position while forested mountains sweep through it is a bit dizzying, but a thought experiment of this sort helps to visualize the reasons for local accelerations in velocity and alterations in flame shape. Fires burn down, out, and (sometimes) up through wildland fuel complexes.

Flame velocity thus describes the rate of flame propagation within the reaction zone, as the flame burns down into rising gases. Rate of spread describes the propagation of the flaming zone as a whole into new fuels. Within the flaming zone, the rate of combustion is described by the *reaction intensity.* But to describe the energy released by the advancing flaming zone as a whole, another measure is needed—*fireline intensity* or Byram's intensity, the product of heat release and rate of spread. A fire's reaction intensity describes the intensity of reactions within its flaming zone; its fireline intensity describes the intensity of the advance of that flaming zone.

In wildland fires, flame shape differs from fire shape. Flames will be distributed differently around a fire. The fire may exist as a point of flame, as a line (or curve) of flames, or as an area of flame. The depth and height of its flaming zone will vary accordingly. Point flames are common among regulated fires such as a candle, and they can persist because new fuel is brought into a confined reaction zone. Although all wildland fires begin as point sources, they cannot long continue in this state; the flame must spread into new fuels to survive and this spread is not confined. This transforms a point of flame into a line, or perimeter, of flame.

The fire grows in size, different mechanisms come into play to control the transfer of heat, and the flaming zone takes on new forms. The flaming perimeter is not so much a line of flame as an ellipse of flame or a chaotic boundary of flames arranged by the complex geometry of the fire environment. Only rarely will the flames along any segment of this perimeter resemble the flames at all other segments. The fire as a whole will burn preferentially in some directions. Some flames will bend forward into new fuels *(heading fires)*, while other flames will bend away from new fuels, or into the old burn *(backing fires)*. Both types of fire spread may coexist around the perimeter of the same fire, but their flame shapes will be different and so will their rates of combustion (Figure 1-7). Fire shape begins to control flame shape, to determine the distribution of heading and backing fires.

For any portion of the flaming zone, moreover, there will be an advancing edge and an expiring edge. For a regulated fire, flame length describes this distance exactly. But for wildland fires, another measure, *flame depth,* is needed. The flaming zone is not only high but wide; it occupies a flaming area, not merely a line of fire. The dimensions of flame depth will depend on the rate of spread and on how long flames persist at one place, their *residence time.* This is the time required for the flaming edge to travel a distance equal to the depth of the reaction zone. Within the flaming zone the time required for complete combustion is referred to as the *combustion period.* That portion which contributes to further propagation is known as the *critical burnout period.* For most wildland fires the critical burnout time and residence time will be equivalent since each applies to the flaming zone. But for those fires in which glowing combustion is significant, the combustion period may be a more useful description of the total burning process.

The critical burnout time and residence time are important determinants of flame depth, and through flame depth, of fire behavior. These variables interact in assorted ways. Normally, rate of spread and residence time are proportioned such that the flaming zone exists as a line or ring around the fire. But if the fire moves rapidly, if the residence time is long, or if the flames should be both rapid and prolonged, then the flaming zone will tend to deepen. A line of fire becomes an area of fire. Flames will increase in height, temperatures will intensify, and new processes may arise that will influence fire behavior—and through fire behavior, combustion.

*Glowing Combustion.* Most wildland fires exhibit flames, and some, like grass fires, may have virtually all flame. Most fires, too, exhibit some degree of glowing combustion. Some, like ground fires in organic soils, tend to make glowing combustion the primary phase of the combustion process. The relationship between flaming and glowing combustion is reinforcing, not exclusive. Many episodes of flaming combustion begin with glowing combustion, and many flaming wildland fires conclude their burning in the glowing state.

Yet the differences between these two phases of combustion are real. Glowing combustion is slow and complete, exemplified by the burning of charcoal. Flaming combustion is rapid—less efficient in the conversion of chemical to thermal energy and characterized by larger residues of un-

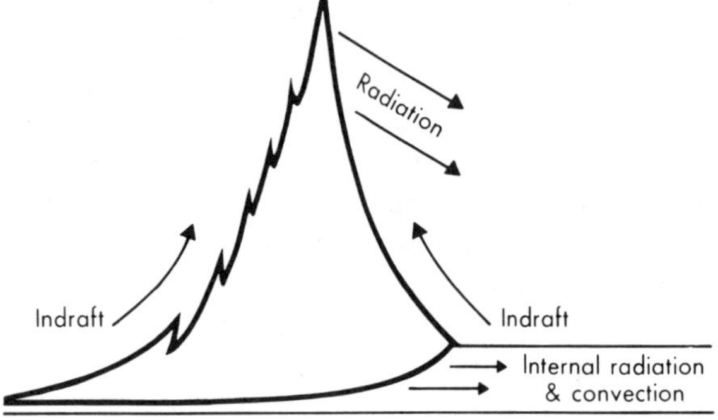
Radiation
Indraft
Indraft
Internal radiation
& convection

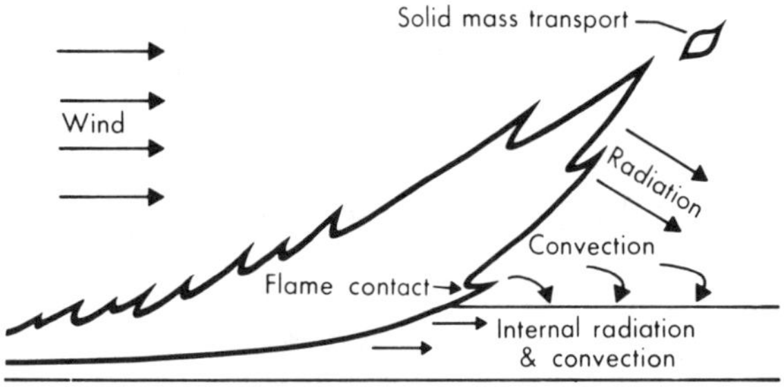
Solid mass transport
Wind
Radiation
Convection
Flame contact
Internal radiation
& convection

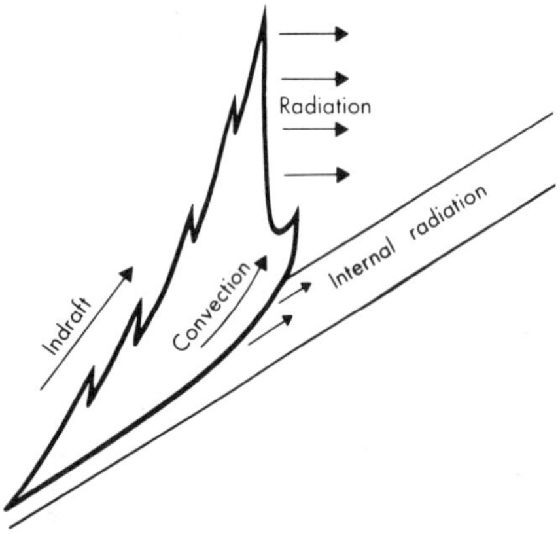
Radiation
Internal radiation
Indraft
Convection

burned material. Glowing combustion, moreover, can occur at lower temperatures. It does not require the high temperatures demanded for rapid pyrolysis and flame. It can thus function like a slow match, holding fire for a long period of time, a significant fact for the ignition process. Most spot fires, for example, begin with glowing combustion. Where the fire moves slowly, where it hovers on the verge of extinction, or where a fire must begin as a spot, glowing combustion can mean the difference between propagation and extinction. If the fire environment changes to more favorable conditions or if that preserved heat can subsequently build up (perhaps because of insulating materials around an ember), then glowing combustion may escalate into flaming combustion.

Glowing combustion contributes to total heat production, even within the flaming zone. But its primary effect is to extend the zone of combustion from the leading edge of the flaming front rearward to include all those fuels which have not yet been reduced to ash. For many fuels and for large fires, this zone may be enormous. Where fine fuels are primarily involved, flaming combustion may consume most of the fuels in a matter of minutes. Where large fuels or deep duff comprise the fuel complex, glowing combustion may persist for days, even weeks, in a slow attrition of fuels and liberation of smoke and energy. As a physical process, glowing combustion adds to the mass and energy equations that describe combustion as a whole. As a biological phenomenon, it contributes significantly to the transformation of a wildland biota into a reservoir of minerals and biochemical compounds whose availability in their new form may be of pertinence to the ecosystem. This is particularly true for the reduction of large fuels and thick litter—only a portion of whose surfaces was exposed during the passage of the flaming front and whose combustibility is limited by slow processes of diffusion and of heat transfer by conduction, but whose biomass may be large. In both cases glowing combustion may persist for long periods of time, slowly mopping up fuels left behind the flaming front. Of the total fuel load consumed by forest fires, glowing combustion will typically account for the mass that is not burned up in the flaming front.

This can make glowing combustion as great a management problem as flaming combustion. On a large fire glowing combustion may provide a persistent source of ignition. It may sustain firebrands that can spot across control lines. It may maintain a reservoir of heat which, under the right conditions, can result in reburns of aerial fuels, a reinitiation of flaming combustion. It may smolder ignominiously for extended periods, manufacturing smoke without the convective heat column needed to ventilate those particulates into the atmosphere. If the area involved in smoldering combustion is large, this enduring smoke may become unbearable from the standpoint of air pollution. And, finally, where glowing combustion fails to clean up the fuels left behind the flaming front, it can inadvertantly promote a long-term cycle of reburns—one whereby the large fuels or green fuels not originally burned may be reduced by successive fires over a period of years or even decades.

*Ideal Fuel Temperature History.* Both pyrolysis and combustion are temperature-dependent. The distribution of temperature within a particle, and within a combustion environment, is critical to the character of the entire burning process. This distribution varies over time. By way of illustration, consider the following idealized thermal history for a woody fuel particle. As flame advances on the particle, its surface is heated. If the particle is small (thin), with a large surface-to-volume ratio, it may be heated more or less uniformly and instantaneously. If it is large, then a thermal gradient will develop between the surface and the interior, and the speed of conduction will depend

◄ *Figure 1-7. Effects of flame shape on fire spread. (Upper) No wind or slope; (middle) wind-driven fire; (lower) slope-driven fire. Photo of slope-influenced fire is for a backing, not a heading fire, though the geometry of flames is identical. For the resulting temperature histories, see Figure 1-10. Schemas from Rothermel (1972), photos courtesy U.S. Forest Service.*

on the intensity of the heat source and the thermal conductivity of the particle.

As heat is absorbed, distillation begins. Some ether extractives will volatilize at temperatures lower than the vaporization point of water. When the particle temperature reaches 212°F (100°C), its fuel moisture begins to boil off as water vapor. This is a critical moment. The specific heat of water is large, and additional energy must be applied to supply the latent heat of vaporization; for dead woody fuels, a fuel moisture content of 24% or so will prevent combustion. If the adsorbed water is expelled, however, then the particle will again absorb heat.

Most extractives will have volatilized by the time particle temperatures reach 390°F (200°C). Most woody constituents volatilize between 570°F and 750°F (300–400°C). By 930°F (500°C) gassification ends. Cellulose shows considerable thermal stability until about 480°F (250°C); after 620°F (325°C), it breaks down suddenly, evolving flammable gases. The rate of gassification increases with increasing particle temperature until about 750°F (400°C). Most flaming combustion occurs between temperatures of 570 and 720°F (300–380°C). The onset of persistent flame begins with the rapid pyrolysis of cellulose around 620°F (325°C). After 840°F (450°C), the rate of gassification slows. By 930°F (500°C) gas evolution is complete. Only char remains. The char may continue to burn by glowing combustion, but it will not flame. The rate of heating does not affect the amount of char yield from cellulose, although it does affect the intensity of flaming combustion (Albini, 1980).

### *Extinction*

If fuel, oxygen, or heat fall below a critical level, the fire goes out. In regulated fires, extinction may be instantaneous—the flame is blown away, eliminating the source of heat; the gaseous fuel supply or oxygen is cut off. On a molecular level, the chain reactions that combustion demands for propagation no longer generate free radicals capable of entering into further reactions but result in stable molecules that leave the system without further chemical activity. Like wildland ignition, however, wildland fire extinction is less an instantaneous event than a graduated process. In ignition, the heat of combustion gradually builds up, the product of successive, multiple ignitions. In extinction, that heat gradually diminishes, particle by particle. With each diminution, the net heat of combustion subsides below the point of glowing combustion, below transient ignition, below the requirements of pyrolysis and dehydration. Because of local insulation, however, heat may persist in certain localities for long periods, gradually lost by conduction as it had gradually built up before.

To extinguish a fire it is necessary to remove fuel or oxygen, to cool the reaction zone, or to interfere with the molecular processes that make the chain reaction possible. All of these techniques are available to wildland fire managers. Fuel is removed in the form of firelines, water and dirt may cool the combustion site, and chemical retardants may intervene in pyrolysis reactions. To extinguish a wildland fire directly and deliberately, however, may require great labor. For control of the fire as a whole, a break in fuel continuity is created around the spreading perimeter. This may be an enormous task, with firelines many inches deep, many yards wide, and many miles long. For control over the flaming front, chemical retardants can be applied, reducing fire intensity. Once the flaming zone is contained, the fire will gradually exhaust itself. To suppress residual burning (often in the form of glowing combustion) direct methods of extinguishment may be brought to bear. Individual fuel particles can be isolated, systematically reduced in size, and cooled. In a fire with large fuels or thick duff, the process is a slow and tedious one. Particle by particle, layer by layer, the fire burns itself out or is deliberately quenched. The sequence of individual ignitions that characterized the buildup of the fire is replaced by a sequence of individual extinctions.

## 1.2 PHYSICS OF WILDLAND COMBUSTION

Fuel chemistry determines the potential energy available for combustion, the reaction mechanisms involved in burning, and the potential range of products that result. But the reactions themselves occur within a combustion environment that includes heat and oxygen as well as fuel. The shape of that environment will be defined by physical processes of heat transfer and diffusion mixing, and the rate of combustion will be set by the rates of heat transfer and air–gas mixing. The combustion environment for wildland fire is not a rigid chamber, but a mobile zone in which physical processes of heat transfer and mixing combine with fuels to shape a site within which chemical reactions occur. That the combustion environment is, in turn, a product of the fire environment only reinforces the importance of physical processes in directing the overall course of burning.

### *Thermophysics of Wildland Fuels*

The physical attributes of a fuel particle and of fuel arrays influence how much fuel burns and by what mode of combustion. For regulated fires such as gas burner, a fuel particle may consist of a molecule. For wildland fires particles will show great variety in size, shape, density, and thermal conductivity, and they will be organized into arrays known as *fuelbeds.* Important particle properties include density, size, and surface-to-volume ratio. Important fuelbed properties include the proportion of fuel particles by size, the depth of the array, and such measures of packing as bulk density and porosity.

The surface area-to-volume ratio of a particle describes its availability for combustion. Fine fuel particles with high ratios become involved quickly. Its thermal conductivity describes the rate at which a particle will transfer heat internally. Unlike metals, wood is a poor conductor, and heat spreads along a thermal gradient from heated surface to cool interior. Large particles thus take much longer to heat than do small particles. Consequently, the surface area-to-volume ratio of a particle is a measure of its availability for combustion. Particles with high ratios become involved quickly since the surface is warmed by rapid heat transfer processes such as convection and radiation; the interior is warmed by the much slower process of conduction. Accordingly, fine fuels are more likely to be involved with flaming combustion, and large fuels with glowing combustion. A particle 3 in. (7.6 cm) in diameter may be considered large. Put differently, large fuels remove a portion of the heat that would otherwise go towards flaming combustion. For the same mass of fuel, fine fuels contribute more to the heat source, while large fuels add to the heat sink. Conversely, large fuels can hold fire for a longer period of time, increasing the residence time of the flaming front and widening the overall zone of combustion.

Similar considerations apply to fuelbeds. Loose packing, effective bulk densities, high porosities —all control the proportion of surface area-to-volume for the array. A fuelbed that is fluffy rather than compact, that consists of fine fuels rather than coarse, that is deep but lightly loaded will heat more rapidly and will more likely flame. The fine fuels, then, not only provide kindling for a fire but contribute proportionately more to its heat source. If the heat of combustion generated by the fine fuels is great, the fire may intensify, involving somewhat larger fuel particles within the flaming zone. On occasion, intense fires may even involve fuel arrays such as tree crowns that are normally removed from a surface fire. Under the proper conditions, it is possible by this rapid but graduated escalation to involve all the particles of the fuel complex, from needles to whole snags.

The physics of fuel particles and their arrangement thus influences how much of the heat source is absorbed by the heat sink, how much of the total fuel load is available fuel, and how much of the total heat of combustion is available as heat of preignition. The important fuel parameters are intimately related to the important fire parameters—the one determining the other. The modeling of fuelbeds is a sophisticated technique that depends on the modeling of fire behavior. Both are examined further in chapters to come.

### *Heat*

The important events of the combustion process —from molecular kinetics to large-scale fire behavior—are temperature-dependent. As a measure of heat, temperature describes the energy dynamics of the fire. Each phase of a fire requires heat input, and the greater the rate of input the more rapid the rates of pyrolysis, combustion, and fire spread. For a fire to grow either in size or intensity, there must be a buildup of heat. Temperature thus controls the intensity of each phase of the fire.

The heat energy of a fire can be analyzed according to how the heat is generated, how it is transferred, and how the liberated heat is absorbed. A fire has a heat source, a heat sink, and mechanisms for transferring heat from source to sink. The heat source is the combustion zone; the heat sink, the unburned fuels within and around the combustion zone; the mechanisms of heat transfer, the familiar processes of radiation, convection, and conduction. In quantitative terms, the heat source can be described by the rate and heat of combustion; the heat sink, by the heat of preignition; heat transfer, by the propagating flux.

*Heat Source.* The *rate of combustion* describes the rapidity with which energy is released during burning. The amount of energy released is known as the *heat of combustion,* and it determines the temperature of the combustion environment. The magnitude of the heat source will depend on the rate and amount of heat flow, fuel flow, and oxygen flow.

Heat of combustion is often used interchangeably with *heat yield,* though the heat yield for a particular fuel will vary with the heat of combustion. The concept of heat yield can be expressed in several ways. *High heat yield* describes the caloric content of a fuel as measured by an oxygen bomb calorimeter; it sets a theoretical maximum on the amount of potential energy available for combustion. The average value for wildland fuels is 4500 cal/g. Wildland fuels burning under natural conditions release less than this value. *Low heat yield* adjusts the high heat yield according to fuel moisture. The heat expended in evaporating adsorbed water cannot be spent in pyrolysis and combustion, thus reducing the amount of energy returned for the heat invested. *Effective heat yield,* or volatile heat content, adjusts the high heat yield for predictable losses due to char formation. It describes that proportion of the fuel which will effectively contribute to combustion, especially to flaming combustion. Effective heat yields range from 34 to 78% of high heat yields. The heat of combustion is thus a variable, not a constant; it is set by the interaction of all the processes that shape the combustion environment.

*Heat Transfer.* Of the potential energy actually converted to heat, some will escape the system, some will be absorbed by heat sinks with little effect, and some will contribute to the propagation of the fire. For the fire to be self-sustaining, a sufficient quantity of heat must be generated and then redirected into preheating new fuels. The success of the transfer will depend on both the character of the heat source, the fire, and the property of the heat sink, the fuels. Not all of the heat released by combustion is effective in preheating fresh fuels, and not all of the fuel load will be available for combustion. The intensity of heat actually transferred to fresh fuels is known as the *propagating flux.* The proportion of fuels that will actually be affected by this transfer, a function of the bulk density of the fuelbed and the particle sizes that make it up, can be described in terms of an *effective heating number.* Porous fuel arrays composed of fine fuels have higher heating numbers than do dense arrays or arrays of coarse fuels (Frandsen, 1973). The amount of heat transferred will help determine whether combustion is sustained; the rate of transfer will help determine the mode of preheating and combustion.

The mechanisms of heat transfer are radiation, convection, and conduction (Anderson, 1969; U.S. Forest Service, 1969). *Conduction* transmits heat within solids, and its rate will vary with the thermal conductivity of the material. But wood is a poor conductor; heat can be brought faster to the surface than it can be transferred to the interior. With char, thermal conductivity becomes even less efficient. Fuels that rely on conduction to ignite will often be out of phase with other fuels that reach their kindling temperature through faster methods of heating. The surface of a large fuel particle, for example, may burn during passage of the flaming front, but unless the residence time of the flaming front is long, the particle will often be left behind to combust outside the primary reaction zone or to extinguish. By allowing heat to build up within larger particles, but not at a rate that sustains combustion, conduction can contribute to the relative size of the heat sink.

*Radiant heating* refers to the transmission of energy along the electromagnetic spectrum. The intensity of radiation varies inversely with the square of the distance between the radiating source and the object it strikes. Thus a small shortening of the distance between flame and fuel can mean geometrically larger influxes of heat by radiation. Since most radiant heat emanates from the flaming zone, the shape of that zone is critical in determining the amount of radiant heat available for sustained combustion.

Most of the heat released by a fire flows by *convection,* a method of heat transfer through the mass movement of fluids. As a fluid is heated, it expands, becomes less dense, and rises; heat is carried with it. In wildland fires flaming combustion depends on the reaction of gases—fuels and oxygen (air); these gases are heated by the energy released by oxidation; they rise through convective processes. This convective flow not only removes heat and combustion byproducts from the reaction zone, but its flow regime helps to shape the flaming zone and it drafts in the fresh air needed to sustain combustion. Thus much as radiant and conductive heat transfer affect the availability of fuels, so convective heat transfer influences the availability of oxygen. *Convective velocity* (or *buoyancy*) describes the rate of convective heat transfer; greater velocity means increased indrafting of oxygen, and the greater turbulence that accompanies higher velocities means improved mixing within the reaction zone. Its velocity also controls the rate of combustion indirectly by purging the reaction zone of combustion byproducts, water vapor, or other noncombustibles.

Convective transfer is not the sole mechanism of diffusion mixing: ambient winds are also important. In heading fires, wind and convective flow act in the same direction. Convective heat contributes to preheating both directly (through contact with fuels) and indirectly (through shaping the flaming zone and improving radiant transfer). Its direct contributions are most felt where fuelbeds are thick and porous. Only a little convective heat is transmitted through a bed of pine needles, more through tall grass, and much more through shrubs or aerial fuels such as tree crowns. Because it rises and because of its quantity, convective flow can enlarge the effective fuel array—incorporating aerial fuels that might otherwise not be involved in a surface fire.

The shape of the flaming zone, of course, will affect the character of heat transfer. Flame shape will, in turn, derive from the overall shape of the fire. In general, the fire may assume three forms: a point, a line, or an area. All fires begin as points, most evolve into lines, and a few develop into areas. Their flame shape, and the intensity of their heat transfer, will vary accordingly. Line fires have larger flaming zones than do point fires, and area fires have larger flames than line fires. If a fuel particle is exposed to a line fire, it will be irradiated from several points, not just one. The amount of radiant heat contributed to the particle will increase, and if the line fire is also a heading fire, convective heat flow will contribute to preheating as well. Area fires, although showing a larger flame, develop strong indrafts that cause the flames to pull away from fresh fuel. The amount of radiant heat transferred to outside fuels may be less because of this shape, and there will be little convective heat transfer. The fire will burn down into larger fuels within the flaming zone, and air–fuel mixing processes, organized by strong upward convective flow, will tend to control overall fire behavior.

Even within the category of line fires, three flame shapes are possible. The flame may rise vertically, it may lean forward into fresh fuel, or it may lean backwards into already burned fuels. A given fire may show all three arrangements, the result of how the flaming zone is distributed around the perimeter of the fire. The *heading fire,* in which the flame bends over fresh fuel, will be the most effective at transferring heat. The flame is lengthened, stretched over the fuels, and may actually touch them. Radiant heat transfer is more efficient, and convective heat flow is possible over or through porous fuelbeds. The *backing fire,* in which the fire leans away from fresh fuels, is correspondingly far less effective. Radiant heating is less efficient, and convective flow will draft in cool air across the preheating fuels. A heading fire will thus burn more rapidly than a backing fire, fashioning a larger and more intense combustion zone and helping to shape the fire as a whole. Between these two arrangements is the neutral, or *flanking fire,* in which the flame rises vertically (see Figure 1-7; Rothermel, 1972).

*Heat Sink.* Of the heat released by combustion, some is lost from the system (entropy), some is expended in preheating new fuels, and some is discharged, mostly by convection, to the atmosphere. Convective heat is not lost since it contributes to preheating and to the diffusion process. But the most critical component of the total heat sink is new fuel. If insufficient heat is transmitted to fresh fuels, they will not combust and the fire will extinguish. If the heat absorbed by fuels is just adequate for combustion, the fire will spread in steady-state fashion. If, however, far more heat can be transferred to fresh fuels than is needed for steady-state propagation, then the fire can build up in total intensity and show different forms of behavior. The necessary heat output, in short, is set by the demands of preignition. In a wildland fire fuel is not a passive heat sink: it must be converted into a heat source if the fire is to continue.

Combustion is a process of successive ignitions, and how much heat it can produce depends, in good measure, on how much heat is required for each ignition. The heat of preignition describes the amount of heat required, while the rate of preheating gives the required rate of fuel flow into the combustion zone. How much of the heat released by combustion actually preheats new fuels depends on the efficiency of heat transfer and the thermophysical properties of the fuel. The efficiency of wildland combustion is not great. Even in special combustion environments, wildland fuels such as wood are less efficiently burned than fuels such as gasoline that have been refined into their most active ingredients and ignite at low kindling temperatures. But the combustion environment of wildland fires plunges even these efficiencies still lower. Considerable portions of the original fuel will only partially combust. In a furnace wood can achieve combustion efficiencies of 80%; in a wildland fire, of perhaps 5%.

Particles with high surface area-to-volume ratios contribute more heat more rapidly than do particles with low ratios. Fine fuels are more readily converted to heat sources, whereas coarse fuels may absorb heat that does not lead to combustion —especially to flaming combustion. In fuel complexes that consist of fuel particles of different sizes, flaming combustion will pursue a path of least resistance—propagating preferentially along those fuels having lower heats of preignition, those that convert most rapidly into heat sources. If fuel particles are mixed, then the finer particles may subsequently inflame the next larger sizes in the fire. If the particles cluster into groups, the flaming front will follow those fuels that react most quickly. An irregular fire perimeter will result. But if the flame residence time is long, then the larger particles may hold fire while the flaming front advances. The fire may become an area fire or it may simply show spots of flame, such as burning logs, left behind an advancing flaming front that moves through pine needles. Moreover, the availability of fuel particles is relative to the transferred heat. Large amounts of heat make more fuel available, reducing the discontinuities of the fuelbed.

*Combustion Environment Temperatures.* Most of the heat of combustion is contained in convection, and secondarily in radiation. Since radiant heat is so important in preignition, and since it can be measured experimentally, consider two illustrations of radiant heat histories. In the first set, the radiant heat histories, or *thermal pulses,* differ according to the properties of the fuelbeds through which they burn. In the second, the temperature profiles differ according to fire characteristics, the shape of the flaming front.

Figure 1-8 gives the results of burning two laboratory fuelbeds of pine needles with equal fuel moisture contents. For a uelbed of ponderosa pine needles, the thermal pulse shows a rapid volatilization of fuel characterized by a high, narrow profile. After this flareup, however, the temperature quickly declines until glowing combustion establishes itself. The initial spike describes the radiant heat produced by the flaming front during the horizontal spread of the fire. The later bulge, produced by glowing combustion, pertains to the downward spread of the fire into the fuelbed. The flaming zone will be narrow, but the total combustion zone large. For a fuelbed of white pine needles, the flaming front shows less rapid but more prolonged volatilization. Greater total energy is released within the flaming zone, and less within the glowing zone. The differences in the physical and chemical properties of the fuel particles and fuel array account for the different thermal pulse profiles (Anderson, 1964).

For fuelbeds of small-sized fuels, the thermal pulse reacts quickly—rising sharply and dropping almost as abruptly. With large-sized fuels, the delay time is longer, but the residence time may also be longer and the heat flux will not fall off so

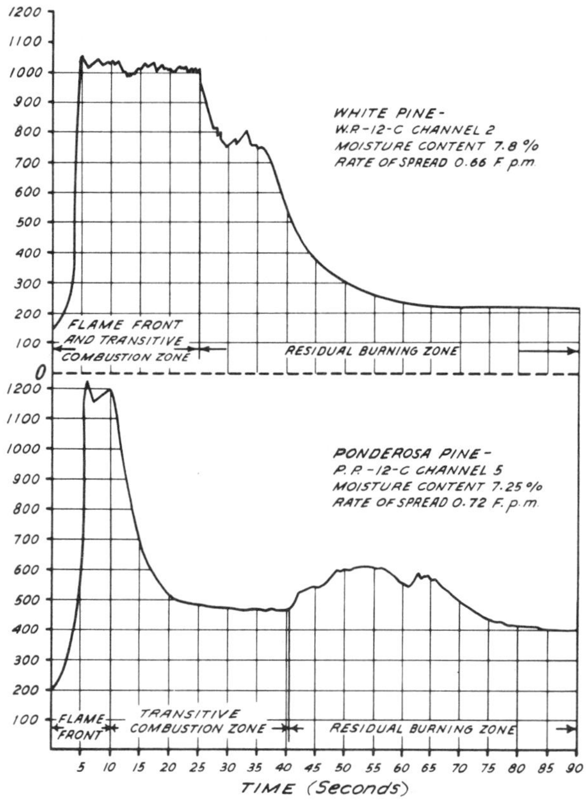


*Figure 1-8. Thermal pulse (combustion temperature profile) for laboratory fuelbeds of pine needles. Upper graph shows white pine; lower graph, ponderosa pine. Temperature in °F. From Anderson (1964).*

rapidly. Fuelbeds of mixed fuel sizes give intermediate thermal pulses (Figure 1-9; Countryman, 1969a, b).

Consider next the differences in radiant heat transfer that result from flame shape (Figure 1-10). Again, the graph pertains to laboratory fires; the fuel type and arrangement is constant, but the fire is heading, backing, or neutral. With a heading fire—flames lengthened and bent over fresh fuels—preheating is virtually instantaneous. Convective heating combines with radiant heating to immerse a surface fuel particle in heat. The range of influence of the flaming zone is relatively large. With a backing fire, however, the effectiveness of radiant heating is reduced. The flames bend back into the burned area, away from fresh

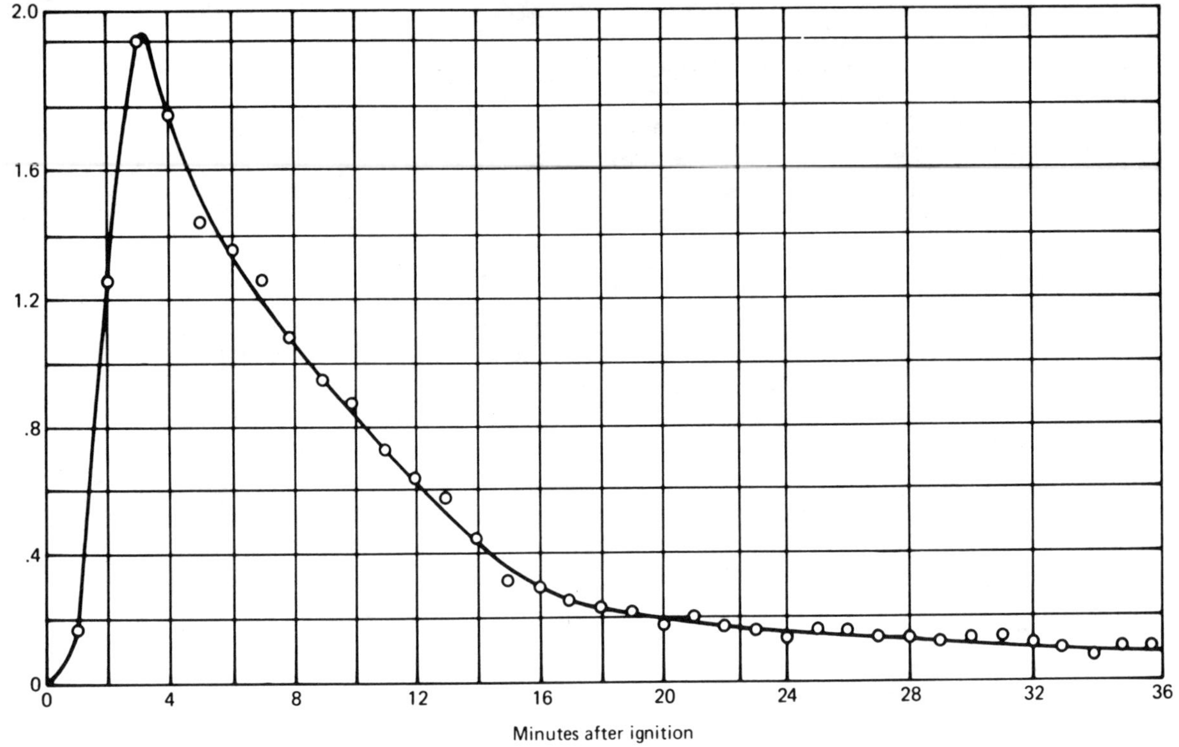


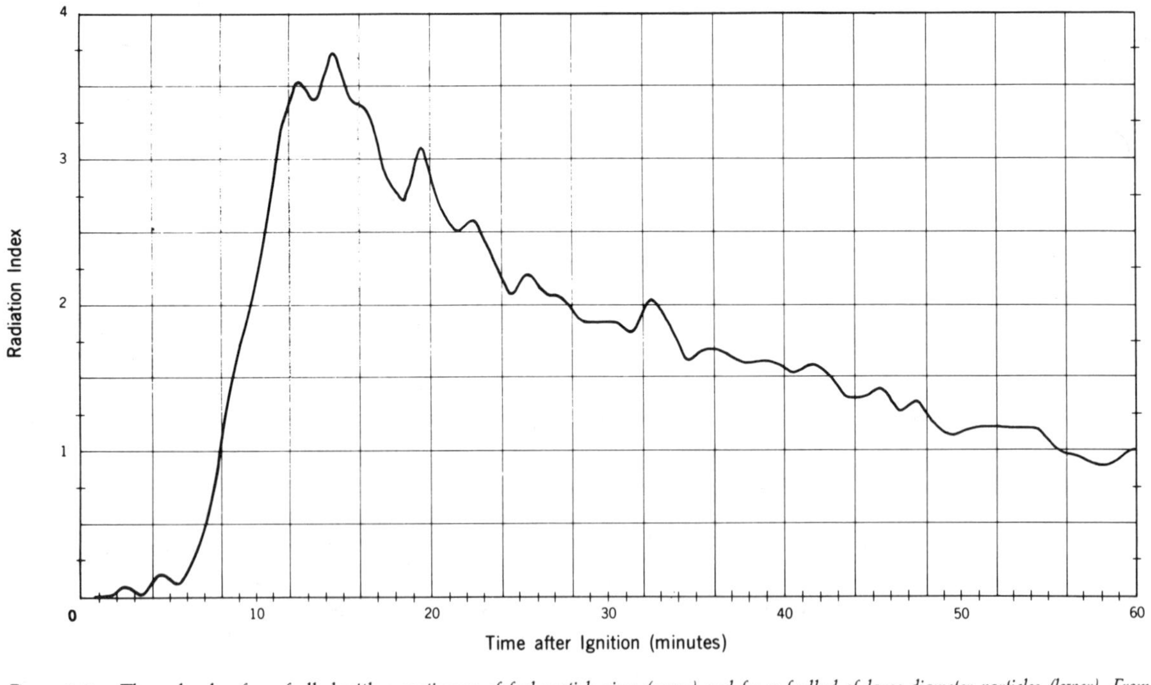


*Figure 1-9. Thermal pulses for a fuelbed with a continuum of fuel particle sizes (upper) and for a fuelbed of large diameter particles (lower). From Countryman (1969a,b).*

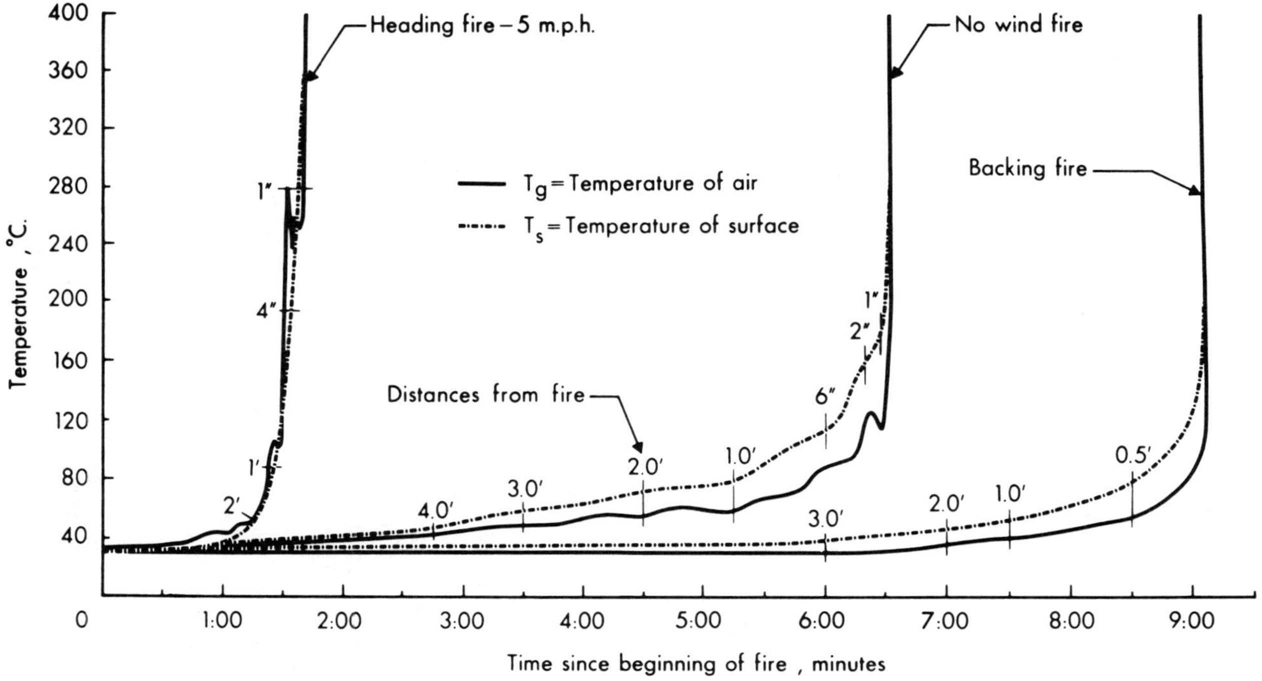


*Figure 1-10.* *Fire spread mechanism and fuel temperature histories. Three cases are illustrated: heading fire, neutral fire, backing fire. From Rothermel (1972).*

fuels. Convective flow does not contribute to preignition, and even retards it by drafting in cool air over the zone of preheating. With neutral fires —neither backing nor heading—the delay time to ignition is less than that for a heading fire, but better than for a backing fire. Heading fires, in brief, are proportionately more successful in transmitting heat to fresh fuels; hence they show higher rates of spread and higher reaction intensities. For a given wildland fire, it is common to have a mixture of heading and backing fires distributed around the flaming perimeter (Rothermel, 1972).

### *Air Flow*

Without oxygen combustion cannot occur. Oxygen must be brought into the reaction zone, it must be mixed with fuel, and noncombustibles (including the products of combustion) must be removed from the scene. Moreover, many secondary pyrolysis reactions require oxygen, and the presence of oxygen will select for one of several alternate pathways. And unless stable products of combustion, or noncombustible products of preheating and pyrolysis, are removed from the reaction site they will dilute the mixing of fuel and oxygen and smother the chain reactions that determine the chemistry of combustion. In regulated fires oxygen may be forcibly injected into the reaction zone, and a combustion chamber may be designed in such a way that the injection of oxygen, the mixing of fuel and oxygen, and the ejection of combustion products occur in sequence, at ideal locations, and with maximum effect. In free-burning wildland fires, oxygen is brought into the combustion environment by wind or convective flow, mixing occurs by the turbulent interaction of air and fuel, and waste products are ejected by wind or convection. The flow regime of air and gases within and around a fire results in mixing through processes of *diffusion.*

Wind and convective flow from the fire together account for diffusion in a wildland fire, but they may be competitive processes. Most fires shape their flaming zone by combining these two flows into one regime. But for some fires one or the other flow dominates. If the wind is very strong, it simply drives convective heat along with it, just as it does other combustion products like smoke. If the convective velocity is greater than the wind velocity, however, then convective flow, organized into a column, will control diffusion within the combustion zone. Strong indrafts bring in fresh oxygen, turbulence mixes oxygen and fuel, and the chimneylike column discharges waste products away from the reaction zone. The rate of diffusion, for wind-driven fires, equates with wind velocity; for convection-driven fires, with convective velocity. Thus intensely burning fires, with high convective velocities, create conditions that encourage increased oxygen flow and mixing.

Diffusion rates can control the rate of combustion. A rapidly moving fire, or fires burning intensely over large areas, can evolve flammable gases faster than diffusion processes can mix them with oxygen. Gases will escape from the combustion zone—some to vanish unburned, and some to flare into streamers of flame at a distance from the fire. In some fires, the inability to purge the reaction zone of noncombustibles may be critical in preventing increased combustion rates. Poorly burning fires (a fire smoldering in pine needles with high fuel moisture content, for example) will not be able to expel the distilled water vapor or other noncombustibles and will not draft in the oxygen they need for rapid combustion. Mixing will be most efficient along the boundaries of the reaction zone. The center, rich with evolved gases, will be somewhat deprived of oxygen, while the outside of the zone will be lacking in sufficient fuels. Where the fuel and air meet and mix, however, combustion will be most intense. That boundary will shape the flaming zone. The flow regime it represents will be far from simple.

Control over diffusion processes is one of the means of manipulating a regulated fire. Even in certain quasi-free-burning but confined fires, control over the flow regime is possible. A fireplace shields a combustion zone from wind and a chimney allows for manipulation of convective flow. In structural fires, some control over diffusion processes—referred to as *ventilation* practices—is possible for those fires that are still confined within a building. By creating openings for the inflow of oxygen and the outflow of combustion products, proper ventilation can manipulate the flow of oxygen, fire behavior, and the direction of fire spread. Conversely, improper ventilation, by suddenly introducing oxygen into a super-heated environment filled with evolved gases, may transform a smoldering fire into a flaming inferno. In wildland fires, there is little control possible over diffusion. Awareness of special fire behavior associated with wind-driven or convection-driven fires, however, and of certain topographic features that act like chimneys to concentrate and organize convective flow is essential to proper fire control strategy.

## 1.3 CHEMISTRY OF WILDLAND COMBUSTION

Combustion is a chemical process that transforms certain substances, known as fuels, into other substances, known as combustion products. In the process, heat is liberated. In regulated fires, the fuels may be refined into their most active constituents, combustion may be complete, and the products predictable. In wildland fires, the fuels are not generally uniform in size or chemistry, the flammable components must be evolved out of these fuel particles in the course of burning, combustion is incomplete, and the final products derive from a great mixture of primary and secondary reactions both within and outside the combustion zone proper.

The physical and chemical attributes of fuel particles, and of the fuel arrays that organize these particles, determine both the potential heat source and the heat sink of the fire. Only a portion of the total fuel load will actually be available for combustion, especially flaming combustion. These available fuels will set both the heat of preignition and the heat of combustion, the heat that must be absorbed for combustion to begin and the heat released by combustion. The relationship is dynamic. The greater the heat of combustion, the greater the quantity of available fuel. Not all of the heat released, of course, is applied to the preheating of new fuels, and not all of the pyrolysates evolved from the fuels are properly mixed and combusted. Heat transfer mechanisms and processes of diffusion intervene. But how fuel and fire interact—through processes of distillation, pyrolysis, flaming, and glowing combustion—will determine the type of byproducts that result.

### *Thermochemistry of Wildland Fuels*

In determining fuel availability, heat yields, and heats of preignition, far more is involved than the physical traits of fuel particles and fuel arrays. Pyrolysis and combustion are chemical processes that are profoundly influenced by the biochemical composition of fuel particles. Different fuel components have different energy potentials and different degrees of thermal stability; they respond in different ways and at different times to preheating, show different caloric contents, contribute differently to the total combustion process (especially to flammability), and result in different products. Its chemical constituency is a fundamental attribute of a fuel particle, and the distribution of chemicals, an important component of a fuelbed.

Considered as a group, wildland (woody) fuels show a certain uniformity of caloric and chemical content in comparison to other groups of fuel. But the differences within that group are significant for the course of wildland combustion and its products. Just as small fuel particles can help ignite larger particles in a fuel array, so a suitable distribution of chemical constituents within an array can make more of the total potential energy available for combustion. Especially significant are those fuel complexes—fine fuel particles with

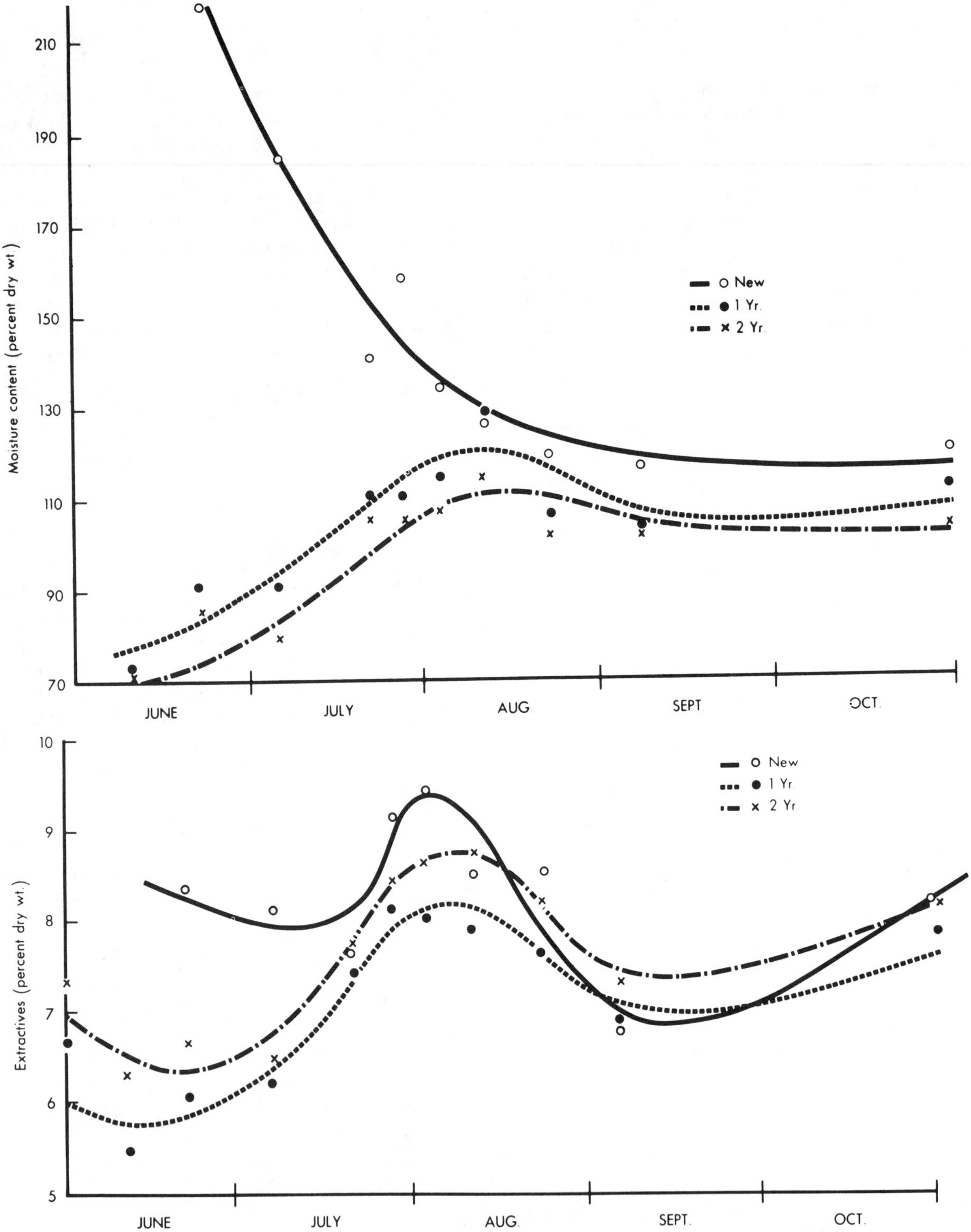


*Figure 1-11. Seasonal changes in fuel moisture content (upper) and ether extractive content (lower) for Douglas-fir needles during the 1968 fire season. Note the coincidence of low FMC and high ether extractive content during August, traditionally a difficult fire month. From Philpot and Mutch (1971).*

readily evolved gases—in which the physical and chemical aspects combine to produce high flammability. Much as fuel arrangements determine what proportion of the fuel load will be available fuel, so chemical composition helps to decide what proportion of the total energy will become available energy. And again the relationship between the thermochemistry of a fuel and the fire —as between fuel thermophysics and a fire—is dynamic. Greater heats of combustion can involve more constituents.

*Chemical Composition.* Most woody fuels consist of cellulose, lignin, extractives, minerals, and water. Its water content is the most variable component of a fuel's chemistry, and in determining the course of combustion, the most significant. Fuel moisture content will vary for dead fuels with precipitation and atmospheric relative humidity; for live fuels, with seasonal growth patterns and drought. Where fuel particles are large, a moisture gradient will develop in dead fuels between the surface and the interior. Where fuel complexes consist of both dead and living fuels, the effective fuel moisture will reflect their combined water content, a reason why most large fires occur in the fall, after the curing of live fuels; in the spring, before flushing; and in droughts, when live fuels are less abundant and contain less moisture (Figure 1-11).

The significance of this adsorbed water is that it acts as a heat sink—its high specific heat drains away much of the heat of combustion, while raising the heat of preignition. All this water must be removed before the particle itself will heat. Unlike wood, water is not a fuel. Its distillation does not lead to combustible products that can add to the heat source. Instead, as a liquid, it slows down pyrolysis by raising the heat of preignition, and as a vapor, it slows down combustion by cooling the reaction zone and by interfering with diffusion mixing. This distilled water, finally, is in addition to the water of reaction—the water that results from the combustion of wood. This, too, must be expelled from the reaction zone, but it does not interfere with preignition processes.

Of the total fuel load, cellulose claims by far the largest proportion, 50–65%. Compared to lignin or extractives, its heat yield is relatively low. But its great abundance, its capacity to generate large volumes of flammable gases, and its susceptibility to influence by small amounts of minerals, all assure it of a critical role in combustion chemistry. That even minute quantities of nonsilica minerals retard its pyrolysis makes cellulose the primary target for the application of fire retardants. By contrast, lignin shows considerable thermal stability, resisting pyrolysis better than cellulose. When it does volatilize, it evolves products of lesser flammability, more prone to glowing than flaming combustion. Lignin comprises 23–35% of softwoods, 16–25% of hardwoods, and up to 65% of punky, decayed wood, resisting biological decomposition as it does thermal degradation. Hemicellulose, a noncellulosic polysaccharide, makes up 15–30% of woody fuels. Figure 1-12 illustrates the differential thermal response of these substances. The primary constituents of wildland fuels, in brief, assume different characteristic forms of combustion: lignin and char favor glowing combustion; extractives and pyrolysates from cellulose, flaming combustion (Tangren, 1976).

Extractives (waxes, terpenes, oils) constitute 0.2–15% of fuel loading. But they are second only to cellulose in the quantity of flammable gases they evolve, and they are the first to volatilize—often at low temperatures, sometimes below the boiling point of water. For small investments of energy by preheating, extractives return large heat energies by combustion. This means that fuels with high proportions of extractives not only have large amounts of high caloric fuels available at low heats of combustion, but that they can function as a kind of biochemical kindling, raising the heat of combustion to levels that induce the pyrolysis of other fuels such as cellulose whose combustion can further intensify the burning process. Ether extractives, in particular, are responsible for high proportions of the total heat yield. In fuels with high ether extractive content, the delay time of ignition diminishes, the

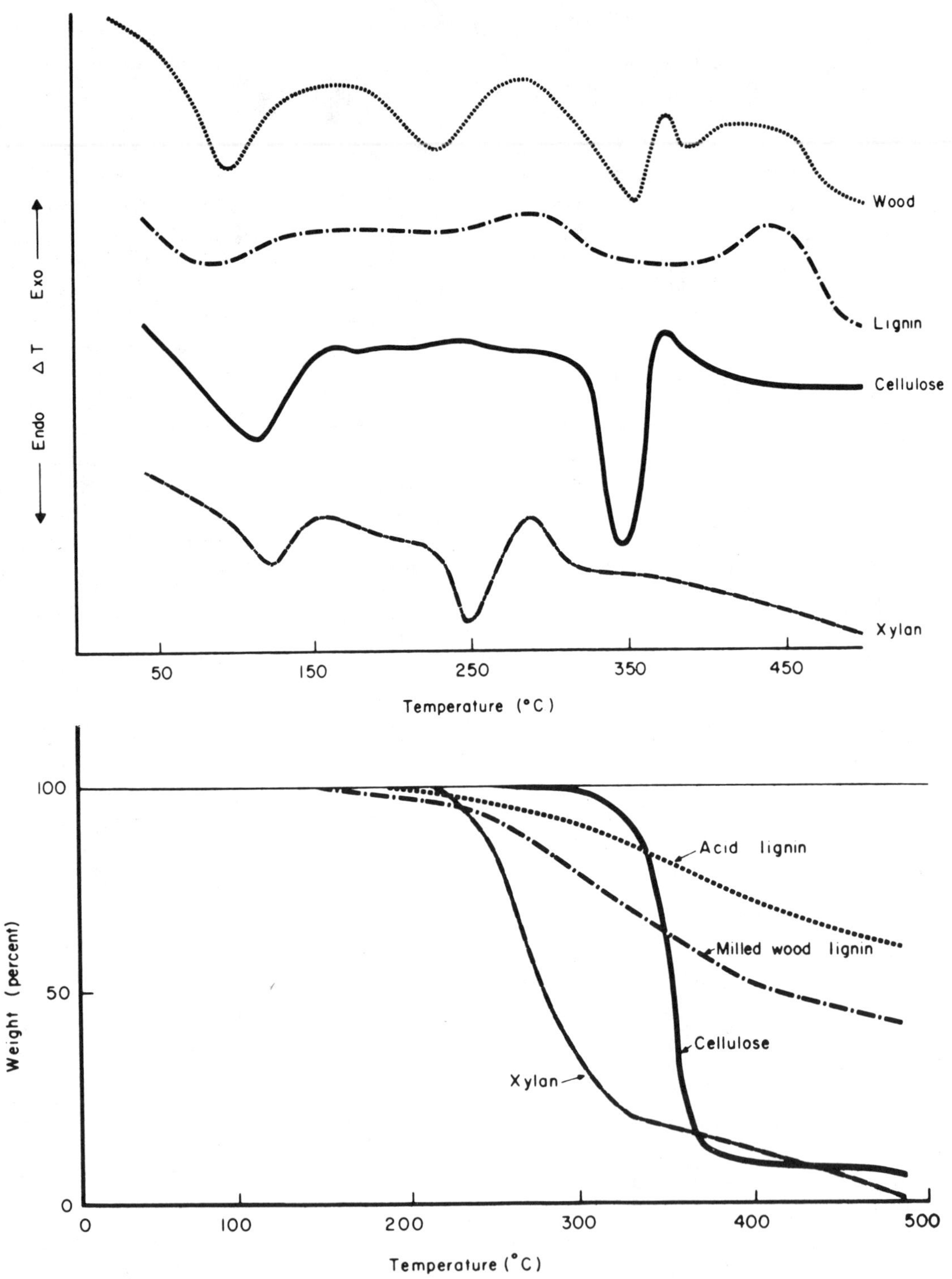


*Figure 1-12. Thermal decomposition of black cottonwood according to its principal chemical components. Upper diagram gives a differential thermal analysis; lower diagram, a thermogravimetric analysis. Note, in particular, the sudden breakdown of cellulose. From Philpot (1971).*

*Figure 1-13. Effect of mineral content on combustion. (Below) The relationship between organic residue and silica-free ash (400°C); (opposite) the effect of moisture and minerals on predicted rates of fire spread. From Philpot (1970), Rothermel (1972).*

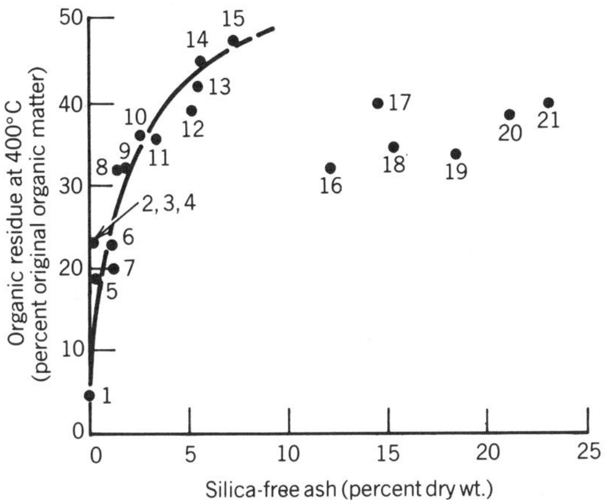


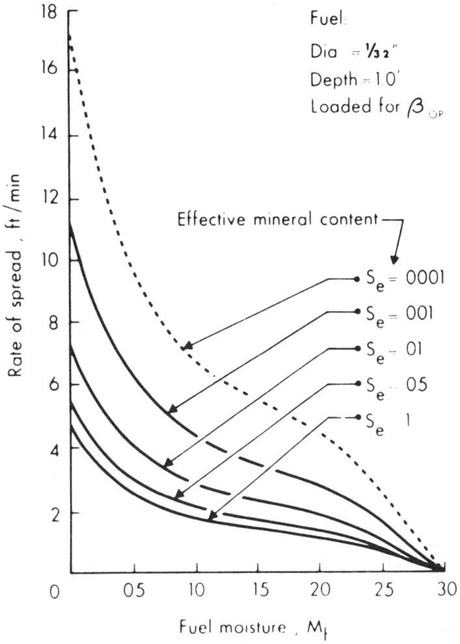


flammability improves, and the heat yield increases. The presence of such extractives within living fuels, localized on needles and leaves, accounts in part for the combustibility of living fuels despite their large fuel moistures. Flammability will show seasonal changes according to the pattern of extractive production (see Figure 1-11; Philpot and Mutch, 1971).

*Heat Yields.* Its various chemical constituents will contribute differently to the total heat yield of a woody fuel. With the heat of combustion set at 750°F (400°C), the relative heat contributions will be

$$\underset{\text{Comb}}{\overset{750°\text{F}}{H}} = (\text{Fuel}) = 3850a_{\text{Cell}} + 5860a_{\text{Lig}} + 7720a_{\text{Ext}} \quad (1)$$

where $a_{\text{Cell}}$ describes the proportion of cellulose, $a_{\text{Lig}}$ the proportion of lignin, and $a_{\text{Ext}}$ the proportion of extractives. The coefficients give the low heat yield for each component—the high yield minus fixed loss due to fuel moisture. Note that the heat yield for cellulose is lower than for the other constituents. But the high percentage of cellulose in woody fuel and its ready gasification during pyrolysis make it the largest single contributor to total heat output. Note, too, that the total heat yield must include both flaming and glowing combustion (Rothermel, 1976).

These tendencies can vary according to the presence of minerals. Some mineral constituents such as silica are inert, irrelevant to the thermochemistry of a fuel. Others such as potassium, sodium, and phosphorus may actively intervene in pyrolysis and combustion reactions. In wildland fuels, silica-free mineral content varies from 0.1 to 3%. The presence of such minerals alters the pathways of pyrolysis, and the quantity of minerals present need not be large in order to be effective (Figure 1-13). Fewer volatiles are generated, more char results, and less heat is released quickly. Glowing combustion replaces flaming combustion, and even glowing combustion may be inhibited as the heat of combustion drops (Philpot, 1970).

The general relationship between mineral content and flammability has been known for some time. It underwrites the use of chemical fire retardants—all of which combine water with selected inorganic salts. By influencing the pyrolysis of cellulose, even small quantities of retardant can help replace flaming combustion with glowing combustion. Moreover, this relationship also underscores the search for plants of low flammability to replace those of high flammability. Desert plants rich in salt content are good candidates.

### *General Reaction Mechanism*

The chemical equations describing the pyrolysis and combustion of wildland fuels are maddeningly complex. The size of fuel molecules and the inconstant fire environment within which these fuels are thermally degraded, oxydized, and resynthesized introduces further variability. The general reaction mechanism is understood, but the elementary reactions are not. Since most woody fuels consist of cellulose, and since cellulose consists largely of sugar, the combustion of glucose sugar can approximate the combustion of cellulosic fuels:

$$C_6H_{12}O_6 + 6O_2 \rightarrow 6CO_2 + 1{,}211{,}000 \text{ Btu} \quad (305{,}172 \text{ kcal}) \qquad (2)$$

One molecule of sugar combines with six molecules of oxygen to yield six molecules of carbon dioxide, six molecules of water, and heat.

The elementary sugar unit for pure cellulose is $(C_6H_{10}O_5)_n$. But woody fuels consist of more than cellulose; they will contain quantities of adsorbed water, and for flaming combustion cellulose does not burn directly, but indirectly through evolved gases. Since oxygen is taken from the air, other constituents of air such as water vapor and nitrogen are brought into the reaction zone to either assist or retard the chain reactions. Nor, under wildland conditions, is combustion complete. The actual heat yield will only represent a fraction of its potential heat yield.

*Chemistry of Flame.* The analysis of combustion chemistry is difficult because the chemistry of burning does not move directly from fuel to combustion product, but passes through a tangle of intermediate products. With wildland fire, the fuel itself must be processed in front of and within the reaction zone in order to combust; combustion is heterogeneous, both flaming and glowing; the reaction zone is not well confined, allowing many constituents to escape in various states of reaction. Especially significant is the chemistry of flame. Its flaming zone is the most important site for chemical activity in a fire, and the most complex.

The *general reaction mechanism* that defines the combustion process sums up many smaller reactions and it includes many reactive intermediates that actually make flaming combustion work. At its foundation are *elementary reactions,* hundreds of which may be present in the breakdown of fuel molecules and the reconstitution of combustion products. Their number increases geometrically with the size of the fuel molecule. For wildland fuels of complex hydrocarbons, the number is very large. Each elementary reaction proceeds at rates that depend on the concentration of reactant molecules and a constant of proportionality, its *rate coefficient.* Because the frequency and speed of molecules upon which the reactions depend vary with temperature, the rate coefficients are temperature-dependent. Higher temperatures increase the rate of reaction. Higher temperatures, in fact, influence all aspects of combustion—the rate of fuel evolution, the mixing of fuel and oxygen, and the speed of reactions.

The elementary reactions involved in combustion are peculiar in that they participate in chain reactions. Ignition is a chain-initiating process, propagation a chain-branching process, and extinction a chain-terminating process. An example of chain-initiation is the reaction $C_3H_3 \rightarrow C_2H_5 + CH_3$. Raised to high temperatures, which invite collisions, an otherwise stable molecule of propane decomposes into two radicals. These radicals become *chain centers* which interact in such a way that more chain centers result, a cascade effect which sets up a chain-branching sequence. Perhaps the most significant of such sequences is that which involves atomic hydrogen and molecular oxygen, from which comes atomic oxygen and a hydroxyl radical: $H + O_2 \rightarrow O + OH$. Propagation is assured because each reaction either maintains or multiplies the number of products available for future reactions. Flame provides an ideal environment for chain-branching processes of this sort. For extinction to occur, the chain-branching process is reversed. Chain centers combine, often in the presence of another chemical, to produce stable molecules. An example is the formation of water by the combination of a hydrogen atom, a hydroxyl radical, and molecular nitrogen: $H + OH + N_2 \rightarrow H_2O + N_2$.

These processes occur at different sites within the combustion zone. Chain-initiation takes place ahead of the flaming zone, chain-propagation within the flaming zone, and chain-termination outside the flaming zone. The flame does not simply determine the distribution of these processes; the arrangement of the processes, as set by fuel properties, diffusion rates, and temperature, determines the geometry of the flame. For chain-branching to occur, combustion must involve flame (Gardiner, 1982).

*Mass Balance and Energy Equations.* The spread of flame within the combustion zone can be imagined in terms of two flows: the flow of mass and the flow of heat. Rates of combustion, reaction intensity, and flame propagation can be equally expressed by the flow rates for fuel and oxygen. Similarly, the process of burning that transforms fuel into combustion products and potential chemical energy into kinetic energy of heat and motion can be equally well expressed in terms of mass balance and the conservation of energy (Figure 1-14).

The mass conserved includes fuel and oxygen. Initially the fuel mass will consist of solids, with a variable amount of free water incorporated into them. At the conclusion to burning, the mass will

**Mass Balance Equation**

$$[O_2] + [M_m] + M_0 = M_{ash} + M_{char} + M_{volatiles} + [M_m] + [O_2]$$

where
$M_{volatiles}$ = Extractives + distillates
$M_{char}$ = Charcoal + unburned fuels
$M_{ash}$ = Mineral solids
$M_m$ = Fuel moisture

**Energy Balance Equation**

$$E_0 = E_{comb} + E_{res} + E_{loss}$$

where
$E_0$ = Total potential energy
$E_{com}$ = Energy released by combustion
$E_{res}$ = Unburned residue
$E_{loss}$ = Entropy

*Figure 1-14. Mass balance and energy equations. After Albini (1980).*

consist of solids, liquids, and gases. The solids left are char and mineral ash. The liquids will be tars or aerosols, and they usually represent poor combustion. Among the gases will be water vapor, both evaporated fuel moisture and the water of reaction; flammable gases evolved during preheating, pyrolysis, and combustion; and gaseous products of combustion, notably water vapor and carbon dioxide. If combustion is complete, only ash and gases will remain. The quantity of gases evolved can be measured by weighing the mass loss of the solid fuel.

The fuel must combine with oxygen to combust. The mass of oxygen needed to combust a mass of fuel can be expressed in terms of a *stoichiometric mass ratio,* a constant fixed by the chemical composition of the fuel. The ratio may take two forms. It may describe the mass of fuel relative to the mass of air, or the mass of combustible gases proportionate to the mass of oxygen. A ton (907 kg) of wood fuel requires approximately 7 tons (6350 kg) of air (175,000 $ft^3$ or 4955 $m_3$) for complete combustion. The actual percentages of air and fuel involved in the reaction will depend on the rates of heat transfer and diffusion that describe the efficiency of pyrolysis, mixing, and combustion (Albini, 1980).

The flow of fuel and oxygen into the combustion zone is complex. In a regulated fire, both fuel and oxygen can be premixed in proper proportion before entering the heated reaction zone. The rate of fuel–air flow is described by flame velocity, and the direction of flow, by flame shape. In wildland fires, fuels are prepared both within and around the reaction zone, and oxygen is brought into the zone by the ambient winds and convective flow that together govern diffusion. Fuel flows from below the reaction zone up into it, and it flows from the sides of the reaction zone inward to it. This complex fuel flow pattern accounts for the complex configuration of the flaming zone. Flame velocity describes the upward fuel flow, and fire spread, the inward flow.

Like any energy transformation, the total energy of combustion is conserved. But not all of it is conserved within the zone of combustion. The total energy of the system will equal the energy liberated during combustion (heat), the energy contained in residue (products of partial combustion), and the energy lost to useful work in the system (entropy). Considered in this way, wildland fires are not efficient. Most of the entropy is lost as radiation, but the magnitude of the loss is not well known. The inefficiency of wildland combustion is largely the result of incomplete combustion, potential energy that is not transformed to heat. Only a portion of fuels ends up as complete combustion products. Many remain as char or intermediate reaction products. Moreover, because wildland fire is heterogeneous, not all of the energy will be released simultaneously. The highest rates of energy release will occur within the flaming zone, but the total quantity of heat liberated must include the lower-intensity but longer-lasting process of glowing combustion.

### *Products of Burning*

The products of combustion will vary according to the chemical constituency of the fuel and the completeness of their burning. The ultimate products of combustion, water vapor and carbon dioxide, will be abundant. But not all fuels will combust. Some of the fuel will pass through the reaction zone more or less intact, as char or particulates, and some of the intermediate products of the general reaction mechanism will escape in various forms from the unconfined combustion environment. Wildland fuels can be made to burn more or less completely in special combustion chambers. In the field, however, pyrolysis is partial rather than total, combustion is incomplete rather than final, and considerable leakage of intermediary products occurs out of the reaction zone. Numerous secondary reactions take place outside the combustion zone proper. The variety of products may number in the hundreds, if not the thousands.

*Preignition Products.* Some products result from simple distillation—water vapor from fuel moisture, volatiles from ether and benzene extractives. The volatiles will combust if they are entrained into the reaction zone and mixed with oxygen. Other products require the pyrolysis of woody components. In general, pyrolysis follows two pathways. Rapid pyrolysis at high temperatures leads to pyrolysates and tars rather than char; slow pyrolysis at low temperatures, to char, water, and carbon dioxide rather than to pyrolysates and tars. The outcome will also differ according to the chemical constituency of the fuel. The pyrolysis of cellulose favors tars and pyrolysates; of lignin, char and aromatics; of extractives, flammable volatiles. Subjected to low, slow heating the fuel will be dehydrated of fuel moisture, volatiles, and pyrolysates. When deliberately controlled in an oxygen-deprived environment, this leaching process produces charcoal. The charcoal can then be burned, but lacking any flammable gases it will only combust by glowing.

*Combustion Products.* The flaming zone is the most important site for pyrosynthesis. Here pyrolysates and volatiles oxidize. Some of the gases will combust completely; others will oxidize only partially. Some lower molecular weight gases will escape to the atmosphere, some will agglomerate into soot, and some will condense into liquid tars. Some of the byproducts represent partial pyrolysis or combustion; others, new chemical compounds generated within the special environment of flame (Figure 1-15). An example of incomplete combustion is carbon monoxide; of pyrosynthesis, soot. Water vapor undergoing condensation and particulates combine to make visible smoke. White smoke indicates relatively high level of water vapor, the product of complete combustion (water of reaction) or of high fuel moistures (distilled water). Black smoke signifies incomplete combustion. The particulates may be particles of solid fuel which broke loose, were entrained in the reaction zone, and passed through the flaming zone with little more than scorching; or they may represent solids newly synthesized around the reaction zone.

Of the pyrosynthesized solids, soot is the most important. Soot is promoted by certain fuels and certain conditions of combustion. Suitable fuels show a low hydrogen-to-carbon ratio among

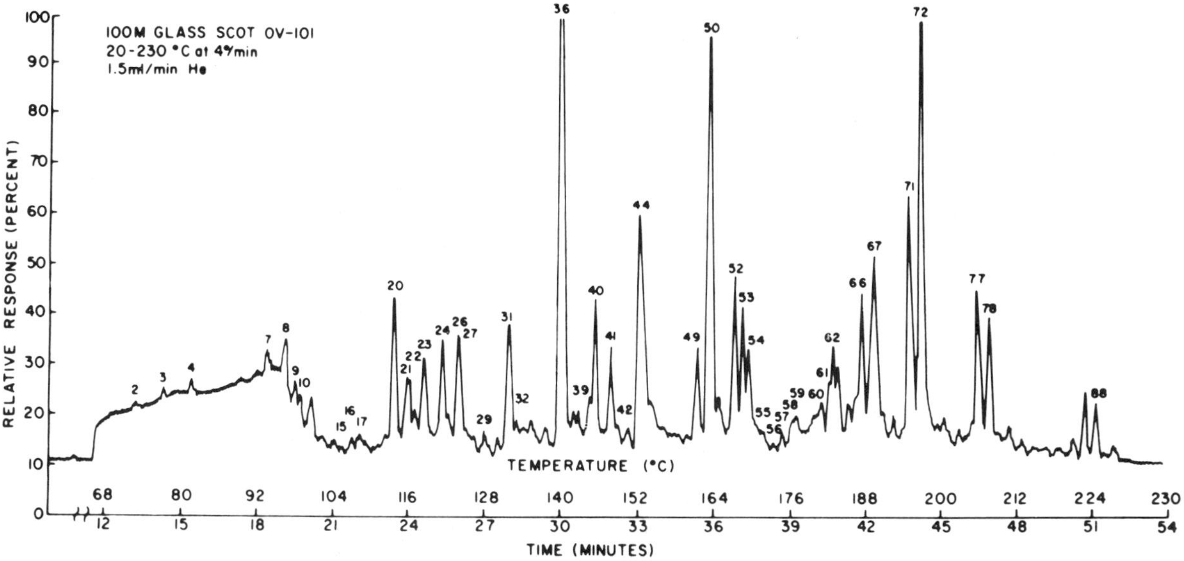


| PEAK | COMPOUND |
|---|---|
| 2 | isopentane |
| 2B | 1-pentene |
| 2C | furan |
| 3 | n-pentane |
| 3A | isoprene |
| 3B | acetone |
| 4A | isopropanol |
| 4B | cyclopentadiene |
| 7 | diacetyl |
| 8 | 1-hexene |
| 8A | methyl vinyl ketone |
| 9 | 2-methylfuran |
| 10 | n-hexane |
| 15 | 2,4-hexadiene |
| 16 | 1,3,5-hexatriene |
| 17 | 3-methylbutanal |
| 20 | benzene |
| 21 | cyclohexane |
| 22 | 4-methylpentene |
| 23 | 2,4-dimethylpentane |
| 24 | 1-trans-2-dimethyl-cyclopentane |

| PEAK | COMPOUND |
|---|---|
| 24A | sorbaldehyde (tent.) |
| 26 | 2,5-dimethylfuran |
| 26A | n-heptane |
| 27 | cis-2-heptane |
| 29 | 2-vinylfuran |
| 31 | 2,3-dimethyl-2-pentane |
| 32 | 2,4-dimethylhexane |
| 36 | toluene |
| 39 | 1-octene |
| 40 | 2,3-dimethylhexa-1,4-diene |
| 41 | n-octane |
| 42 | 2,3,5-trimethylfuran |
| 44 | furfural |
| 49 | ethyl benzene |
| 50 | p-xylene |
| 50A | 2-propionylfuran (tent.) |
| 52 | styrene |
| 53 | o-xylene |

| PEAK | COMPOUND |
|---|---|
| 54 | n-nonane |
| 54A | 2-isopropylfuran |
| 55 | anisole |
| 56 | 2-methyl-5-isopropenylfuran |
| 57 | cumene |
| 58 | n-decane |
| 59 | camphene (tent.) |
| 60 | n-propylbenzene |
| 61 | m-ethyltoluene |
| 62 | p-ethyltoluene |
| 66 | 1-decene |
| 67 | benzofuran |
| 71 | m-diethylbenzene |
| 72 | limonene |
| 77 | p-α-dimethylstyrene |
| 78 | n-undecane |
| 88 | n-dodecane |

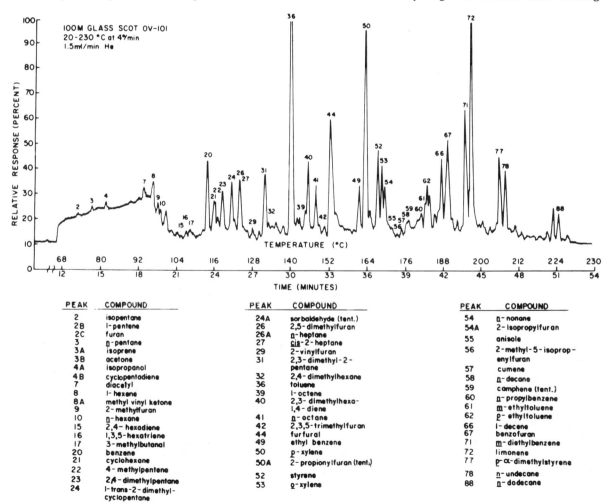

*Figure 1-15. Chromatogram of organic vapors in loblolly pine smoke. Each peak represents a separate compound. Smoke is also rich in particulates. From Tangren (1976).*

their compounds. A favorable combustion environment exhibits poor mixing—the product of too much fuel relative to available air or of poor diffusion mixing between fuel and air. Out of these circumstances soot forms in three stages. Instead of oxidizing, as occurs in complete combustion, the final molecules polymerize instead, forming polyacetylenes and polyaromatics. Some of these molecules, known as soot precursors, disappear. The aromatics persist, however; they are joined with aerosols (hydrogen and trapped hydrocarbons) and shaped into spheres. The aerosols are finally subjected to two competing processes: they oxidize in the flame and they agglomerate. If the oxidation rate is the faster process, the flames glow yellow and produce little black smoke. If agglomeration proceeds most rapidly, then black smoke forms. The agglomerated particles may continue to burn, provided the temperature remains high and the oxygen plentiful. But in wildland fire they are not confined within the reaction zone, and as soot they are expelled from the system (Gardiner, 1982).

The inefficiency of wildland fire has important consequences for fire ecology and fire management. As with fuels, so it is with biomass: fires tend to be selective in their effects, not total. The fire does not combust all fuels, and only a portion of the combusted fuel ends up as heat, water, and carbon dioxide. Much will remain, and much will only be reconstituted. For an ecosystem this residuum of fire can be an important source of nutrients, and the newly constituted fire environment an important factor in the quality of the site or habitat. For humans, however, the vast smoke plumes and towering convective columns that accompany large fires can become an annoying atmospheric pollutant. For fire managers, the staggering quantity of char and partially combusted fuels left behind the combustion zone presents serious problems. Any residual combustion must be mopped up to extinguish the fire. At the same time, the char and rearranged fuels make fires in many forested environments a part of long-term cycles rather than singular events. A landscape of charred snags, soot-encrusted litter, and standing conifers bearing limbs of dessicated needles is often only the initiation of a series of reburns, which like a slow parody of the combustion process, extends over years, decades, or even centuries.

## 1.4 FIRE RETARDANTS

Mankind can create heat sources of great magnitude, but not heat sinks on an equal order. The only sink immense enough to absorb the energy generated by large-scale combustion is the atmosphere, where most of the heat and kinetic motion of a fire eventually go. Part of the problem is that fire propagates, but fire-retarding agents dissipate. Except in fires of limited size and intensity, the ratio of propagation of dissipation is unfavorable and direct extinguishment is impossible. Nonetheless, fire intensities can be reduced and fire effects more easily controlled if retarding agents are present. Retardants slow down the reaction rates of a fire, encourage pyrolysis pathways that favor charring over flaming, and in special circumstances suppress fires. Retardants can be useful in all aspects of fire management, but they are best used to complement other operations, not to substitute for them.

Fire retardants fall into two general groups. Physical agents influence heat and diffusion processes. Chemical agents affect fuels, modifying the course of pyrolysis and combustion. Physical retardation relies primarily on water, though in limited areas dry powder (including dirt) is effective. Chemical retardation depends on chemical additives applied to the surface of fuel particles. Because their effectiveness vanishes with the evaporation of water, most physical agents are *short-term retardants;* by contrast, the persistence of chemical additives, even after the loss of a watery matrix, makes chemical agents into *long-term retardants.* Because they act selectively on one phase of combustion—flame—most wildland fire retardants belong to that category of fire inhibitors known as *flame retardants.*

### *Physical Agents*

Because of its high specific heat and natural abundance, water makes a superb extinguishing agent. Applied to the reaction zone, water absorbs heat and lowers the heat of combustion; applied to fuels, it raises the heat of preignition. And once converted to steam, it smothers the reaction zone even as it lowers the amount of heat produced, thereby diminishing the convective flow that might expel it. It is particularly effective in urban environments where, with ventilation under control, the expanding, penetrating, and smothering powers of steam can be exploited. In wildland environments steam, much as the case with foam, is less useful; it cannot be confined.

The efficiency of water can be improved by additives that reduce its surface tension (*wetting agents)* or that increase its viscosity (*thickening agents)*. By reducing surface tension a wetting agent increases the amount of surface area available for a given volume of water, improves the ability of water to coat rough surfaces, and diminishes the viscosity of the water. The increase in surface area means that more heat can be absorbed more rapidly. Reduced surface tension (the tendency of water to bead) allows treated water to penetrate into deeply furrowed logs burning by glowing combustion and to insinuate into deep mats of grass and needles. Reducing viscosity enhances the movement of water through hoses, with more water moved for less work.

But these assets come at a cost. Making water thinner improves its rate of heat absorption, but reduces its heat absorbing capacity. Where high heat absorption is needed, thickening agents are used. A wetting agent can reduce the height of a layer of water by half; thickening agents can congeal that layer up to 20 times its normal height. The increased viscosity which results means that thickening agents must be combined with water at the nozzle, not pumped through a hose. But they have advantages. Viscous fluids adhere better to fuel surfaces, coat and smother reaction surfaces, absorb greater quantities of heat, and penetrate better into the reaction zone. When delivered by aircraft such mixtures will not so readily dissipate during their descent and will better penetrate forest canopies. Such mixtures are known as gels or slurries, and the thickening agent is a clay (like bentonite) or a gum (like Gelgard). The rheological properties of gum-thickened slurries are generally superior, but the mixtures more costly.

Foam comprises a special category of thickeners. Foams both cool and smother a fire. The capacity of foam to expand means that more retardant by volume will be produced than from a similar weight of water alone, and the structure of foam allows it to resist quick dissipation. Most foams are restricted to structural fires, where, as with steam, their expansion properties can be put to best use. But some wildland foams have been developed for initial attack by ground tankers and for the protection of equipment and personnel trapped by a fire (Ebarb, 1978).

The different properties of water additives make particular mixtures useful at different stages of suppression. Water treated with viscous agents is most effective for the pretreatment of fuels and for initial attack, where much heat must be absorbed. Water treated with wetting agents is most suitable for mopping up smoldering fires in logs, grassy mats, and heavy duff where penetration by water would otherwise be difficult.

The means of delivering water to a fire are many. Water can be brought by hand, with buckets or backpack pumps; by portable power pumps, either hand-carried or mounted on vehicles; by locomotives, outfitted with tank cars and pumps; by vehicle, from ground tankers to engines; by aircraft, from helicopters to air tankers. Common to all techniques of application is the desire to direct water as a spray so as to maximize its surface area, to direct it at the base of the flames, and to apply it in sufficient quantities and at appropriate rates. High volume pumping is needed to control flaming combustion; low volume discharge, to mop up glowing combustion. For many areas, attack by water is the primary mode of suppression, the means by which to integrate rural and wildland fire protection, and an important technique for prescribed fire. In grasslands, wetlines can substitute for the firelines traditionally bared to mineral soil. In organic soils, flooding with water is the only practical method of extinguishment.

### *Chemical Agents*

There are various methods by which to chemically retard a fire. The chemical structure of the substance may be changed, the reaction surfaces coated, or flame-inhibiting agents injected into the combustion zone. Changes in structural chemistry are possible when dealing with a synthetic environment. The flammability of an object might be revised without destroying other properties, much like the manufacture of an artificial sweetener. Other fuels can be coated with special agents, such as borate or tumescent paints, that seal off the reaction surface when they are heated. Oxygen cannot enter and pyrolysates cannot escape, so combustion fails. In wildland situations such manipulation is not possible. To some extent certain fuels can be substituted for others, and certain structural arrangements of fuels can replace other arrangements. In general, however, chemical retardants must be applied to fuel surfaces.

The flame-retarding agents used in wildland fire control are mixtures of water and selected inorganic salts. Laboratory studies indicate that a long-term retardant, when fully dried out, is nearly twice as effective in reducing rate of spread as a short-term retardant which still retains two-thirds of its original moisture. The density of retardant application depends on the same fuel and fire properties as combustion; the fire and the retarding chemicals compete for the same fuel. Those fuel complexes with heavy fuel loads and a high proportion of exposed surfaces require more retardant. Retardants are most effective when applied in advance of a fire, on fuels that have good surface exposures, and on fires whose intensity is low. Retardants are best used, in brief, during initial attack.

The retardant mixtures, or *slurries,* cool the fuels, alter the routes and products of pyrolysis, quench the heat of combustion, and interfere with diffusion and oxidation in the reaction zone. Different retardants work on different phases of the combustion process. Some primarily influence pyrolysis, promoting char rather than combustible gases. Some interfere with flaming combustion, providing traps for the chain centers that are necessary to sustain chain-propagation. Some interrupt glowing combustion, coating and cooling the reaction zone until the heat of combustion drops below the point of extinction. Phosphorus compounds lower the ignition temperature of the fuel to which they are applied. The result is a decomposition process which—because it occurs at lower temperatures—does not evolve flammable gases but produces larger quantities of char and smoke. Glowing combustion replaces flaming combustion, reducing total heat output by as much as 35% (Figure 1-16). This, indeed, is the general mechanism of wildland fire retardants. They change the preignition pathway from hot, fast pyrolysis to cooler, slow pyrolysis, from the rapid evolution of flammable gases to the slow leaching of pyrolysates and the production of water, carbon dioxide, tar, and char. Since cellulose is the primary fuel constituent, and since it is especially sensitive to the presence of minerals, a small amount of the proper salts can produce proportionately large effects.

The chemical elements most responsible for retardation are phosphorus, antimony, chlorine, bromine, boron, and nitrogen. These have been formed into a great variety of compounds, marketed under a number of trade names. For wildland fire control, the most popular retardants are ammonium sulfate (AS) [$(NH_4)_2SO_4$] and diammonium phosphate (DAP) [$(NH_4)_2H_2PO_4$]. The latter combines the properties of both phosphorus and nitrogen, and there is evidence that a synergistic effect occurs when both are introduced simultaneously. AS is marketed under the trade name Fire-Trol; DAP, as Phos-Chek. Since the early 1960s, these two compounds have dominated fire retardant use—in part because of their

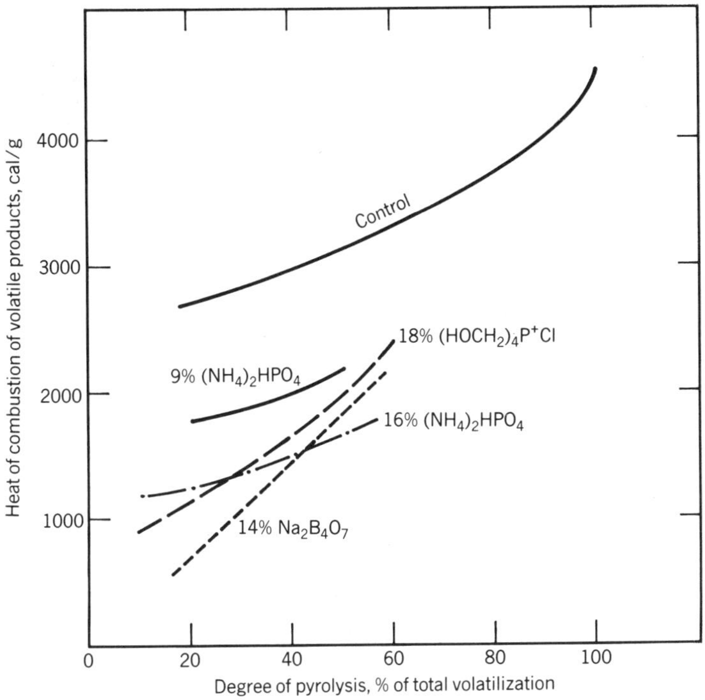


*Figure 1-16. Effect of chemical retardants on the heat of combustion. Volatile pyrolysis products are given as a function of degree of volatilization for ponderosa pine. Altering the pathway of pyrolysis also diminishes the total heat release as much as 35% or so. From Lyons (1970).*

intrinsic effectiveness, and in part because their similarity to agricultural fertilizers has assured their abundance, relative cheapness, and environmental compatibility.

These fire retardants interfere with both the physical and chemical processes of combustion. Its water matrix cools and its thickened gel coats reaction surfaces; the inorganic salts favor certain pathways of pyrolysis. By generating nonflammables rather than flammable gases, retardants affect the amount, rate, and type of mass lost. With fewer gases to combust, heat output is reduced, so the rate of fire spread diminishes. Moreover, retardants affect the mode of burning, promoting glowing over flaming combustion.

All of these effects vary with the type of retardant used. For example, ammonium phosphates are effective against both flaming and glowing combustion; sulfates, primarily against the flaming phase. The difference in effectiveness is on the order of 1.5 times, with only about two-thirds as much DAP required to do the work of a given quantity of AS. This success comes at a cost, however. By altering the pathways of pyrolysis, DAP reduces flame and heat but increases the production of particulates and residues. Figure 1-17 shows the effect of DAP and AS on weight loss rate and particulate production. Note that the effect in all cases is to slow the rapidity of mass loss. With DAP the proportion of particulate matter, however, increases significantly; much of this is tar. The value of DAP for prescribed burning—where smoke management may be important—is thus less than that of AS (George and Blakely, 1972).

Figure 1-18 shows the relative influence of DAP and AS on the rate of spread of laboratory fires in fuelbeds of pine needles. Rate of spread slowed with increased concentrations of retardant. Except for very thin or porous fuelbeds, however, increasing the retardant cover will not extinguish the fire. Only the surface layer will be affected. Flaming will be inhibited, but glowing combustion can continue beneath the retardant cover. For any fuelbed, then, there is an optimum coverage level (George and Blakely, 1972).

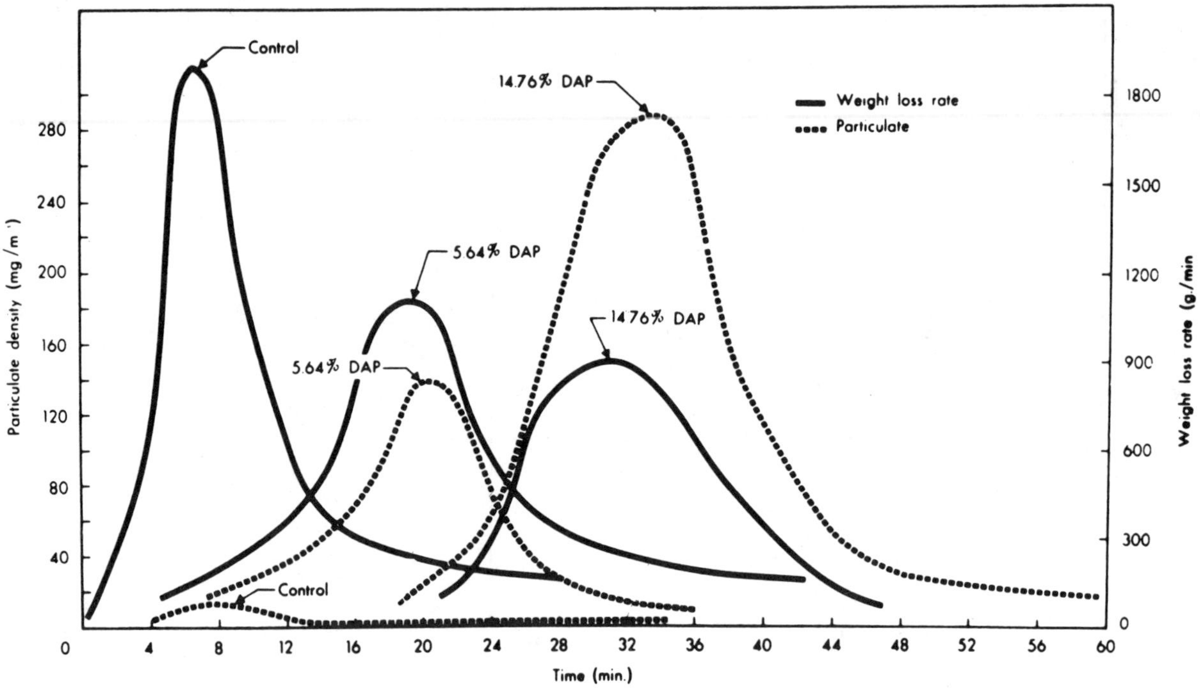


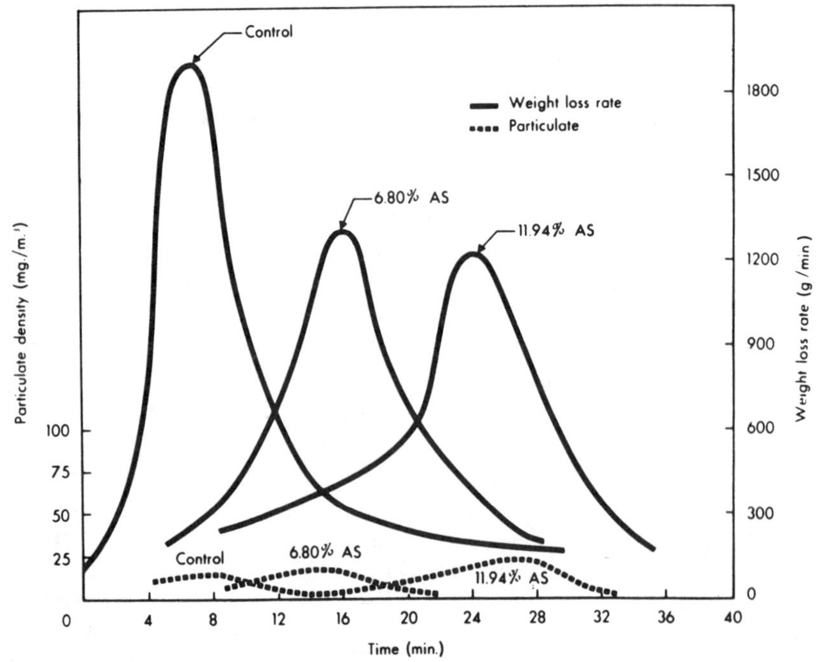


*Figure 1-17. Effect of retardants on mass loss and particulate production. Upper graph gives three treatment levels for DAP; lower graph, for AS. From Philpot et al. (1972).*

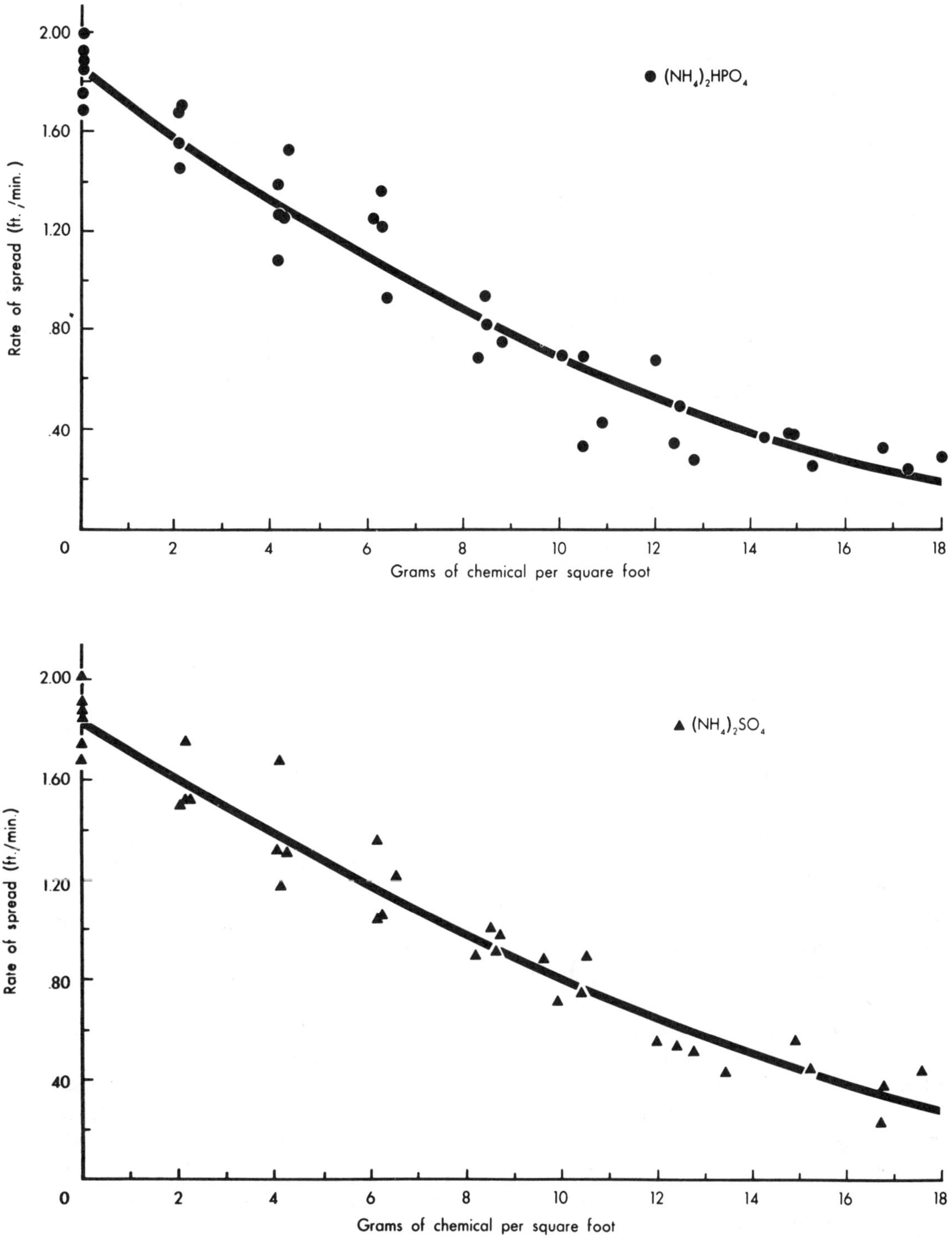


*Figure 1-18. Effect of retardants on rate of spread in pine needle beds. (Upper) Effect of DAP; (lower), of AS. From George and Blakely (1972).*

Slurries consist of more than salts and water. Thickening agents are normally added, and will be mandatory if the mixture is to be dropped free-fall from an aircraft. The salts are corrosive against some metals used in mixing and storage, so inhibitors must be introduced to prevent excessive deterioration. Ammonium phosphates will corrode aluminum, copper, and brass; ammonium sulfates will also corrode steel. When the mixture will be stored for any length of time, flocculants must be added to prevent settling and bactericides to retard spoilage. Dyes are helpful in showing where the mixture has been applied. For ground tanker operations that rely on retardants, some of these problems are solved by using liquid concentrates rather than dry powder as a source.

Its chemical properties determine the suitability of a slurry as a flame inhibitor but its physical properties may determine its method of application. Its rheological properties are especially important when the slurry is to be dropped by air tanker. Its effective viscosity determines the size and penetrating capability of the individual droplets that atomize upon release, and the coherence and momentum of their descending cloud. Its effective shear largely controls the rate of evacuation from the holding tanks; this flow rate shapes the cloud and the relative densities characteristic of its ground pattern. Its solubility measures not only how readily a retardant will mix with water, but how rapidly it may leach away when on the ground (Swanson et al., 1976).

One problem with fire retardants in general has been their toxicity. In the case of wildland fire retardants, an early version (sodium calcium borate) was abandoned after it was found to sterilize ground to which it was heavily applied. The active chemicals in modern retardants, by contrast, are fertilizers. But like any fertilizer, especially phosphates, they can damage an environment when applied too heavily or indiscriminately. Dumped into waterways, the phosphates can encourage eutrophication and, in some cases, contribute to fish kills. The active ingredient seems to be free ammonia ($NH_3$), which can affect certain invertebrate aquatic organisms (such as shrimp) and the egg-sac fry of certain vertebrates (such as Coho salmon and rainbow trout; Van Meter and Hardy, 1975). In sensitive areas, retardants may be prohibited. In other areas, their use will be circumscribed by the likelihood that they will enter reservoirs in sufficient quantities to become a pollutant.

## *REFERENCES*

In wildland fires combustion and fire behavior are so interrelated that a distinction is not always easy, or relevant. General overviews of wildland combustion can be found in Browne (1963), "Theories of the Combustion of Wood and Its Control" and, though somewhat dated, in Byram (1959), "Combustion of Forest Fuels." Summaries that examine combustion primarily from the physical aspect include U.S. Forest Service (1969), "Intermediate Fire Behavior," and Countryman (1969a), "Project Flambeau." For an overview of the chemistry involved, see the various articles contained within the "Southern Forestry Smoke Management Guidebook." For a mathematical synthesis of fundamental importance—since it forms the basis for a popular fire behavior model —see Rothermel (1972), "A Mathematical Model for Predicting Fire Spread in Wildland Fuels."

Summary works on pyrolysis include Shafizadeh (1968), "Pyrolysis and Combustion of Cellulosic Materials," and Philpot (1971), "Pyrolysis Products and Thermal Characteristics of Cottonwood and Its Components." More technical, but full of observations, is Albini (1980), "Thermochemical Properties of Flame Gases from Fine Wildland Fuels." For ignition, see Anderson (1970), "Forest Fire Ignitability." For flaming combustion, several good summaries exist. Gardiner (1982), "The Chemistry of Flames," and Fristrom and Westenberg (1969), *Flame Structure,* complement each other and describe the general parameters of flame analysis. Albini (1980) outlines the general framework for flame description. Rothermel (1972) consolidates the important fire behavior variables.

There are several fine summaries on the subject of fire retardants. Lyons (1970), *Chemistry and Use of Fire Retardants,* gives a comprehensive survey including the research on wildland fire retardants up to 1970. The NFPA (1967) second edition of "Chemicals for Forest Fire Fighting" is thorough on both theory and application. Considerable research has been conducted since both these works were published, however; a summary up to 1976 can be found in C.W. George et al. (1976), "Forest Fire Retardant Research. A Status Report."

Excellent summaries of smoke content can be found in Sandberg et al. (1979), "Effects of Fire on Air. A State-of-Knowledge Review"; and Tangren et al. (1976), "Contents and Effects of Forest Fire Smoke," in the "Southern Forestry Smoke Management Guidebook."

For a more detailed set of references on fire behavior, consult Chapter 2. For the modeling of fuels, see Chapter 3. For smoke dispersion, see Chapter 10. For air flows, see Chapter 4. And for the application of retardants, see Chapter 9.

Albini, F.A., 1980, "Thermochemical Properties of Flame Gases from Fine Wildland Fuels," U.S. Forest Service, Research Paper INT-243.

Anderson, Hal E., 1964, "Mechanisms of Fire Spread. Research Progress Report No. 1," U.S. Forest Service, Research Paper INT-8.

———, 1968, "Fire Spread and Flame Shape," *Fire Technology* **4**(1): 51–58.

———, 1969, "Heat Transfer and Fire Spread," U.S. Forest Service, Research Paper INT-69.

———, 1970, "Forest Fire Ignitability," *Fire Technology* **6**(4): 312–322.

Broido, A., 1973, "Flammable—Whatever That Means," *Chemical Technology* **3**(1): 14–17.

———, 1964, "Some Problems in Fire Research," *Pyrodynamics* **1**: 27–40.

Brown, A.A. and Kenneth P. Davis, 1973 rev., *Forest Fire: Control and Use,* 2nd ed. (New York: McGraw-Hill).

Browne, F.L., 1963 rev., "Theories of the Combustion of Wood and Its Control. A Survey of the Literature," U.S. Forest Service, Forest Products Laboratory Report No. 2136.

Byram, George, 1959, "Combustion of Forest Fuels" and "Forest Fire Behavior," in Kenneth P. Davis, *Forest Fire: Control and Use* (New York: McGraw-Hill).

Countryman, Clive M., 1969a, "Project Flambeau...An Investigation of Mass Fire (1964–1967). Final Report—Volume I," U.S. Forest Service, Pacific Southwest Experiment Station.

———, 1969b, "Fuel Evaluation for Fire Control and Fire Use," in U.S. Forest Service, "Intermediate Fire Behavior," TT-81 (Washington, D.C.: U.S. Forest Service).

Ebarb, Pat, 1978, "Texas Snow Job," *Fire Management Notes* **39** (3): 3–5.

Frandsen, William H., 1973, "Effective Heating of Fuel Ahead of a Spreading Fire," U.S. Forest Service, Research Paper INT-140.

Fristrom, R. M. and A. A. Westenberg, 1969, *Flame Structure* (New York: McGraw-Hill).

Gardiner, William C., Jr., 1982, "The Chemistry of Flames," *Scientific American:* 110–124.

George, C.W. et al., 1976, "Forest Fire Retardant Research. A Status Report," U.S. Forest Service, General Technical Report INT-31.

———, 1977, "Evaluation of Liquid Ammonium Polyphosphate Fire Retardants," U.S. Forest Service, General Technical Report INT-41.

Hartong, Allan L., 1971, "An Analysis of Retardant Use," U.S. Forest Service, Research Paper INT-103.

Johansen, Ragnar W. et al., 1976, "Fuels, Fire Behavior, and Emmissions," pp. 29–44, in Southern Forest Fire Laboratory Personnel, "Southern Forestry Smoke Management Guidebook," U.S. Forest Service, General Technical Report SE-10.

Kelsey, Rick G. et al., 1979, "Heat Content of Bark, Twigs, and Foliage of Nine Species of Western Conifers," U.S. Forest Service, Research Note INT-261.

Kerr, J. W. et al., 1971, "Nuclear Weapons Effects in a Forest Environment— Thermal and Fire," The Technical Cooperation Program, Panel N-2 Report N2:TR 2–70 (Santa Barbara, Calif.: Department of Defense Nuclear Information and Analysis Center).

Lyons, John W., 1970, *The Chemistry and Use of Fire Retardants* (New York: Wiley).

Martin, S., 1964, "Ignition of Organic Materials by Radiation," *Fire Research Abstracts and Reviews* **6**: 85–98.

National Fire Protection Association, 1967, "Chemicals for Forest Fire Fighting," 2nd ed. (Boston: National Fire Protection Association).

———, 1969, *Fire Protection Handbook,* 13th ed., (Boston: National Fire Protection Association).

National Wildfire Coordinating Group, 1981, "Airtanker Base Planning Guide."

Philpot, Charles W., 1970, "Influence of Mineral Content on the Pyrolysis of Plant Materials," *Forest Science* **16**(4): 461–471.

———, 1971, "The Pyrolysis Products and Thermal Characteristics of Cottonwood and Its Components," U.S. Forest Service, Research Paper INT-107.

Philpot, Charles and Robert W. Mutch, 1971, "Seasonal Trends in Moisture Content, Ether Extractives, and Energy of Ponderosa Pine and Douglas-fir Needles," U.S. Forest Service, Research Paper INT-102.

Philpot, Charles et al., 1972, "The Effect of Two Flame Retardants on Particulate and Residue Production," U.S. Forest Service, Research Paper INT-117.

Rothermel, Richard C., 1972, "A Mathematical Model for Predicting Fire Spread in Wildland Fuels," U.S. Forest Service, Research Paper INT-115.

———, 1976, "Forest Fires and the Chemistry of Forest Fuels," pp. 245–259, in Fred Shafizadeh et al. (eds.), *Thermal Uses and Properties of Carbohydrates and Lignins* (New York: Academic).

Rothermel, Richard C. and Hal E. Anderson, 1966, "Fire Spread Characteristics Determined in the Laboratory," U.S. Forest Service, Research Paper INT-30.

Rothermel, Richard C. and Charles E. Hardy, 1965, "Influence of Moisture on Effectiveness of Fire Retardants," U.S. Forest Service, Research Paper INT-18.

Rothermel, Richard and C.W. Philpot, 1975, "Reducing Fire Spread in Wildland Fuels," pp. 369–403, in "Experimental Methods in Fire Research," Proceedings of a Meeting to Honor Clay Preston Butler (Stanford, Calif.: Stanford Research Institute).

Rundel, Philip W., 1981, "Structural and Chemical Components of Flammability," pp. 183–207, in H.A. Mooney et al. (eds.), "Fire Regimes and Ecosystem Properties," U.S. Forest Service, General Technical Report WO-26.

Sandberg, D.V. et al., 1979, "Effects of Fire on Air," U.S. Forest Service, General Technical Report WO-9.

Shafizadeh, Fred, 1968, "Pyrolysis and Combustion of Cellulosic Materials," pp. 419–477, in Melville L. Wolfrom and R. Stuart Tipson (eds.), *Advances in Carbohydrate Chemistry* (New York: Academic).

Shafizadeh, Fred and William F. DeGroot, 1976, "Combustion Characteristics of Cellulosic Fuels," pp. 1–17, in Fred Shafizadeh et al. (eds.), *Thermal Uses and Properties of Carbohydrates and Lignins* (New York: Academic).

Shafizadeh, Fred et al., 1977, "Effective Heat Content of Green Forest Fuels," *Forest Science* **223**(1): 81–89.

Simms, D. L., 1963, "On the Pilot Ignition of Wood by Radiation," *Combustion & Flame* **7**(3): 253–261.

Stockstad, Dwight S., 1976, "Spontaneous and Piloted Ignition of Cheatgrass," U.S. Forest Service, Research Note INT-204.

———, 1978, "Spontaneous and Piloted Ignition of Pine Needles," U.S. Forest Service, Research Note INT-194.

———, 1979, "Spontaneous and Piloted Ignition of Rotten Wood," U.S. Forest Service, Research Note INT-267.

Susott, Ronald A. et al., 1975, "Heat Content of Natural Fuels," *Journal of Fire and Flammability* **6**: 311–325.

Swanson, D.H. et al., 1976, "Air Tanker Performance Guides: General Instruction Manual," U.S. Forest Service, General Technical Report INT-27.

Tangren, Charles D. et al., 1976, "Contents and Effects of Forest Fire Smoke," pp. 9–22, in Southern Forest Fire Laboratory Personnel, "Southern Forestry Smoke Management Guidebook." U.S. Forest Service, General Technical Report SE-10.

U.S. Forest Service, 1969, "Intermediate Fire Behavior." TT-81, U.S. Forest Service.

Van Meter, Wayne and Charles E. Hardy, 1975, "Predicting Effects on Fish of Fire Retardants in Streams," U.S. Forest Service, Research Paper INT-166.

*Chapter Two*

# Fire Behavior

Combustion supplies the energy for a fire, but the behavior of the fire as a whole is not identical with the process of combustion. In part this reflects a problem of scale. Combustion responds to chemical processes that occur on the scale of molecules, while fire responds to physical and meteorological processes on the order of meters or kilometers. Moreover, the question of scale involves more than area. Fires grow in intensity as well as size. As wildland fire increases its dimensions, different mechanisms of heat transfer and different modes of flame propagation come into prominence. The combustion history of a fuel particle from ignition to steady-state flaming to extinction is only broadly analogous to the history of a fire. Of equal significance to those processes that occur within the reaction zone are those processes that account for the evolution, arrangement, and spread of the reaction zone as a whole. Practically speaking, this means the history, shape, size, and mode of propagation of the flaming zone. Collectively these topics constitute the subject of fire behavior.

In well-ordered combustion chambers, fire spread and fire growth are determined by strictures on fuel flow and oxygen mixing and the physical confinement imposed by the chamber itself. Fire behavior is synonymous with the combustion of new fuels introduced into a fixed reaction zone. In wildland environments, no such constraints exist. The behavior of the fire will depend on its larger environment of fuels, weather, topography, and other fires. The prediction of how the flaming zone will spread from a point of origin, how the heat released from combustion will be organized, and what flame characteristics it will acquire all follow from considerations of the behavior of the fire as whole. On these properties will depend the characteristics of the combustion process, and from the interrelationship of combustion and fire behavior, the potential for fire management. No aspect of fire management cannot be integrated with fire behavior measurements, and no program will long succeed without such considerations.

## 2.1 FIRE GROWTH

In describing fire behavior, fire growth has two meanings. It can refer to an increase in intensity or to an increase in size. Intensity describes the buildup of heat within a fire, both in amount and in rate of transmission—a function of heat release. According to their heat buildups, fires show stages of growth, a kind of life cycle. All fires experience a stage of persistent ignition at their origin; many undergo a first order transition to a self-sustaining steady-state fire; a small fraction undertake a second order transition to the category of large fire. Size increases, by contrast, result from rates of spread. All fires begin as points, expand to circles, and, if they escalate in intensity to steady-state propagation, assume the broad form of ellipses with differential rates of spread for the heading, backing, and flanking fires distributed around their perimeter. In some respects, in the earlier stages of a fire, fire growth as a function of rate of spread and fire growth as a function of intensity are complementary. During steady-state propagation the two coexist. But as second order transitions become possible, they become increasingly competitive. For fires of very high fireline intensities, energy may derive from either high rates of spread or high rates of heat release.

In either expression growth is seldom uniform. Like combustion, fire is a multiple not a singular event. Fires do not grow once, from a solitary point, but often, from an array of points and at different times, from the same arrays. This is in part a product of, and in part an analogue to, the phenomenon of ignition. The ignition process appears in every stage of fire growth: it initiates the fire as a point source, it governs the steady-state spread of a fire from particle to particle, and it allows large fires to multiply their areas through spotting by firebrands. Similarly, fires do not often spread continuously or with regularity. Even when burning within a range of intensities that share common modes of propagation, the flaming front will surge, flare intermittently, and compound thermal pulse with thermal pulse, like the mixing of waves in the sea, into a plexus of flame. The pattern will vary along different portions of the flaming perimeter, and for a given portion of the flaming front at different times of fire history or burning cycles. Fire behavior is probabilistic, not deterministic; irregular rather than uniform; variable in intensity as well as in size.

### *Fire Growth as Intensity*

In the combustion history of a fuel particle there is a period of heat absorption followed by a period of heat release. Combustion occurs only if the temperature of the particle, the net heat budget, reaches certain values. If the particle is large, the periods of heat absorption and release may be prolonged. Something analogous occurs with a fire. One difference is that heat flow is not continuous but discontinuous, sustained by successive new ignitions of fresh fuel particles. Another difference is that synergistic effects become important as many particles ignite, with the outcome that the flaming zone as a whole shows properties distinct from the sum of its individual particle flames. What both processes share is the need to raise temperature. For a fire to sustain itself from a point of ignition, there must be a heat buildup, and this heat must be organized in such a way that it is capable of carrying the fire into new fuels. The buildup of heat is measured by the fire's intensity, and the organization of that heat flux will determine whether the fire can make a first order transition to steady-state spread.

For industrial fires and for structural fires, this initial buildup of heat within the reaction zone is expressed by *time–temperature curves.* Since the combustion chambers are regulated, the buildup of heat can occur at predictable rates. The heat released by combustion, by and large, remains in the chamber to further raise the heat of combustion. By this feedback mechanism the temperature of the reaction zone can be elevated rapidly until the desired heat of combustion is reached. Even in the case of structural fires, standard time–temperature curves exist for particular environments. Deviations are possible, even normal, since the fire environment is not completely confined; if the structure is not well ventilated, heat may build up without combustion, delaying flashover. Yet the buildup of heat is predictable. Such curves provide guidelines for the response time of urban fire suppression units (Figure 2-1).

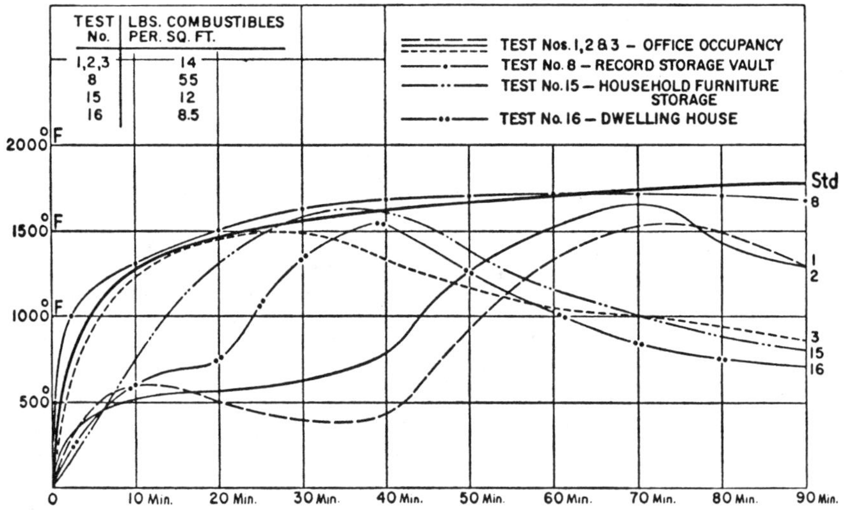


*Figure 2-1. Time–temperature curve for structures. The standard curve is indicated by the bold line. The leveling of the line marks the onset of steady-state burning. The actual buildups will depend on fuels and ventilation. From National Fire Protection Association (1969).*

*Initiating Fire.* Wildland fires do not occur within enclosed chambers or semiconfined structures. The heat released during the early stages of combustion is not retained automatically, not redirected into the reaction zone to raise the heat of combustion. Winds, diffuse convection plumes, and poorly organized flaming zones all may allow much of the heat to be dissipated. In a steady-state fire, each point of ignition is surrounded by other points, more or less simultaneously ignited, and by a flaming zone which contributes to the heat flux that leads to ignition. But at its origin a wildland fire does not enjoy such an environment. The *initiating fire* begins as a point source. Gradually, the point spreads outward into new fuels. As it increases the area under flame, the fire can, at this early stage, proportionately increase its heat output and accelerate, or it can lose its rate of combustion and decelerate into extinction. Which option occurs will depend on the combustion period and the critical burnout time.

The issue is how soon the center burns out relative to the spreading perimeter of flame (Figure 2-2). If the burnout is rapid, then a perimeter of fire results, with the center no longer flaming. In this circumstance, some fires, especially those burning in fuels near the moisture of extinction, will witness a gradual reduction in flame size as the perimeter expands. But if the burnout is slower, then a larger area of flame results, the residual heat from the center combines with the heat along the flaming edge to increase the heat of combustion, and proportionately greater heat is available for propagation along the perimeter. At some critical point, even while the center burns out, the flaming zone along the perimeter acquires sufficient heat to sustain itself. The critical combustion rate at which this transition occurs is on the order of 10–30 Btu/min (151–453 kcal/sec). Note that it is the rate of heat release, not simply the total quantity of heat, that is fundamental. The initiating fire becomes a *steady-state fire.*

Because this buildup of heat occurs in free-burning circumstances, it does not follow a regular time–temperature curve (Figure 2-3). It may occur quickly, with the fire rapidly accelerating into steady-state burning or into extinction. Or it may occur over hours, days, even weeks. A point of ignition may smolder for long periods without becoming an initiating fire. Wood is a poor conductor of heat, but a good insulator. Heat from an ignition source may build up very slowly, eventually bursting into flame after considerable time has elapsed. Moreover, during the course of its

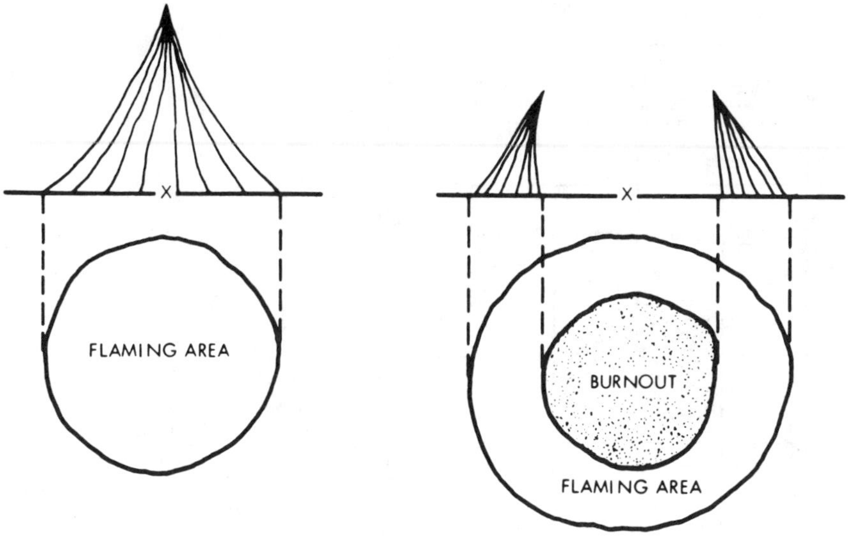

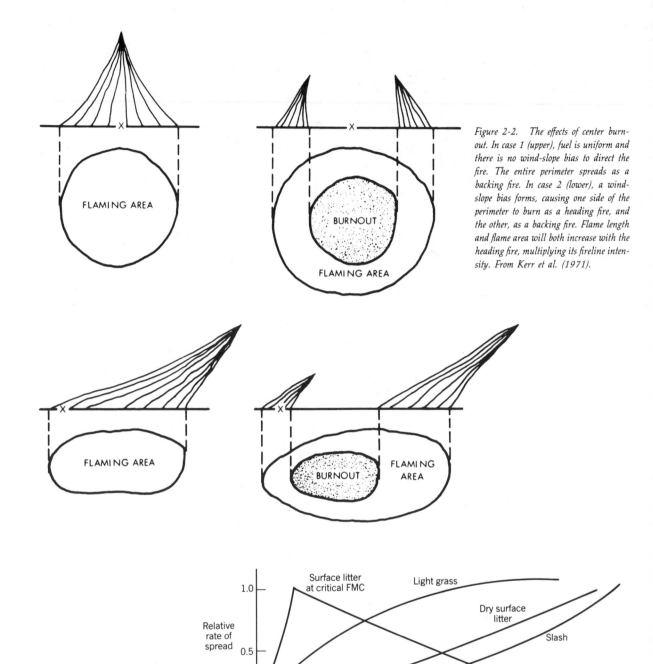


*Figure 2-2. The effects of center burnout. In case 1 (upper), fuel is uniform and there is no wind-slope bias to direct the fire. The entire perimeter spreads as a backing fire. In case 2 (lower), a wind-slope bias forms, causing one side of the perimeter to burn as a heading fire, and the other, as a backing fire. Flame length and flame area will both increase with the heading fire, multiplying its fireline intensity. From Kerr et al. (1971).*

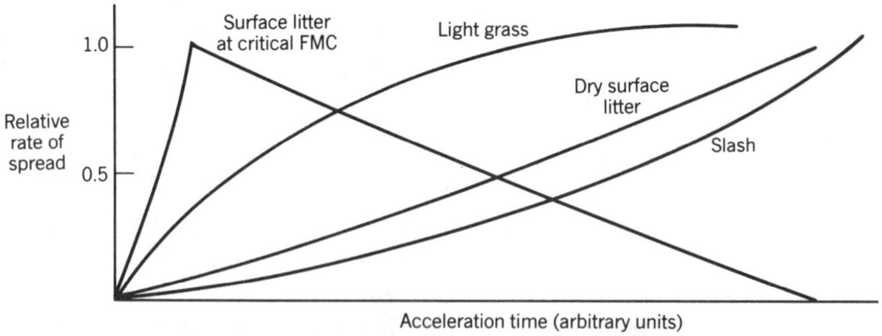


Surface litter at critical FMC: failure to reach steady-state after center burnout; maximum flame height, 10 ft.
Dry pine litter: fuel loads, 4 tons/acre; maximum flame height, 8 ft.
Slash: fuel load, 50 tons/acre; maximum flame height, 50 ft.
Light grass: fuel loads 2 tons/acre; maximum flame height, 6 ft.

*Figure 2-3. Idealized time-spread curves for wildland fuels. The curves show the acceleration from initiating to steady-state fire. With so many environmental variables uncontrolled, the time units remain arbitrary. Note that maximum flame heights, a measure of fire intensity, vary dramatically. From Kerr et al. (1971).*

history, a wildland fire may make this transition once or many times, from a single point of origin, or from many points distributed around a smoldering perimeter. The controlling variables will relate to fuels, primarily to fuel moisture. With decreasing fuel moisture, a given amount of heat can lead to proportionately greater heat releases.

*Steady-State Fire.* Ideally, an initiating fire burns in a circular pattern, its flames rising upward and inward from convective flow. But as the center burns out, this pattern cannot be sustained. Nor does such a pattern convert as much of the total heat released into the propagating flux along the perimeter. If flames can be bent, however, or if the fire can be given a direction in which to concentrate its convective flow, then more of the released heat can be directed onto new fuels. Rather than dissipate its heat in all directions, the heat is focused. This is precisely what occurs when wind or slope act on a fire. Unless the fuels are exceptionally dry and the heat of combustion unnaturally high, some such bias in fire shape and heat flow stimulates the fire into steady-state spread. The fire no longer shows a uniformly distributed flaming zone, but one that is much more developed on one side of its perimeter. The effect is to create a surplus of heat on this side, raising the heat of combustion within the reaction zone and generating still further heat releases. At this stage the fire exhibits three varieties of flaming zones and three rates of heat release: that which burns downwind or upslope (heading fire), that which burns upwind or downslope (backing fire), and that which burns between these zones (neutral or flanking fire).

The expression steady-state fire is used loosely. It does not mean that such a fire burns day and night, through fine fuels and coarse, with a constant rate of spread or reaction intensity. Where the fire environment is constant, such a fire will burn with a quasi-steady-state flow. But this environment is constantly changing in ways both large and small, and the steady-state fire ceaselessly accelerates and decelerates with changes in wind, topography, and fuels. Rare is the fire that burns under true steady-state conditions, and common the fire that shows local accelerations, pulsations, flareups, and surges. Considered as a whole, a fire burns with different rates of spread around its perimeter at any one time, and at each point along that perimeter with different rates as the fire experiences changes in its environment. How closely a fire approximates a steady-state flow, strictly defined, depends on the scale of time and space chosen. Virtually every fire is an amalgamation of discrete ignitions, and a composite of smaller pulsations.

Rather, the steady-state fire describes a stage of fire growth—those fires whose intensities range between the precarious heat release of the initiating fire and the massive intensities of very large fires. These are fires whose behavior is primarily controlled by environmental conditions that determine its flame geometry, and fires whose propagation follows certain mechanisms of continuous fire spread through surface fuels. Steady-state fires are self-sustaining, show a perimeter of flame (usually an ellipse), grow in size and shape but not in intensity, and spread as surface fires through more or less constant heat transfer by radiation with additional heat transfer by convection for heading fires.

An initiating fire is, for practical purposes, a *point fire.* Inevitably that point will spread outward into a ring of flame. If the fire survives this transition to steady-state behavior, its flaming front acquires the properties of a *line fire.* This means that there is a certain ratio between flame width and flame height and that the fire as a whole is large enough that the spread of any portion is independent of influences from the opposite portion of the perimeter. The actual dimensions of such a perimeter will vary with fuel properties, but the important concept is that the propagating flux does not derive from a single point or collection of points, but from a larger unit, a line. Each fresh fuel particle receives heat from the flames to the side of it as well as from those to its immediate front.

This geometry, too, may change with further increases in fire intensity. To say that steady-state fires do not change in intensity is not exactly true. They change, though like rates of spread the changes do not show a constant acceleration and do not significantly change the modes of propagation characteristic of a steady-state fire. Yet circumstances can lead to further heat buildups well in excess of that needed for sustained spreading. Typically this is manifested by an enlargement in the depth of the flaming zone. The flaming zone does not increase merely in terms of its total perimeter length but in the thickness of that perimeter. Such an increase may occur due to high rates of spread or long residence times or a combination of the two. A result is that the unit geometry of the flaming zone changes from that of a line to that of an area. Fireline intensity (Byram's intensity) increases. For a given unit of fire, more heat is released; for a given unit of time, more fuel units are involved in combustion. The depth of the flaming zone assumes an importance equal to its height and width. No longer is one portion of the flaming perimeter unaffected by other portions. A line fire becomes an *area fire.*

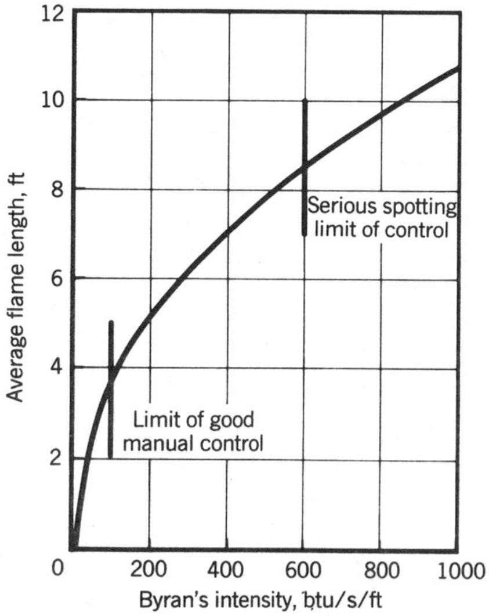


*Figure 2-4. Relationship between fireline intensity and flame length. From Albini (1976).*

*Large Fire.* With such increases in intensity, the steady-state fire makes a second order transition to the status of large fire. New modes of fire propagation become prominent. Depending on whether the fireline intensity derives from high rates of spread or high heat release/area, whether the fire is wind-driven or convection-driven, the category of large fires is sometimes subdivided into *conflagrations* and *mass fires.* Like the transition from initiating fire, a steady-state fire may make this second order transition once or several times in its total history. The number of fires that make this transition is small, but their effects are large and disproportionate.

The growth of a fire as measured by intensity has important management implications. Obviously, the earlier a fire is detected in this evolution the simpler it is to control. An initiating fire can be suppressed more easily than a steady-state fire, and a steady-state fire much more readily than a large fire. Figures 2-4 and 2-5 illustrate well the relationship between fireline intensity, flame length, and the prospects for perimeter control of the fire. During their high-intensity phase, large fires are uncontrollable by any means under human manipulation. Only very broad changes in the fire environment—fuels and weather, principally—significantly modify the behavior of such a fire; only when the large fire has returned to steady-state conditions, can it be manipulated. The strategy of fire control is thus to intervene early in the period of heat buildup. Equally, it is possible to accelerate this buildup, a useful tool for many forms of prescribed burning. By igniting fires with various flaming zones—multiple lines of fire, or even areas of fire—heat buildup can come rapidly, escalating through slower, less predictable stages, shaping the geometry and distribution of the flaming zone. By putting down successive lines of fire beyond the flaming perimeter, it is possible to increase the effective rate of spread of a steady-state fire without changing its important properties. Conversely, under proper circumstances, it is possible to go from initiating fire to mass fire within minutes.

| Burning index | Fireline intensity | Flame length | Narrative |
|---|---|---|---|
| | *Btu/s/ft* | *Ft* | |
| 0-28 | 0-50 | 2.8 | Most prescribed burns are conducted in this range |
| 38 | 100 | 3.8 | Generally represents the limit of control for manual attack methods. |
| 78 | 500 | 7.8 | The prospects for control by any means are poor above this intensity. |
| 96 | 700 | 9.6 | The heat load on people within 30 feet of the fire is dangerous. |
| 108 | 1,000 | 10.8 | Above this intensity, spotting, fire whirls, and crowning should be expected. |

*Figure 2-5. Common flame measurements and their interpretation. From Deeming et al. (1978).*

### *Fire Growth as Size and Shape*

For an industrial fire burning within a combustion chamber, a time–temperature curve may be an adequate description of growth. Once a stage of steady-state combustion is reached, the important measure of the flaming zone is its flame velocity, a product of the rates of fuel and oxygen brought into the chamber. With changes in the rates of fuel and oxygen, the geometry of the flaming zone may show some modifications (becoming larger or longer) until a new size is achieved appropriate to its new steady-state flow.

*Influences on Growth.* In wildland fires, however, the fuel and oxygen flow are not concentrated into a single site. The site itself grows, as the flaming perimeter searches out new fuels. The flaming front shows not one velocity, but three; not one flame geometry, but three; not a simple propagation of flame, but a growth of the flaming zone itself. Used in this sense, fire growth refers to the growth of the flaming perimeter and the growth of the fire area. What determines these growth rates for steady-state fires is the development of a directional bias, a *head.* The fire perimeter then consists of the heading, the backing, and the flanking fires, each of which has a characteristic mode of propagation and different rates of spread, with the heading fire significantly faster than the others. This means that the flaming zone will spread more rapidly in one direction than in others, and that the shape of the fire will less and less resemble the circular form characteristic of an initiating fire.

For wind and slope, which principally determine the growth of steady-state fires, fire shape will resemble an ellipse (Anderson, 1983). More precisely, considering that both heading and backing fires are at work, the fire shape will resemble a double ellipse, with heading fire and backing fire each circumscribing a semiellipse along a shared minor axis. The more powerful one factor becomes, the more the fire perimeter approximates this shape. If the fire environment remains constant, the double ellipse will show a constant length-to-width ratio.

*Fire Shape.* The double ellipse is a good approximation of fires burning under a constant wind velocity. Knowing the forward rate of spread (heading fire) and the wind speed at midflame height, one can predict the shape and rate of spread of the flaming perimeter with reasonable accuracy (Figure 2-6). But few fires show a final form of this type. The fire environment is not constant—winds shift frequently, topography and wind may compete rather than complement each other, and fuels change in load and type. The double ellipse describes a single episode of sus-

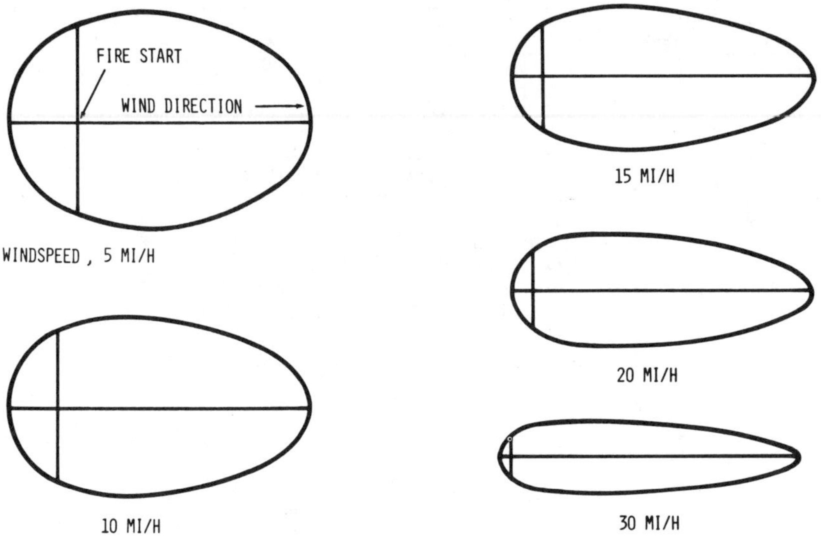


*Figure 2-6. Fire shapes under different wind regimes. Note how the backing fire becomes less influential as windspeed increases. The sizes shown are arbitrary. From Albini (1976).*

tained forward spread, a *run.* The final size, as often as not, represents a composite of many episodes, not all of which begin as point sources.

Fires tend not to grow uniformly, but spasmodically. Their final perimeter may be fashioned by many episodes around a single ring of fire or a succession of episodes occuring on a single fire. For a steady-state fire, whose perimeter tends to behave as a composite of "lines" few of which are dynamically related to the others, the possibility of multiple heads is very likely. In mountainous topography, separate canyons can develop individual heads, and the passsage of a cold front perpendicular to the main axis of an elliptical fire may spawn several heads with the wind shift. Only with very intense fires does the fire behave as a single entity, with a single head and a shape close to a simple double ellipse.

In the field fires are not commonly described in terms of their geometry. Instead there exists a vocabulary of fire anatomy (Figure 2-7). The area inscribed by the heading fire is the *head,* the area circumscribed by the backing fire is the *rear,* and the sides are the *flanks.* Since the perimeters of most fires are irregular, reflective of the fire's history and environment, the shape is likely to be a ragged ellipse or a compound of ellipses. Small extensions of the perimeter that radiate out from the main fire are called *fingers,* and the unburned area between fingers and main fire, *pockets.* If the pockets are completely isolated within the burn, they are called *islands,* or, for biological purposes, *refugia.* The point where the fire begins is its *origin.* Small fires ignited by firebrands outside the main burn are *spot fires.*

*Management Considerations.* The implications of fire spread rates and shape are many. The most significant measure of velocity is forward rate of spread, and the most significant growth rates are those for perimeter and area. Perimeter and area are not equal in their growth rates or their effects. Perimeter growth is proportional to time; area growth, to the square of the time. Perimeter is significant because fires are controlled along their perimeter, with larger perimeters increasing the difficulty of control. For large fires with high forward rates of spread, the ratio of length-to-width is large, and the perimeter proportionately larger than the inscribed area. For a given investment in suppression, less area is protected. Since backing fires and flanking fires have lower rates of spread and intensity, the suppression of large fires often begins with control over the rear and flanks, grad-

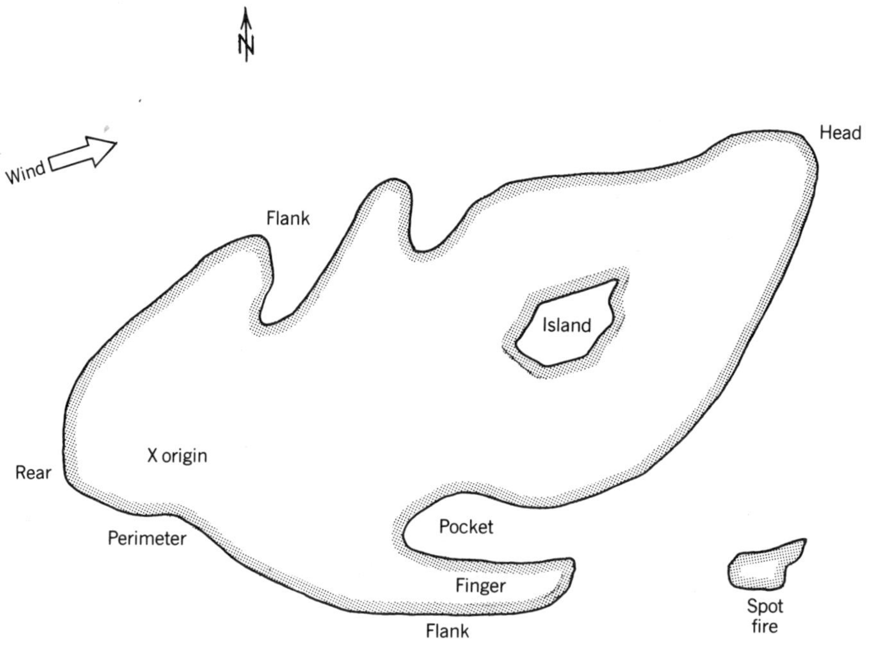


*Figure 2-7. Fire anatomy with common terms. For mathematical analysis, the shape can be considered as a double ellipse. The stronger the wind or slope, the closer the total fire shape will approximate the ellipse. When the fire burns slower, influenced by several factors, more ragged shapes result, for which common terms are necessary.*

ually pinching off the head. In terms of perimeter geometry, such a control strategy increases the length-to-width ratio (Figure 2-8). If, however, the head can be controlled first, the total size of the fire will be dramatically reduced.

Fire area, by contrast, measures damages; burned area increases with the square of the time involved. Obviously, fires with a high rate of forward spread will burn larger areas than fires with lower velocities, but not proportionately larger. In terms of minimizing burned area, it is important to keep one fire run from multiplying into others —by preventing the explosion in the first place, by confining the fire perimeter immediately after a run is exhausted, by cutting off fire spread from topography that would increase the number of heads, or by protecting firelines against wind shifts that could transform a flank into new heads.

The very considerations that shape fires and regulate rate of spread can be used deliberately in the service of prescribed burning. The processes of growth may be reversed; lines of backing fires, burning downslope or against the wind, can substitute for a single heading fire; the fire can be burned from a carefully prepared perimeter inward. Instead of trying to fashion a perimeter to contain a fire's spread, the spread rate is adjusted to conform to a predetermined perimeter. In a wildfire, strong winds, for example, may render the fire more uncontrollable; in a prescribed burn, they may actually enhance control over the fire by reducing the fire behavior options open to the system, by making it behave in predominantly one way.

### *Fire Characteristics Curve*

In the vernacular, fires tend to be categorized by the stratum of fuel in which they burn. There are *ground fires,* which burn in organic soils, *surface fires,* which burn on surface fuels like litter, and *crown fires,* which burn into the aerial canopy. Any given fire may show combinations of these fires, one leading to another under proper circumstances of weather and fuel. The smoldering ground fire may flare into a flaming surface fire for a period of time, and the surface fire may torch aerial fuels

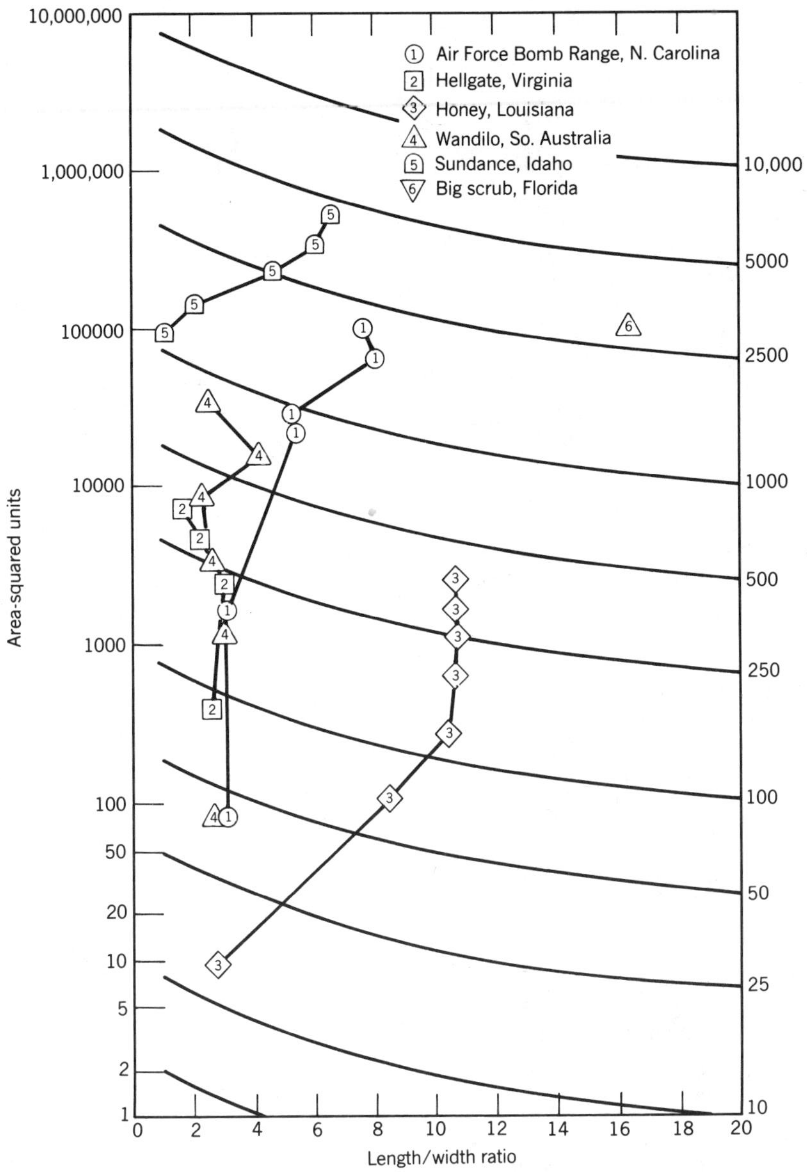


*Figure 2-8. Changes in fire size and shape during the history of observed fires. As measured by length–width ratio, the shape varies until major runs occur. The effect of suppression is to increase the length–width ratio—how rapidly depends on whether the fire is attacked at the head or along the flanks. From Anderson (1983).*

and spread in tandem with a crown fire for some distance. If the fire runs, or shows exteme convective development, it is often called a *blowup fire.*

Another way to characterize wildland fires, however, is by considering their two most significant dynamic properties—rate of spread and heat release/area. Plotting one against the other organizes fires along a hyperbolic curve known as the *fire characteristics curve.* For a given fireline intensity, the two properties are inversely related (Andrews and Rothermel, 1982). Although most fires occupy a transitional zone, showing medium values for both rate of spread and energy release, large fires favor one property or the other. Fires with a very large rates of spread are called *runaway fires,* or *conflagrations,* and fires with very large reaction intensities are known as *mass fires,* or *firestorms.* Figure 2-9 (upper) shows the distribution of experimental fires of a given intensity and (lower) projected curves for a range of intensities.

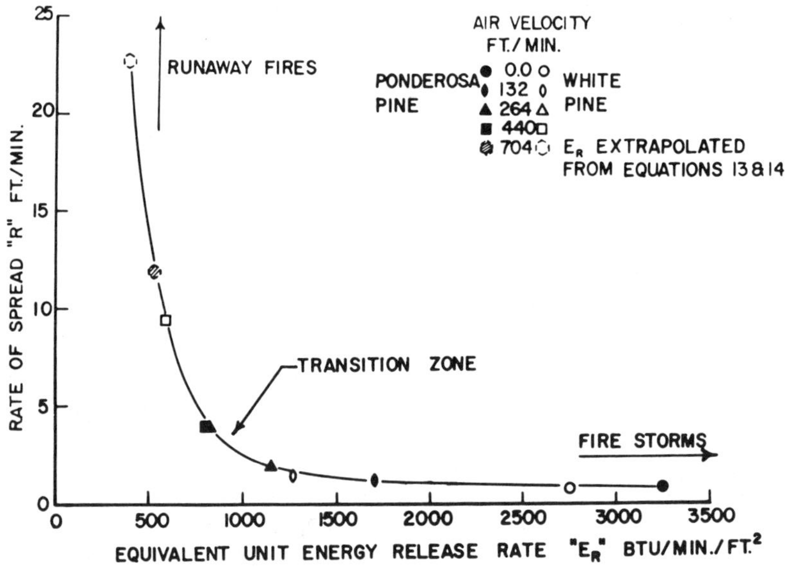


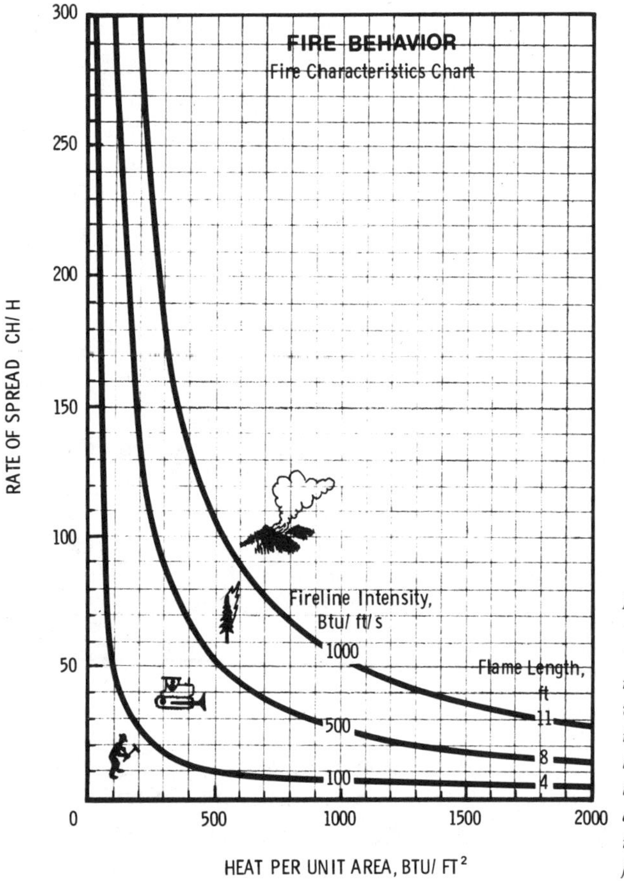


*Figure 2-9. (Upper) Fire characteristics curve from experimental fires. The "firestorm" values indicate fires burning down into thick fuelbeds; the "runaway" values, fires that burn over the surface of fuelbeds. From Rothermel and Anderson (1966). (Lower) Fire characteristics chart for interpreting wildland fire behavior measurements. Fireline intensity and flame length are indicated, as well as appropriate methods of control. The firefighter indicates the range of manual control and the bulldozer, of mechanical control; the torching tree describes the realm of discontinuous fire behavior and the convective column, the domain of large fires. Note how, with higher fireline intensities, the sharp distinction between convective-driven and wind-driven fires tends to smooth out. From Andrews and Rothermel (1982).*

Intuitively, one might expect all fires with high energy releases to show high rates of spread. That some do not follows from the operation of different processes that control the fire environment. The processes that direct heat, fuel, and oxygen may be competitive rather than complementary. Diffusion may follow from convection or ambient wind; fuel flow, from horizontal or vertical spread; heat transfer, from convection or radiation; flame propagation, from continuous spread or discontinuous saltation (spotting). Most fires show combinations of all these processes, but in certain circumstances one set can dominate. Conflagrations show wind-driven diffusion, horizontal fuel flow, and convective heat transfer. The fire spreads rapidly, engulfing more area simultaneously within the flaming zone but with lower heat output per unit area. Mass fires, by contrast, have convection-driven diffusion, vertical fuel flow, and radiant heat transfer. Strong indrafts caused by high convective velocities prevent outward spread; the fire burns downward into large fuel particles and heavy fuelbeds; spread, if it comes, tends to be discontinuous, more a process of firebrand spotting and fire coalescence than of continuous outward propagation. High winds prevent the organization of convection columns. And once established, a convective column may not be readily responsive to ambient wind flow. Because these processes tend to be exclusive, large fires favor one form over another.

More is involved than simply flow regimes, however. For a convective column to develop, there must exist high energy release rates. This can only occur with appropriate fuel loadings. Fine fuels, like grasses, cannot generate the reaction intensity needed to create mass fires. Nor can fires burning in heavy fuels, with scant quantity of fine fuels intermixed, show rapid spread rates. Large diameter fuels can, of course, become involved—often do become involved—but only where there are abundant fine fuels to carry the flaming front. In such circumstances, the involvement of the large fuels is typically temporary, an instance of transient ignition. Only when fuel properties can sustain high reaction intensities does the opportunity exist for a convection column to become organized, and only if the energy of that column is greater than the energy of the ambient wind regime can that column persist. In most circumstances, a complex pattern of interactions is the result.

## 2.2 FIRE SPREAD

Steady-state fires propagate with continuous flow. That flow may not be continuous in velocity, but full of local accelerations, pulses, and surges; and it may not be continuous on the lowest levels, being rather a composite of an infinite number of ignitions. But the geometry of these ignitions is such that each follows in proximity to the other; the fire spreads from particle to adjacent particle. On the scale of normal fires the process may be considered suitably continuous. Fire spread equations assume such uniformity, adjusting rate of spread to such environmental variables as fuel, topography (slope), and wind. Correspondingly, the modes of propagation are those mechanisms of heat transfer, convection, and radiation which transmit heat from the flaming zone to adjacent fuels. Fire shape is dominated by the formation of a head.

Yet there are other modes of propagation that rely on other mechanisms of heat transfer, some of which are discontinuous, and that characterize the behavior of large fires. Some modes such as crown fires relate to apparent discontinuities in the fuel complex. Some such as spotting by firebrands involve ignition that is not adjacent to the main fire, that spawns multiple new fires which may, in turn, interact in striking ways with the main fire. Some, such as firewhirls, organize the flaming zone in particular ways; others, such as convective columns, organize the plume that discharges the byproducts of combustion; still others, such as horizontal roll vortices, organize the ambient wind flow. A large fire will typically exhibit several such mechanisms, interacting in complicated, often singular combinations.

### *Influences on Rate of Spread*

Consider, first, a steady-state fire burning under uniform conditions. The Rothermel fire model gives a useful approximation of rate of spread by considering the ratio of heat source to heat sink —of the reaction intensity, as modified by environmental conditions, to the heat of preignition, as modified by fuel properties. The heat source can be increased by improving the fuel properties, a change which converts more of the reaction intensity into effective heat flux; by increasing ambient winds, which improves both convective and radiant heat transfer; by increasing the angle of slope, which has similar consequences to an increase in wind. Moreover, all of these changes increase the heat source without similarly increasing the heat sink. As these variables increase, so does the rate of spread. Commonly, the fire, or portions of the fire, will encounter local differences in fuels, topography, and wind that will result in local changes in rate of spread and local deformations in the shape of the flaming perimeter.

Observed by itself, each variable induces certain predictable behaviors. But only rarely is one variable dominant. Instead several are present, and they interact in complex, often unique ways. Sometimes they combine to accelerate a fire; sometimes they compete; always they influence. Only in the case of very large fires does one variable effectively override all others. Within the range of steady-state fires, different variables affect the fire in different ways and at different times, and even the variables themselves change during the history of the burn.

*Fuel.* The influence of fuel on fire behavior comes through its contribution to the combustion process. Particle size, shape, and chemistry—and the organization of particles into a fuel array—will all influence rate of spread (Figure 2-10). Fuel

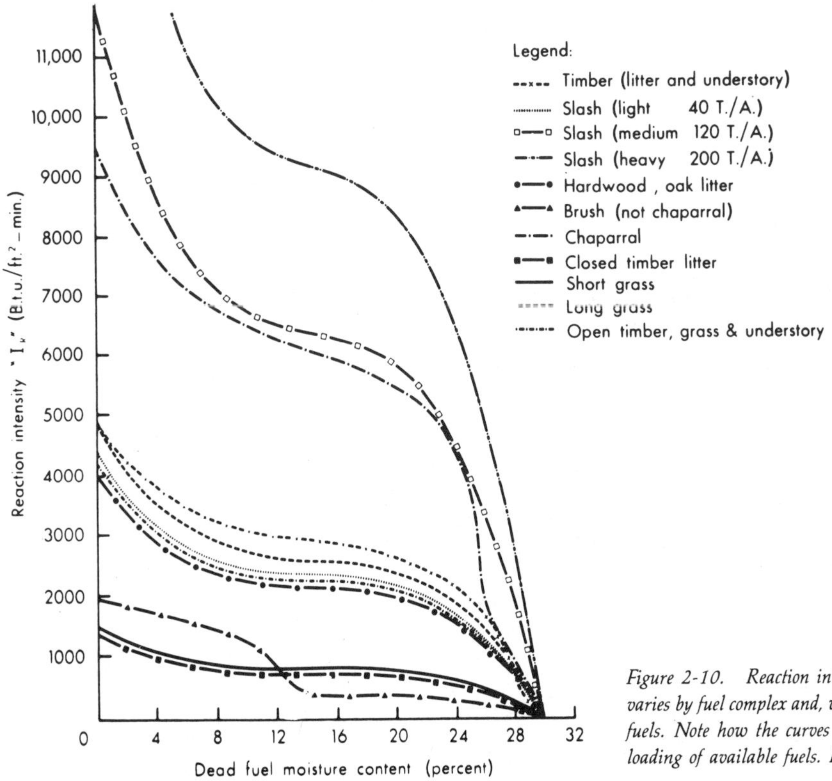


*Figure 2-10. Reaction intensity of typical wildland fuels. The intensity varies by fuel complex and, within each complex, by the fuel moisture of dead fuels. Note how the curves tend to fall into three groups according to the loading of available fuels. From Rothermel (1972).*

complexes with high effective bulk densities will show faster spread rates than complexes with low bulk densities. Fuels with high fuel moistures or fuel complexes with high proportions of live fuels will show slower rates than complexes of dead and dry fuels. Fuels rich in extractives will show more explosive rates of propagation. Where mixtures of fuels occur, the rate of spread of the flaming perimeter will vary accordingly. Other factors aside, the flaming zone will proceed rapidly through fine fuels and perhaps stall at concentrations of coarse fuels. The shape of the flaming zone will reflect the properties of the fuel complex. For purposes of fire behavior forecasting the relevant fuel parameters are expressed as models, each model typical of a natural array of fuels.

*Wind.* Ambient winds act on flame in various, sometimes contrary ways. Wind improves diffusion by introducing a steady flow of oxygen and encouraging turbulent mixing. It improves heat transfer by directing the convective heat flux towards surface fuels and by reducing the distance between flame and fuel, causing flames to bend forward. Winds increase forward rate of spread, amplify fireline and reaction intensities, and improve the likelihood of spotting by windborne firebrands. With higher intensities and longer flames, more of the fuel complex becomes available fuel. In mode of propagation and shape, the forward progress of the fire assumes the properties of a fire head.

Yet most fires have a perimeter of fire, of which the heading fire, driven by winds, is only a portion. For the backing fire, wind works to opposite effects. It cools the new fuels, separates convective heat flux from the propagating flux, and bends the flames away from the fuels, diminishing the efficiency of radiant heat transfer. The backing fire, too, takes the shape of an ellipse, but one with a much smaller length-to-width ratio than that characteristic of the heading fire. The effects of wind on flanking fires are mixed, with the flanking fire showing more forward advance than outward spread. Within the range of steady-state fires, rate of spread increases with increased wind velocity (Figure 2-11).

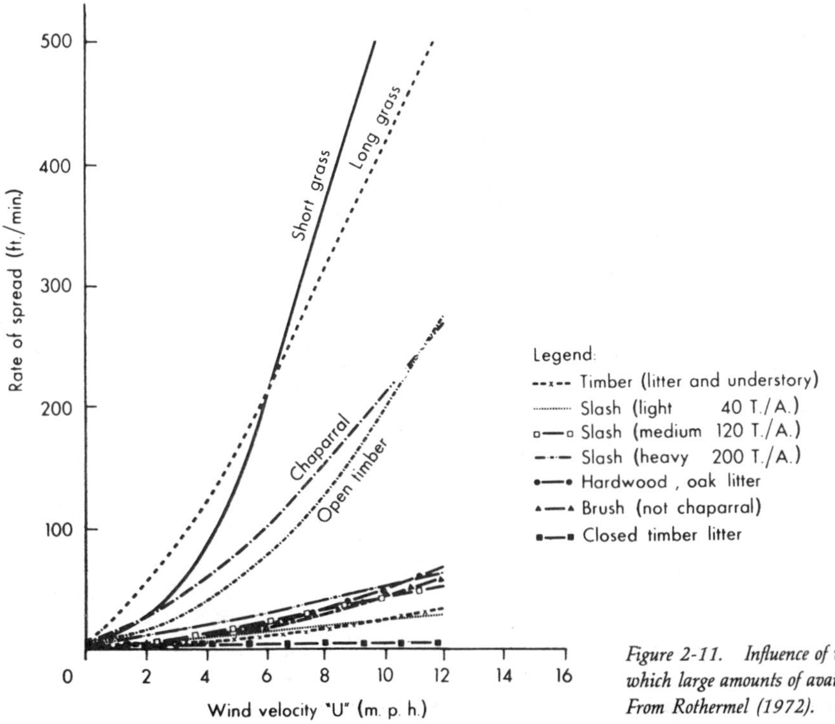


*Figure 2-11. Influence of windspeed on rate of spread. Fuel complexes for which large amounts of available fuel are exposed to the wind respond best. From Rothermel (1972).*

There are limits to the effectiveness of wind. For fires of very small combustion rates (such as initiating fires), the effect of wind is not proportionate to its speed. Similarly, above certain velocities (approximately 35 mph; 56 km/hr) further increases in wind speed do not stimulate additional increases in rate of spread or fireline intensity. On the contrary, by promoting advance spot fires, strong winds may actually work against the fire by causing new fuels to burn out before the main fire can advance upon them. The effective wind speed, moreover, is the wind velocity at midflame height. Obviously this speed will vary —considerably in those forested or mountainous environments where turbulence is the norm (Figure 2-12). And the effectiveness of wind, like other aspects of burning, will vary with fuel properties. Fine fuels and fuelbeds organized into loose, well-ventilated complexes are far more responsive to wind than coarse fuels or opaque, densely stocked woods.

The prevalence of wind makes it a universal component of fire spread calculations, the variability of wind makes it among the greatest problems in the prediction of fire behavior. That variability appears as changes in both speed and direction. Wind affects different portions of a fire differently at any one time, and the same portion differently at different times. Its gustiness can cause a quiet fire to flare and spread rapidly. Its shiftiness can make a flank fire into a head fire. The complex perimeters typical of most steady-state fires reflect, more than anything else, the complicated history of winds experienced by the fire.

*Topography.* Many topographic features influence fire indirectly through the impact of landforms on fuel and weather. Topography affects fuels through such local climate considerations as elevation, which controls temperature and precipitation; aspect, the orientation of a site relative to the

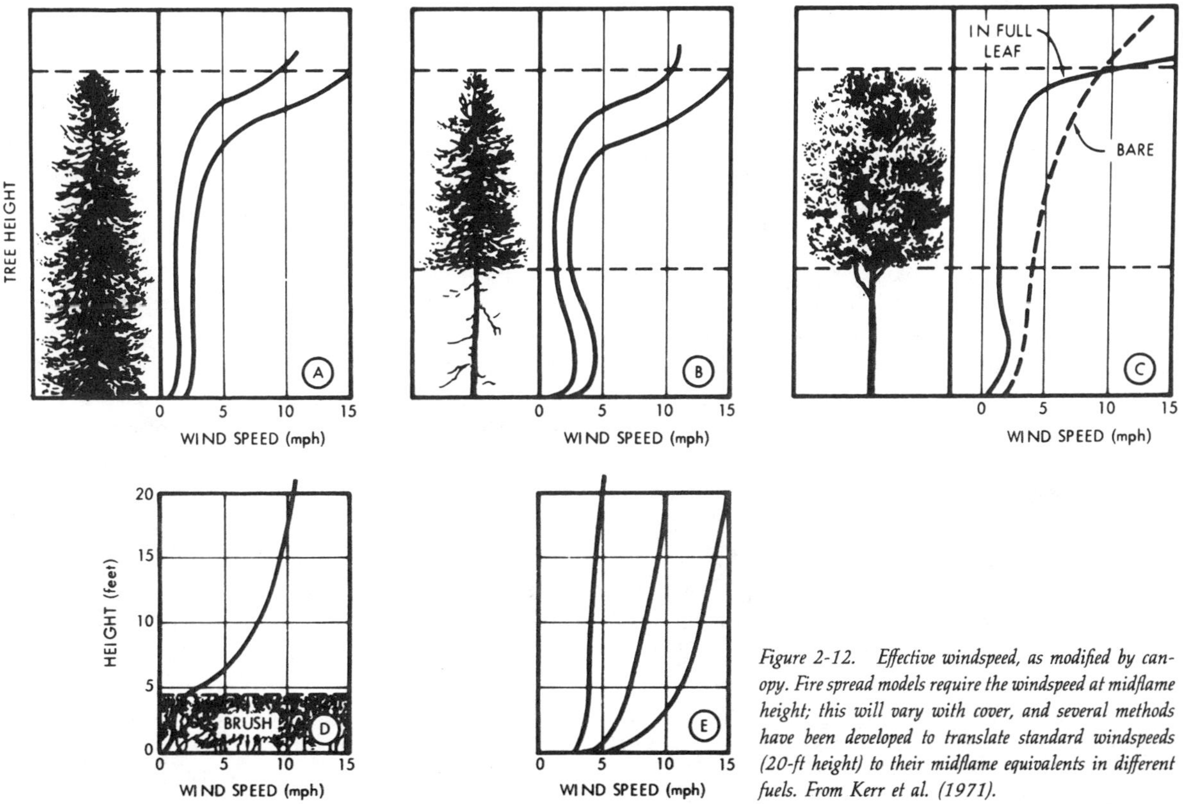


*Figure 2-12. Effective windspeed, as modified by canopy. Fire spread models require the windspeed at midflame height; this will vary with cover, and several methods have been developed to translate standard windspeeds (20-ft height) to their midflame equivalents in different fuels. From Kerr et al. (1971).*

sun; irregular terrain, such as canyons, that shapes microclimates supportive of different fuels. The type of fuel prevalent and its fuel moisture will vary according to topographic considerations. Topography affects weather because of its ability to fashion many microclimates and because it perturbs the wind field. Ridges cause an acceleration of wind across their summits; saddles, an acceleration through their passes. Rugged topography induces turbulence, ranging from local gusts to large roll eddies on the lee side of mountains. A diurnal (daily) cycle of winds develops on slopes, upslope during the day and downslope at night. Cold air collects in vaileys at night, leading to inversions.

But the impact of relief on fire behavior can be more direct. Deep, narrow canyons are more likely to burn out as a unit—the radiant heat and firebrands generated by one side tend to ignite the other. Sculptured mountain flanks and deep canyons can encourage the formation of convective columns above fires, acting like a chimney by channeling heat into relatively narrow funnels. By concentrating the plume, the convective velocity increases, and with it, the rate of combustion. These effects combine with another, the most common influence of terrain on fire behavior: slope. A fire burning on an incline has its flaming zone bent in a way identical to that produced by winds (Figure 2-13).

Fires burning upslope show the same mechanisms of heat transfer as heading fires driven by wind; fires burning downslope, those of backing fires. The fire perimeter assumes the same shape as wind-driven fires. One reason is that convective heat flow sets up a stream of wind into the fire that helps drive it upslope. Another is that, during the daytime, it is common for upslope winds to develop simply because of solar heating. In both cases the slope, and the winds induced by it, create a fire with properties nearly identical to those for wind-driven fires. Slopes, moreover, encourage spotting. Ahead of the flaming front, spots develop from firebrands carried upward by convective wind; below the front, by rolling debris. Left alone, those slopes tend to burn out as

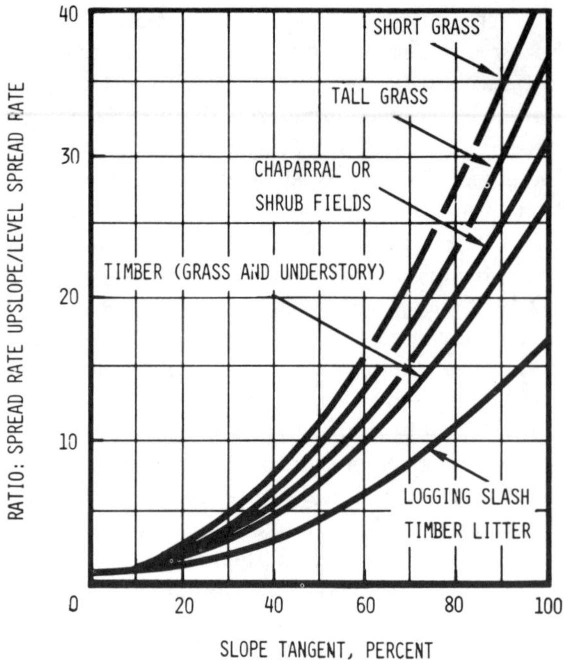


*Figure 2-13. Influence of slope on fire spread. Slope introduces still other complexities in that burning debris may roll down the slope, igniting fires below the backing fire. From Albini (1976).*

a unit, and whole drainage basins form natural burning blocks (Figure 2-14).

Topography cannot be changed by fire management agencies, and unlike winds it does not change its configuration rapidly of its own accord. The changes in fire behavior that it stimulates come about as a result of the fire perimeter entering new terrain. These facts have important management implications. Topography is predictable in ways that wind is not. Firelines for the control of a fire perimeter will therefore conform to topographic features—avoiding terrain that is likely to cause accelerations in fire behavior. A number of fire casualties have occurred from fire behavior dominated by topographic considerations such as steep slopes and chimneys. At the same time, by sculpturing natural burning units, topographic features outline sites for fuelbreaks, barriers for preattack planning, and burning blocks for prescribed fire. Again, the features that make a wildfire burn upslope make it easy to control a fire burning downslope.

*Figure 2-14. Topographic influences on fire spread. In the upper photo (Southern California), the narrowing of the canyon created a chimney and helped to develop a convective column. In the lower photo (northern Arizona) the topographic configuration made the basin into a natural burning unit, though some pockets (islands) remained. Courtesy U.S. Forest Service.*

### *Modes of Propagation*

One of the characteristics of large fires is that new modes of propagation supplement, or substitute for, the modes of flame spread that typify steady-state fires. Each mode affects different processes of combustion and shapes the flaming front in different ways, but all bring a special degree of organization to that process and distinctive shapes to the fire as a whole.

*Head Formation.* An initiating fire cannot remain as a point. To escalate into a steady-state fire means that it will behave as a line fire, and for a line fire to move, the perimeter will normally develop a head. 'Head" is both an anatomic as well as dynamic term. Descriptively, it refers to that most active portion of a fire perimeter shaped by a heading fire. But the shape of the fire approximates a *double* ellipse, and there is also a head formed to the rear by the action of the backing fire. In this sense, *head formation* describes a process by which fire perimeters successfully grow outward and the shape that such a process inscribes. The perimeter does not spread evenly, but with a bias. Heat is concentrated at the center of a line rather than on the ends; the center spreads slightly more rapidly, generating more heat which is again concentrated preferentially at the center. The line of fire narrows; a head forms. The process applies equally to heading and backing fires. The initial bias results, in part, from differential radiant heat transfer. More often, for heading fires, it comes from the relative redirection of convective heat by wind or topography and, for backing fires, by the differential dissipation of heats by wind and slope.

Head formation gives organization to the flaming front. Head formation is a primary means of fire spread, and therefore of fire propagation, for steady-state fires. It characterizes fires whose principal fuel flow is horizontal and whose available fuel loads are not large. Where wind is prominent even large fires will develop heads—often become large through the mechanisms of head formation. For many large fires, the process of head formation controls the fire's history. Moreover, for fires that are not point sources, it is possible for multiple heads to develop simultaneously. This can occur with wind shifts acting over long lengths of fire perimeter or where mountainous terrain allows portions of a flaming perimeter to enter several drainage basins. A single fire becomes many, loosely connected fires.

*Convection Column Formation.* For most fires the discharge of heat and other combustion byproducts takes the form of a *plume.* While hot air causes the plume to rise, but the configuration of the plume is the product of the air mass surrounding it. Ambient or slope-induced winds will push a diffused plume ahead of the fire. But in certain cases the plume assumes a special degree of organization, forming a *convective column* (Figure 2-15). Where multiple columns exist, the separate columns can coalesce, with important outcomes for fire behavior.

In general, the development of a column signifies a series of fundamental transformations—changes in the fire's properties, with an acceleration in its combustion rate; in the fire's mixing processes, a shift from wind-driven to convection-driven diffusion; in its fuel flow, from horizontal to vertical flow; and in its mode of propagation, from continuous spread by head formation to discontinuous spread by spotting and firewhirls. For a given fireline intensity, the establishment of a column changes the position of a fire on the fire characteristics curve. More commonly, it means that the fire has moved to a new range of intensities altogether (Figure 2-9).

Head formation and column formation are, in some respects, competitive processes. In theory, they propose exclusive means of organizing the air flow around a fire. Large fires show some features of large storms. The basis for the comparison is the analogy between the formation of towering cumulus clouds and the development of convection columns. The heat released by a large fire initiates a process of lifting; whether lifting continues depends on the structure of the surrounding air mass. For both cumulus cloud development and fire-induced convection, strong

*Figure 2-15.* Convection columns. (Upper) Diffuse smoke plume, the product of a wind-driven fire; (middle) well-developed convection column, with many large storm features; (lower) bent convection column, the result of strong wind shear above the surface. Even though the column has been severely flexed, it maintains its integrity. The situation is ideal for saturation showering by firebrands. Courtesy U.S. Forest Service.

winds retard development and unstable air promotes it. Whether development is possible depends on the ratio of the energy system of the wind to that of the fire.

In practice, wind and convective columns show degrees of interaction—stiff winds commonly bending the column like an elastic object, for example. The interaction of ambient winds and convective columns gives rise to other modes of propagation such as firewhirls and long-range spotting by firebrands. And the presence of multiple columns encourages another mode of propagation, should the upper portions of the columns encounter each other and combine. The outcome of *coalescence* is analogous to combining two small diameter hoses into one larger diameter hose: the velocity of fluid through the smaller hose increases. When convective columns merge, fire intensity increases, indrafting winds accelerate, and the fires are drawn together. This form of behavior can be deliberately exploited in setting backfires, allowing the main fire to draw the backfire into it.

Column formation and interaction give an added degree of organization to the fire as a whole. Rather than burning as a complex of geometrically connected but dynamically unrelated heads, the fire acquires a more singular organization. All parts contribute to the burning of other parts. Within this special environment of extremely intense burning, other forms of fire, wind, and flame organization may occur.

*Crown Fire.* When a fire flames through a canopy of aerial fuels, it is known as a *crown fire.* The expression can refer to several conditions, however. In geometry and dynamics, the crown fire may represent only a difference in size from the surface fire, not a difference in process. A surface fire, for example, may simply engulf aerial fuels within its flaming zone. In this case the involvement of crown fuels is a restatement of the proposition that fuel accessibility is relative to flame size and heat flux. A crown fire, moreover, may spread through aerial fuelbeds independently of the surface fire. In this case the dynamics of crown fire propagation are identical with those of surface fires. Crown fire becomes merely a vernacular expression for unremarkable processes.

Yet there are few things unremarkable about a crown fire, and in reality crown fires show complex linkages with surface fires; they relate integrally to the behavior of large fires in selected fuel complexes, and they cooperate with other, often dramatic modes of fire propagation. In closely stocked jack pine and red pine forests, in mature chaparral brushlands, and in dense eucalypt woodlands, the normal mode of large fire behavior is the crown fire. Its intimate connection with the mechanisms of spotting, mechanics of horizontal roll vortices, and processes of large wind-driven fires make crown fires worthy of independent study.

Crowning fires include a range of phenomena, from the torchings of individual trees to actual fire spread through the crowns. *Torchings* may be thought of as a case of transient ignition. The surface fire transmits heat to a cluster of vertical fuels: the fuels ignite, flame, and then expire as the heat source passes on. For persistent ignition to result, the fuels must be suitable; the heat generated, of sufficient amount; and the heat transferred, of adequate rate to support sustained combustion. The fire and fuel criteria are essentially the same as for surface fires, with a few allowances. For one thing, the vertical dimension to the heat flux—its height above the source—is more important than in surface fires. The heat of ignition is applied to the base of the fuelbed not to its surface. For another, the original heat source remains within the fire environment, much like the pilot light of a furnace. Surface and crown fires can, and generally do, interact, and the persistent presence of the surface (pilot) fire can reignite the crown fire over and over again as conditions favor. The crown fire may become an integral part of a single flaming zone or an episodic phenomenon erupting spasmodically out of a surface fire (Figure 2-16).

Three categories of crown fire have been identified on the basis of three criteria for fire initiation and spread (Van Wagner, 1977). The mass fuel flow, the horizontal heat flux through the

*Figure 2-16. Crown fire, in profile and area. Courtesy Petawawa National Forestry Institute, Canadian Forest Service.*

crown, and the rate of flame spread furnish the criteria. To produce a fire all three must exceed certain threshholds. But the extent to which the surface fire supplies the requisite heat or fuel to the flaming zone of the crown fire will decide whether the crown fire is passive, active, or independent.

In the case of the *passive crown fire,* the heat flux is sufficient but the rate of spread and the rate of mass fuel flow are not. Some fuel is supplied to the flaming zone of the crown by the surface fire, so the rate of spread of the surface fire determines the rate of spread of the crown fire. An example would be sparse crown foliage, or foliage otherwise unavailable, perhaps because of high fuel moisture. The two fires, surface and crown, remain coupled. In the case of the *active crown fire,* the mass fuel flow and the rate of spread are adequate, but the propagating flux is not. The surface fire makes up the heat deficit. An example would be a dense crown whose heat output is insufficient, or whose transfer of heat is inadequate, perhaps because of low winds. Again, surface fire and crown fire are linked to mutual benefit. The enlarged flaming front generates more heat for the propagation of both fires. Only the *independent crown fire* can burn without assistance from the surface after the point of initial ignition. The actual mechanics of such propagation are not well understood, though strong winds, which bend the flame over and drive flame through crown foliage, seem to be essential. Such fires, moreover, do not sustain themselves for long, suggesting that the conditions for their existence are both rigid and transient. Most crown fires show a mixture of these three types of behavior.

Common to all categories of crown fire, however, is the presence of strong winds. Crown fires are large, wind-driven fires with huge flaming zones. Like other wind-driven fires, crown fires thus exhibit head formation; and if the wind field collapses, so does the crown fire. But fire field and wind field interact. Air above the flaming zone rises from convection, while ambient air flows horizontally. At one extreme, convective heat is simply entrained within the wind field with little organization; at the other, the development of a convection column resists ambient wind, which tends to flow around it as though it were an elastic solid.

But where convective flow and ambient wind can interact without one or the other dominating, wind shear results. Like other aspects of large fire behavior, the shear may assume special organization—in this case as a *horizontal roll vortex.* One of a variety of vortical motions associated with wildland fires, horizontal roll vortices occur with crown fires, perhaps accounting for a peculiar pattern of unburned strips of canopy amidst burned-out fields. Wind flow along the fire boundary is helical, usually confined to one side of the fire or the other. The horizontal roll vortex may also account for sudden surges and shifts in crown spread along the fire flank (Haines, in press).

*Spotting By Firebrands.* Among the products of a fire are burning particles. Those particles that are capable of igniting fuels are known as *firebrands.* Certain particles have properties that make them good candidates for firebrands, but whether they ignite new fires depends on their mechanism of transport and the fuel complex to which they are applied (Figure 2-17). The fires ignited by brands are known as *spot fires,* and the process of ignition, as *spotting.* Spotting introduces the possibility of discontinuous fire spread, not so much as a process of growth as of cloning. Spotting can initiate new, distant fires independent of the source fire; but more often spot fires and source fires interact in ways that may either assist or retard the source fire. Spotting accompanies most large fires, and a few large fires depend on spotting as a primary mode of propagation.

Many natural fuels make suitable firebrands: cone scales, grass clumps, bark flakes, parts of branchwood, and moss. To be effective a firebrand must continue to combust during its transport; most often this means glowing combustion, though flaming combustion contributes to more rapid ignition after settling in new fuels. A firebrand must also have suitable properties for aerial transport. Its size, shape, volume, and mass all

(This table is based on data compiled by Hal E. Anderson, USDA Forest Service, Intermountain Forest and Range Experiment Station, Northern Forest Fire Laboratory, for Fire Behavior Officer training (National Interagency Fire Training Center, Marana Air Park, Arizona).)

| Fuel moisture content | | | Relative ease of chance ignition, likelihood of spotting, general burning conditions |
|---|---|---|---|
| Relative humidity* | 1/2-inch fuel stick | Forest litter | |
| - - - - - - Percent - - - - - - | | | |
| | | >25 | Little or no ignition[1] |
| >60 | >15 | >20 | Very little ignition;[2] some spotting may occur with winds above 9 mi/h[3] |
| 45-60 | 12-15 | 15-19 | Low ignition hazard - campfires become dangerous;[1] glowing brands cause ignition when relative humidity <50 percent[3] |
| 40+<br>30-45 | 7-12 | 11-14 | Medium ignitibility - matches become dangerous;[1] "easy" burning conditions[4] |
| 26-40 | | 8-10 | High ignition hazard - matches always dangerous;[1] occasional crowning, spotting caused by gusty winds;[2] "moderate" burning conditions[4] |
| 15-30 | 5-7 | 5-7 | Quick ignition, rapid buildup, extensive crowning; any increase in wind causes increased spotting, crowning, loss of control;[2] fire moves up bark of trees igniting aerial fuels; long distance spotting in pine stands;[3] dangerous burning conditions[4] |
| <15 | <5 | <5 | All sources of ignition dangerous;[1] aggressive burning, spot fires occur often and spread rapidly, extreme fire behavior probable;[2] critical burning conditions[4] |

*Relative humidity is a surrogate for "fine" fuel moisture content.

[1]Gisborne (1936)
[2]USDA Forest Serv., Northern Region (1973)
[3]Florida Div. For. (1973)
[4]Barrows (1951)

*Figure 2-17. (Upper left) Torching tree, a common mechanism for lofting firebrands; (upper right) long-range spotting, from fire in heavy fuels burning right to left; (lower left) spotting potential as related to fuel moisture content. From Albini (1979), photos courtesy U.S. Forest Service.*

determine its aerodynamic success as measured by drag, buoyancy, and settling velocity. What complicates the analysis is that these properties vary as the particle burns. Mass loss derives primarily from flaming combustion, and volume loss from glowing combustion. Loss of either mass or volume reduces the terminal velocity of the firebrand, so that the more the particle burns the more likely it is to travel far. The maximum effective distance from the source fire thus depends on two competing processes: the firebrand improves its distance by growing smaller, but growing smaller reduces its ability to ignite fuels when it finally comes to rest (Albini, 1979).

Distance depends, too, on how that particle is entrained in the wind field. It is the nature of spotting that it requires winds. But within the wind field firebrands may be entrained along the surface, as with tumbleweeds and cattle dung; or somewhere above the level of the flaming zone, typical of firebrands lofted by the process of torching; or at some height above the surface, lifted up within the convective column. They may result from flareups, transient pulses of flame that tear off and eject debris, or they may come from sustained, high-intensity sources such as slash piles. Similarly, the wind that entrains the particle may represent transient vortices such as dustdevils or firewhirls, sustained vortices such as the convective column, or broad wind fields, further broken by topographic configurations into rolls and eddies. How far the firebrand travels depends on the interaction of its aerodynamic properties and the wind profile into which it is introduced. Higher wind velocities (and longer brand trajectories) result from particles placed at higher levels of the wind profile. A general prescription for spotting recommends midflame winds over 10 mph (16 km/hr), fine fuel moisture under 10%, and fireline intensities greater than 700 Btu/ft/sec (579 kcal/m/sec).

The source of the spotting process may be isolated or massive, a single torching tree or a crown fire; stationary or mobile, a flaming slash pile or a flaming front. Several mechanisms operate, but common to all of them is the lifting of firebrands by the convective buoyancy of the flaming zone. The convective updraft generated by the fire lofts particles upward where they become entrained in the ambient winds. The interaction of particle with convective plume, and of plume with the ambient winds, is complex, but three versions have been investigated: the torching of a single tree or clump of trees, sustained flaming associated with slash piles, and the flaming typical of a wind-driven line fire. Torching, or the sudden rush of fire upward through a tree crown, increases fire intensity and convective buoyancy, breaks loose small debris, and like a thermal catapult ejects the particles into the wind profile above the canopy. The process is transient, and for spotting to be characteristic of a fire the episode must be repeated many times with new trees. Slash fires, by contrast, offer sustained flaming and convective lift. But such flames are not truly constant: they pulse or flicker, ejecting bubbles of heated air upward.

Line fires resemble slash fires in that they too pulsate, with bubbles of heated air constantly breaking off from the flaming front. The interaction of the ambient winds with the line of convection rising above the fire is exceedingly complex. But the general effect is to add another source of transport to the firebrand. Rather than merely lifting a particle upward and then releasing it to the wind regime, the convective flow from a wind-driven fire carries the particle some distance away from the fire before releasing it for transport by the ambient winds (Albini, 1979, 1981, 1983).

The characteristic result of such processes is spotting of two sorts. One is a somewhat random spotting, often at considerable distance from the source fire. This occurs when a suitable firebrand has been hurled upward high into the wind profile. The other outcome is a systematic showering of firebrands close to the source. This often occurs as embers fall out of a bent convective column or when burning debris is driven along close to the ground by strong surface winds. Distant spotting tends to ignite new, distinct fires that may or may not subsequently interact with the source fire. Saturation spotting, however, does interact with the source fire. The spotting process is broadly analogous to the transport of sand by saltation along a stream bed, with the important difference that in a fire the transfer of energy from particle to particle does not come by the instantaneous transfer of momentum but by an intervening process of ignition, combustion, and the production of more firebrands.

This distinction is basic to an understanding of how spotting relates to the behavior of the fire as a whole. Between deposition and combustion—the period of ignition—there exists a delay time; between combustion and the buildup of heat to support the production of more firebrands, there is another delay. For the firebrand to ignite a distant fire of its own these delay times must be brief, the distance from the source fire long, and the rate of spread of the source fire relatively slow. Otherwise the source fire will arrive at the scene of the spot before the onset of sustained combustion. The effect of spotting in this case will be to reduce the delay time involved in the ignitions that occur along the flaming front. The source fire absorbs these spots and the spots help to speed the fire along. Large, wind-driven fires with close-range spotting belong in this group. For such fires spotting is a common phenomenon, but not a distinctive means of spread.

There are cases, moreover, for which spotting can become a significant mode of propagation. Isolated, long-range spotting can produce independent fires miles from the source fire. Under proper circumstances, these fires can themselves behave like another source fire, growing from a few points. The history of these spots will depend on how they interact with the source fire. With wind-driven fires, spotting may deprive the source of new fuels; with convection-driven fires, it may add new sources. For wind-driven fires, they may become inhibitors. When a large fire encounters the spot, the fuels will be at least partially burned out. In this way spotting may account in part for the exhaustion of large fires, the self-arresting character of runs. For convection-driven fires, however, spotting may be an essential mechanism for survival. The reason is that their strong indrafts prevent horizontal spread; the fire burns down and out. If the column is bent, as it might be under the impact of winds or slope, the firebrands that it contains will fall out before they burn up. The deposition may take the form of an ember shower. The particles simultaneously ignite an area, another intense fire results, and this new fire effectively enlarges the area burned. A convection-driven fire may thus spread laterally, even developing the appearance of a head.

Spot fires are a major problem in fire control. Long-range spots multiply the number of fires under attack, reduce the effectiveness of perimeter control by firelines, and endanger crews. More than once crews have been trapped by unseen spot fires that burned below them on slopes. Yet, once again, the ability to ignite many fires simultaneously and to have these fires interact in various ways is a useful technique for many programs of prescribed fire. Interestingly, as with their natural counterparts, spot ignition is best suited for aerial delivery.

*Firewhirls.* The subordering of wind and flame has perhaps no more striking example than the firewhirl. Firewhirls share with dust devils and tornadoes certain originating mechanisms, and they share with other fire-related vortices the special environment posed by large fires. Like all vortices, firewhirls require as conditions of development that there exist a source of energy, a generating eddy, a fluid sink within the eddy, and upper and lower boundary limits to the system. Within the environment of the fire—an environment that includes the ambient wind field flowing around the fire—combustion supplies the energy source; diffusion, the heat sink; the ground and the upper limits of flame bouyancy, the boundary conditions. The firewhirl gives these loosely connected phenomena an intense degree of organization. For most firewhirls this organization resembles a dust devil in size and mechanics, rising upward from within or around the flaming zone. For a few it simulates a tornado, spiraling downward from the lee side of a bent convective column. The fire creates circumstances that favor the development of firewhirls, and in turn the firewhirl locally orders the flow of wind and structure of flame, accelerates fire intensity, and disseminates spot fires to such an extent in some cases that it becomes a distinctive mode of fire propagation (Byram and Martin, 1970).

The triggering mechanism, the generating eddy, can appear in different areas of a fire and from different causes (Countryman, 1971). Wherever such a wind exists in conjunction with an appropriate rate of heat release, a firewhirl is possible. Eddies tend to appear in zones of low pressure where some shearing occurs. Within the flaming zone, vorticity may result from the interaction between indrafts and convection, the shear produced by mixing horizontal and vertical flow around cells of intense burning. Outside the flaming zone, eddies result from the interaction of ambient winds and the convective column, concentrating on the lee side of the column, and from the interplay of wind and topography, favoring zones of low pressure such as those found on the lee side of ridges (Figure 2-18). The eddy must show some stability, usually by the presence of surrounding "curtains" which enclose the zone of vorticity. In laboratory experiments curtains have been created by mechanical enclosures, rings of compressed air, or circles of thermal jets; in na-

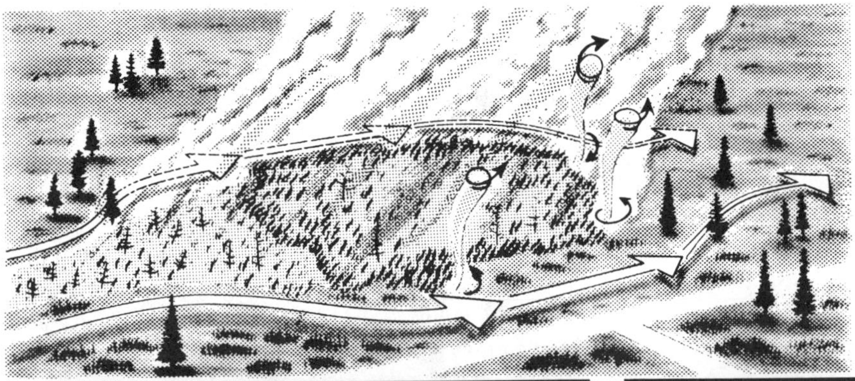

*Figure 2-18. Firewhirls. A common site for eddy location is the lee-side of a convective column, but any substantial obstruction to ambient wind flow will do. Photo shows a firewhirl generated in the lab (left) and one in the field (right). From Countryman (1971), photo courtesy U.S. Forest Service.*

ture, they take the form of topographic features, quasi-elastic convection columns, and distinctive columns that develop around cells of very intense burning.

The vortex organizes a vertical flow of air. That flow results from temperature differences between the air and a heated surface. An unstable atmosphere will assist this movement, but a sufficiently intense heat source will result in upward lifting regardless. Thus firewhirls may develop over level terrain or under evening inversions as well as in such eddy-rich environments as mountains or within unstable air masses. In dynamics the firewhirl and dust devil show similarities and both may be found in fires. The firewhirl resembles the dust devil in that it derives its energy from a heated surface and develops from the ground up—in contrast to tornadoes, whose heat source results from condensation in clouds and which develop downward. It differs from the dust devil in that it has higher heat sources. For actively burning fires firewhirls are common during flaming stages; dust devils, on the blackened, solar heated surfaces left in the burned zone.

Once it initiates a firewhirl, the generating eddy collapses into a condensed, high organized form with $\frac{1}{10}$ to $\frac{1}{20}$ its original diameter. It is this particular organization that distinguishes the firewhirl from other forms of flaring, torching, or fire flashing. But its originating conditions change. The flaming zone moves into new fuels and new topography, and those firewhirls that form around the perimeter of the flaming zone may themselves break loose to roam, driven by the ambient winds, and dissipate. Should it so separate from its energy source or enter terrain that perturbs its tightly ordered vorticity, its structure will break down and the firewhirl vanish.

The transition to firewhirl is abrupt, with sudden increases in flame height, noise, rate of combustion, and indrafting winds. The increase in combustion rate is significant. In laboratory experiments with liquid fuels, the combustion rate tripled; with solid fuels, the rate increased two to five times. Firewhirls do not merely associate with intense fires: they further intensify the source fire. Convective velocities increase dramatically, the product of higher heat releases and the shrinking of the vortex diameter. The pattern of indraft winds follows this reorganization of the vortex. Nearly all air enters from along the lower surface boundary, and the convective velocity reaches speeds of from a few tens of feet per second to the velocities of small tornadoes. Flame height rises in a column 10–50 times the diameter of the vortex.

Yet these properties merely intensify processes already at work. The character of the fire changes, but firewhirls become a distinctive mode of propagation by other means. They often wander, particularly around the head of the fire, a pillar of flame charged with firebrands. And they are a likely mechanism to emit firebrands, not simply to carry them. With its high convective velocity and its tall, if spindly flame column, the firewhirl can wrench loose suitable firebrands and propel them into higher wind velocities. Especially troublesome are fire-related whirls, including dust devils, that form over smoldering areas of the fire or blackened regions heated by solar radiation. A source of firebrands, such phenomena can prolong fire suppression efforts into those final stages of a fire normally designated for mopup.

### *Fire Behavior Models*

Many models exist, and many types of models can exist, to describe fire behavior. The models may be descriptive or quantitative, empirical or theoretical. They may describe any or several aspects of fire behavior—history, spread, intensity, fuel consumption, smoke production, or fire effects. There are models for area and perimeter growth, for torching and spotting, for duff burnout, for crown fires, and for mass fire, both in wildland and urban settings. Proposed in 1959, the Byram fire model developed a theoretical description for fire behavior which allowed for scaling to various fire sizes and intensities.

Within the United States today, the expression fire model refers principally to the Rothermel fire model; the Rothermel model, to a mathematical model of fire behavior; fire behavior, to the spread of steady-state surface fires. Already a hierarchy of mathematical models (with the Rothermel model at its foundation) has developed into a general fire information system (Figure 2-19). Any information needs that require fire behavior forecasts can be integrated within this network. Long-term fire planning, short-term presuppression strategies, fireline operations, and estimates of fire effects all depend on fire behavior characteristics; in the United States all rely on the Rothermel fire model to convert environmental data into numerical predictions about those characteristics.

*Rothermel Fire Model.* Published in 1972, the Rothermel model is a mathematical model for the prediction of fire spread. Like all fire behavior models, it first describes the free-burning combustion of wildland fuels, which defines the properties of the reaction zone, and then it accounts for the behavior of that zone as determined by fuel, weather, and slope. For inputs the model requires certain kinds of information about fuel moisture, slope, wind (velocity at midflame height), and the fuel complex, or more properly, about the surface fuelbed. The relevant fuel parameters may be entered directly into the model; more commonly, normal values for typical complexes of wildland fuels have been synopsized into fuel models. The fuel model can then interact with the fire model. To get information requires information, and the resolution of the model depends on the accuracy of its fuel data.

The fire model assumes uniform environmental conditions, and it applies to fires that spread uniformly. By combining its predicted outputs for such combustion measures as reaction intensity with other fire-related models, the Rothermel fire model can assist in the prediction of other manifestations of fire behavior. It can indicate the potential for large fire behavior, for example,

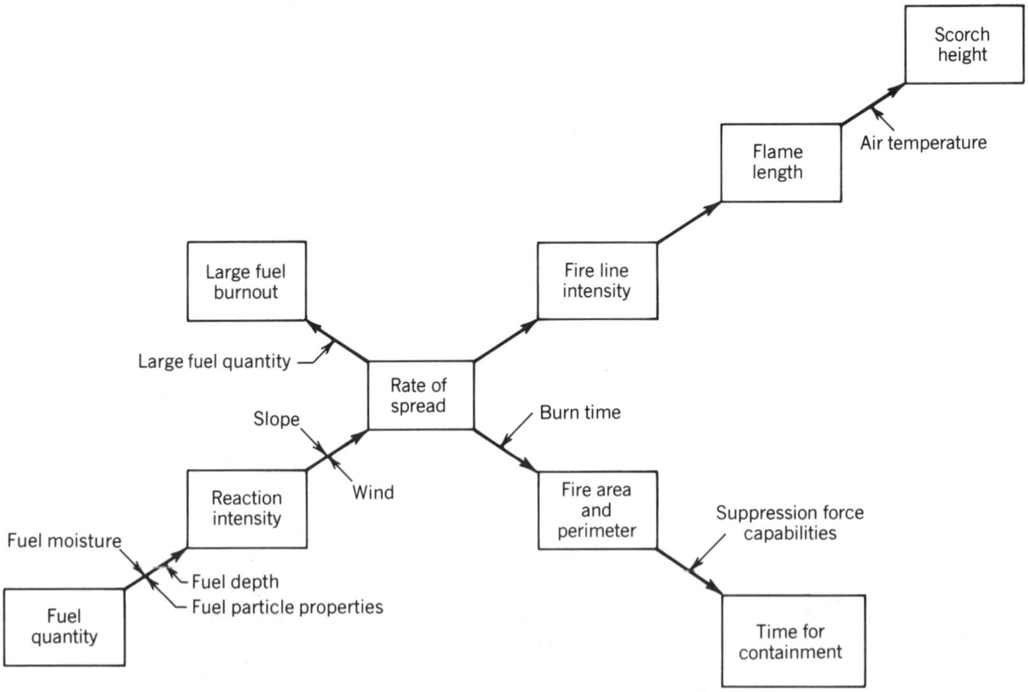


*Figure 2-19. Surface fire model capabilities. The reaction intensity describes combustion; slope and wind, the behavior of the flaming zone. Other models build on these calculations. From Rothermel (1980).*

though it does not forecast how large fires will behave. The fire model developed out of extensive experimentation, and it has been subsequently tested both in the laboratory and, insofar as it underwrites various systems for the prediction of fire behavior, in the field. Refinements are extending the range of the model into nonuniform fuels and certain expressions of nonuniform fire behavior.

*Fire Information Systems.* The applications of the fire model are varied and growing. Virtually no system of fire management that seeks quantitative information fails to use the Rothermel model for fire behavior analysis, and few systems of fire intelligence fail either to contribute to its input needs or to extrapolate from its outputs. At all levels of fire planning, and increasingly at all levels of fire operations, fire behavior information is mandatory. For the most part, the necessary calculations derive from the Rothermel fire model. The character of the fire model, in turn, has stimulated the demand for other, compatible models that can simplify the problem of entering good input data, translate fire behavior output into other management programs, and improve the resolution of the model itself. The presence of the fire model not only encourages the production of other models, but defines the kind of information they should contain. At present the variability of weather and discontinuities among natural fuel complexes are greater limitations on the use of the model than its internal workings. Accurate real-time firecasting is thus restricted to periods of 12–24 hours—about the same as the weather forecasts upon which it depends. Most model "failures," in brief, result from poor input or the inappropriate application of output, from using the model in situations and for purposes other than those for which it was designed.

In general, the model is applied to three kinds of fire situations: possible fire situations, potential fire situations, and actual fire situations (Figure 2-20). Although the processing unit, the model

| Objective of System | Consequences of Decision | Resolution | | Accuracy of Inputs | |
|---|---|---|---|---|---|
| | | Space | Time | Fuels | Weather |
| Possible fire situation (long-range planning) | Significant in relation to resource response, economics, and operational costs | Diverse—drainage to Regional or Forest level | Long—years | Good—should have capability of reflecting changes caused by planned activities | Moderate—climatologic data |
| Potential fire situation (prefire planning) | Significant in relation to operational costs | Large—Regional or District | Moderate—seasonal | Moderate—used as base to indicate effects of weather | Good—weather changes are the primary driving force |
| Actual fire situation (operational plans and decisions) dispatching and escaped fires | Significant operational costs and human life risk | Site-specific—10 to 100 acres | Short—minutes to hours | Very good to excellent—must be keyed to site and strata expected to carry fire | Best possible—fire weather forecast or on-site meteorologist |

*Figure 2-20. Scope of fire behavior systems and input needs. From Rothermel (1980).*

itself, is unchanged for all of these uses, the resolution of input and output does vary. Possible fire situations include hypothetical fires that are needed for long-term planning. By specifying certain situations, as defined by environmental inputs, planners can compare resulting fire behavior to strategies of protection, fuels management, or habitat manipulation. Instead of fighting the fires of the past, planners can project, through simulation and gaming, the fires of the future. Potential fire situations include such short-term planning needs as fire danger rating, prescribed fire projects, and smoke management. In many ways, the fire model works better for prescribed fire than for wildfire. The environmental variables can be controlled, ignition and spread timed, and more uniform conditions promoted with prescribed burning; in a wildfire, all of these circumstances are, by definition, uncontrolled.

Actual fire situations provide perhaps the most demanding use of the model and the highest skills for its application. Real-time fire behavior forecasting (firecasting) for specific sites and particular events requires high resolution input and knowledgeable interpretation of the output. Actual fires include wildfires, for which initial attack is planned; escaped fires, for which policy requires a special analysis of alternatives; and prescribed fires, for which continual monitoring is necessary. For initial attack, the fire model forms part of an information system that assists the dispatcher. Examples include the Firecasting module used in Southern California and the initial attack management system (IAMS) developed by the Bureau of Land Management. In both cases, automatic weather stations provide much of the site-specific data needed by the model. For project (escaped) fires, portable systems are available for processing the data, such as hand calculators and nomographs; intensive formal training with fire prediction methods has led to a special staff position, the fire behavior officer. For prescribed fire, the process of monitoring site data and measuring predicted fire behavior against actual fire spread is fundamental to the execution of the burn. Predictable fire effects require predictable fire behavior. Without some fire behavior model—especially without one that can make numerical forecasts—prescribed natural fire programs would be difficult to execute. With them logic trees can be designed to help guide management decisions.

## *2.3 LARGE FIRES*

Large fires are relatively rare but immeasurably important. Even with aggressive fire protection, fewer than 5% of all fires account for more than 95% of all burned acres. Large fires are more intense, inflicting proportionately greater effects, regardless of whether the episode is considered as an ecological process or an economic event. The need to understand large fires has determined much of fire research; the need to cope with large fires, much of the structure of fire managment in the United States: its planning, financing, and strategy of fire control and fire use. Large fires show higher fireline intensities, different modes of propagation, and a different distribution pattern than steady-state fires. On the basis of these criteria large fires may be classified, and through manipulation of these mechanisms, managed.

### *Types of Large Fires*

The distinctive feature of all large fires is that they have high fireline intensities. Large fires show much higher intensities than steady-state fires. Within the realm of high fireline intensities, however, important differences exist. Recall from the fire characteristics curve (Figure 2-9) that a given intensity may be expressed in either of two extreme forms: as high rate of spread or high rate of heat release. The former characterizes wind-driven fires, large in area; the latter, convection-driven fires, which only become large in area if they interact in some way with wind or topography. A classification of large fires begins with this distinction. In reality, most large fires show some degree of interaction between these two dominant processes, manifested by assorted modes of propagation. Topography intervenes in important ways. In the form of long, steep slopes, topography leads to fires whose processes are indistinguishable from wind-driven fires. Rugged terrain can multiply heads, much as variable winds do. In the form of narrow chimneys, topography assists the development of convection-driven fires. The effect of terrain is to fashion a fire environment that favors either wind-driven or convection-driven fires, or some combination of the two processes. But the influence is not so much determinant as suggestive; very strong winds, for example, can override the potential effects of land sculpture.

A summary classification of fires based on the interaction of wind and convection systems is given in Figure 2-21. The table is meant to be a suggestive description, not an exhaustive inventory. It summarizes some common types of large fires; in nature, many more forms are possible. All of these fires exhibit large fireline intensities. Types I–III characterize fires with well-developed convection columns. Types IV–V apply to fires whose convection column is distorted by ambient winds. Type VI describes a fire dominated by ambient winds. Types VII and VIII represent fires with multiple systems, with topography as a primary intervening variable. They include fires with multiple heads or convection columns, each of which may represent one of the previous types of large fire and which, collectively, may or may not interact among themselves (Figure 2-22).

For Types I–V the convection column controls the rate of combustion. Unless the fire enters additional fuels laterally, it will soon exhaust its downward fuel supply and burn out. This necessary horizontal movement may come in several ways. If light winds can push the entire column (Type I), then the flaming zone can spread outward as well as down. The same effects can result from slope (Type II), though the ridgetop imposes a limit on how far spread can take place. With

| No. | Type | Dominant Features |
|---|---|---|
| I | Towering convection column with light surface winds | Moderate to rapid fire spread persistent until changes in the atmosphere or fuel |
| II | Towering convection column over a slope | Rapid short-term spread with convection cutoff at ridge crests |
| III | Strong convection column with strong surface winds | Fast, erratic spread with short-distance spotting |
| IV | Strong vertical convection cutoff by wind shear | Steady or erratic fire spread with occasional long-distance spotting |
| V | Leaning convection column with moderate surface winds | Rapid erratic spread with both short- and long-distance spotting |
| VI | No rising convection column under strong surface winds | Very rapid spread driven by combined fire and wind energy; frequent close spotting |
| VII | Strong surface winds in mountain topography | Rapid spread both up- and down-slope with frequent spotting and area ignition |
| VIII | Multiple head fires (mostly types I through V | Broad fire front with two or more independent convection columns |

*Figure 2-21. Summary of large fire types. From Kerr et al. (1971)*

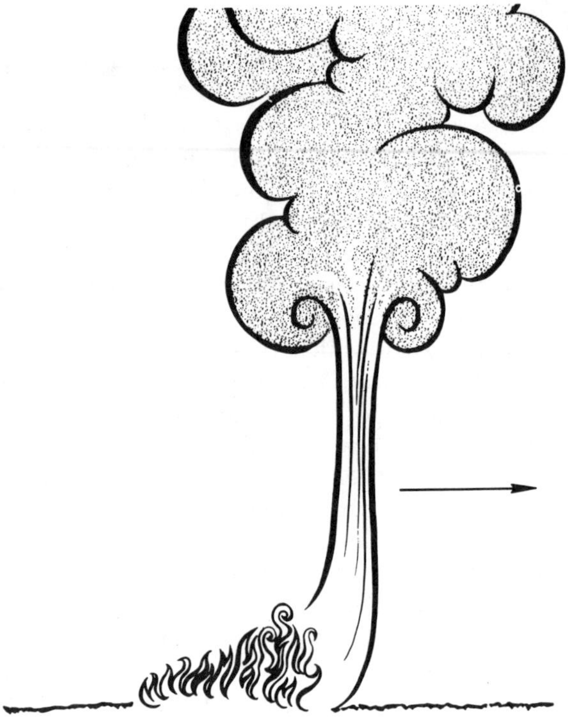

Type I

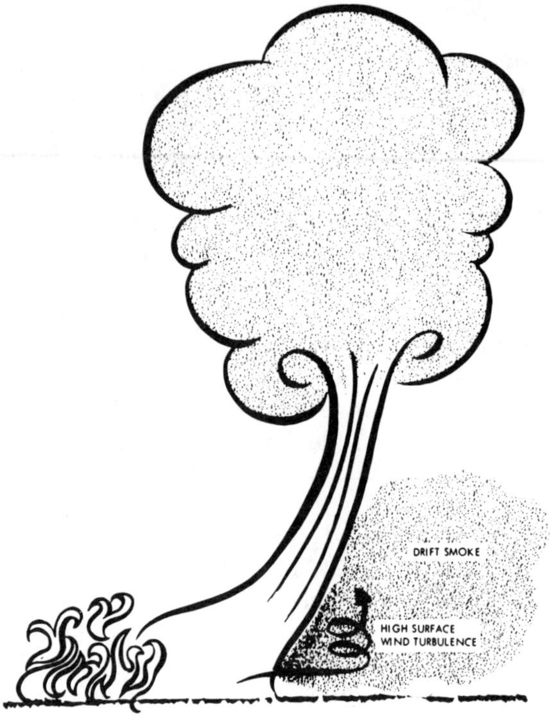


Type III

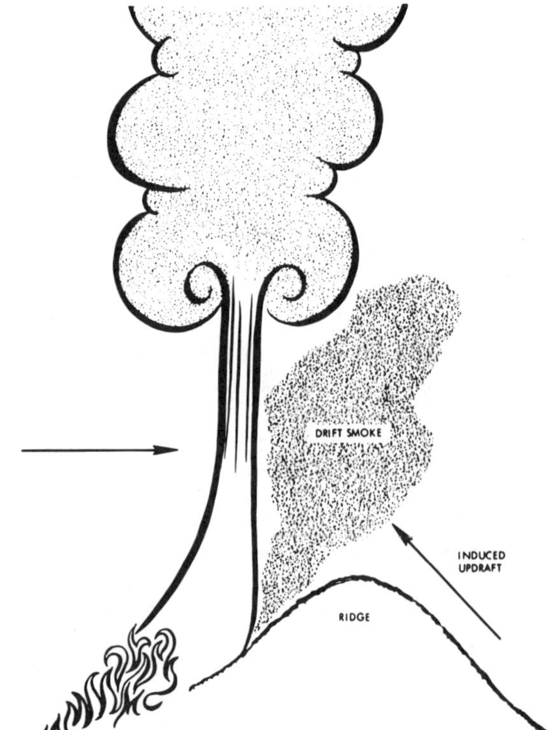


Type II

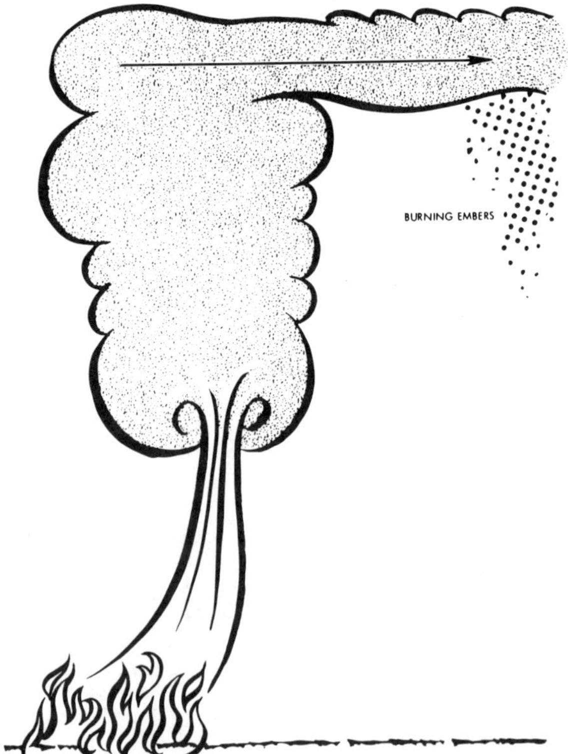


Type IV

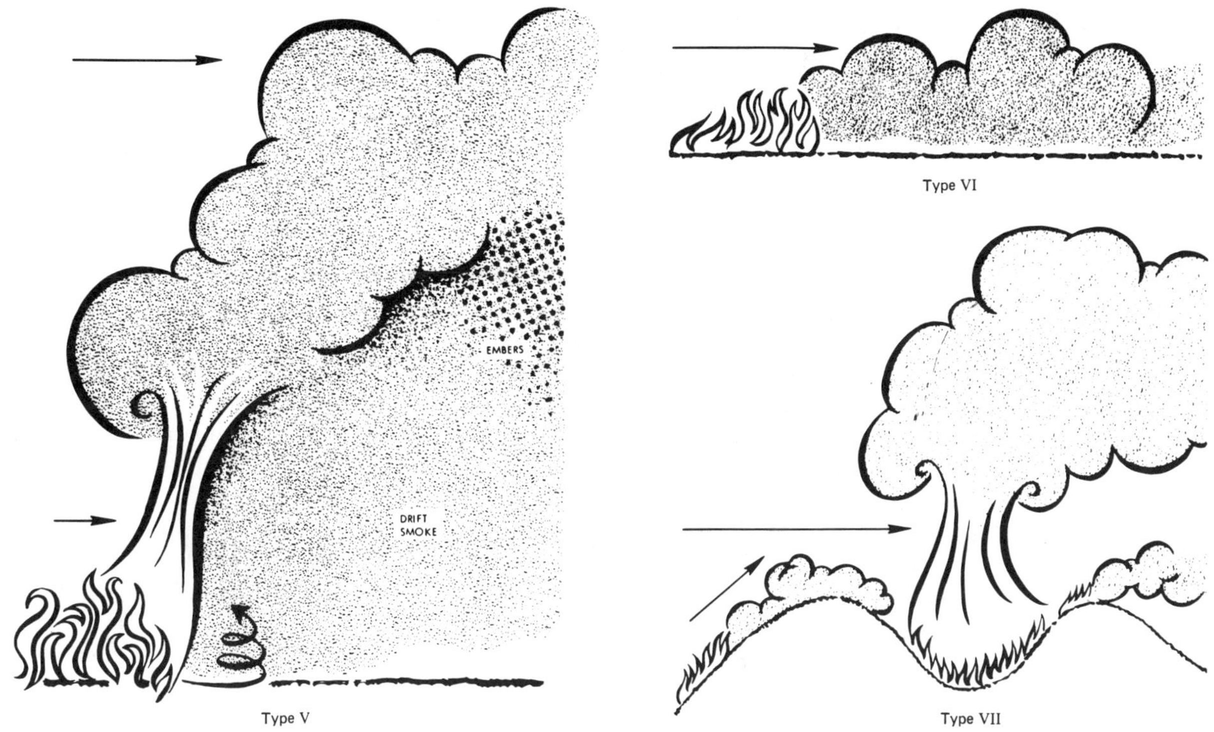


*Figure 2-22. Large fire types. From Kerr et al. (1971)*

surface winds (Type III), short-range spotting and firewhirls can occur on the lee side of the column, intensifying burning and increasing the area under fire. If the winds aloft are stronger, the convection column may either be cut off (Type IV) or bent (Type V). Spotting is likely; to the extent that truncating the convection column distorts convective diffusion, the flaming zone may show erratic behavior. In all these cases, wind does not act directly on the flaming zone but on the convection column and its immediate environs, and the influence of wind is communicated to the fire through modification of that column.

Type VI describes the familiar wind-driven large fire. Convection-column development is negligible, or so confined to the ground by strong wind that its effects are indistinguishable from a wind-driven fire. Ambient winds enter the flaming zone directly. Types VII and VIII refer to complex, but common combinations of wind-driven and convection-driven fire systems. These are most prevalent in rugged topography. In Type VII wind determines the overall shape and dynamics of the fire, though portions partially sheltered from the wind may develop convection columns. Type VIII lumps together the phenomenon of multiple large fires, characterized by multiple convection columns or multiple heads.

That fires burn with a spreading perimeter of flame means that in most landscapes the fire environment will change. Separate portions of the perimeter may develop independent heads, burning up particular canyons, for example. Even in the case of convection-driven fires occupying large areas, there seems to be a limit on how much area a single convective column can organize. Combined with the exhaustion of fuels, the result may be a proliferation of heads or columns. But against this process of divergence, there is also a process of convergence. Large fires absorb small fires; large convective columns incorporate small columns; multiple heads burn out intervening fuels and merge. Under these circumstances, the presence of other fires becomes a significant component of the fire environment.

### *Concept of a Mass Fire*

The expression *mass fire* refers to that species of large fire characterized by high reaction intensities, vertical fuel flow, and convection-dominated diffusion. As a physical event mass fire received notoriety as a result of the saturation firebombing of German and Japanese cities during World War II and the prospect of even greater destruction through the use of thermonuclear weapons. But mass fire is abundant in wildlands. It requires only suitable fuel loads, a pattern of area fire, and an air mass that will not prevent the development of a vigorous convection column (Figure 2-23).

There is no apparent intrinsic size at which a mass fire begins or at which its structure collapses. Minimum area requirements are as little as 100,000 ft$^2$ (9,290 m$_2$), and minimum heat output is on the order of 5,000–8,000 Btu/ft$^2$/min (226–362 kcal/m$^2$/sec). Its initiation requires simultaneous ignition over an area or a broadening of the flaming front such that the flaming zone is deep relative to flame height and width. Appropriate atmospheric conditions include instability that will encourage convective transfer upward and a wind field that will not break up organized convective flow. Once initiated the buildup to a mass fire follows a process of positive reinforcement. Greater combustion rates increase the amount of available fuel. Larger fuel loads promote a dramatic increase in burning intensity, signified by the development of a convection column. Greater intensities transform the mechanisms of fire propagation from processes of continuous spread to discontinuous ignition by firebrand spotting and marauding firewhirls. The combustion period is long and the combustion rate high. A mass fire may develop once or several times during the history of a single fire. The Countryman model gives a qualitative description of mass fire (Figure 2-24; Countryman, 1969).

| Fuel Bed (Wood Fuel): | Specifications for a mass fire: |
|---|---|
| Surface to Volume Ratio ($\sigma$) | 25.11 ft.$^2$/ft.$^3$ |
| Porosity ($\gamma$) | 23.85 x 10$^{-2}$ ft.$^3$/ft.$^2$ |
| Fuel Loading | > 17 lbs./ft.$^2$ |
| Kindling Fuel | 10 x 10$^{-2}$ lbs./ft.$^3$ |
| Fuel Bed Area | > 1700 ft.$^2$ |
| Ignition Points | 1/300 ft.$^2$ of fuel bed area |
| Fuel Moisture | <10 percent |
| Fire: | |
| Fire Area | > 100,000 ft.$^2$ |
| Fuel Bed Spacing | <25 ft. |
| Terrain | Level (Approx.) |
| Weather: | |
| Ambient Wind 10 ft. level | <10 ft./sec. |
| Ambient Wind 5000 ft. level | <20 ft./sec. |
| Relative Humidity | <30 percent |
| Temperature | > 50° F. |
| Lapse Rate | Neutral to unstable |

*Figure 2-23. Prescription for mass fire. From Countryman (1964).*

What the transition to mass fire brings is heightened organization. Mass fire is a pattern phenomenon. Its transition represents the triumph of organizing forces over dispersing forces; it consolidates fire rather than disseminates it. Instead of an amalgamation of points and lines of fire loosely strung out around a perimeter, there is a concentration of fire activity within a circumscribed area. Instead of a convective plume, there develops a column; instead of wind diffusion, convective indrafts control mixing; instead of diffused arrays of flames, firewhirls arise. Though the burning zone may appear as a solitary flame, it is actually a well-orchestrated ensemble of smaller, intensely organized cells for which the mass fire as a whole provides a suitable environment.

Perhaps what most characterizes a mass fire is its convective column. That the same atmospheric properties which support thunderstorm development also sustain convection column formation gives mass fire some of the characteristics of severe storms. That column is not homogeneous, but heterogenous; it is not solid, but porous. It can interact in complex ways with the surrounding air mass. This porosity begins with the combustion

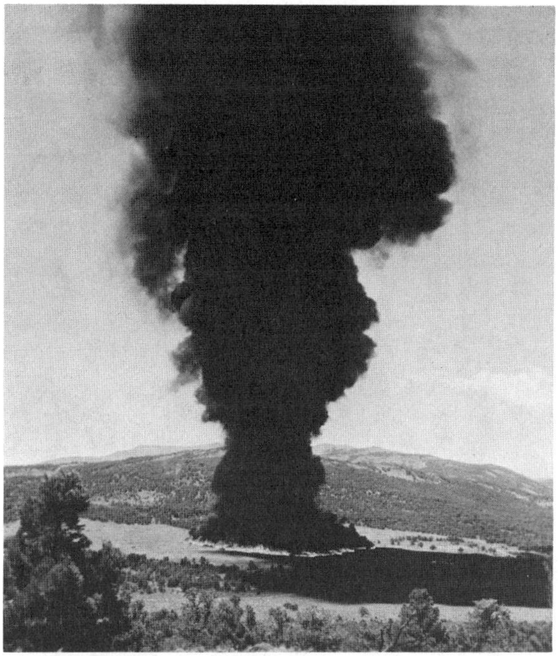

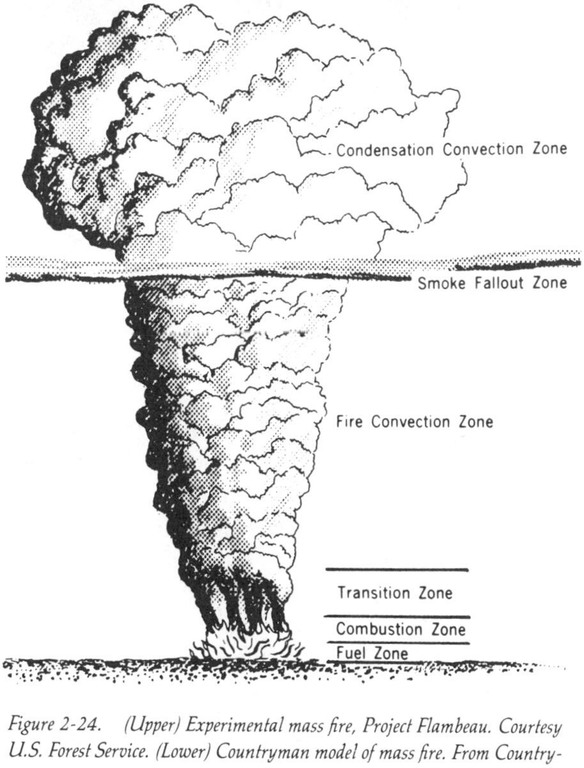



*Figure 2-24. (Upper) Experimental mass fire, Project Flambeau. Courtesy U.S. Forest Service. (Lower) Countryman model of mass fire. From Countryman (1969).*

zone. Within the flaming zone there develop cells or miniature columns of intense burning. This leaves pockets of reduced intensity into which air may be introduced. The process of injection occurs at high velocities and with great turbulence. Winds associated with large fires can reach gale-force velocities, cause massive blowdowns of timber, and render fire control efforts ineffective. Although some accounts report winds flowing out of a mass fire, the awesome winds of historic record were indrafts (Figure 2-25).

When the question of mass fire was first addressed after World War II, it was believed that the fire behaved like a gigantic vortex, a hurricane or a colossal firewhirl. This allusion was fortified by loosely referring to mass fire as a *firestorm.* It was believed, too, that as a firestorm spread slowly outward, the center of the burn would experience oxygen starvation, analogous perhaps to the eye of a hurricane, or fuel burnout, collapsing the area pattern demanded by the fire. Since most mass fires are relatively stationary, it followed that there was a maximum limit for firestorm size, a value set by the need for oxygen in the central core.

But although indrafts do—and must—exist, the process of drafting air into the combustion zone is much more complex than this early image suggested. Only a portion of air comes from the surface, and much from upper levels of the atmosphere through momentum transfer. The pattern of induction does not resemble a well-defined vortex so much as a perforated container, in some parts impermeable and in others porous. The airstreams resemble a tangle of jets rather than a normal vortex. Above the flaming zone, individual columns appear which rise like ring vortices. Entrainment of ambient air can come through induction into these numerous vortices as well as by indrafting through the porous spaces between them directly into the center of the fire. The boundary between convection column and fire is continuous, and the zone represents the site of column merger. Internal mixing is apparently good. The greatest velocities of indrafted air occur within the combustion zone proper.

*Figure 2-25. Blowdown from mass fire-induced winds, Sundance fire, Idaho. Courtesy U.S. Forest Service.*

Surface wind does not generally enter the combustion zone, but flows around it. With respect to control over diffusion processes, the fire determines where and in what quantities air will be drafted into the combustion zone. Yet the relationship between fire and wind cannot remain exclusive. Paradoxically, it is often wind that initiates the transition to mass fire. Surface winds stimulate an increase in burning rates, promote saturation spotting, and deepen the flaming zone. The same winds, of course, disperse the elements of a fire that a mass fire attempts to organize. But if the wind velocity is not excessive, if the fuels are large enough and their loading heavy enough to sustain a long combustion period, if the structure of the atmosphere can support vertical development, then wind may trigger the transition, only to find its influence overpowered by the structure of the mass fire.

Once established, the column no longer behaves like a thermal plume, but a thermal jet, showing some properties of an elastic, slightly porous solid. Strong winds may deflect it, causing it to bend or twist, but do not disperse it. Air flow around the column resembles fluid flow around a solid object, like wind around a large tree trunk. On the windward side, some air will accelerate into the fire system; on the flanks, some will be absorbed through turbulent entrainment; on the lee side, eddies develop, with some indrafting into the burn. The eddies may in turn be organized into firewhirls. Whereas a conflagration is the product of strong winds, a mass fire is a producer of winds. Driven by winds from outside the flaming zone, a conflagration spreads; confined by winds drafted inward, a mass fire remains stationary.

Yet a mass fire must spread or it will burn out

as its large fuels are exhausted. The fact is that mass fires do burn out, much as wind-driven heading fires do. Large fire behavior, regardless of form, seems to be self-arresting. Nonetheless, the tendency of a mass fire to consolidate and concentrate a field of fire rather than to disseminate it makes its exhaustion rapid unless some mode of propagation other than wind is available. Such modes exist in the form of saturation spotting, firewhirls, and convection coalescence. Collectively, they spread mass fire more by a process of cloning than of growth; new mass fires are started from old ones, then annexed to the source fire (Figure 2-26). Here, again, the relationship to ambient winds is paradoxical. Without some winds acting on the column, spotting and firewhirls would be unimaginable.

Coalescence describes the process of integration, and it is not limited to mass fires. Even with steady-state fires (in reality a complex of particle fires) there exists a coalescence phenomenon whereby flames begin to interact and fires start to pull together. Until the flames actually merge, the rate of spread, flame height, and fire intensity all increase; at merger, they reach a maximum. The same process applies to multiple large fires that coexist within proximity to each other. In the case of mass fires, the convective columns can merge at some distance above the ground, as much as 2000 feet or more (610 m). This is similar in principle, though not in scale, to the coalescence characteristic of the column above the flaming zone, with its ensemble of intense cells. The effect is to accelerate convective velocities of both columns and thereby the combustion rates of their source fires. If the fires are uneven in size, the smaller fire will be most affected. Where the columns converge—and if the combined energy at the point of convergence is sufficiently high—then the convergence point will descend toward the surface. The fires approach each other and meet. This effect may be apparent with fires a mile or more apart. Beyond this distance the tendency is to develop multiple mass fires, each more or less complete in itself.

The coalescence process is an important mechanism for the propagation of mass fire because it can escalate spot fires into mass fires and then annex such spot fires into the main burn. Since the distribution of spot fires and firewhirls follows from ambient winds, an apparent head can develop as the fire incorporates a succession of new flaming zones. "Head," as used here, is an anatomical term, not a mode of propagation.

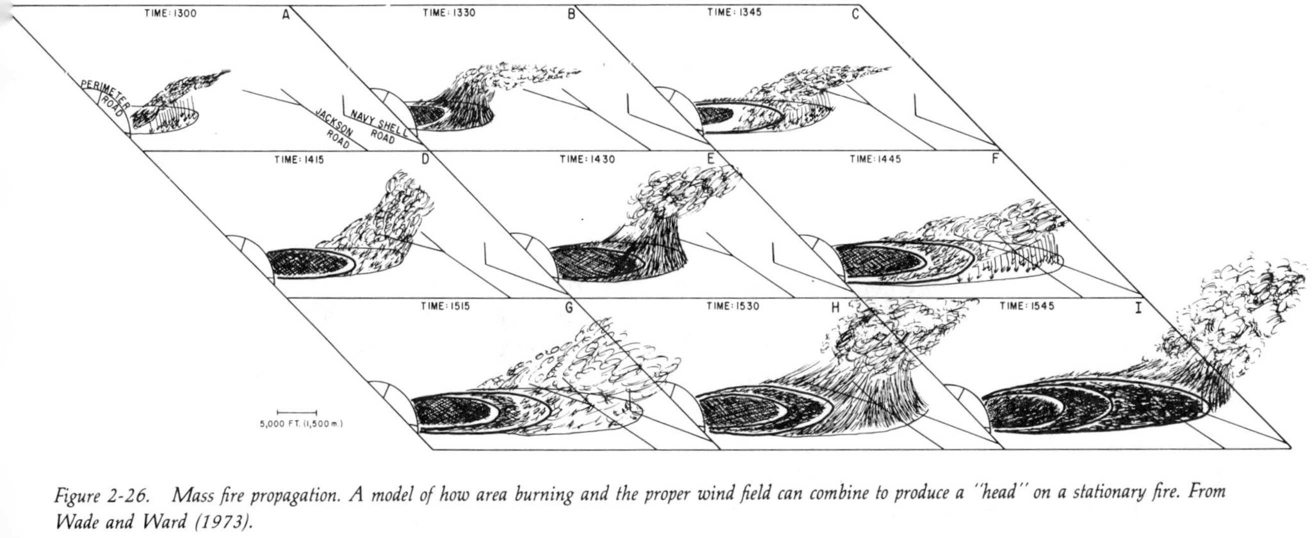


*Figure 2-26. Mass fire propagation. A model of how area burning and the proper wind field can combine to produce a "head" on a stationary fire. From Wade and Ward (1973).*

### *Distribution of Large Fires*

Large fires do not result from one cause, but many. Once started, a wind-driven fire will increase its rate of spread in a way that is roughly linear with increases in wind velocity. But wind alone cannot make a fire large. Though one component may dominate fire shape or intensity, a large fire burns within a fire environment, and a large fire only happens if all the components of that environment are suitable. One component must occur with another, which must compound with still others. The fire environment is the product of many small events. Their individual probabilities may be high but their product is necessarily low. The distribution of fires by largeness, whether defined as size or intensity, is thus not a linear function but a logarithmic one. As a natural event, wildland fires show a frequency distribution based on fireline intensity much as earthquakes do on the basis of energy release or floods on the basis of discharge. Fires are episodic, like floods, not cyclic, like growth rings. One can speak of 10, 20, 50, or 100-year fires as one can floods. There are a large number of small events, and an exponentially smaller number of large events.

Large fires are rarely singular. Instead they are wide in geographic extent and long in time, typically manifested by complexes of fires and cycles of reburns. Historically, large fires occurred together as whole regions felt the impact of drought or high winds. More large fires are found proportionately in certain seasons than in others. And once initiated, a large fire in an old growth forest can set a series of reburns that might last for decades, with each successive fire feeding on the charred residue of its predecessor. When a large fire will occur is more problematic than where it will occur: the prediction of place is simpler than the prediction of time. In this regard, too, wildland fires resemble floods. Just as its history of flooding shapes a river's profile, so its fire history shapes a region's fuels. Particular fuels show particular patterns of fire behavior, a proneness to certain types of large fires, just as the properties of a drainage basin make its stream predisposed to certain types and frequencies of flooding.

From the standpoint of fire protection large fires are the big problem. The strategy of fire control, simply put, recognizes that big fires grow out of little fires, that little fires are more easily suppressed than big ones, and that the surest way to control big fires is to control the small fires that precede them. That transition from small to large fire may come quickly, but for many free-burning fires it will come slowly, if at all. That almost every fire has the potential to become a big fire argues for the rapid suppression of all fires, and that even steady-state fires begin as point sources, initiating fires, suggests that suppression at the earliest stage is the simplest form of control.

Large mass fires mock control efforts because of the magnitude of their heat release. No anthropogenic heat sink can absorb this intensity and extinguish the fire. Large-sized fires resist containment because of their long perimeters, all of which must be surrounded by firelines, and they delay control because the area in need of mopup or burnout increases geometrically with time. During a major run, large fires are simply uncontrollable. Only along the rear and flanks, against other than heading fires, can control lines be effective. Such fires must be stopped before they make their transition or immediately after a run, before conditions favor yet another episode. With the adoption of aggressive programs of fire protection, the proportion of fires that have made the transition to large fires has shrunk. By 1980 only 2–3% of all fires became large fires, as measured by size. Of 200,000 wildland fires in all fuel types, only 4,000 exceeded 100 acres (40 hectares). Any further reduction does not seem likely through better systems of fire control. Large fires exhaust themselves when they encounter new fuels or weather, and some modification of the fire environment seems to be necessary to diminish further the proportion of large fires. This has led to a strategy of fuel modification.

Large fires, even mass fires, can be used under controlled conditions. Deliberately set large fires have been used in warfare, in fire research, and for forest management, notably for slash removal and site preparation. Mass fires are especially useful for disposing of large diameter fuels and for delivering smoke high into the surrounding air mass, improving dispersion. The processes of coalescence often provide the terms under which backfires can be successfully set, the main fire drawing the backfire into it. The very properties that give mass fire its implacable strength delineate the means by which it may be consciously manipulated.

## 2.4 SELECTED LARGE FIRES: BEHAVIORAL CHARACTERISTICS

The following case studies offer brief synopses of certain behavioral characteristics associated with historic fires from four different regions of the United States. Each fire occurred under fuel complexes and weather patterns typical of large fires in its region, and each displayed a more or less typical process of growth by intensity and size.

### *Sundance Fire, Idaho, 1967*

The Sundance fire was part of a large complex of lightning fires that struck the Northern Rockies during the droughty summer of 1967. The initial fire smoldered on the Kaniksu National Forest for several days before becoming visible; it was caught at 35 acres (14 hectares) on August 24. Late on August 29 it began a run that increased its size to 2000 acres (809 hectares), then 4000 (1619 hectares). It was at 4000 acres on the morning of September 1 when it began its major run. Within 9 hours the fire increased in size by more than 50,000 acres (20,234 hectares), shaping an ellipse a full 16 miles (28 km) long and 4 miles (6.4 km) wide and taking the lives of two firefighters (tractor operators) caught in its path. Ambient winds ranged from 20 to 50 mph (32–80 km/hr), increasing towards afternoon, but indraft velocities induced by fierce burning led to timber blowdowns and estimated local velocities on the order of 80–120 mph (129–193 km/hr). A convective column swelled to 35,000 feet (10,668 m), though the wind field bent it forward with considerable short and long-range spotting as the predictable result. Spot fires ignited 10–12 miles (16–19 km) in advance of the main burn (Figure 2-27).

The fire advanced in surges. At its maximum, energy release reached 474 $\times$ $10_6$ Btu/sec (119.45 $\times$ $10_6$ kcal/sec) and fireline intensities, 22,500 Btu/sec/ft (18,603 kcal/sec/m). The fire burned cross-grain to steep topography, its shape dominated by the interaction of wind and convection column. The rugged terrain hardly perturbed the shape, all the more remarkable considering that the run began over a large area, not a point. The relief did result in pockets of more or less intense burning, however. In some places, wind and slope cooperated to accelerate the fire's rate of spread, and in other places, they competed. In one valley, a mass fire resulted from saturation spotting and slight sheltering from the ambient wind.

The fire ceased when large-scale fuel and weather conditions changed. Fuels thinned as the fire entered an old logging site and spread into areas already stripped of fine fuels by spot fires. As evening deepened, temperature dropped, humidity rose, and winds subsided, diminishing both the combustion rate and the probability of spotting. The fire made its remarkable run in advance of a cold front. The surface winds typical of approaching fronts drove the fire, while the unstable air that normally sustains thunderstorm development along the front created favorable conditions for convection column formation. Despite the passage of the cold front, with a pronounced wind shift from southwest to northwest, control efforts successfully prevented another outbreak (Anderson, 1968).

### *Air Force Bombing Range Fire, North Carolina, 1971*

Rather different fuels stoked the Bombing Range fire that burned in pocosins and marshlands along the coastal plain. Topographic relief was negligible. Yet the pattern and mechanisms of spread were remarkably similiar to those of the Sundance fire. The fire began on the morning of 22 March, from a practice bomb on an Air Force

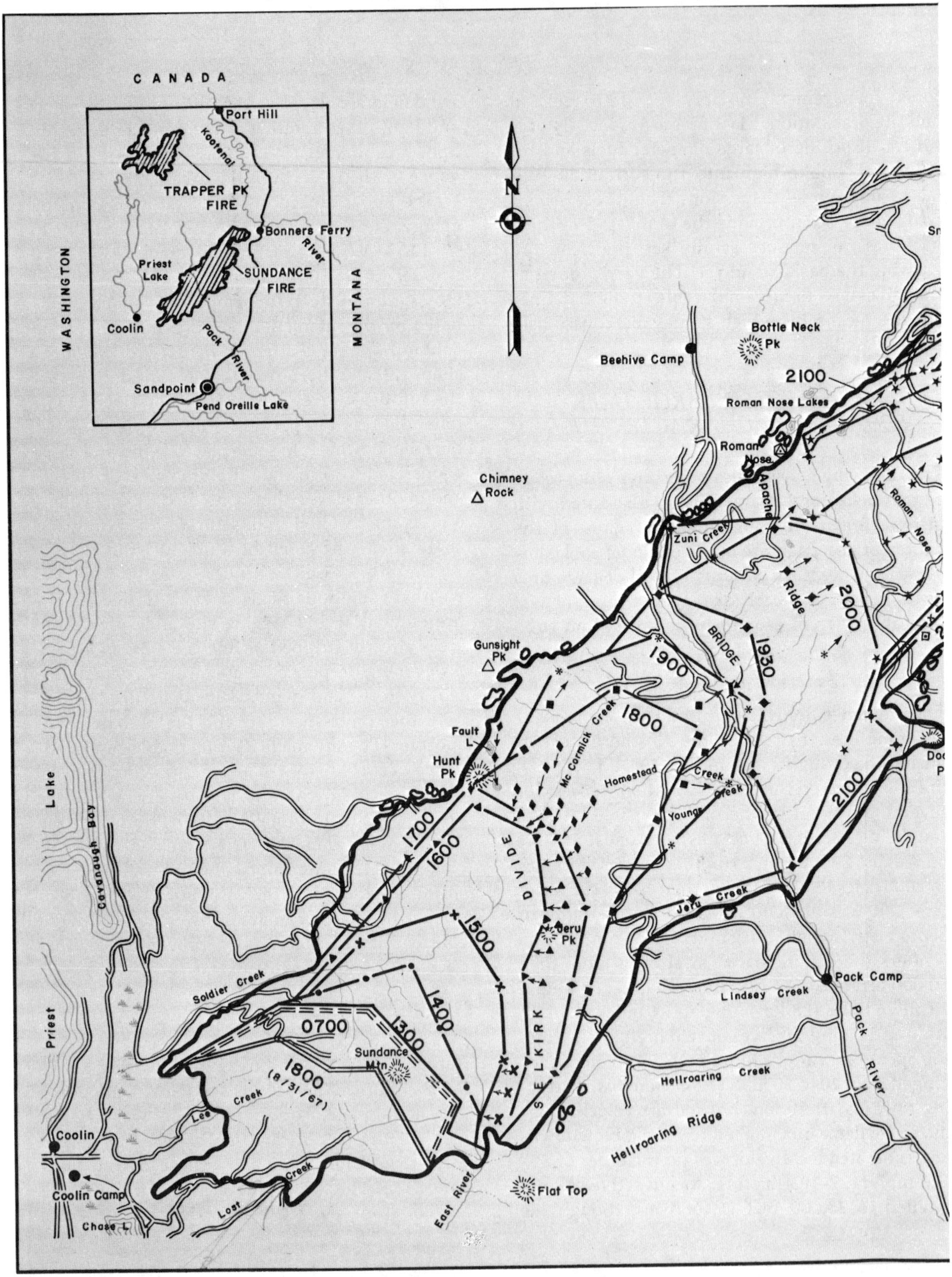
CANADA
Port Hill
Kootenai
TRAPPER PK FIRE
Bonners Ferry
River
Priest Lake
SUNDANCE FIRE
WASHINGTON
MONTANA
Coolin
Pack
River
Sandpoint
Pend Oreille Lake
N
Bottle Neck Pk
Beehive Camp
2100
Roman Nose Lakes
Roman Nose
Chimney Rock
Apache
Zuni Creek
Roman Nose
Ridge
2000
Gunsight Pk
1900
BRIDGE
1930
1800
Fault L.
Hunt Pk
Mc Cormick Creek
Homestead
Creek
Youngs Creek
2100
1700
1600
DIVIDE
Jeru Creek
1500
Jeru Pk
Pack Camp
Soldier Creek
Lindsey Creek
1400
0700
1300
Sundance Mtn
SELKIRK
1800 (8/31/67)
Hellroaring Creek
Lee Creek
Hellroaring Ridge
Coolin
Coolin Camp
Creek
Lost
East River
Flat Top
Chase L.
Lake
Cavanaugh Bay
Priest
Pack
River

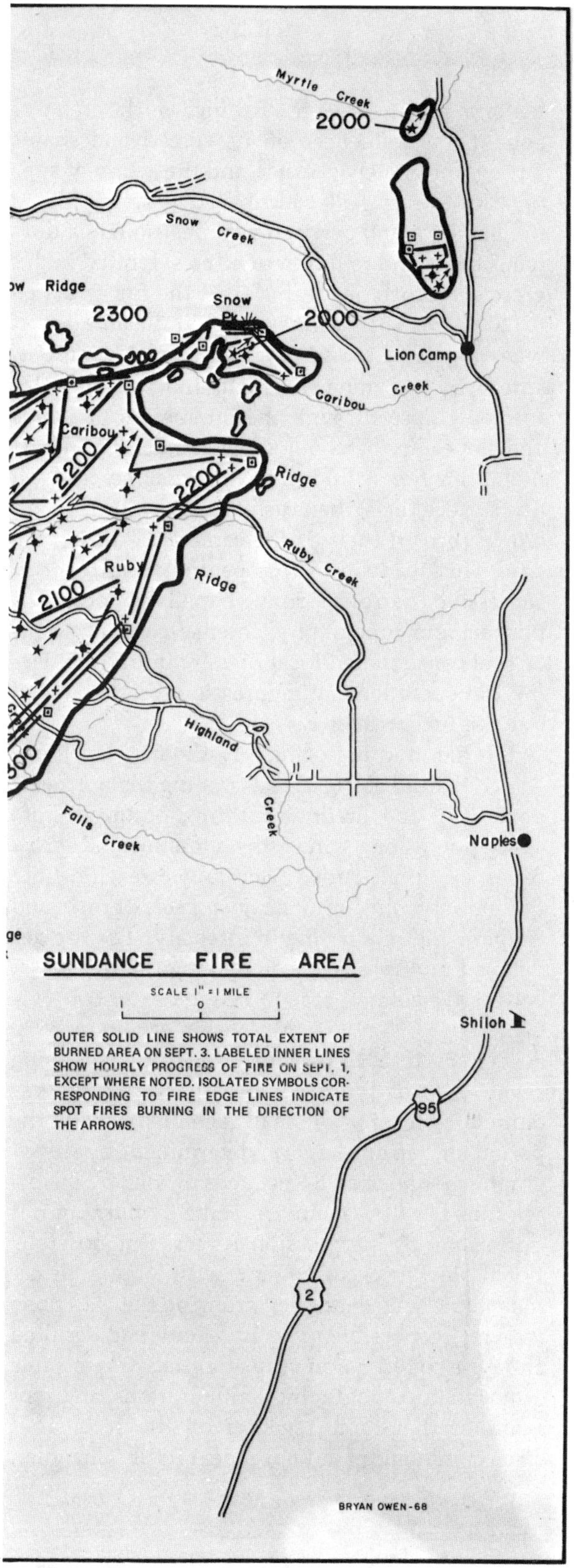


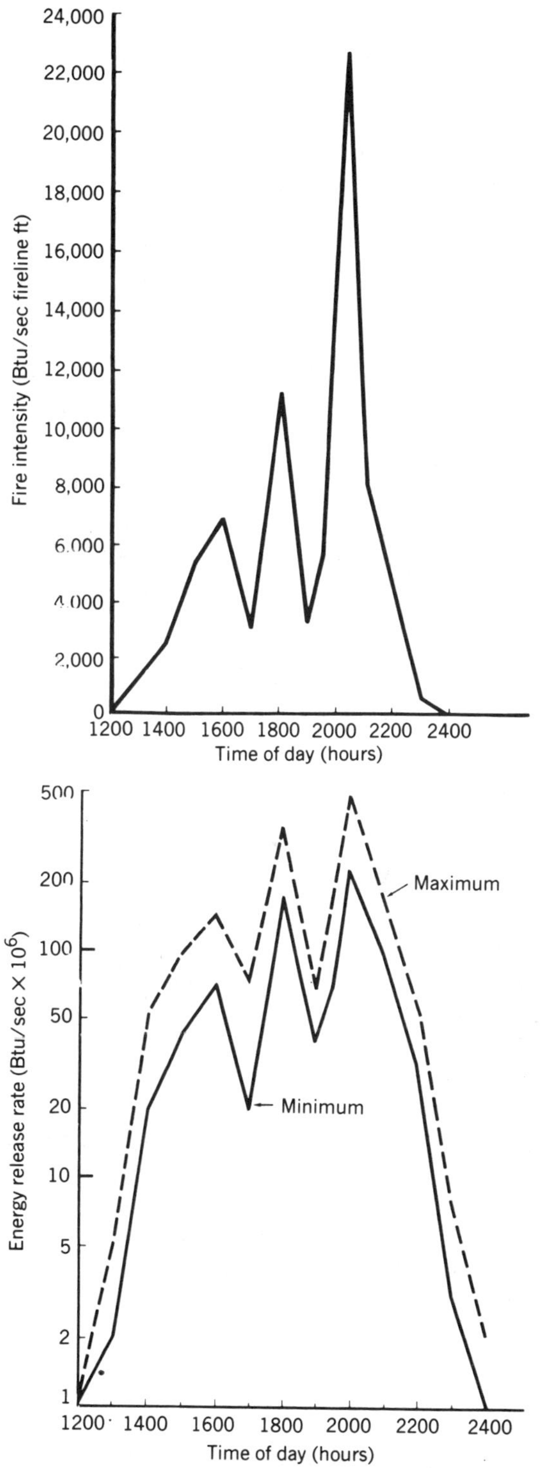


*Figure 2-27. Sundance fire. Map of fire history. Note how, during the blowup, the entire shape of the fire came to more closely approximate an ellipse. (Above) Fireline intensities and rates of energy release. From Anderson (1968).*

range. Though the water table was high, surface fuels were in good conditon for burning and the fire built up heat rapidly. Ambient winds were consistently fresh, blowing an average of 20 mph (32 km/hr). The fire soon made its major run to the northeast along a major axis 14 miles (23 km) long. During the four hours of peak burning time, the rate of forward spread ranged from 2 to 5 mph (3–8 km/hr); corresponding rates of energy release reached $1.144 \times 10_8$ Btu/sec ($2.883 \times 10_7$ kcal/sec) and fireline intensity, 18,000 Btu/ft/sec (14,882 kcal/m/sec). A convection column developed, though winds bent it low, keeping it under 15,000 feet (4,572 m). Like the Sundance fire, the Bombing Range fire showed a tendency to surge or pulsate, with each surge during the burning period growing in size and intensity. The flaming front burned as a crown fire. Short-range saturation spotting was pronounced, probably accounting for the surge process and leaving a peculiar pattern to the burned vegetation. Successive strips of unburned fuel—ghostly outlines of the flaming front—were preserved amid an otherwise fire-scoured landscape (Figure 2-26).

Like the Sundance fire, the Bombing Range fire burned along the boundary of an approaching cold front. But unlike the Sundance fire, control of the perimeter was all but impossible. With the abrupt shift in wind speed and direction, the southern flank developed two heads (Figure 2-28). The fire expired when it burned into wetlands or open water or against backfires set by suppression forces; eventually rains extinguished the burn entirely. Its final shape reflected its complex history: a basic elliptical shape, distorted along the northern flank by suppression backfires and swollen along the southern flank by the gradual transformation of flanking fires into heading fires, culminating in the development of two independent heads (Wade and Ward, 1973).

### *Basin Fire, California, 1961*

The Basin fire began during the afternoon of July 13 on the Sierra National Forest. What makes its history particularly interesting is the relative equality of influences on its spread and shape. Topography, fuel, weather, and the action of suppression forces—all contributed prominently to fire behavior and perimeter shape, though at different times and to different effects. Ignited in dry grass and brush on steep slopes, the fire escalated to 200 acres (81 hectares) within 20 minutes. It was largely directed by slope considerations, spread by crowning in brush, and shaped a deformed ellipse through the interaction of winds and slope. Fire behavior showed a daily cycle. At night the fire would creep downslope through fine surface fuels, then rush back upslope through crown (brush) fuels of the same area during the day. This readvance upslope took the form of successive, narrower runs shaped by local fuel pockets and topography. Intense concentrations of fuel consolidated local fires by indrafting. Firing-out operations by suppression forces rounded out the fire perimeter.

On the morning of July 16 the fire developed an active head along a road framing the northeast portion of the fireline. Spotting occurred from gusty downslope winds, the spots merged into a series of uphill runs, eventually breaching the fireline, escaping into heavier fuels (brush and timber), and redoubling in intensity. The fire developed multiple heads along major drainages. A convective column rose to heights of 20,000–30,000 feet (6,096–9,144 m). Within six hours 5000 acres burned (2023 hectares). Over the remaining seven days of the fire's history, fire shape was defined by a series of small increments along the perimeter, again largely determined by topographic boundaries. Some areas burned by incipient runs from the main fire, some by burning-out operations by suppression forces. But no additional blowup occurred, and with large-scale atmospheric subsidence beginning on the 18th none was really possible. The breakout had resulted from the combination of heavy fuels, an unstable atmosphere, topographic configurations, and spot ignitions that came together on the 16th. That same combination did not repeat itself (Chandler, 1961).

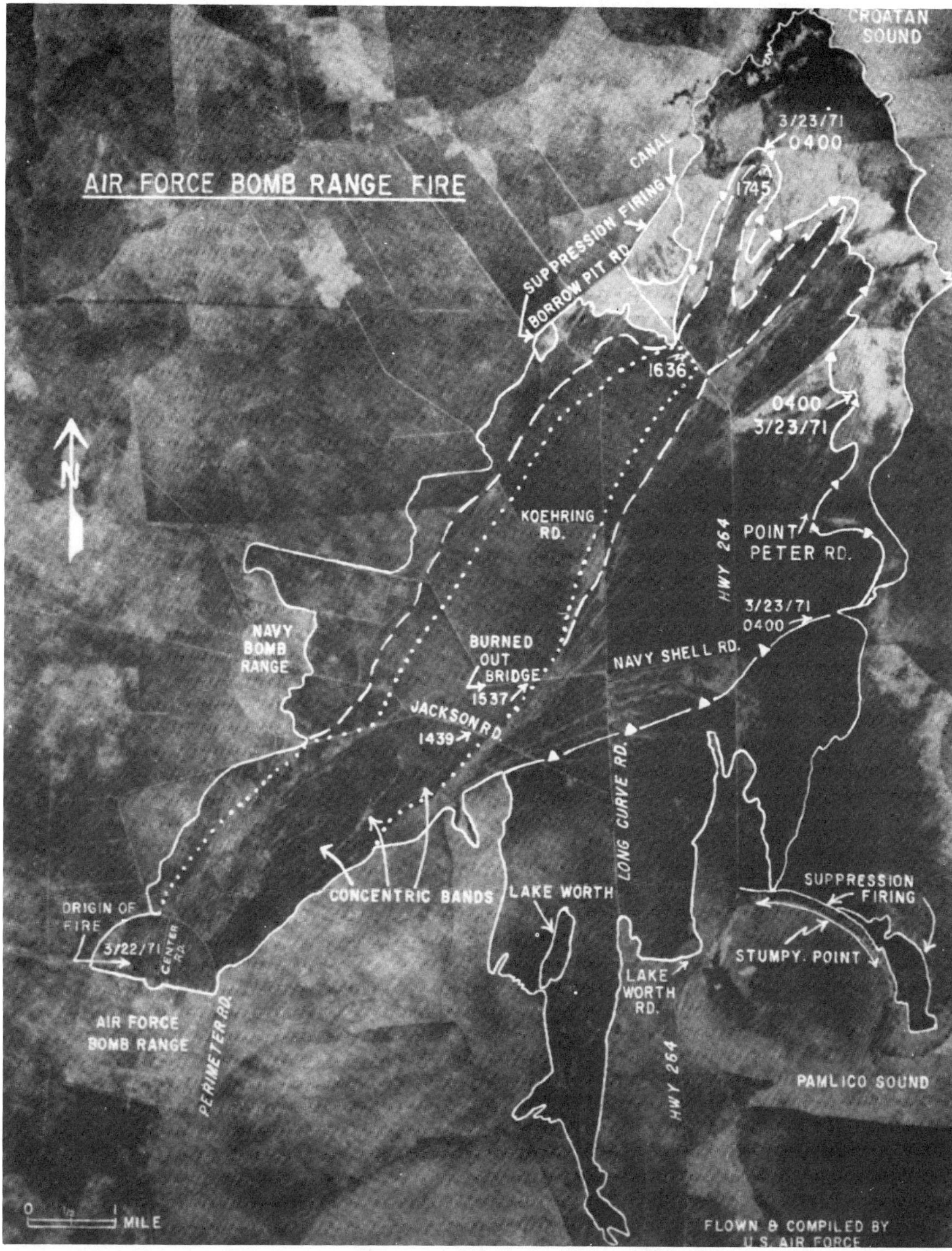


*Figure 2-28. Bombing Range fire. Note the development of several heads along the east flank of the fire following passage of the cold front. From Wade and Ward (1973).*

### *Mack Lake Fire, Michigan, 1980*

A prescribed burn for slash disposal and habitat improvement on the Huron National Forest, the Mack Lake fire was ignited on the morning of May 5. Within 2 hours spotting had spread the fire into dense stands of jack pine on more exposed hillsides; within 10 more hours, and most of that confined to 6 peak hours of burning, the fire had exploded over 25,000 acres (10,117 hectares) and released energy on the order of $3.5 \times 10^{12}$ Btu ($8.82 \times 10^{11}$ kcal), the equivalent of 100 average thunderstorms or 10 Hiroshima-size atomic bombs. In the process one firefighter (a tractor-plow operator) died and fire swept through the town of Mack Lake, consuming 44 residences.

Circumstances that favored the blowup included heavy fuel loads, especially fine fuels both on the ground and in the tree canopy; low fuel moisture, both surface and foliar; suitable atmospheric conditions, notably strong winds and unstable air along the advancing edge of a cold front. The principal mode of propagation was a head and crown fire, assisted by the interaction of a convection column with the ambient wind field. Near the surface the column bent at 45°, suggesting the relative equivalence of wind and convective energies; then it rose more steeply to 60° and finally ended in cumulus cloud development up to 15,000 ft (4572 m). Horizontal roll vortices apparently account for the residual pattern of unburned strips and burned bands in the crown fuels. The fantastic energy releases were sustained entirely from fine fuels, primarily needles.

During its major run the fire traveled 7.5 miles (12 km) in 3.5 hours, an average forward rate of spread of 2.1 mph (3.4 km/hr). Surges occurred, with velocities of 6–8 mph (10–13 km/hr). Fireline intensity for this period averaged 9,300 Btu/ft/sec (7689 kcal/m/sec), with peaks estimated at 31,000 Btu/ft/sec (25,631 kcal/m/sec)—the highest computed for any comprehensively studied fire. During the wind shift associated with frontal passage, the southern flank spread more vigorously, developing several small heads. With completion of the frontal passage, the onset of poorer evening burning conditions, and the exhaustion of jack pine fuels, the fire eventually expired. The increase in final size from the size attained during the major run was negligible (Simard et al, in press).

## *REFERENCES*

Good summaries of existing fire behavior models are given in Albini (1976a), "Computer-Based Models of Wildland Fire Behavior: A User's Manual," Albini (1976b), "Estimating Wildfire Behavior and Effects," and Rothermel (1980), "Fire Behavior Systems for Fire Management." The National Wildfire Coordinating Group has developed three courses for practical training in firecasting: S190, "Introduction to Fire Behavior"; S390, "Fire Behavior"; S590, "Fire Behavior Officer." Student workbooks, films, and instruction manuals for the first two courses are available through the National AudioVisual Center. The U.S. Forest Service has produced an excellent training manual, "Intermediate Fire Behavior" (TT-81), though it is becoming dated.For its outstanding replacement—a how-to manual that synopsizes the important concepts—see Rothermel (1983), "How to Predict the Spread and Intensity of Forest and Range Fires."

The Rothermel fire model is described in Rothermel (1972), "A Mathematical Model for Predicting Fire Spread in Wildland Fuels." A summary of tests made on the Rothermel model is given in Andrews (1980), "Testing the Fire Behavior Model." Refinements on the model are described by Frandsen (1973a), "Effective Heating of Fuel Ahead of Spreading Fire"; Frandsen and Andrews (1979), "Fire Behavior in Nonuniform Fuels"; and Frandsen and Schuette (1978), "Fuel Burning Rates of Downward Versus Horizontally Spreading Fires." A useful synopsis of the model is given in Kessell et al. (1978), "Analysis and Application of Forest Fuels Data." The various uses that the model has been put to are summarized in Rothermel (1980), "Fire Behavior Systems for Fire Management." For the use of the model with a hand programmable calculator, see Burgan

(1979), "Fire Danger–Fire Behavior Calculations with the Texas Instruments TI-59 Calculator: A User's Manual." For descriptions of the combustion component of the model, see the references given in Chapter 1. For fuel models associated with the model, see Chapter 3.

Many other fire models exist, especially to describe aspects of large fire behavior. Among the most important are Byram (1959), "Forest Fire Behavior," and (1966) "Scaling Laws for Modeling Mass Fires"; Countryman (1964), "Mass Fires and Fire Behavior," and (1969), "Project Flambeau"; Byram and Martin (1970), "The Modeling of Fire Whirlwinds"; and Van Wagner (1977), "Conditions for the Start and Spread of Crown Fire." Much of the mass fire modeling has been done for urban landscapes, but summaries of wildland models are given in Kerr et al. (1971), "Nuclear Weapons Effects in a Forest Environment—Thermal and Fire," and TCP (1969), "Mass Fire Symposium."


Albini, Frank, 1976a, "Computer-Based Models of Wildland Fire Behavior: A User's Manual," U.S. Forest Service, Intermountain Station.

———, 1976b, "Estimating Wildfire Behavior and Effects," U.S. Forest Service, General Technical Report INT-30.

———, 1979, "Spot Fire Distance From Burning Trees—A Predictive Model," U.S. Forest Service, General Technical Report INT-56.

———, 1981, "Spot Fire Distance from Isolated Sources—Extensions of a Predictive Model," U.S. Forest Service, Research Note INT-309.

———, 1983, "Potential Spotting Distance from Wind-Driven Surface Fires," U.S. Forest Service, Research Paper INT-309.

Anderson, Hal E., 1964, "Mechanisms of Fire Spread. Research Progress Report No. 1," U.S. Forest Service, Research Paper INT-8.

———, 1968a, "Sundance Fire: An Analysis of Fire Phenomena," U.S. Forest Service, Research Paper INT-56.

———, 1968b, "Fire Spread and Flame Shape," *Fire Technology* **4**(1): 51–58.

———, 1969, "Heat Transfer and Fire Spread," U.S. Forest Service, Research Paper INT-69.

———, 1983, "The Shape of Wind-Driven Wildland Fires," General Technical Report INT-xx.

Anderson, Hal E. et al., 1966, "Mechanisms of Fire Spread. Research Progress Report No. 2," U.S. Forest Service, Research Paper INT-28.

Andrews, Patricia L., 1980, "Testing the Fire Behavior Model," *Proceedings, Sixth Conference on Fire and Forest Meteorology*. American Meteorological Society and Society of American Foresters.

Andrews, Patricia L. and Richard C. Rothermel, 1982, "Charts for Interpreting Wildland Fire Behavior Characteristics," U.S. Forest Service, General Technical Report INT-131.

Barrows, Jack, 1951, "Forest Fires in the Northern Rocky Mountains," U.S. Forest Service, Northern Rocky Mountain Experiment Station Paper 28.

Brown, A.A. and Kenneth P. Davis, 1973, *Forest Fire: Control and Use*, 2nd ed. (New York: McGraw-Hill).

Brown, James K., 1972, "Field Test of a Rate-of-Fire-spread Model in Slash Fuels," U.S. Forest Service, Research Paper INT-116.

Burgan, Robert, 1979, "Fire Danger–Fire Behavior Computations with the Texas Instruments TI-59 Calculator: A User's Manual," U.S. Forest Service, General Technical Report INT-61.

Byram, George, 1954, "Atmospheric Conditions Related to Blowup Fires," U.S. Forest Service, Southeast Experiment Station Paper 35.

———, 1959, "Combustion of Forest Fuels" and "Forest Fire Behavior," Chapters 3 and 4, in Kenneth P. Davis, (ed.) *Forest Fire: Control and Use* (New York: McGraw-Hill, 1958).

———, 1966, "Scaling Laws for Modeling Mass Fires," *Pyrodynamics* **4:** 271–284.

Byram, George and Robert E. Martin, 1970, "The Modeling of Fire Whirlwinds," *Forest Science* **16**(4): 386–399.

Byram, George and Ralph Nelson, 1966, "The Modeling of Pulsating Fires," *Fire Technology* **6**(2): 102–110.

Byram, George et al., 1966, "Final Report. Project Fire Model. An Experimental Study of Model Fires," U.S. Forest Service, Southeast Experiment Station.

Chandler, Craig C., 1961, "Fire Behavior of the Basin Fire. Sierra National Forest, July 13–22, 1961," U.S. Forest Service, Pacific Southwest Experiment Station.

Countryman, Clive M., n.d., "The Fire Environment Concept," U.S. Forest Service, Pacific Southwest Station.

———, 1963, "A Study of Mass Fires and Conflagrations," U.S. Forest Service, Research Note PSW-22.

———, 1964, "Mass Fires and Fire Behavior," U.S. Forest Service, Research Paper PSW-19.

———, 1969, "Project Flambeau...An Investigation of Mass Fire (1964–1967). Final Report— Volume I," U.S. Forest Service, Pacific Southwest Experiment Station.

———, 1971, "Fire Whirls—Why, When, and Where," U.S. Forest Service, Pacific Southwest Experiment Station.

Finklin, Arnold I., 1973, "Meteorological Factors in the Sundance Fire Run," U.S. Forest Service, General Technical Report INT-6.

Frandsen, William H., 1971, "Fire Spread Through Porous Fuels from the Conservation of Energy," *Combustion and Flame* **16:** 9–16.

———, 1973a, "Effective Heating of Fuel Ahead of a Spreading Fire," U.S. Forest Service, Research Paper INT-140.

———, 1973b, "Using the Effective Heating Number as a Weighting Factor in Rothermel's Fire Spread Model," U.S. Forest Service, General Technical Report INT-10.

Frandsen, William H. and Patricia L. Andrews, 1979, "Fire Behavior in Nonuniform Fuels," U.S. Forest Service, Research Paper INT-232.

Frandsen, William H. and Richard C. Rothermel, 1972, "Measuring the Energy-Release Rate of a Spreading Fire," *Combustion and Flame* **19:** 17–24.

Frandsen, William H. and Robert D. Schuette, "Fuel Burning Rates of Downward Versus Horizontally Spreading Fires," U.S. Forest Service, Research Paper INT-214.

Gaylor, Harry P., 1974, *Wildfires. Prevention and Control* (Bowie, Maryland: Robert J. Brady Co.).

Haines, Donald A., 1982, "Horizontal Roll Vortices and Crown Fires," *Journal of Applied Meteorology,* 21(6): in press.

Haines, Donald A. and Gerald H. Updike, 1971, "Fire Whirlwind Formation over Flat Terrain," U.S. Forest Service, Research Paper NC-71.

Johansen, Ragnar W. et al., 1976, "Fuels, Fire Behavior, and Emissions," pp. 29–44, in Southern Forest Fire Laboratory Personnel, "Southern Forestry Smoke Management Guidebook," U.S. Forest Service, General Technical Report SE-10.

Kerr, J.W. et al., 1971, "Nuclear Weapons Effects in a Forest Environment—Thermal and Fire," The Technical Cooperation Program, Panel N-2 Report N2: TR 2-70 (Santa Barbara: Department of Defense Nuclear Information and Analysis Center).

Kessell, Stephen R. et al, 1978, "Analysis and Application of Forest Fuels Data," *Environmental Management* 2(4): 347–363.

Kourtz, P.H. and W.G. O'Regan, 1971, "A Model for a Small Forest Fire...To Simulate Burned and Burning Areas for Use in a Detection Model," *Forest Science* 17(2): 163–169.

Luke, R.H. and A.G. McArthur, 1978, *Bushfires in Australia* (Canberra: Australian Government Publishing Service, 1978).

National Fire Protection Association, 1967, *Chemicals for Forest Fire Fighting,* 2nd ed. (Boston: National Fire Protection Association, 1967).

———, 1977, *Fire Protection Handbook,* 14th ed. (Boston: National Fire Protection Association, 1977).

National Wildfire Coordinating Group, "S190, Introduction to Fire Behavior," 'S390, Fire Behavior," and "S590, Fire Behavior Officer" (Washington: National AudiVisual Center), continually revised.

Rothermel, Richard C., 1972, "A Mathematical Model for Predicting Fire Spread in Wildland Fuels" U.S. Forest Service, Research Paper INT-115.

———, 1974, "Concepts in Fire Modeling" U.S. Forest Service, Northern Forest Fire Laboratory.

———, 1976, "Forest Fires and the Chemistry of Forest Fuels," in Fred Shafizadeh et al (eds), *Thermal Uses and Properties of Carbohydrates and Lignins* (New York: Academic Press, 1976), pp. 245–259.

———, 1980, "Fire Behavior Systems for Fire Management," *Proceedings, Sixth Conference on Fire and Forest Meteorology.* American Meteorological Society and Society of American Foresters.

———, 1983, "How to Predict the Spread and Intensity of Forest and Range Fires," U.S. Forest Service, General Technical Report INT-143.

Rothermel, Richard C. and Hal E. Anderson, "Fire Spread Characteristics Determined in the Laboratory" U.S. Forest Service, Research Paper INT-30.

Simard, Albert J. et al, 1983, "The Mack Lake Fire," U.S. Forest Service, Research Paper NC-00.

Taylor, Dee F. and Dansey T. Williams, 1968, "Severe Storm Features of a Wildfire," *Agricutural Meteorology* 5: 311–318.

Technical Cooperation Program, *Mass Fire Symposium,* 2 vols. Panel N2, Working Group J (Maribyrnong, Victoria: Defense Standards Laborataories, 1969).

U.S. Forest Service, 1969, "Intermediate Fire Behavior," TT-81.

Van Wagner, C.E., 1977, "Conditions for the Start and Spread of Crown Fire," *Canadian Journal of Forest Research* 7: 23–34.

Wade, Dale D. and Darold E. Ward, 1973, "An Analysis of the Air Force Bomb Range Fire" U.S. Forest Service, Research Paper SE-105.

*Chapter Three*

# Wildland Fuels

A fire begins and ends with its fuels. The presence of a proper fuel is mandatory for the smallest of flames, and its absence can shut down the largest of conflagrations. As a physical and chemical process, a wildland fire converts stored chemical energy into kinetic energy of heat and motion. As an ecological process, fire is one variety of biological decomposition; it reassembles solid organic fuels into other forms. In any ecosystem biomass is being produced and consumed, but wherever the rate of production exceeds the rate of consumption by biological agents, then fire is possible. In temperate and boreal zones, where the ratio of production to decay is high, fire is probable. No aspect of the fire environment is more subject than its fuels to biological processes, and none is more susceptible to anthropogenic manipulation.

Yet fuel is not synonymous with biomass. Only a portion of total biomass, its phytomass, is potentially available as fuel, and not all of the phytomass is available for combustion. Nor, for purposes of combustion and fire behavior, is it significant that this fuel has a biological origin. Its important parameters are physical and chemical. Biota that show the same fuel chemistry and geometry will burn alike, whether or not they contain the same organisms. The organization of fuel particles into a fuel complex, moreover, is not uniform. Particles exist in different sizes, and these size classes will be distributed differently in different fuel complexes. Particles show different fuel chemistries, and this chemistry, too, will be distributed in different ways and, within a single complex, at different times. A single process—like rain or flame—will affect different portions of a fuel complex differentially. And the complex shows changes over time. Some changes, like those due to aging are systematic, and some, like the aftermath of storms, are stochastic. The distribution of fuel moisture within a complex is the most variable and, at the same time, the most directly relevant trait for combustion. But there are others. Brush, reproduction, and understory rough put high caloric fuels near the surface; seasonal changes in live fuels as they flush, grow, and cure affect the ratio of live to dead fuels; and simple aging transfers matter from live to dead and back again to live fuel categories.

The description and classification of fuel complexes is not simple. Since it is not the potential fuel as such that is of concern, but that proportion of the total fuel which actually burns, the fire determines the important parameters of the fuel. Fuel models organize the information needed by fire models. The fuel complex must be inventoried, but it must also be appraised—a process of translating fuel data into fire behavior data, from which can be derived information on fire effects. One of those effects is a transformation of the fuel array. With sophisticated computer programs, information on fuel particles can be assembled into fuelbeds, and fuelbeds into fuel models that describe, for fire behavior purposes, the fuel complex.

It is through fuels that a fire influences an ecosystem and that humans most influence fire. Suppression most commonly results from the deliberate removal of fuels from in front of a spreading perimeter. Few studies of fire ecology can avoid considering the biota as a fuel complex, and few fire management programs lack a separate work unit dedicated to fuels management. The wholesale mangement of fuels has become a fundamental strategy of fire management. Through manipulation of the fire environment—fuels—fires can be prevented, fire intensity abated, fire control simplified, and fire effects managed. Wildfire may be more easily controlled, and prescribed fire better promoted. Whether changed prior to a fire or after one, it is through the medium of fuels that ecological manipulation by a fire must come. The Rothermel fire model depends for its accuracy on the accuracy of its fuels data; fuel models are thus as ubiquitous, and essential, as the fire model itself. Plans to replace wildfire with prescribed fire have, as their administrative cohort, plans for fuels management—to replace natural fuel complexes with more intensively managed complexes.

## 3.1 FUEL MOISTURE

Fuel moisture intersects the two most variable components of the fire environment, fuel and weather. Fuel moisture content (FMC) thus integrates many processes, and it influences all aspects of combustion and fire behavior. As adsorbed water, moisture content affects the availability of fuels by increasing the heat sink at the expense of the heat source. More of the propagating flux goes towards distillation of fuel moisture, and less towards pyrolysis and combustion. As a vapor, fuel moisture inhibits combustion reactions through a combination of smothering in the reaction zone and of cooling within the flame.

The processes of moisture exchange differ according to whether the fuel is alive or dead, to whether the moisture is liquid or vapor, and to whether the moisture gradient is large or small—a function of particle size, the intensity of the gradient, and the duration of exposure. Live fuels show different mechanisms than dead fuels. Dead fuels vary according to particle size, and arrays of dead fuels according to their bulk density and distribution within the fuel complex as a whole. Fine fuels respond more quickly than coarse, and coarse fuels act as a moisture sink just as, in the case of combustion, they act as a heat sink.

The exchange of fuel moisture is in some ways analogous to heat transfer. Both follow from a differential between the particles and their surrounding environment, heat and water traveling from regions of excess to regions of deficiency. Both act on the surface of the fuel particle, and those fuel particles with high surface-to-volume ratios and those fuel complexes with large proportions of fine fuels and high porosity for their packing will show higher rates of transfer. Large particles become moisture and heat sinks. Within a large particle, a gradient will develop between the surface and the interior. The rate of moisture transmission depends on the size of the initial gradient, the diffusivity of the particle, and the length of time the gradient persists.

Fuel moisture is measured by weight, a percentage of the ovendry weight of the fuel particle. For live fuels FMC may range from 70 to 200%. For dead fuels, FMC will vary according to the mechanisms of moisture transfer: immersion in water leading to figures of 200–300%, and transfer by vapor pressure giving an effective range of 1.5–30%. FMC can be measured directly by drying fuels in an oven or, indirectly, through mathematical models that describe the transfer of moisture for known sizes of fuels and given environmental conditions. A popular compromise is to use fuel moisture sticks (0.5 in. wooden dowels), which are weighed as part of fire weather records. The sticks give a FMC for 10-hr timelag fuels.

The duration factor may be more important than the gradient. Persistent high humidities or fog, for example, are more effective in wetting fuels than a sudden downpour, most of which washes away without more than surface contact. To continue the analogy between moisture and heat exchanges, the rainstorm may be compared to the flaming front, and moisture duration to flame residence time. The effect of a rainstorm on large fuels or deep fuelbeds may be negligible, and heat or moisture exchanges may end with the passage of the storm or flaming front. When exposures are prolonged, as with drought, the larger fuel particles and the deeper fuelbeds will dry out and become available for combustion. Because it competes with fire preferentially for the same fuels, FMC is fundamental to fire behavior considerations.

Fuels may be living or dead, and the processes of moisture exchange and the resultant amounts will differ accordingly. With live fuels, physiological processes can actively work to accelerate or retard the exchange of moisture between a particle and its environment; with dead fuels, the response is passive. Live fuels show higher quantities of moisture, resist moisture deficiency better, and exhibit seasonal changes in moisture in accordance with physiological processes such as spring flushing and fall curing. A living fuel, moreover, may contain extractives that increase the total flammability of the particle—substances leached or volatilized away after death. Dead fuels are chemically and physiologically simpler, though the cellular structure that the dead fuel inherits from its living state makes the process of moisture exchange more complicated than a simple diffusion gradient might suggest. Moisture is acquired and shed in stages, not through a continuous process like heat diffusing through an iron bar.

### *Live Fuel Moisture*

The moisture content of living fuels results from a mixture of physical and physiological processes. Although there exist moisture gradients between the plant and the atmosphere, and between the plant and the soil, a variety of physiological processes intervene to keep plant moisture higher than it would be if the plant were dead. Dead fuels exchange moisture passively; live fuels can, up to a point, actively acquire or shed moisture. Physical factors account for the general gradient between water supply and water demand. Precipitation brings water into the system, and soil stores it. Evaporation back to the atmosphere completes the cycle, accounting for the loss of surface moisture to the atmosphere. If one places a tree into this system, the tree becomes another mechanism of moisture exchange between soil and atmosphere, absorbing water from the soil by its root system and transpiring water to the air by its leaves or needles. The process would seem to be a simple input–output arrangement for the mass balance of water, driven by a gradient between the stored water in the soil and the humidity contained in the air mass above it.

In a gross sense, it is. But intervening between water supply and water demand are a series of physiological processes that both pump and store moisture. Each process responds to water deficits within a particular portion of the tree, and each modulates the magnitude of the water gradient (Figure 3-1). Root uptake results from a deficit in the root compartment relative to the available water in the surrounding soil; tissue recharge, from the differential between sapwood water and available root water; sapwood water exchange, between extensible and mature sapwood; and stomatal control (as manifest by leaf resistance to moisture loss), by the moisture differential between sapwood and the atmosphere. This last process is critical: transpiration losses drive the entire operation and the stomata are the immediate organs in control of transpiration. Their resistance to loss derives, in part, from their great surface area, and, in part, from processes that can regulate water flow directly, to some degree, according to the needs of seasonal growth patterns (Running, 1978).

Not all live fuels behave according to this simple model. For purposes of assessing the flammability of a fuel complex, a number of distinctions

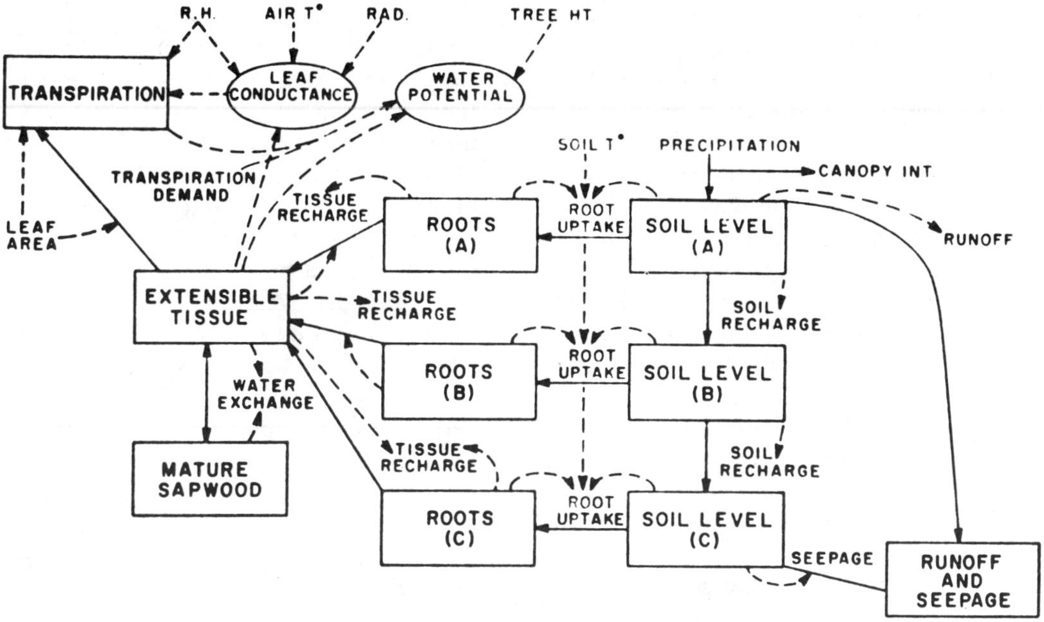


*Figure 3-1. A compartment flow diagram for water transport. From Running (1978).*

among the total live fuels are made. Live fuels may be *woody* or *herbaceous.* Woody vegetation includes tree reproduction and brush; herbaceous vegetation, the grasses and forbs. Most woody fuel exists as *surface fuel,* though, in select fuel complexes prone to crown fires, the amount of needles and branches within the canopy, the *aerial fuels,* is relevant. Herbaceous fuels are further subdivided into *annuals* and *perennials,* with a fuel complex characterized by one or the other depending on which type is in the majority. The reason for this distinction is that annuals are more sensitive to seasonal changes, especially drought, and they complete their life cycle well within a normal growing (and fire) season; perennials show fewer fluctuations.

All live fuels show predictable changes over time (Figure 3-2; Burgan, 1979). Moisture is highest with new foliage and at the time of emergence; lowest with older (more than one-year old) foliage and at the onset of death or dormancy. The life cycle of moisture for woody fuels is expressed in three stages: rapid growth, maturing growth, and severe drought or dormancy. The annual cycle of moisture for herbaceous fuels follows three analogous stages: greenup, maturity, and curing. Prior to emergence or *greenup,* most fine, live fuel can be considered as dead fuel. During greenup, moisture increases rapidly to a seasonal high, perhaps two to three times the dry weight of the plant. Even needles or fine branchwood, which had plummeted to moisture contents of 80–100% during dormancy, suddenly flush with moisture, recharged sap, and luscious new growth. For both new and old foliage, however, this greenup rate slows quickly, so that by autumn the plant either dies or becomes dormant; material which had been considered as live fuel now becomes dead fuel, or live fuel with significantly reduced moisture content. The process is

*Figure 3-2. Seasonal changes in fuel moisture. Some of the change is the result of moisture acquired or shed by dead fuels; some, by the transfer of live fuels to dead fuels with curing. Upper diagram shows the changes for 1-hr TL fuels and grasses; lower diagram, for 1000-hr TL fuels and live woody vegetation. From Burgan (1979).* ➤

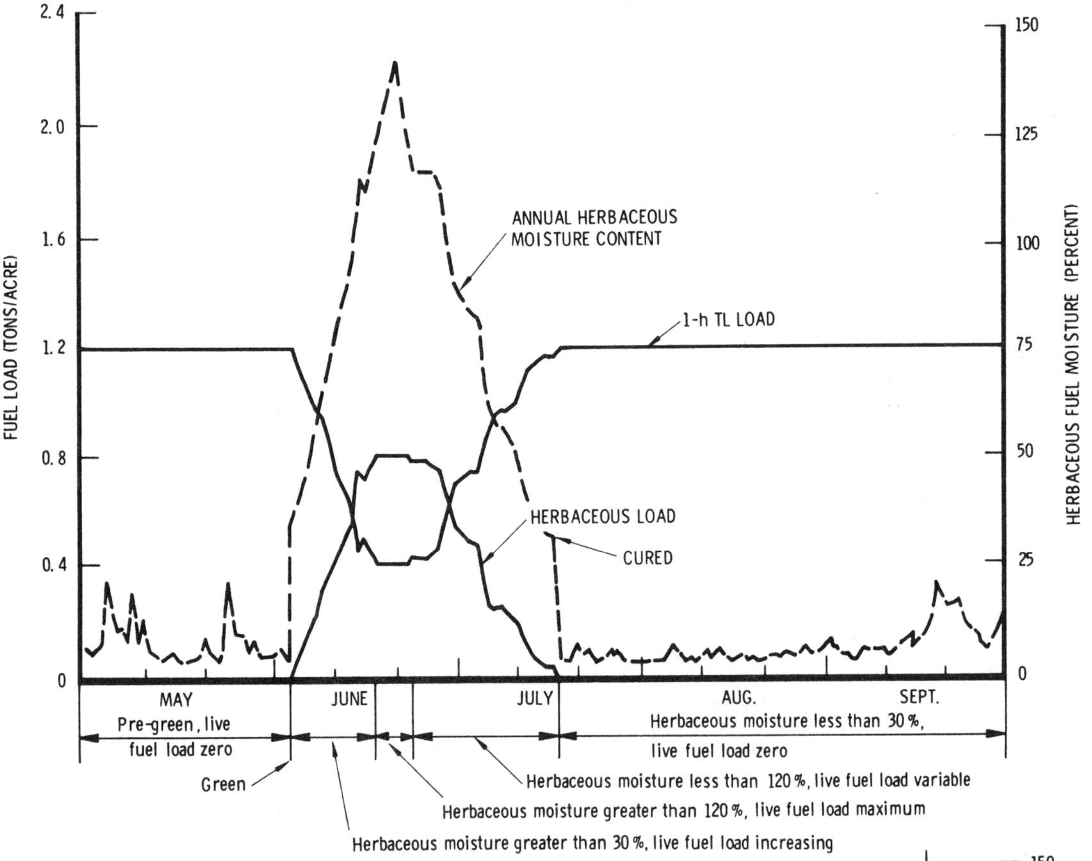

2.4
2.0
1.6
1.2
0.8
0.4
0
FUEL LOAD (TONS/ACRE)
150
125
100
75
50
25
0
HERBACEOUS FUEL MOISTURE (PERCENT)
ANNUAL HERBACEOUS MOISTURE CONTENT
1-h TL LOAD
HERBACEOUS LOAD
CURED
MAY
JUNE
JULY
AUG.
SEPT.
Pre-green, live fuel load zero
Herbaceous moisture less than 30%, live fuel load zero
Green
Herbaceous moisture less than 120%, live fuel load variable
Herbaceous moisture greater than 120%, live fuel load maximum
Herbaceous moisture greater than 30%, live fuel load increasing


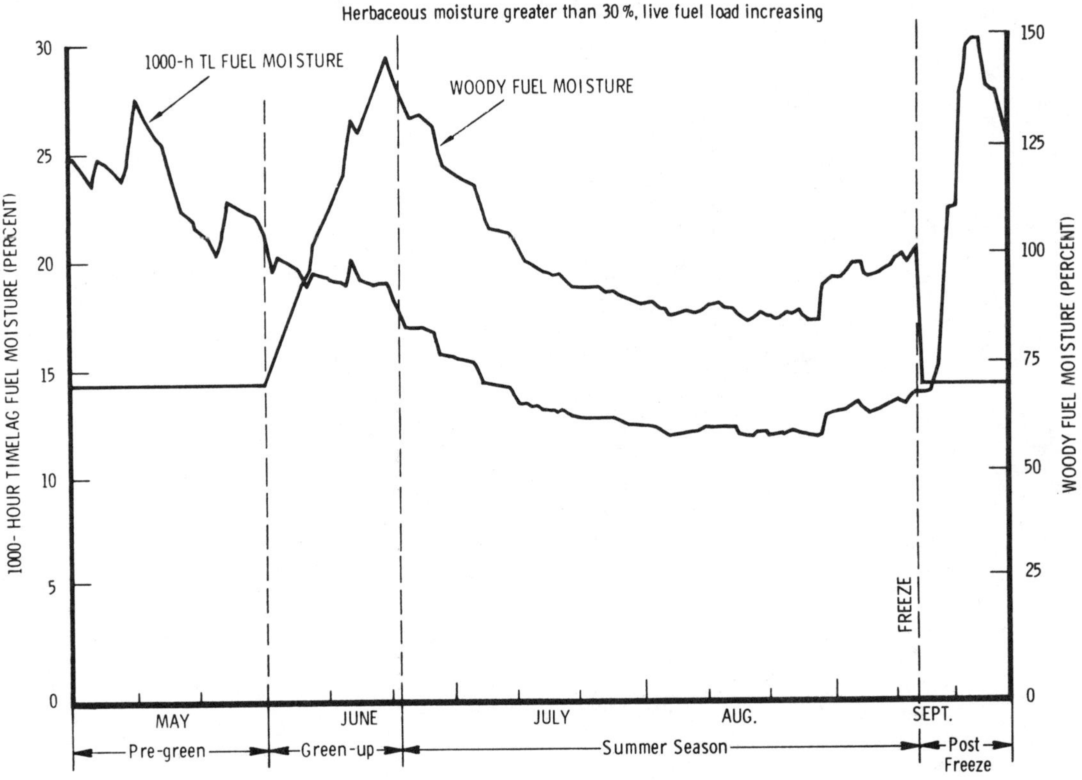

30
25
20
15
10
5
0
1000-HOUR TIMELAG FUEL MOISTURE (PERCENT)
150
125
100
75
50
25
0
WOODY FUEL MOISTURE (PERCENT)
1000-h TL FUEL MOISTURE
WOODY FUEL MOISTURE
FREEZE
MAY
JUNE
JULY
AUG.
SEPT.
Pre-green
Green-up
Summer Season
Post Freeze

commonly known as *curing.* For annuals, curing means the death of the plant; for deciduous woody fuels and perennial herbaceous fuels, it means the death and loss of foliage; for evergreen shrubs and trees, it results in partial death of the foliage, leaving a mixture of live and dead fine fuels in the crown.

This cycle of fine fuel moisture is intimately connected with fire patterns. Most grass fires occur before greenup or after curing, and most crown fires burn prior to spring flushing or after fall dormancy and dieoff. Since deciduous plants grow all new foliage each year, their crown fuel moisture is high. Evergreen shrubs and conifers, conversely, add perhaps only 20% new foliage to the crown each year, keeping their total crown moisture lower.

To these seasonal changes must be added episodic ones. Plants respond differently to drought. Prolonged drought gradually leaches away moisture from the plant, but plants adapted to arid environments cope better. Consider the case of the chamise chaparral, a woody evergreen shrub common in Southern California. As an adaptation to frequent drought, the organism may cease to grow by midseason, reducing its foliar moisture to 40–50%. If rains return, foliation renews. But if drought persists, the plant may expire in whole or in parts. Herbaceous fuels in such landscapes accommodate common drought with varied grow-

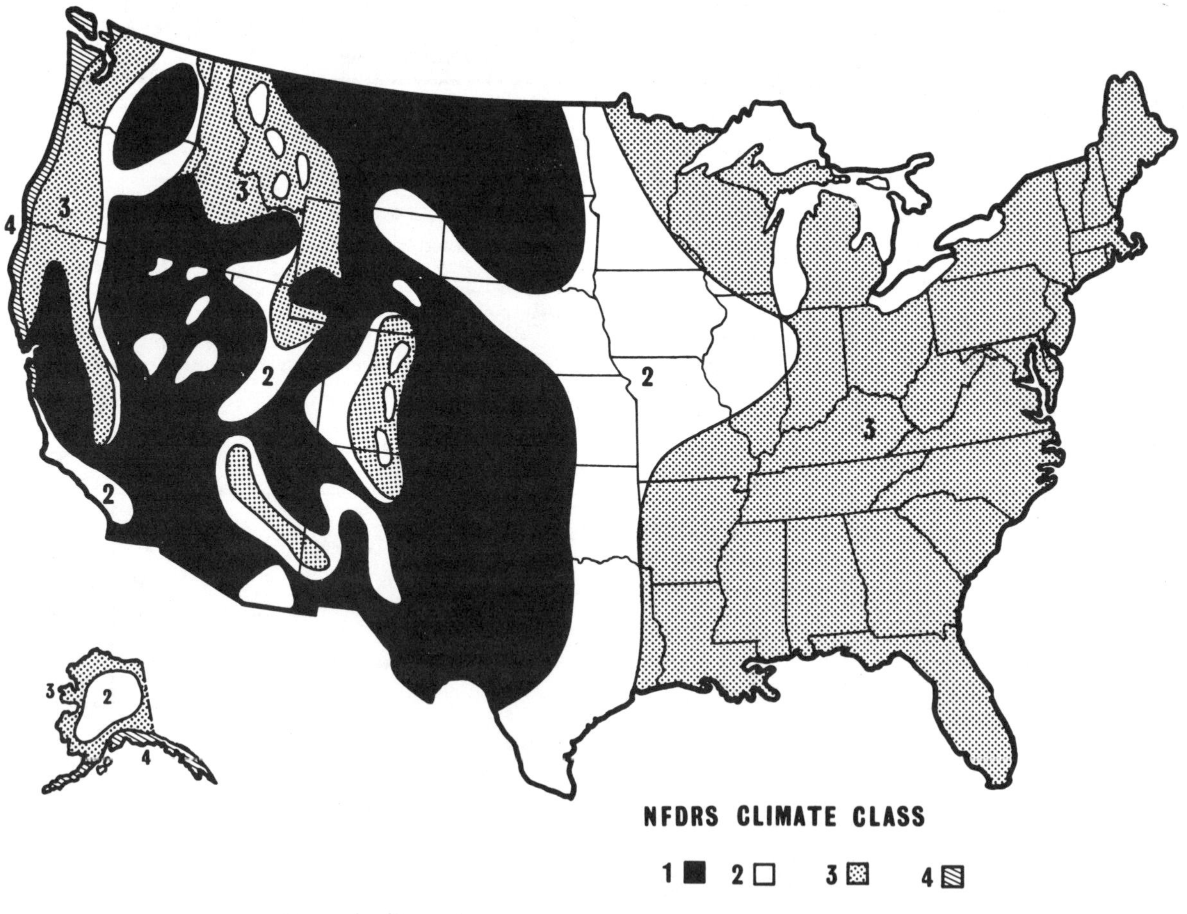


*Figure 3-3. Climate classes, as used by the NFDRS. From Deeming et al. (1978).*

ing cycles. Accordingly, live fuels are further discriminated on the basis of their likely adaptability to drought. The concept of *climate class* identifies four regimes for use with the National Fire Danger Rating System (Figure 3-3).

The important measurements of live fuel moisture include the fuel moisture (%) of live fuel particles, and the ratios of live fuels to dead fuels and of herbaceous to woody live fuels (by volume). Increasing the percentage of live fuels, especially herbaceous live fuels, effectively raises the fuel moisture of the entire fuel complex. Whether or not such particles enter into combustion, they act as heat sinks, draining heat away from the burning process. These ratios are not fixed. Dead fuels increase at the expense of live fuels, and live fuels continue to resprout. In the case of aerial fuels, a large percentage of fine fuels will be live.

Live fuels are thus a paradox. Moisture is a retardant, and the high moisture content of live fuels would seem to make them inaccessible to combustion, a colossal heat sink. Yet because such fuels are living, they also contain chemical constituents (notably extractives and fats) whose heat content and flammability is high. This is most true for the leaves and needles of woody live fuels, a reason for distinguishing them from herbaceous live fuels. Some live fuels are thus both a heat sink, because of their moisture content, and a heat source, because of their thermochemistry. Which tendency predominates depends on the intensity of the fire, the total chemistry and physics of the fuel elements, and the actual level of fuel moisture. Conifer needles may show a strong tendency to ignite before and after their principal growth season, but once cast to the ground, their special combustibles, and flammability, leach or volatilize away. Needles on the surface may burn with less ferocity than needles still on the tree.

### Dead Fuel Moisture

*Wetting and Drying Processes.* Moisture exchange by dead fuels follows different processes (Schroeder and Buck, 1970). Basically, two situations exist, depending on the state of water: that in which fuel directly contacts liquid water and that in which it contacts water vapor. When the fuel is in direct contact with liquid water, water is absorbed into the porous structure of the material and can reach moisture contents equal to that which the plant had when it was living. This occurs with precipitation on the fuel particle or immersion in snowmelt, and it is most pronounced with punky logs or broken fuels with irregular surfaces that will not shed the ponding liquid.

When the exchange of moisture occurs with water vapor, the process is more complicated, involving adsorption rates that depend on the hygroscopic properties typical of cell walls. Driving the whole process is a moisture differential between the vapor pressure of the atmosphere and the vapor pressure of the fuel particle interior. When the vapor pressure of the atmosphere is greater, then moisture moves into the fuel; when the vapor pressure of the fuel is greater, then moisture moves out. The process begins with *adsorption,* or chemical bonding, to the cell walls. Water vapor becomes *bound water.* The bond is greater than the vapor pressure of the bound water, reducing the vapor pressure inside the particle and maintaining a high vapor pressure gradient between the particle and the ambient air. The adsorption process continues until the cell walls are saturated (Figure 3-4). For woody fuels this equals a fuel moisture of about 30%.

Once bound water covers the cell walls, and only after hygroscopic bonding covers all walls, it is possible for *free water* to move into interstitial voids. After saturation, moisture transfer proceeds by the osmosis of liquids rather than by vapor pressure. Free water passes through cell walls, spreading like a contagion from saturated to unsaturated pores. Only direct contact with liquids, not vapor pressure, can move free water. Excess water is simply shed from the particle surface.

The process of drying fuels reverses this sequence. If the fuel has been filled with free water above its fiber saturation point, then evaporation first expels the free water within interstitial pores.

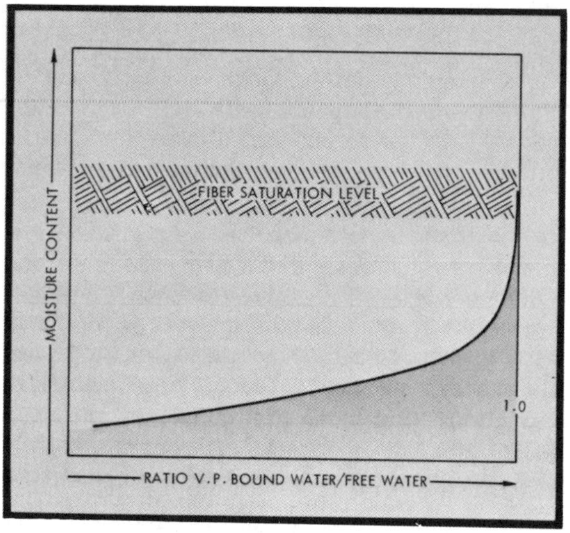


*Figure 3-4. Vapor pressure and moisture content. Prior to saturation the vapor pressure of bound water is less than that of free water; at saturation, the vapor pressures equal each other. From Schroeder and Buck (1970).*

The mechanism is that characteristic of any free water exchange; it shows a rate proportional to the outward vapor pressure gradient and it continues until only bound water remains. The rate of drying by this process is constant, and this stage of drying is often designated the *constant-rate period.*

The problem of disposing of bound water proceeds in two additional stages. First, the amount of bound water is reduced to the level of fiber saturation, and then the amount drops to equal the vapor pressure (relative humidity) of the ambient air. The rate of drying is initially variable, then falls rapidly. Accordingly, the second phase of fuel drying is referred to as the *decreasing-rate period* and the third phase as the *falling-rate period.* The factors affecting the decreasing-rate period are complex, and the processes are not reversible; the mechanisms apply to drying but not to wetting. For the falling-rate period, however, the rate depends on the differential vapor pressures between the bound water and the atmosphere. Without a continuous coating of bound water, this vapor pressure develops a strong gradient. How pronounced it becomes depends, in part, on the size of the fuel particle. The transmission of moisture within a particle, a process of internal diffusion, progresses more slowly than the exchange between particle surface and atmosphere.

In a sense, the two rates create a double moisture gradient. There is one gradient within the fuel and another gradient between the fuel surface and the atmosphere. The process of drying below fiber saturation thus requires either small fuel particles with high surface-to-volume ratios or large, sustained gradients such as are produced by drought or direct fuel heating. Without a steep gradient, the process slows down and asymptotically approaches a point of moisture equilibrium. For the first two stages of drying, wind speed affects the rate of evaporation, with windspeed and evaporation increasing or decreasing together; for the final stage, dominated by the internal moisture gradient and the process of internal diffusion, wind effects are negligible except as they affect fuel surface temperatures.

A condition of *moisture equilibrium* exists when the vapor pressures outside and inside the fuel particle are equal. The vapor pressure of the atmosphere depends on the temperatures and moisture content of the air; that of the bound water in the fuel, on the temperature and moisture content of the fuel particle. Neither pressure is constant, so moisture equilibrium in nature is virtually impossible. Instead *equilibrium moisture content* is defined as the value that actual moisture content would approach if the fuel particle were exposed to constant atmospheric conditions for an indefinite period of time (Figure 3-5).

This relationship is itself variable: the different processes involved in drying and wetting proceed at different rates. The rate of exchange during the falling-rate period is logarithmic—10 times as rapid when the fuel moisture is 10% than when it is 1%. The rate of moisture exchange for this case depends on the magnitude of the gradient between the particle and the atmosphere. As the fuel moisture of the particle approaches equilib-

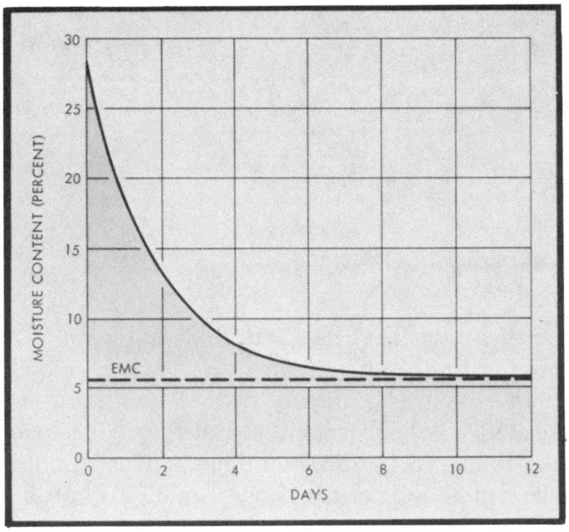


*Figure 3-5. Equilibrium moisture content. A drying curve for a 2-in. layer of litter for an EMC of 5.5%. From Schroeder and Buck (1970).*

rium moisture content, the gradient lessens and the rate of exchange slows. Because of the chemical bonding involved in adsorption, it is easier for a dry fuel particle to acquire moisture from a moist atmosphere than for a moist fuel element to surrender moisture from a dry atmosphere. Fuels wet more readily than they dry, and fuel moisture rises more rapidly than it drops.

*Timelag Fuels.* To simplify this variable rate of drying, a *timelag principle* is used. For bound water, the approach to equilibrium follows a logarithmic rate for which the time required to reach equilibrium moisture content can be divided into periods. Each *timelag* (TL) *period* describes the moisture exchange to the amount of 63% (1–1/e) of the departure from equilibrium. The actual duration required to achieve equilibrium moisture content depends on the properties of the fuel, including its size and diffusivity (Figure 3-6). Laboratory determinations apply standard conditions of 80°F (27°C) and 20% relative humidity. By the end of five or six timelag periods, actual moisture content would very closely approximate equilibrium moisture content. For fine fuels, the change may take only minutes; for large fuels, days or weeks. For the exchange of moisture to continue effectively, the duration over which the gradient persists must be proportionately long.

For purposes of inventorying fuel complexes and designing fuel models, four categories of fuel particles are recognized based on their timelag periods: 1-hr TL, 10-hr TL, 100-hr TL, and 1000-hr TL. Fuels are committed to one of these categories as a function of their diameter, in the case of particles, or their depth, in the case of fuelbeds. The 1-hr TL category includes fuels up to .25 in (.64 cm) diameter and fuelbed surfaces; the 10-hr TL category, fuels from .25–1 in (.64–2.5 cm) and fuelbeds down to .75 in (1.5 cm); the 100-hr TL category, fuels from 1–3 in (2.5–7.6 cm) and fuelbeds from .75–4 in (1.6–10.2 cm); and the 1000-hr TL category, fuels from 3–8 in (7.6–20.3 cm) and fuelbeds from 4–12 in (10.2–30-5 cm) (Figure 3-7). Fuelbeds exchange moisture much like large diameter fuel particles. Each has a "cel-

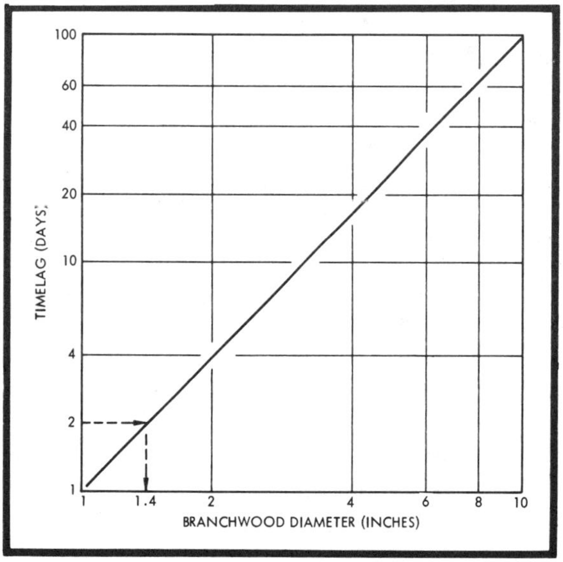


*Figure 3-6. Timelag and fuel particle diameter. From Schroeder and Buck (1970).*

| | Time lag class | | | |
|---|---|---|---|---|
| | *1 hr* | *10 hr* | *100 hr* | *> 100 or 1,000 hr* |
| Time lag class interval | 0–2 | 2–20 | 20–200 | > 200 or 200–2,000 |
| Equivalent fuel dimensions: | | | | |
| Roundwood | < ¼″ | ¼ to 1″ | > 1 to 3″ | > 3″ |
| Litter and/or duff | Surface | Surface to ¾″ | > ¾ to 4″ | |

*Figure 3-7. Timelag fuel equivalencies. From Martin et al. (1979).*

lular″ structure with interstitual voids, each communicates moisture along a gradient, and each has a gradient between fuel surface and atmosphere and between fuel surface and soil. In this way a multitude of gradients may exist, surface and internal both (Figure 3-8).

The ratio of fuel surfaces to fuel volume is as critical for moisture exchange as for heat exchange. Particles with large surface-to-volume ratios, and fuel complexes with large amounts of such fuels or high effective bulk densities, respond rapidly. Large particles and fuel complexes composed of large TL fuels respond more slowly. At the same time, high fuel loads act as big sources and sinks for both heat and moisture. Both the quality and the distribution of fuel particles by size class are important in determining how much moisture and heat will be exchanged and at what rates. The small particles (1-hr TL and 10-hr TL categories) and the upper crust of the surface fuelbed (under 2 in, or 5 cm) control fire spread and power even high intensity fires. Many of the same fuel parameters that describe

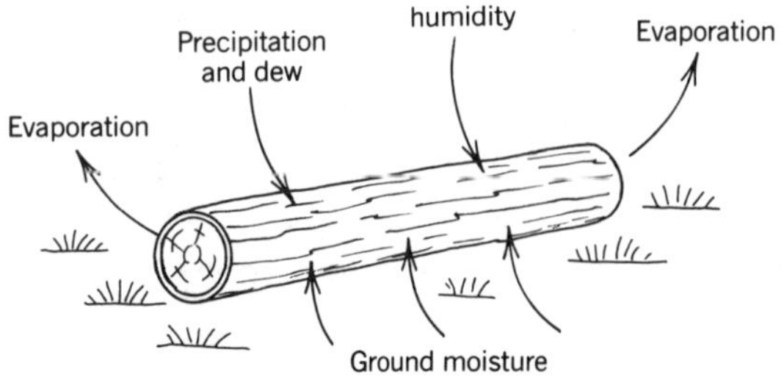


*Figure 3-8. Moisture exchange in wildland fuels. From NWCG, S390.*

the effectiveness of heat transfer account equally for moisture exchange. In brief, because a fuel complex consists of many sizes of fuel particles, different portions of the complex will become available as fuel at different times, will experience different periods of fire danger, and will contribute in different ways to fire spread.

The largest fuel category, 1000-hr TL, includes fuel particles with diameters of 3–8 in (7.6–20.3 cm) and fuelbeds from 4 to 12 in (10.2–30.5 cm). The actual FMC of 1000-hr TL fuels derives almost wholly from their initial moisture content at the onset of fire season and the amount of precipitation that falls on them during the season (Figure 3-9). Low FMC in this size class indicates in a general way an increase in fuel availability over the complex. For many fuel complexes, with abundant fuels in the 1-hr and 10-hr classes, the 1000-hr fuels are marginally involved in combustion and are not figured directly into fire behavior calculations. Yet combined with the appropriate climate class, the 1000-hr fuel moisture relates to live fuel moisture and furnishes a useful drought index. For many dense conifer fuel complexes, with most of their fuel loading concentrated into large downed boles, the 1000-hr FMC is important in predicting total energy release and the likelihood of large fires.

How fuels of these dimensions contribute to fire spread is somewhat problematic. They do not perform so much directly as heat sources as they do indirectly by reducing (through low FMC) the size of their potential heat sink. This increases the amount of heat flux available for propagation; the overall combustion rate improves. Fireline intensity derives more from heat release (heat output/

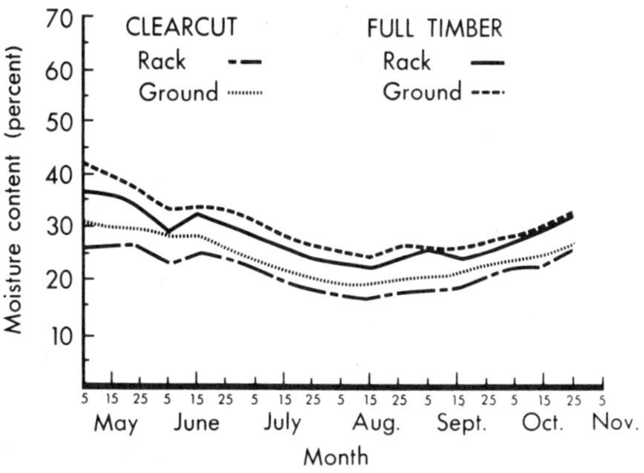


*Figure 3-9. Seasonal fuel moisture changes for large diameter fuels, Idaho. From Brackebusch (1975).*

unit area) than from rate of forward spread. For steady-state fires the 1000-hr fuels are important in that they burn—when they burn well—with long residence times, helping to make the transition from line fire to area fire. For mass fires, where fire spread is principally downward, large-diameter fuels (or deep fuelbeds) are important to maintain the flow of fuel. For certain conifer biotas of infrequent but high-intensity fires, successful combustion of the 1000-hr (and greater) fuels is critical for getting the fire into the canopy, where it can spread as a crown fire. Large fuels, moreover, contribute to residual combustion, either flaming or glowing, complicating mopup operations and the persistence of smoke. Their biological significance, of course, is large since 1000-hr and greater fuels contain large amounts of otherwise inert biomass undergoing very slow decomposition.

### *Drought*

Drought is a condition of moisture deficiency that is large in area and long in time. It induces moisture stress in live fuels and deepens the moisture gradients of dead fuels and fuelbeds. Up to a point, live fuels can resist drought, maintaining a relatively high moisture content because of physiological processes. Even dead fuels may hold on tenaciously to some moisture. Because it is so difficult to purge water from the interior of large diameter fuels or deep fuelbeds, a gradient can persist in dead fuels for considerable periods of time. But the long-term consequences of drought on FMC can be pervasive. Drought affects equally the gradients between particle and atmosphere and between particle and soil, parching dead fuels on all sides.

Several means exist by which to measure drought relative to fire danger. The *climate class* concept recognizes that drought is a relative condition, that many organisms show different adaptations to moisture stress. Two months without rain may constitute a drought in the Southeast, but it is normal, annual event in the Southwest. Live fuels, moreover, respond differently than dead fuels to similar meteorological conditions. A more quantitative assessment results from consideration of the FMC of *1000-hr TL fuels.* A practical indicator of short-term drought, the 1000-hr TL calculation describes the moisture content of large fuels, deep fuelbeds, and live fuels. For prolonged droughts, the *Palmer drought index* and the *Keetch-Byram index,* both based on soil moisture deficits, are available.

The Keetch-Byram index measures drought in terms of the amount of precipitation necessary to recharge the soil, which it conceives as a reservoir, to full capacity. The index assumes that the moisture regime will range from 0 to 8 in. (0–20 cm) of water (8 in. being saturation), organizing this range on a scale of 0–800. The lower the number, the greater the deficit. Originally designed with the organic soils of the Southeast in mind, the Keetch–Byram index gives an accurate portrait of soil field capacity and, since the depth of such fires corresponds directly with height of the water table, a good assessment of fire behavior. The index does show aberrations early in the fire season, however. The ground may be saturated from winter rain or snow, but have surface grasses and fine fuels in prime condition for fire. This difficulty highlights the general problem of determining just when a fire season begins, and it is not unique to determinations of drought. Grass and even crown fires in the spring are likely before the onset of flushing.

The Palmer drought index similarly examines soil moisture stress, but it divides the soil profile into two layers—an upper and a lower—and it normalizes its scale so that the index represents a departure from average. This departure may be either positive (with a wet cycle) or negative (with drought). The effective range is between + 4.00 and -4.00. Conceived for crops, the Palmer index is routinely reported by the National Weather Service and the Department of Agriculture. But like the Keetch–Byram index it suffers from what might be termed an uncertainty of statistical generalization. The Keetch–Byram index relies on point source data for computation; the Palmer index, on areas. The Keetch–Byram index is thus good for particular localities, typical of fire sites, but suffers when generalized over a range of microclimates. The Palmer index, conversely, typifies a region, but may not be helpful for smaller areas on the scale of fires. As a result it is common to see both indices and the 1000-hr TL fuel calculation cited at the same time. For a graphic comparison, see Figure 3–10.

The interpretation of drought indices is not easy. Drought is neither a necessary nor a sufficient cause for large fires. Droughts do not result in spontaneous combustion any more than large fuel accumulations do. Both the Keetch–Byram and Palmer indices, moreover, most accurately describe the condition of ground fuels, like organic soils, whereas fires burn through surface fuels. Even within the realm of surface fuels the role of coarse fuels—1000-hr TL and larger—is problematic. Most surface fires spread through fine fuels (1-hr and 10-hr TL), and low fuel moistures for these size classes do not require drought. Coarser fuels can lengthen residence time and augment total heat output, but they do not sustain the flaming front. Instead they often improve the effective heat source by diminishing the heat sink. By drying out large fuels, a drought improves the total heat output: more heat is applied to combustion and more of the total fuel load is brought into the category of available fuel. Similarly, drought transfers live fuels to dead fuel categories, again transforming heat sinks into heat sources.

Prolonged drought reduces the variability in the fuel complex as a whole; all fuel size classes

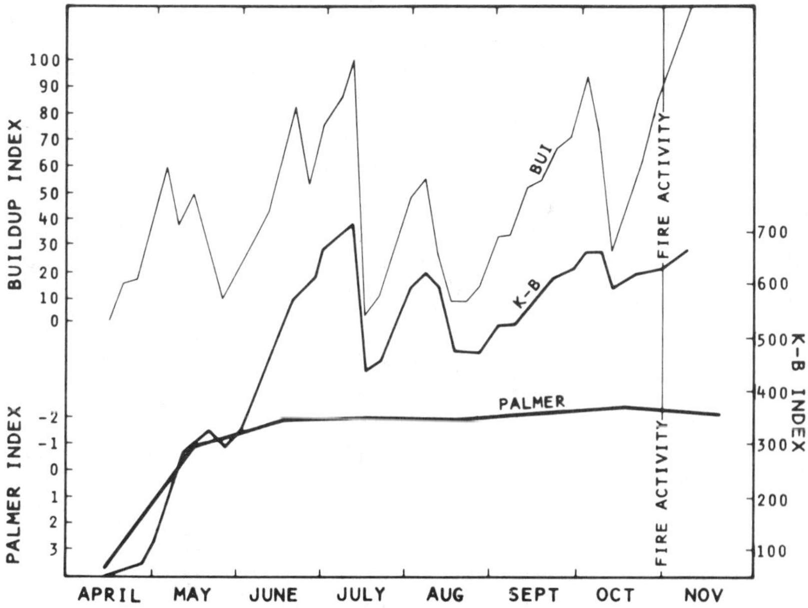


*Figure 3-10. Drought indices compared. Palmer indicates the Palmer drought index; K–B, the Keetch–Byram index; BUI, the buildup index, an early version of the burning index. The values shown are for the 1952 fire season in Missouri. From Haines et al. (1975).*

become potential heat sources rather than partial heat sources and partial heat sinks. Nearly all large historic fires have been associated with droughts. The differential moistures that characterize the fuel complex as a whole disappear. Instead of a mosaic of microclimates, the landscape is uniformly parched. Everywhere fuel moisture plunges, and the probability that the fine fuels which occupy any single site will be dry increases. The likelihood improves that ignition and low fuel moisture will coincide, and the prospects rise for making the transition to large fire status.

## *3.2 FUEL COMPLEXES AND FUEL CYCLES*

Arrangements of fuel show patterns over space and time. In general, there are two ways to describe these arrangements: as biological systems and as physical systems. The biological approach defines the fuel complex as a community of organisms. Organisms make up the fundamental units of composition, and the dominant species give their name to the whole complex—identified as a *cover type, habitat type,* or *fire type.* Similar assemblages tend to show common fuel and fire characteristics, and these characteristics will be different from those typical of other biotas. The advantage of this method is the ease with which a fuel complex may be classified and the extent to which the method merges with ecological analysis. Its disadvantage is its lack of quantitative information about those fuel parameters essential to fire behavior.

By contrast, *fuel models* describe biotas strictly in terms of their relevant thermophysical and thermochemical properties. Particles are assembled into fuelbeds, and fuelbeds into fuel arrays or models. Several methodologies of fuel modeling exist, but all (in the United States) synopsize the fuel parameters that the Rothermel fire model demands. In some cases, fuel models have been developed that incorporate changes over time. The changes may be systematic, the result of growth and decay, or they may be episodic, the result of massive disturbances like logging or wildfire.

### *Fuel Arrays*

The elementary fuel unit is the particle or cell. The *fuel particle* describes the smallest discrete object in a fuel complex, and the *fuel cell,* the smallest volume of fuel particles that has sufficient mass to be statistically representative of the complex. The unit may be classified in several ways. Biologically, it may be alive or dead, woody or herbaceous, annual or perennial. Chemically, it can be characterized by its low heat yield, mineral content (minus silica), and fuel moisture. Physically, its important properties are its loading, density, and surface-to-volume ratio. Because both moisture and heat are exchanged across surfaces, particle size is further subdivided according to four timelag categories: 1-hr, 10-hr, 100-hr, and 1,000-hr TL fuels.

When combined, fuel particles and cells show several degrees of organization. A first order structure transforms elementary units into *fuel strata* or *fuelbeds.* Fuelbeds are described according to the proportions of the various fuel particles present in them—the ratio of live fuel types to one another, the ratio (by volume) of live fuels to dead fuels, the depth of the strata, particle packing within the strata, bulk density and bulk loading, and moisture content. The most important strata for most fuel complexes is the surface fuelbed. The fuelbed can be characterized according to the distribution and arrangement of the different particle size classes that make it up; if the fuelbed is not uniform, then unit fuel cells may be used, or the different fuel components weighted to make an average stratum. The depth of the fuelbed may be classified by appeal to timelag categories analogous to those used for fuel particles (see Figure 3–8).

A fuel complex may contain several strata. A layer of organic soils (*ground fuels)* and a stratum of foilage (*aerial fuels*) may frame a bed of *surface fuels* (Figure 3-11). Models of these strata will emphasize slightly different parameters, and the strata will be organized such that they burn with different fire behavior characteristics (Figure 3-12). Not measured directly, but implied for all these strata,

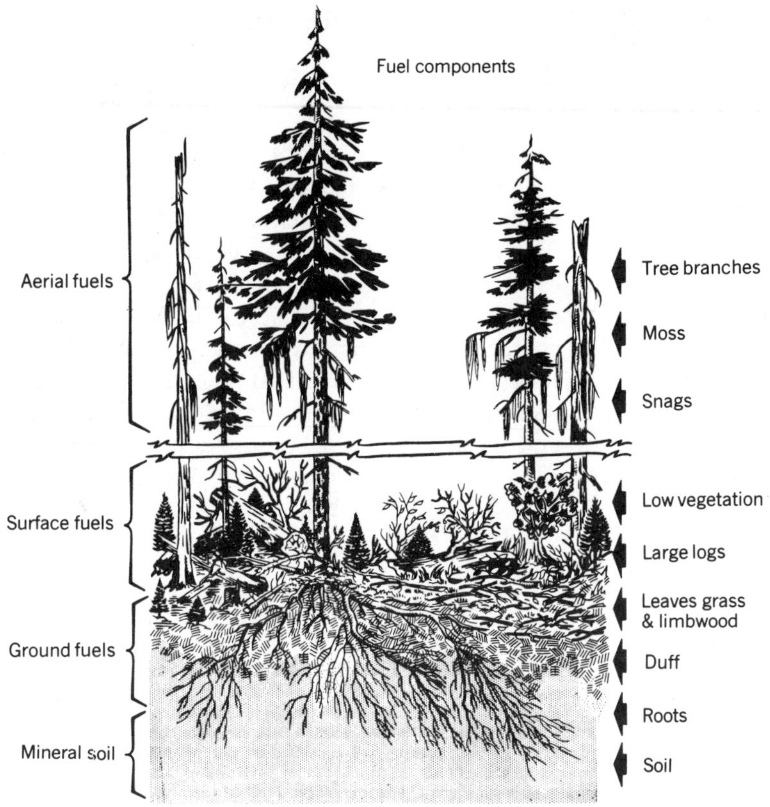


*Figure 3-11. Principal components of ground, surface, and aerial fuels. Note the organization of fuels into stratified fuelbeds. In some instances—dense tree reproduction, for example—the fuel complex links one stratum to another; at other times, a large flaming zone binds the surface fuelbed to aerial fuelbeds. From U.S. Forest Service (1969).*

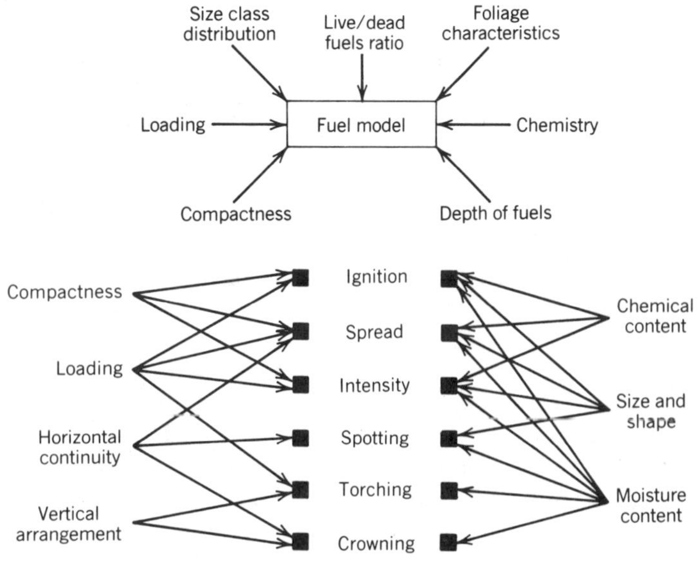


*Figure 3-12. Fuel model considerations and characteristics. From NWCG, S390.*

is a sense of continuity. Fuel models, like fire behavior models, assume uniformity. For all strata, too, fuel moisture is the most important variable. Fire behavior in ground fuels is determined largely by the depth of wetting, and in crown fuels, by the load, bulk density, and fuel moisture of the foliage. In surface fuels the controls over fire behavior are more complicated, but the proportion of live to dead fuels and the FMC of the principal timelag categories are fundamental. The special importance of the surface fuelbed is that fires originate and initially propagate through it. Only if a fire grows in intensity does flame progress into other fuel strata, and then only rarely does such a strata propagate fire independently of the surface fire. The surface fire sustains other fires, and through its flames the surface fuelbed is coupled to other strata.

A second order organization occurs in the vertical dimension. All fuelbeds show depth as well as extent; fires burn down as well as over fuelbeds; and fuelbed depth determines the residence time, a measure of the relative size of the heat sources and heat sinks. By vertical dimension, however, is meant the upward development of fuel strata, the stacking of one fuelbed onto another. In this case fires burn up as well as down. Not all fuel arrays show much vertical development, and not all fuels that are present in canopies can be considered as available fuel. But when several fuelbeds are integrated together, or when only one fuelbed is present, the result is a *fuel complex.* The medium of integration may be the fire or it may be the fuels. If the fire intensifies and shows considerable flame height, aerial strata of fuels may become involved. If the medium of integration is fuel, such that there exists vertical continuity between several fuelbeds, then the arrangement is referred to as *fuel ladders.* Even in the absence of high-intensity burning, the fire in such a fuel complex can propagate upward as well as outward.

Ladder fuels are a serious hazard. For a relatively small increase in fuel load, they multiply the quantity of fuel available to the fire by incorporating aerial fuel strata (especially fine fuels) that are normally beyond the reach of surface fires. The dense understory that develops beneath southern pines, the rough, is perhaps the best known example. But similar situations can be found where heavy conifer production congeals into dense understories, or where brush communities blossom to great heights with foliage rising from ground to canopy. Special fuel models have been constructed to account for such arrays. For the most part, however, fuel models include only surface fuels because these fuelbeds incorporate the parameters demanded by the Rothermel fire model, and the model pertains to steady-state surface fires. Where atypical understories of ladder fuels have developed, the fuel hazard will be much greater than a model of surface fuel alone would suggest. The transition to crown fire or to large fire can be made more easily.

Rare is a fuel array that shows unbroken continuities over large areas. Just as fuel particles make up a fuel cell, so fuel complexes contribute to larger associations called *fuel mosaics.* For many fuel complexes to be involved in a fire, the fire must be large, and fire behavior models, which set the important parameters for fuel models, do not apply to large, intense fires. The fuel mosaic is thus more often used as a biological description than as a fuel description. To model very large complexes, it is necessary to use several fuel models, much as, in constructing fuelbeds to model nonuniform fuels, it is often necessary to use fuel cells, with each cell weighted for the proportion of the fuelbed it represents.

### *Fuel Histories*

All of these descriptors pertain to the geography of fuels. Yet from particles to complexes, fuels also show a history; they change over time. The magnitude, character, and organization of these historical changes lead to the concept of a *fuel cycle.* And from this concept comes another, the *fire cycle.* As their names imply, for these concepts historical changes are considered to be regular and at least roughly periodic. In fact, they are neither. In general, only where humans intervene with fire and fuel management practices is there an approximate cycle or an apparent periodicity.

Fuels change in their amount, a measure of the general production and decomposition of biomass. They change in their character, too—acquiring and shedding moisture, reorganizing their chemistry as they grow and die, transferring their status from live fuel to dead, from small particle to large and large to small, from surface fuel to canopy and back to surface. With each season come changes in the heat yield values of fuel particle, in the ratio of live-to-dead fuels within a complex, and in the moisture content of both live and dead fuels. With respect to fuel moisture there exists a hierarchy of scales to express routine changes: diurnal, or the daily cycle of temperature and humidity; synoptic, coinciding with the movement of air masses on the scale of 1–10 days; seasonal, appearing on the order of months; and planetary, corresponding to astronomical periods, years, and long-term secular developments.

Similarly, fuels change in their arrangement. Old strata acquire or lose loading, density, and compactness. New strata or ladder fuels are added, greatly expanding the range of available fuels. Old ladder fuels may grow to maturity, segregating aerial fuels from surface fuels. The mechanisms of fuel accumulation and attrition are predominately biological, and the concept of fuel cycles suggests that the patterns of biological change will set the pattern of fuel loading and fuel arrangement. After all, the matter in question does exist as biomass, not simply as inert fuel, and as an ecosystem, not solely as a fire environment. One might expect fuels to show diurnal, seasonal, annual, and successional changes comparable to those evident in organisms. And out of this predictable fuel cycle could develop a normal fire cycle. For one version of this concept, the *reburn cycle,* historical change may take two forms. It may describe the time required to build up fuels to the point at which they can again carry fire, or the time required by successive fires to clean out the residue of a former burn. In either case the resulting fire cycle shows a predictable periodicity.

Only in a few special environments, however, does fuel accumulate or diminish with regularity. Examples include those fuel complexes dominated by brush (such as chamise) or by single conifer species (such as ponderosa pine or lodgepole pine). As the species increases its proportions of fine available fuels, so increases the flammability of the biota. For a few systems, which are composed of a limited number of species and not subject to disturbances, biomass shows developments over time that translate into increased fire size and intensity (Figure 3-13). But not all biomass is fuel, and not all fuel is available for combustion. Even increases in surface litter do not automatically mean increases in the fuel available for surface fires. Few important systems, moreover, are restricted to dominance by one or two species; rather, successional competition among the various species subtly shapes complicated associations of organisms and complex arrays of fuels. And, finally, few sites remain undisturbed for long. Natural disturbances from biological processes such as disease, from meteorological processes such as drought and high winds, or from geological processes such as flooding are not periodic, but episodic.

So it is with fire. Fires occur from a compounding of biological and meteorological events, and they show a frequency–intensity distribution similar to natural outbursts of energy like floods. The relationship is logarithmic, not linear. Nor is a fire singular in its effects or deterministic in its outcome. Fires may produce more fuels than they consume; the net fuel loading after a fire will depend, in good measure, on the character of the fire, and a wildfire is a stochastic compounding of many events, not an inevitable byproduct of a fuel complex. Disturbances such as fires, moreover, have different effects on a site as a function of the site's prior history of disturbances: each additional disturbance does not initiate another identical cycle, but comes as a unique event, one dependent on circumstances shaped by previous events.

Fuel histories based solely on the growth of individual organisms and on the successional patterns of groups of organisms ignore, too, the fundamental influence of humans. In most systems, the patterns of fuel are the product of human activity, including fire practices. The character of that intervention may itself be episodic, such as the conversion of forest to farm or of brush to grass; or it may be periodic, typical of programs of controlled burning, fire suppression, and the rotational harvesting of timber. The human presence thus works in two ways. On one hand, it magnifies the tendency for episodic phenomena to dominant fuel history by coupling fuels to the idiographic nature of human history. On the other, it shapes fuel history into patterns with greater predictability and periodicity than would ever be found in natural conditions.

The conceptual problem has perhaps more to do with a philosophy of history than with the measurement of fuel parameters. But the important points are that fuels do not show simple or regular rates of accumulation and attrition, that fuel history, like fire history, is increasingly under

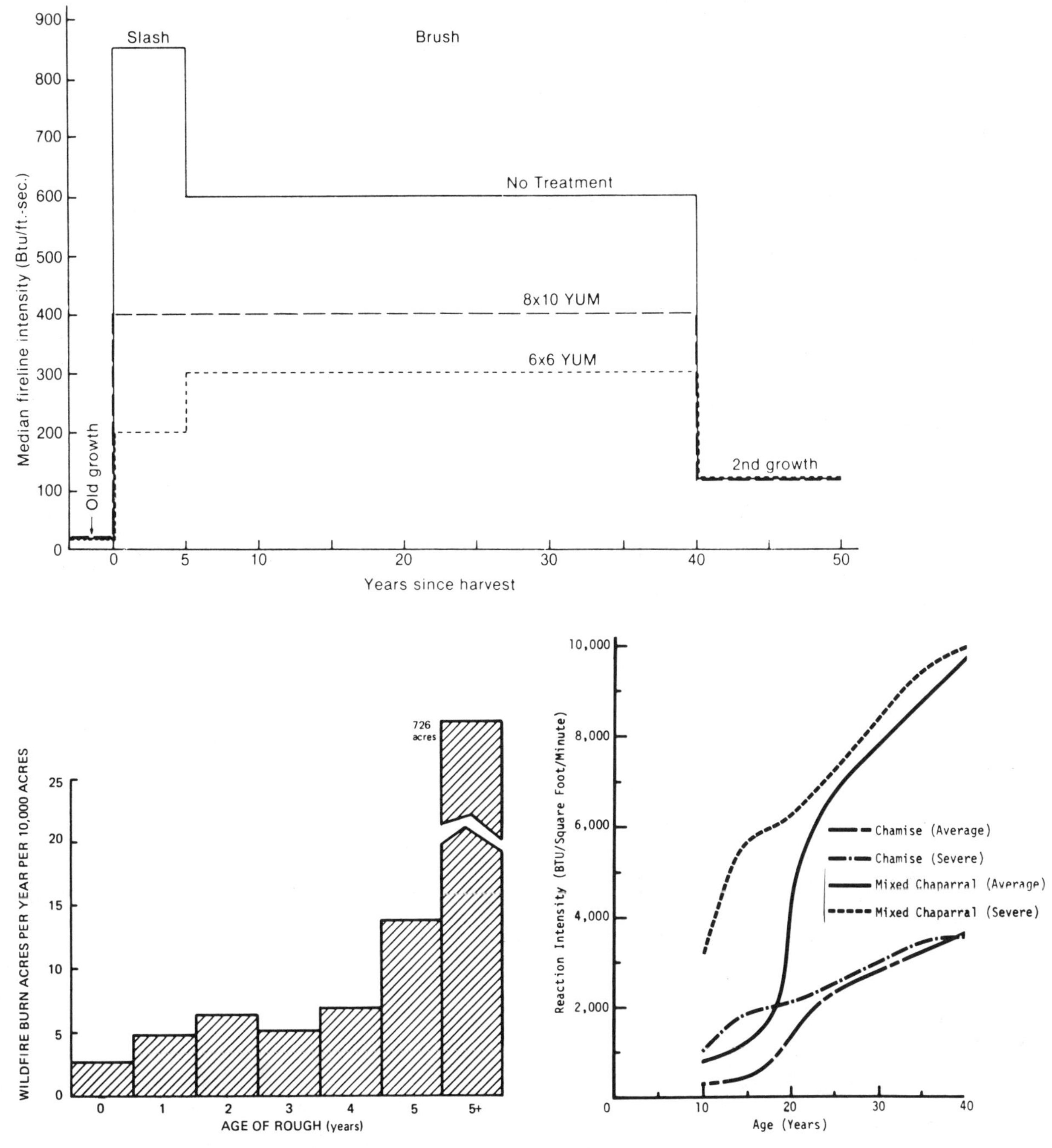


*Figure 3-13. Fire and fuel relationships over time. The graph opposite compares fireline intensity and time for slash; the lower left graph, fire size and age of rough; the lower right graph, reaction intensity and age for chaparral. From Hirsch et al. (1980); Cooper (1976); and Philpot (1971).*

the control of human agents, and that the relationship between fuel and fire is stochastic (Figure 3-14). Fuels support fires but they do not cause them. Large fuel loads do not, of themselves, result in frequent fires or intense fires. Hazard must combine with risk, fuel with ignition, and fire with a suitable environment. The fuel cycle and fire cycle concepts are convenient ways to characterize the fact that fuels and fires have histories, but erroneous designations of how those histories evolve and interrelate.

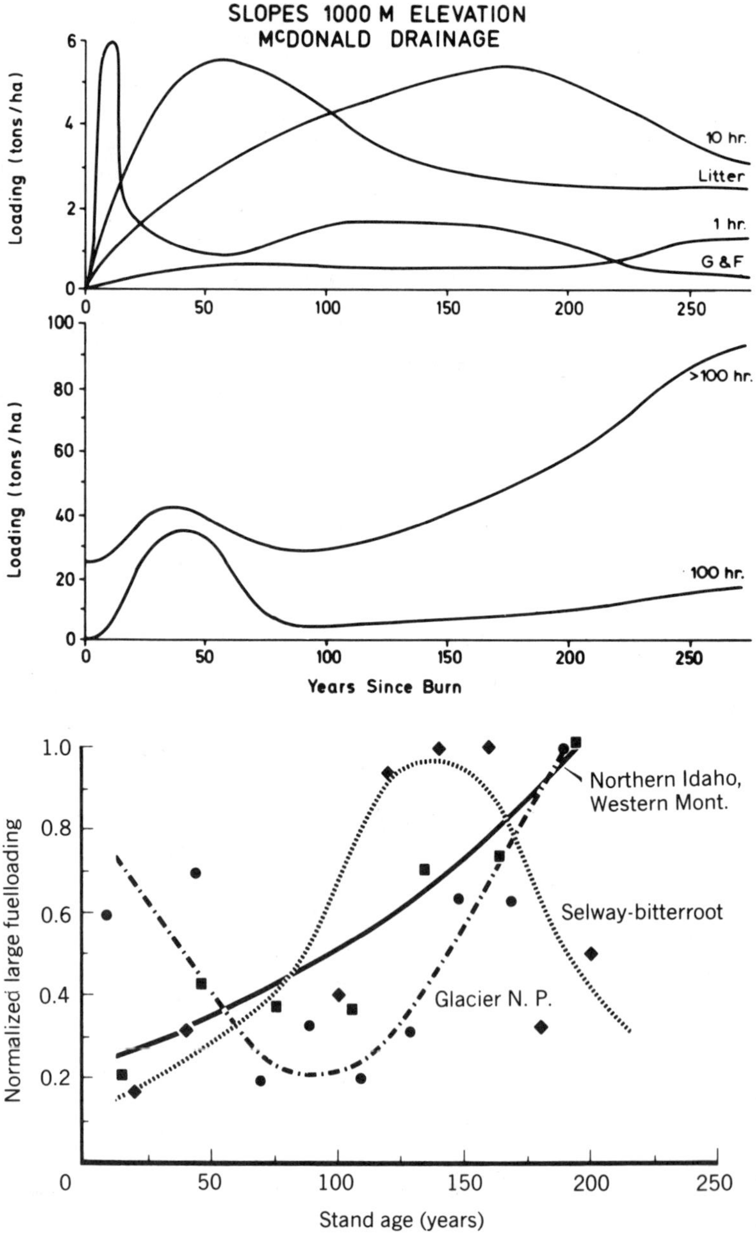


*Figure 3-14. Fuel loading over time. The upper graphs present an idealized history for fuels in six categories as they developed since the last fire. The lower graph gives normalized loading of large downed woody fuel in lodgepole stands of varying ages. Both studies describe the Northern Rockies. From Kessell (1976) and Brown and See (1981).*

### *Fuel Appraisal*

The classification of fuels involves processes of characterization, inventory, and appraisal. *Characterization* identifies the fundamental physical and chemical parameters of fuel particles, fuelbeds, and fuel complexes; *inventory*, the measurement, or sampling, of these properties; *appraisal*, the estimation of the hazard posed by these fuels based on the likelihood of their ignition and the fire behavior properties they would exhibit were they to burn. Curiously, perhaps, it is the appraisal process that determines the characterization process. The important fuel parameters are defined by the needs of fire behavior models (Anderson, 1982).

*Characterization.* There are certain biologically obvious components to any fuel description. The surface litter layer, the duff, dead woody material on the surface, grasses and forbs, living woody shrubs and trees (including canopies), and reproduction—all are easily discernible elements within a fuel complex. Depending on what type of vegetative cover dominates, the fuel complex may be divided into four general groups: grass, shrub, timber, and slash. To convert a general characterization such as these into fire behavior information, however, certain important physical and chemical traits for each component must be identified, quantified, and then assembled into fuelbeds. Each of the important traits contributes to fire behavior.

*Fuel loading* (weight) by size classes, *size class distribution* of the load, and *arrangement* of the classes in terms of compactness or bulk density all influence ignition. There must be ample quantities of fine fuel for kindling. *Horizontal continuity* decides whether a fire can spread outward and whether its rate of spread will be steady or spasmodic. *Vertical continuity* (or arrangement) and fuel loading describe the presence of very deep fuelbeds or ladder fuels. This will influence flame size, residence time, torching, and, under the proper conditions, the prospects for crown fire. With good vertical arrangement, several fuel strata may be integrated into a single fuelbed. *Fuel moisture content* affects all aspects of fire behavior—growth, spread, intensity, and the prospects, through spotting, of large fire behavior. *Chemical content* includes caloric values, especially easily volatilized oils and waxes, and mineral content, which retards pyrolysis and combustion. These values will show seasonal and annual changes.

Basic fuel characteristics combine in various ways to determine the flammability of a fuel complex. To organize these separate components into fuelbeds, and the fuelbeds into a fuel complex, is the task of fuel modeling. To describe how seasonal changes affect the way in which the various characteristics combine is the job of fire danger rating. At times, components may work at cross-purposes. Fuel load and arrangement may encourage high fire intensity, but fuel moisture and chemical content may dampen fire behavior; ignition and spread will be unlikely. At other times, many of the components may combine to support quick ignition, rapid rates of spread, and high fire intensities. High fuel loads in fine fuel size classes, low fuel moisture contents, abundant easily volatilized oils—all may coalesce at the same time to promote high fuel hazards.

*Inventory.* Measurement of these characteristics makes up a fuel inventory. Curiously, attempts to convert standard forest inventory data based on biomass (timber) directly into fuel inventories for the complex as a whole have had limited success. Phytomass is not synonymous with fuel, much less with available fuel; structural considerations, not simply load, are significant. But direct inventories for specific sites are costly and laborious. Instead special methods for sampling the necessary fuel components have been developed. According to the procedures used to assess them, these components fall into six groups: standing trees, shrubs, herbaceous vegetation (grasses and forbs), forest floor litter, forest floor duff, and downed woody material (Brown et al., 1982).

For standing trees, normal forest mensuration techniques are applicable, though particular interest resides in tree reproduction close to the ground and special techniques have accordingly been developed. Two methods are used for shrubs. One relates biomass to stem diameters, and the other relates to canopy area and canopy volume. Herbaceous vegetation can be inventoried by clipping and weighing, or by assorted procedures that combine estimation techniques with some selective sampling. Estimation may involve guesses of actual weights for sample plots, or guesses involving the weights of plots relative to the known weight of a standard plot. For duff and litter, bulk density is the critical property; but it is a parameter more easily identified than measured. Duff depth can be sampled, but the determination of litter depth requires trained judgment. Instead a version of the relative weight-estimate technique is employed, adapted from herbaceous fuel sampling. For downed, woody surface fuels a planar intersect technique is used to estimate the volume of fuels according to size classes. Fuel loads are then calculated from volume by applying estimates of the specific gravity for the various woody materials present (Brown, 1974).

Based on this latter methodology, the U.S. Forest Service has promoted a National Fuels Classification System. To reduce the labor of separate samplings, techniques have been developed to standardize fuel complexes and to consolidate fuels data. By matching a specific site to inventoried sites, known fuels data can be applied to new settings. To assist this process, the Forest Service has established a National Fuels Inventory Library (NFIL). The NFIL contains a Fuels Inventory Database and a suite of software programs to process the raw data (Bevins and Roussopoulos, 1980).

*Appraisal.* The appraisal of fuels requires that fuels data be converted into fire data. Fuels inventories, the raw fuel data for the relevant fuels parameters, is converted into fuelbeds or fuel models. The fuels are ultimately expressed as an index of fire hazard, or expected area burned per year. Again, the U.S. Forest Service has developed a standard process by which to determine this index. The National Fuel Appraisal Process combines fuel modeling with fire behavior modeling to project probable fire characteristics (Roussopoulos, 1980). These are in turn processed through a decision analysis program. Whether a fire occurs, and under what weather conditions, requires probabilistic estimates. Ignition likelihood is determined from historic records, stored as a National Fire Occurrence Library; fire weather, from historic records stored in the National Fire Weather Data Library and processed through various programs that project the distribution of weather and fire behavior indices over a season (Firefamily). The appraisal process is best developed for activity fuels, less so for natural fuels.

It is fire, not simply fuels, that is the problem for fire management. The fuels appraisal process translates known fuel data into probabilistic fire behavior data. Based on fire behavior risks, it assists in deciding how much fuel information is needed and it weighs the probable outcomes of various fire (and fuels) management programs. Information costs money, detailed information demands very costly sampling, and not every site demands high resolution fuels data. The alternatives are expressed as a decision tree. In some instances, little fuels knowledge may be needed and little formal treatment. In other cases, where ignition is likely and anticipated fire behavior intense, considerable investment in acquiring fuels data and in executing fuels treatment programs may be warranted. The fuels appraisal process provides a quantitative methodology for making fuels decisions.

## *3.3 FUEL MODELS*

### *Historical Synopsis*

From its inception organized fire protection in the United States has found it desirable to classify fuels. In gross terms, the biotic groupings developed early in the twentieth century are the same as those used in the 1980s, and they are applied to the same purposes. The function of fuel classification was to estimate fire behavior, and the prediction of fire behavior was fundamental to fire danger rating, fire suppression, and fire planning. For each revolution in fire planning, new fuel classifications have been proposed. Originally, the fuels were typed according to whether grass, brush, or timber dominated the biota. Gradually, these cover types evolved into descriptions more specifically related to problems of fire control. With the development of mathematical models of fire behavior, and with the adoption of fuels management programs as a major strategy of fire management, the contemporary array of fuels models has appeared (Anderson, 1982).

The process began with Coert duBois' *Systematic Fire Protection in the California Forests* (1914). Dubois identified three *cover types*: grass, brush, and timber. Combined with fire weather information, each type expressed a certain ignitability and rate of spread. Based on these estimates (largely derived from primitive fire reports) a first order fire suppression organization could be planned. The cover types were multiplied and refined. When S.B. Show and E.I. Kotok transformed duBois' concept of systematic fire protection into the concept of hour control (1929–1930), they identified nine cover types for northern California. For planning purposes, each cover type was characterized by a particular rate of spread, a value loosely determined through field trials and fire reports. Moreover, each type had an associated index that characterized its value as a protected resource. Other researchers adapted and extended this classification scheme to other regions. The process of defining fuel properties in terms of fire properties, and of separating fuel traits from biological traits, was underway.

By the 1930s the concept of cover type evolved into the concept of a *fuel type.* Each fuel type exhibited two primary characteristics: a rate of spread and a resistance to control. Each of these two traits could be assigned one of four values: low (L), medium (M), high (H), or extreme (E). Rate of spread was estimated by experienced fire officers, and it was calculated for "average worst" burning conditions. Resistance to control described the difficulty of constructing and holding a fireline through the fuel complex. Fire control was thus imagined as a two-dimensional competition between the perimeter increase of a fire and the line-construction capabilities of the control forces mustered against it. As elaborated by L.G. Hornby (1936), the Hornby fuel classification system underwrote extensive fire planning for the national forest system. Fuel type maps became widespread. A fuel type of H-H, for example, meant that the fuel showed high rate of spread, and offered high resistance to control; a fuel typed as L-H, that rate of spread was low but the difficulty of putting in a fireline high; and H-L, that rate of spread was rapid but firelines could be constructed quickly. An example of a L-H fuel type would be a dense, damp forest; of a H-L fuel type, a grass. The estimated values were based on statistics contained in fire reports and on the best judgment of fire experts. The development of a fire danger rating meter introduced some refinements, and the adoption of the 10 AM Policy eliminated the association of cover types with timber resources under protection.

But different strategies of fire control demanded greater refinement, and new developments in fire danger rating and fire behavior modeling made new fuel classifications possible. It became necessary, first, to describe fires in terms of their intensities, not merely their rates of spread; then, to model fires suitable for precision prescribed burning, and to model fuels for the specific changes that a fire could induce. Control required not only that a fireline be cut but that it be held, and for this it was necessary to measure the difficulty of control directly against the properties of the flaming front, not indirectly against its fuels. Both the National Fire Danger Rating System (NFDRS) and the Rothermel fire model appeared in 1972. Both continued the process of transforming vegetative cover into fuel, and fuel traits into fire behavior traits. Other quantitative fuel-related properties could be added. For each fuel model, for example, there was assigned an *emissions factor,* which described the production of particulate matter, and a *coverage level,* which expressed the suitable density for fire retardant. The quantification of fuels was underway. Indeed, even biological approaches to classification followed suit. Biotas were expressed as fire regimes and cover types as fire types, identifying complex ecosystems solely on their responsiveness to fire.

### *Fuel Modeling*

Given the requirements of the Rothermel fire model, three approaches exist for the modeling of wildland fuel complexes. One is actual on-site inventorying. The resolution of fuel data is high, but the process is costly and the outcome site-specific. For some sites, however, the information is worth the investment. Another approach resorts to inferential or statistical fuel modeling. For the most part, such techniques apply to natural rather than to anthropogenic fuel complexes, to areas in need of high-resolution data which can be derived from existing ecological information. Gradient modeling of environmental data, with appropriate conversion equations, for example, can transform ecosystems into fuel complexes. Another approach—the most prevalent—employs sets of stylized, or averaged, fuel models (Roussopoulos, 1980).

The averaged fuel models may, in turn, take three forms: static, dynamic, or stylized. *Static fuel models* define fuel parameters for only a short period of time, such as a given fire season. *Dynamic fuel models,* by contrast, accomodate elements of growth and decay, predicting systematic changes in fuel parameters over the course of several years. To date, three such models exist: one for slash decay, one for the growth of palmetto–gallberry fuel complexes, and one for chamise chaparral. The slash decay model describes more the rearrangement of fuels than their loss from the system, but the transfer of available fuel to unavailable categories has the same effect as their removal from the system. The other dynamic models work because the fuel complex is unusu-

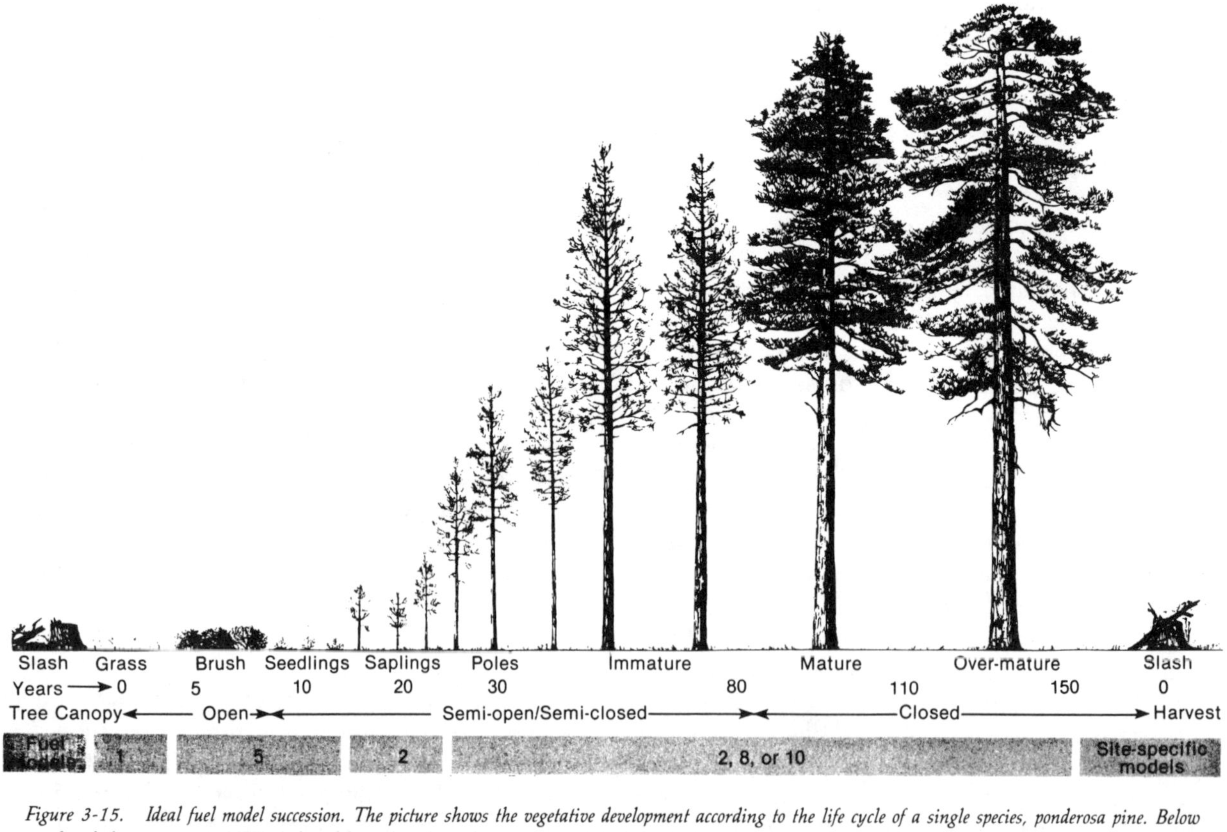


*Figure 3-15. Ideal fuel model succession. The picture shows the vegetative development according to the life cycle of a single species, ponderosa pine. Below are listed the appropriate NFFL fuel models to describe each stage. From Hirsch et al. (1981).*

ally dependent on the growth properties of a single species; the history of the complex is closely intertwined with the life cycle of this species. Most complexes, however, consist of many competing species, and to describe changes over time it is more useful to match particular stages of development against discrete stylized fuel models (Figure 3-15). All of the dynamic models project fuel changes for simple systems over relatively short periods of time—5 years for slash, 25 years for palmetto–gallberry, and 50 years for chamise. All assume that no other disturbances intervene within this period. For some disturbances, that occur once, such as fire or logging, it is possible, however to predict the fuel complexes that result.

*Stylized fuel models*—as a specific form of modeling rather than as a generic category—average the net growth and decay of fuel complexes into a characteristic ensemble. Two sets of stylized fuel models are in common use. The NFDRS includes 20 fuel models, and the Northern Forest Fire Laboratory (NFFL) developed a slightly different set of 13. As biotic associations, both sets fall into four groups: grass, brush, timber, and slash. As fuel complexes, they are distinguished according to the distribution of fuel loads by size classes, by the depth of the fuelbed, and by their moisture of extinction (Figure 3-16).

The distinction between fuel load and fuel depth divides the NFFL models into two broad groups (Figure 3-17). One group, grass and brush, is oriented vertically. Increased fuel load means increased fuelbed depth, and increased depth means an increase in available fuels. The other group, timber and slash, is more horizontally oriented. Increasing fuel loads in this case leads to compaction, only slowly increasing fuelbed depth and fuel availability. Within each group, a range of values exists. Short grass, for example, has different loads, depths, and moisture of extinction than long grass; hardwood litter differs from closed conifer litter; light slash from heavy, and rough from Southern chaparral (Figure 3-18).

Moreover, there are some differences between the NFDRS and NFFL fuel models (Albini, 1976). The NFDRS models incorporate changes associated with weather and, to a limited extent, seasonal cycles of growth. Live fuels flush and cure, the ratio of live-to-dead fuels changes, dead fuel moisture varies with atmospheric conditions

| Fuel model | Typical fuel complex | Fuel loading | | | | Fuel bed depth | Moisture of extinction dead fuels |
|---|---|---|---|---|---|---|---|
| | | 1 hour | 10 hours | 100 hours | Live | | |
| | | Tons/acre | | | | *Feet* | *Percent* |
| | **Grass and grass-dominated** | | | | | | |
| 1 | Short grass (1 foot) | 0.74 | 0.00 | 0.00 | 0.00 | 1.0 | 12 |
| 2 | Timber (grass and understory) | 2.00 | 1.00 | .50 | .50 | 1.0 | 15 |
| 3 | Tall grass (2.5 feet) | 3.01 | .00 | .00 | .00 | 2.5 | 25 |
| | **Chaparral and shrub fields** | | | | | | |
| 4 | Chaparral (6 feet) | 5.01 | 4.01 | 2.00 | 5.01 | 6.0 | 20 |
| 5 | Brush (2 feet) | 1.00 | .50 | .00 | 2.00 | 2.0 | 20 |
| 6 | Dormant brush, hardwood slash | 1.50 | 2.50 | 2.00 | .00 | 2.5 | 25 |
| 7 | Southern rough | 1.13 | 1.87 | 1.50 | .37 | 2.5 | 40 |
| | **Timber litter** | | | | | | |
| 8 | Closed timber litter | 1.50 | 1.00 | 2.50 | 0.00 | 0.2 | 30 |
| 9 | Hardwood litter | 2.92 | .41 | .15 | .00 | .2 | 25 |
| 10 | Timber (litter and understory) | 3.01 | 2.00 | 5.01 | 2.00 | 1.0 | 25 |
| | **Slash** | | | | | | |
| 11 | Light logging slash | 1.50 | 4.51 | 5.51 | 0.00 | 1.0 | 15 |
| 12 | Medium logging slash | 4.01 | 14.03 | 16.53 | .00 | 2.3 | 20 |
| 13 | Heavy logging slash | 7.01 | 23.04 | 28.05 | .00 | 3.0 | 25 |

*Figure 3-16. Description of NFFL fuel models. From Anderson (1982).*

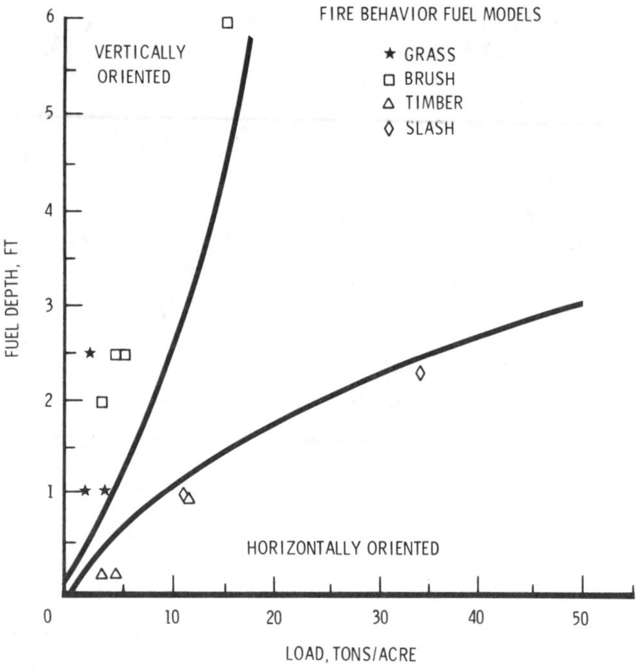


*Figure 3-17. Fuelbed dimensions for four general fuel groups. Grasses and shrubs are oriented vertically; timber, litter, and slash, horizontally. Grass fires have high rates of spread and low short residence times, whereas timber, litter, and slash have comparatively slower rates of spread and longer residence times. From Anderson (1982).*

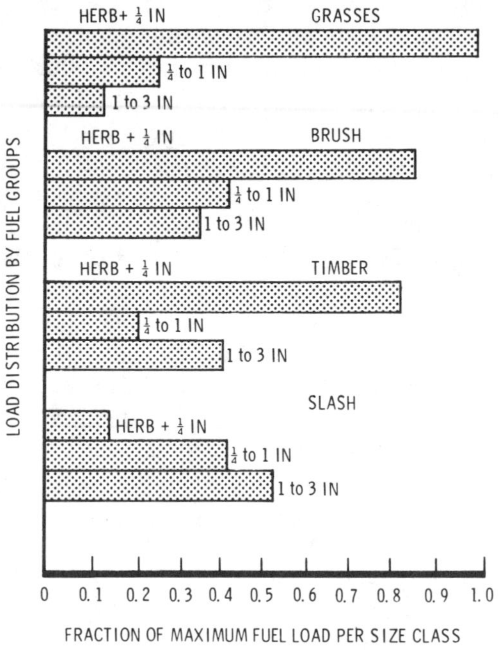


*Figure 3-18. Distribution of maximum fuel load by size class for the four general fuel groups. From Anderson (1982).*

and the distribution of size classes. The NFFL models are more specific: site-specific in that they require on-site fuel data, and time-specific in that they apply to peak burning conditions. The NFDRS models were intended to assist with the forecast of high fire-danger periods, as judged by probable fire behavior; the NFFL models, with real-time fire behavior forecasts for actual fires. For greater specificity without needlessly multiplying the number of fuel models, photo series have been developed. Matching the appearance of the photographed fuel complex with the site in need of fuel data gives approximate values for the fuel input data needed as input for fire behavior models.

### *Fuel Model Selection*

With the proliferation of fuel models has come a complementary elaboration of means by which to select an appropriate model, to match obvious biotic communities to more esoteric fuel models, and to interrelate the various published models. Two common methods employ keys or photo series. *Keys* were produced for fuel typing, and new versions have been developed for fuel modeling. Essentially, the keys follow the same logic used in biological systematics. A series of yes–no decisions steers through a logic tree, the end result of which is the proper typing of the object in question. The keys refer to the criteria that apply at each stage in the decision evolution. The NFDRS includes a key to help users determine an appropriate fuel model. The Fahnestock keys assemble 36 distinct combinations of fuel characteristics that describe rate of spread and another set that assesses the likelihood of crown fire. Use of the Fahnestock keys, however, requires knowledge of a special terminology—one that accounts for fuel properties in descriptive, qualitative language rather than in quantities (Fahnestock, 1970).

DATA SHEET — Residue descriptive code 2-PP-4-PC

**LOADING**

| Size class (inches) | Weight (tons/acre) | Volume ($ft^3$/acre) |
|---|---|---|
| 0.25-1.0 | 2.2 | 132 |
| 1.1-3.0 | 3.4 | 220 |
| 3.1-9.0 | 2.9 | 234 |
| 9.1-20.0 | 0 | 0 |
| 20.1+ | 0 | 0 |
| Total | 8.5 | 586 |

**OTHER MEASUREMENTS**

| | | |
|---|---|---|
| Average residue depth | (feet) | 0.2 |
| Ground area covered by residue 1/4-inch diameter and larger | (percent) | 43 |
| Average duff and litter depth | (inches) | 0.8 |
| Sound residue 3.1-inch diameter and larger ponderosa pine | (percent) | 100 |
| ______ | (percent) | |
| ______ | (percent) | |
| Rotted residue 3.1-inch diameter and larger | (percent) | 0 |

**HARVEST INFORMATION**

| | |
|---|---|
| Gross volume cruised (M fbm/acre) | 3.6 |
| Net volume cruised (M fbm/acre) | 3.4 |
| Average stems/acre cut | 17 |
| Average d.b.h. of stems cut (inches) | 16 |
| Stand age (years) | 110 |
| Cutting prescription | Tree selection |
| Yarding method | Tractor |
| Slash treatment | None |
| Period since cut or treatment (months) | <12 |

**PRECOMMERCIAL THINNING INFORMATION**

| | |
|---|---|
| Stems cut/acre | |
| Stems remaining/acre | |
| Basal area/acre before | |
| Basal area/acre after | |
| Average d.b.h. before (inches) | |
| Average d.b.h. after (inches) | |
| Thinning method | |
| Slash treatment | |

**FUEL RATING**

U.S. Forest Service Region 6 fuel type identification: ML

**REMARKS**

*Figure 3-19. Sample photo series. Photo shows selectively logged ponderosa pine in the Pacific Northwest. From Maxwell and Ward (1976).*

Other keys, a residue of fuel typing, are still published in many fireline handbooks.

The *photo series* can match visual scenes directly to fuel models or it can provide an intermediate form of fuel modeling. The photo series may publish its own fuel data—a kind of synoptic fuel inventory. This raw data may be entered directly into fire models, or additional handbooks can be consulted that give important fire behavior properties for a particular entry in the photo series (Figures 3–19 and 3–20).

Since all fuel models relate to the same fuel complexes, it is possible to correlate the various fuel models—each of which emphasizes particular fuel data—among themselves. The NFDRS and NFFL fuel models have been so correlated (Albini, 1976). For essentially the identical fuel complex, the best model, or suite of models, for the purposes at hand may be selected. Ensembles of models usually provide the best fit. Much as heterogeneous fuel cells, with suitable weighting, can be combined into a single fuelbed, so groups of fuel models may be combined to better approximate the actual fuel complex. One model may best describe rate of spread, and another, intensity. Two or even three models may be mixed, in some proportion, to obtain the fire behavior predictions that best fit the fire. Since fuel mosaics usually result from biotas of different ages, the effect is to substitute geography for history, to

| FUEL MOISTURE | RATE OF SPREAD: EFFECTIVE MIDFLAME WIND (MI/H) | | | | | | | | | FLAME LENGTH: EFFECTIVE MIDFLAME WIND (MI/H) | | | | | | | | |
|---|---|---|---|---|---|---|---|---|---|---|---|---|---|---|---|---|---|---|
| | 0 | 2 | 4 | 6 | 8 | 10 | 12 | 14 | 16 | 0 | 2 | 4 | 6 | 8 | 10 | 12 | 14 | 16 |
| PERCENT | CHAINS PER HOUR | | | | | | | | | FEET | | | | | | | | |
| 2 | 1 | 4 | 9 | 14 | 19 | 25 | 31 | 41 | 58 | 2 | 4 | 5 | 7 | 8 | 9 | 10 | 11 | 13 |
| 3 | 1 | 4 | 8 | 12 | 17 | 22 | 27 | 36 | 51 | 2 | 4 | 5 | 6 | 7 | 8 | 9 | 10 | 12 |
| 4 | 1 | 3 | 7 | 11 | 15 | 20 | 24 | 32 | 46 | 2 | 3 | 5 | 6 | 7 | 7 | 8 | 9 | 11 |
| 5 | 1 | 3 | 6 | 10 | 14 | 18 | 22 | 29 | 41 | 2 | 3 | 4 | 5 | 6 | 7 | 8 | 9 | 10 |
| 6 | 1 | 3 | 6 | 9 | 12 | 16 | 20 | 27 | 38 | 1 | 3 | 4 | 5 | 6 | 6 | 7 | 8 | 9 |
| 7 | 1 | 3 | 5 | 8 | 11 | 15 | 19 | 25 | 35 | 1 | 3 | 4 | 4 | 5 | 6 | 6 | 7 | 9 |
| 8 | 1 | 2 | 5 | 8 | 11 | 14 | 18 | 23 | 33 | 1 | 2 | 3 | 4 | 5 | 5 | 6 | 7 | 8 |
| 9 | 1 | 2 | 5 | 7 | 10 | 13 | 17 | 22 | 31 | 1 | 2 | 3 | 4 | 4 | 5 | 6 | 6 | 7 |
| 10 | 1 | 2 | 4 | 7 | 10 | 13 | 16 | 21 | 30 | 1 | 2 | 3 | 4 | 4 | 5 | 5 | 6 | 7 |
| 11 | 0 | 2 | 4 | 7 | 9 | 12 | 15 | 20 | 26 | 1 | 2 | 3 | 3 | 4 | 4 | 5 | 5 | 6 |
| 12 | 0 | 2 | 4 | 6 | 9 | 12 | 15 | 19 | 27 | 1 | 2 | 3 | 3 | 4 | 4 | 5 | 5 | 6 |
| 13 | 0 | 2 | 4 | 6 | 9 | 11 | 14 | 18 | 26 | 1 | 2 | 2 | 3 | 3 | 4 | 4 | 5 | 6 |
| 14 | 0 | 2 | 4 | 6 | 8 | 11 | 13 | 18 | 25 | 1 | 2 | 2 | 3 | 3 | 4 | 4 | 4 | 5 |
| 15 | 0 | 2 | 4 | 6 | 8 | 10 | 13 | 17 | 24 | 1 | 1 | 2 | 2 | 3 | 3 | 4 | 4 | 5 |
| 16 | 0 | 2 | 3 | 5 | 8 | 10 | 12 | 16 | 23 | 1 | 1 | 2 | 2 | 3 | 3 | 3 | 4 | 4 |
| 17 | 0 | 2 | 3 | 5 | 7 | 9 | 12 | 15 | 22 | 1 | 1 | 2 | 2 | 2 | 3 | 3 | 3 | 4 |
| 18 | 0 | 1 | 3 | 5 | 7 | 9 | 11 | 14 | 20 | 1 | 1 | 2 | 2 | 2 | 2 | 3 | 3 | 4 |
| 19 | 0 | 1 | 3 | 4 | 6 | 8 | 10 | 13 | 18 | 0 | 1 | 1 | 2 | 2 | 2 | 2 | 3 | 3 |
| 20 | 0 | 1 | 2 | 4 | 5 | 7 | 9 | 11 | 16 | 0 | 1 | 1 | 1 | 2 | 2 | 2 | 2 | 3 |
| 21 | 0 | 1 | 2 | 3 | 4 | 6 | 7 | 10 | 14 | 0 | 1 | 1 | 1 | 1 | 2 | 2 | 2 | 2 |
| 22 | 0 | 1 | 2 | 3 | 4 | 5 | 6 | 8 | 11 | 0 | 1 | 1 | 1 | 1 | 1 | 1 | 2 | 2 |
| 23 | 0 | 1 | 1 | 2 | 2 | 3 | 4 | 5 | 8 | 0 | 0 | 1 | 1 | 1 | 1 | 1 | 1 | 1 |
| 24 | 0 | 0 | 1 | 1 | 1 | 2 | 2 | 3 | 4 | 0 | 0 | 0 | 0 | 0 | 0 | 1 | 1 | 1 |

*Figure 3-20. Fire behavior and control information calculated for the fuel shown in Figure 3-19. The fuel parameters were processed through the Rothermel fire model. From Ward and Sandberg (1981).*

describe historical changes with spatial ensembles of stylized models. Even when the fuel complex shows large changes over time, it is more common to select one stylized fuel model after another, at appropriate intervals, than to develop dynamic fuel models that describe these evolutions internally.

The selection of a fuel model is part of the appraisal process. How detailed the model need be will depend on the uses to which it is put, and the comparative costs of getting high-resolution data (Radloff et al., 1982). Perfect knowledge of a system can only come by destroying that system —the act of measurement will alter the complex it assesses. Inevitably forms of sampling are used; some approximations are necessary. When the information gained exceeds the information lost, or when the value of more refined data drops relative to management needs and other uncertainties in the system, then the best level of fuel data has been reached. Proper ensembles of existing fuel models are probably adequate for most situations. Appeal to published photo series can add another level of refinement to areas which show great variability in their fuel mosaic. And for selected high-value areas, special site-specific inventories may be justified. For the present, however, greater uncertainties are associated with weather data than with fuels data; and high-intensity fire behavior (which presents the greatest management problem) is not yet modeled with the sophistication achieved for the spread of steady-state surface fires.

## 3.4 SELECTED FUEL COMPLEXES: FIRE CHARACTERISTICS

The effect of fuels on a fire, and of fire on fuels, can be summarized in terms of the three principal strata of fuels: ground, surface, and aerial. The surface stratum is the most fundamental, and it is customarily subdivided into four major biotic groupings: grass, brush, forest, and slash. Each of these groups may be characterized by fuel models, each of which differentiates the fuel complex on the basis of certain structural properties, such as its size class distribution. Disregarding such environmental variables as fuel moisture and wind, the behavioral characteristics of fires will differ according to these structural properties. For grass fuels, nearly all of the surface fuel is available fuel; for brush, from 5 to 95%; for forest fuels, 5 to 25%; for slash, from 10 to 70% (Figures 3–21). Nuances in the particular complex and variables like fuel moisture will determine the actual quantity consumed.

The relationship between fire and fuel has many meanings. Wildfire feeds on fuels, but so does prescribed fire. One way to control wildfire is to modify its fuels, either around its perimeter or over its entire area. Ecological manipulation by fire can only come through the fuel complex—making fire, like fuel, a biological phenomenon. Changing fuel properties is one method of changing fire properties, and the use of fire for the purposes of fuel modification means that the fuel complex will be affected in certain ways. By way of example, consider the mutual interaction of fires and fuels for the four major biotic groupings. For each group the principal stratum of the fuel complex is the surface fuelbed, but its parameters, and its relationship to other fuelbeds, is different in each case. Figure 3–18 summarizes the fuel information parameters for each group as synopsized for the NFFL fuel models.

### *Grass*

Grass may exist by itself or in combination with brush and timber. Wherever it flourishes, however, it becomes the primary fuel for carrying surface fires. Grasslands consist exclusively of 1-hr TL fuels; their principal structural parameters are fuel depth (grass height), fuel loading, and the ratio of living to dead fuels. The live-to-dead fuel ratio varies annually, of course, and most fires occur either before spring greenup or after fall curing. The moisture of extinction is also directly related to fuel depth, short grasses have low values (12%) and tall grasses have higher values

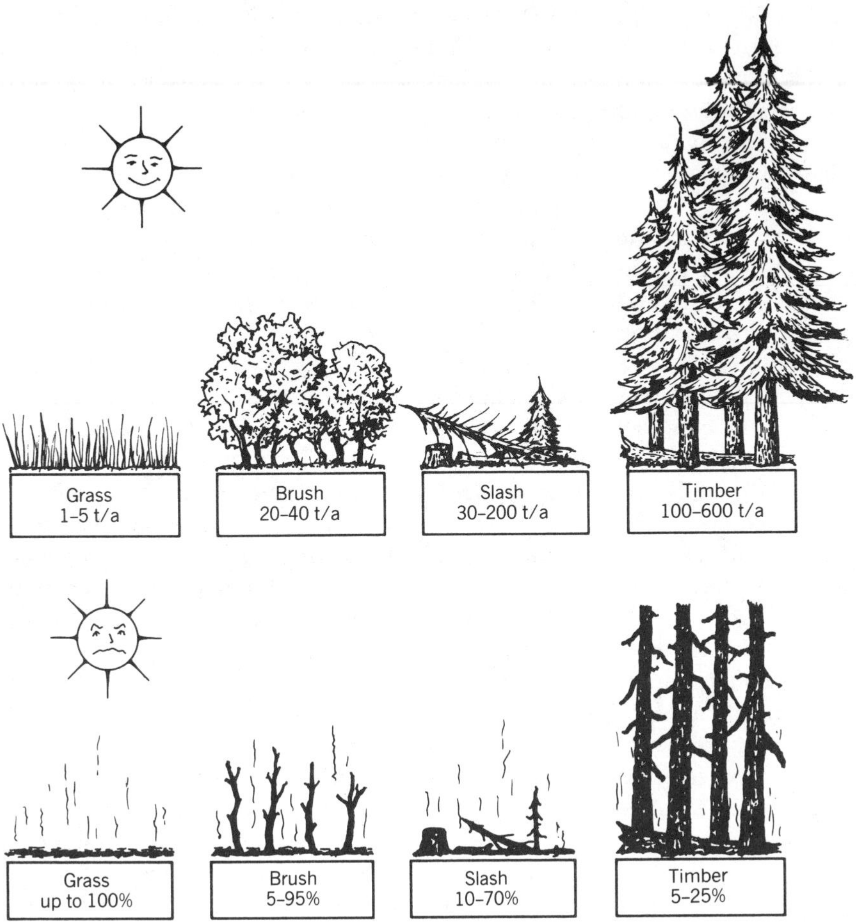


*Figure 3-21. Fuel loading and fuel availability. From NWCG, S390.*

(25%). Its small diameter fuels, high surface-to-volume ratios, and low bulk density make grasses quickly responsive to heat and moisture transfer; grass fires can show high fireline intensities as a result of high rates of spread. Flame height and velocity will vary with fuel depth and loading, the one increasing steadily with the other (Figure 3-22). For large, dense grasses, flames may soar to 20–30 ft (6.1–9.1 m) and, running before strong winds, may exhibit even greater flame lengths.

Fire effects vary with the intensity of the fire, but the tendency is to consume virtually all surface fuels. Even with short residence times, the flaming front may clean out nearly all standing fuels, leaving perhaps only residual ground clumps to smolder. The grass returns with the next growing season and the burn may be repeated. Without some form of removal, however, the productivity of the grass drops off sharply. Fuel loading does not increase at a steady rate year by year; most of the fuel load is restored in the year following a burn. Where grassy fuels are burned or cropped for fuel management purposes, fire is applied on a cycle of one to three years.

### *Brush*

Brush includes a wide range of fuels, from blueberries to the southern rough to chaparral. All have in common that they enlarge the depth of

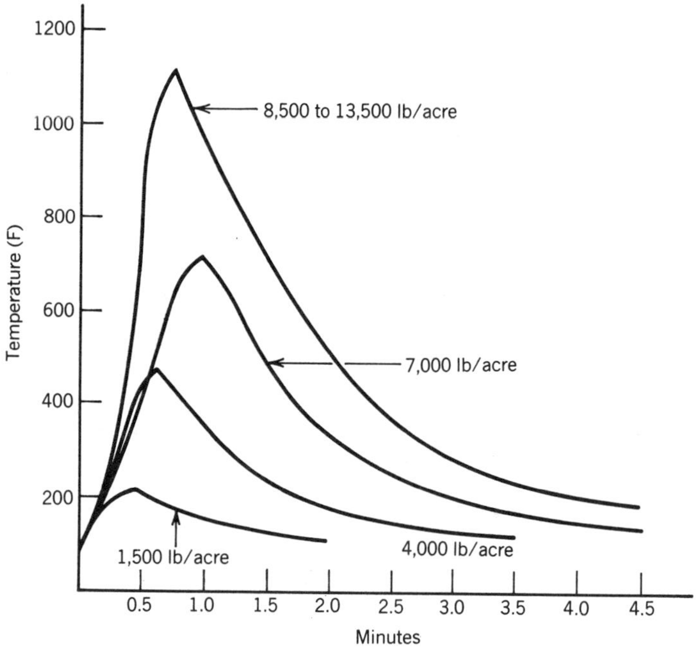


*Figure 3-22. Fuel loading and grass fires. Virtually all fuels are available fuels, so fire intensity rises accordingly. From Wright and Bailey (1982).*

the surface fuelbed and introduce a chemistry of flammable live fuels to compound with surface dead fuels. The fuels introduced as brush are small, loosely compacted, and available. The results can be a significant increase in fireline intensity and flame height, with further escalation into a stratum of aerial fuels should a forest canopy be present. Despite its continuity with surface fuels, high brush burns with many of the characteristics of a crown fire. Fire effects vary with fire intensity, and the discrete nature of chamise fuels makes it possible for pockets of fuel to escape burning; fire spreads from shrub to shrub rather than along a continuous fuel stratum punctuated by flareups. For high intensity fires, however, consumption is total.

Consider the case for chamise chaparral. For a fuel complex dominated by 6-ft (1.8 m) chamise, there are fuels in the 1-hr, 10-hr, and 100-hr TL fuel classes, with heavy loadings for the 1-hr and 10-hr TL fuels. The surface-to-volume ratio is high, the packing ratios low, and the depth large. The mixture of live to dead fuels is favorable, with the individual plant increasing the proportion of dead branches to live within its crown over time. The live fuels, moreover, have a high proportion of extractives, ebbing and flowing on an annual cycle. At 20% the moisture of extinction is relatively low. Under proper wind and fuel moisture conditions, the chamise chaparral biota is notorious for high fireline intensities and high rates of spread.

Following a fire, the plant resprouts from its roots. The buildup of live and dead fuel for a particular plant is predictable, and dynamic fuel models can be constructed for biotas dominated by chamise (Figure 3–23). Initially, the live fuels build up rapidly, then slow down. Conversely, the dead fuels accumulate slowly at first, then accelerate. By 15 years of age, only about 15% of chamise is dead; by 30, over 20%; by 50, over 50%. With further aging dead fuels drop to the ground and provide at least a spotty layer of dead surface litter. Decomposition by means other than fire is negligible.

Because of this peculiar fuel cycle, it is possible to speak of a fire cycle of sorts. The flammability of chamise increases with age. Prior to 15 years of

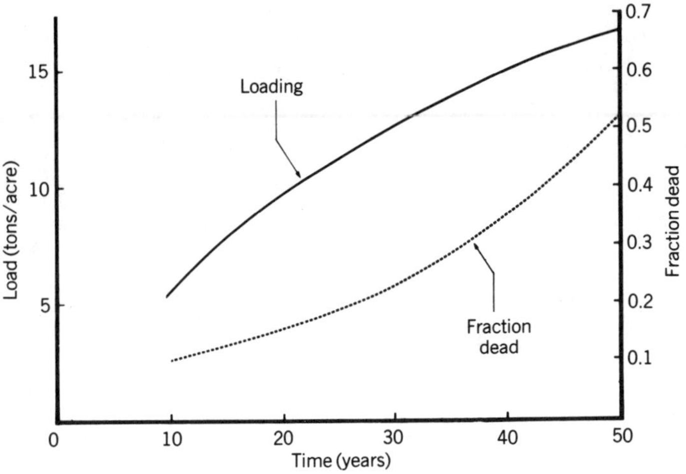


*Figure 3-23. Fuel loading and chaparral fires. The increase in fuel load is a function of time. The opposite graph shows the rising fraction of the total load that is dead. The lower left graph shows the changes in rate of spread that occur over a season, adjusted for different ages of chamise. The lower right graph shows changes in rate of spread for mixed chappral as a product of aging and windspeed. From Philpot (1971) and (1977).*

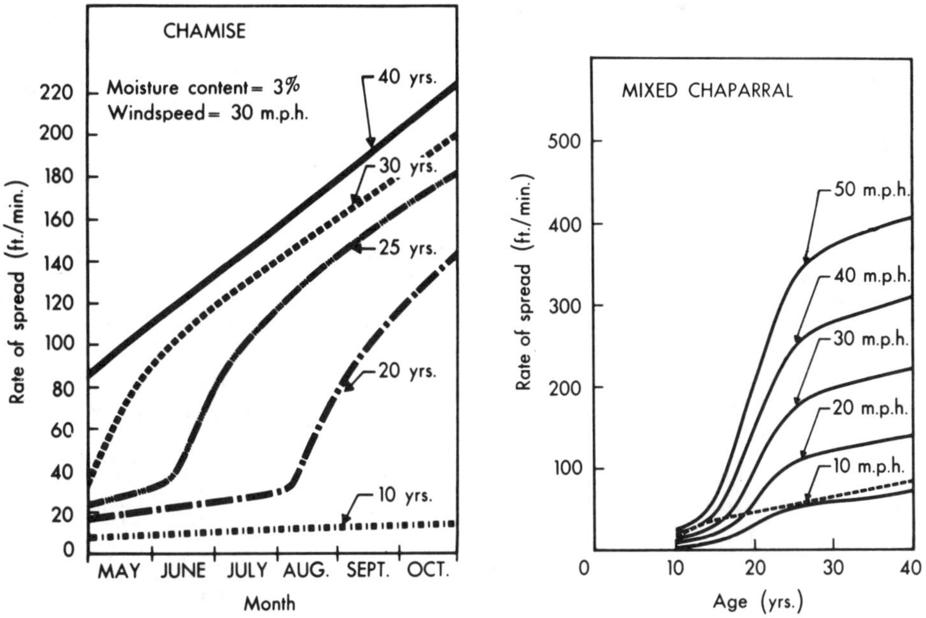


age, it will not burn; after 25 years it can burn readily. Nearly all large fires occur in the older fuels, and the prospect of fire control by fuel management is very attractive. Of the chamise chapparal burned on the Angeles National Forest since 1950, 1% was in the 0–10-year age class, 2% in the 10–19-year class, 4% in the 20–29-year class, 12% in the 30–39-year class, and 80% in the over-40-year class (Minnich, 1983). It is likely that broadcast fire will be employed in some capacity in any large-scale program of fuel reduction.

### *Forest*

Forest fuel complexes differ from pure grass complexes in having a mixture of fuel size classes, with proportionately heavy loads in the larger size classes. Some have a grassy or brushy understory; of special interest are those that concentrate their fine fuel particles into litter and duff. Of its total quantity of fine fuels, only the litter and a portion of the duff—its crust—is available fuel. Most of the surface litter consists of small diameter branchwood. Total fuel depth is typically shallow; much of the total fuel load is either buried or locked up inside large dead boles; burning tends to be heterogeneous rather than concentrated into the flaming front. Glowing and residual combustion will be common in duff and large logs. Rate of spread is typically less than for grass and brush, and the shielding effect of the forest further reduces effective wind velocity; fireline intensity will depend more on heat release/area. Forest fires thus burn in complex ways, a function of their mixed fuels.

The variability in fuels matches the variability in fires, and together they make an accurate determination of mutual effects difficult. Most surface fires in forest fuels effectively reduce the 1-hr and 10-hr TL fuels, and make headway against any duff and large diameter logs that are present. The flaming front consumes the former, and residual combustion, the latter. The Rothermel model and the fuel models that feed into it normally consider the 1-hr and 10-hr TL fuels; for a few fuel complexes, for which there exists considerable deposition of branchwood, the 100-hr TL fuels enter into the reaction. Wildfire tends to consume more fuel along the flaming front than does prescribed fire, but it also produces more dead fuels by killing the live fuels; the net result is often ambiguous. In general, fuel history shows two periods of excess accumulation: early and late in the life cycle, though even this generalization assumes a single initiating event when, in fact, the events are multiple.

Consider, for example, two fires in the ponderosa pine fuel type of Arizona. A wildfire (Coconino National Forest) reduced litter fuels from 15 to under 1 ton/acre (5.5–0.4 metric tons/hectares; 90–100%) reduction and woody fuels from 18 to 4.5 tons/acre (6.6–1.7 metric tons/hectares; 75%). A prescribed burn nearby reduced litter fuels from an average of 15 tons/acre (5.5 metric tons/hectares) to 5.5 tons/acre (2 metric tons/hectares; 63%) and woody fuels from 8.2 tons/acre (3 metric tons/hectares) to 2.8 tons/acre (1 metric ton/hectares; 66%). Most of the reduction came through residual burning. Differences in their initial fuel loads make a comparison between the two situations tricky, but the trend is typical. Reduction on the order of 78% for particles under 3 in. (7.6 cm) is a good average (Harrington, 1981).

Surface fuels build up in sufficient quantities to support fire after about five to seven years. The possibility of controlling this accumulation through prescribed broadcast fire is excellent, though the great variability in actual fuel complexes makes the construction of precise, general prescriptions perplexing. Fuel managment projects, moreover, tend to differentiate between the periodic reduction of surface litter and the occasional reduction of large diameter fuels (over 1000-hr TL) and deep duffs. Large fuels are eliminated by burning at times when small fuels will not become involved and spread will be slight; fine fuels, by broadcast burning at times when the large fuels and deep duffs are unlikely to do more than scorch and residual burning will not be a problem.

### *Slash*

*Activity fuels,* commonly known as *slash,* can assume several forms. They may be left in situ, shaped into piles or windrows, or mechanically crushed or chipped. Their structural properties will vary accordingly. These properties are subject to direct manipulation. As a rule, slash shows high proportions of large diameter fuels and, so far as needles remain on branches, of fine fuels. The mixture can be explosive. A large slash field shows many of the structural properties of brushfields, with the addition of very large (greater than 1000-hr TL) fuels. To the degree that fine fuels remain available, fire may spread rapidly; to the extent that the larger fuels become available, mass fire is possible. Slash fuels are treacherous

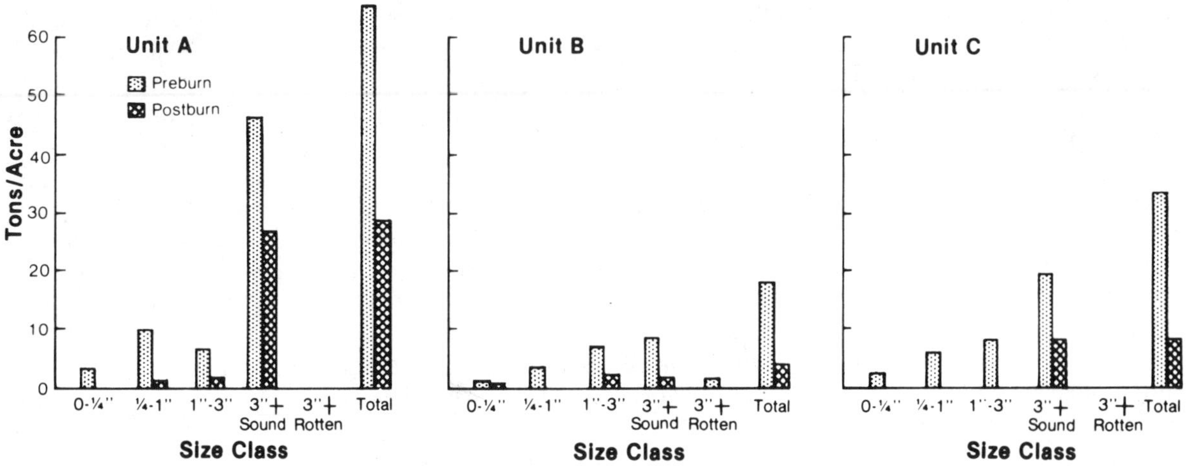


*Figure 3-24. Slash fuel consumption by size class for prescribed burns. From Zimmerman (1982).*

sites for wildfire. Considerable efforts will be taken to prevent fires from entering them, including hazard reduction by both mechanical means and fire. Where fuel reduction calls for fire, the general strategy is to create fires with high reaction intensities, even to the point of stimulating miniature mass fires. This improves the consumption of larger diameter fuels while simultaneously assisting in control over fire spread through the development of a convective column. Sample patterns of fuel consumption are given in Figure 3–24, based on burns in larch–Douglas-fir clearcut slash.

The accumulation of slash is practically instantaneous. Its decomposition rates will largely depend on its fire history. For the first couple of years the structural properties of in situ or piled slash make it extremely hazardous, diminishing only with suitable weather or fuel moistures. But then chemical changes transform fine fuels, such as needles, from volatile live fuels to less volatile dead fuels, and mechanical breakage reduces the availability of needles and branchwood by depositing them on the ground. Concurrently, large live boles will cure into large dead fuels, though their contribution to fire spread will be far less than their total fuel load might suggest. Gradually, the sensitivity of the site to fire decreases, and fire effects on it diminish. The site becomes less vulnerable to wildfire, but less suitable for prescribed fire. For silvicultural preparation, the large diameter fuels must be eliminated. Prescribed fire must be introduced relatively early into the slash fuel cycle to be effective, and its effectiveness must be enhanced by measures to prepare the fuel.

### *Duff*

Surface fuels grade downward into ground fuels, much as they grade upward into aerial fuels. This lower, subsurface stratum of fuels is known as *duff.* Typically, it does not interact with the flaming front, though duff fires are coupled in various ways to the surface fire. Considered somewhat differently, duff is sometimes taken to include all the dead fuels resting on the ground. In this case, the upper layer that contains surface fuels is known as the L-layer (litter); the remaining duff is divided between the F-layer (fermentation) and H-layer (humus) which underlies it. All fires require surface fuels. A few fires, like those in chamise, burn primarily through surface fuels that elevate some distance from the ground and show some properties of aerial fuels; a few others, like peat fires, burn along the surface of organic soils having all the properties of the H-layer of duff. But if duff rarely contributes to the flaming front, it adds measurably to the total heat production and fuel consumption of the fire through residual combustion, and it has considerable biological importance.

By its nature duff shows little internal structure. Its measurements are depth and load, and the chief determinant of fire consumption is the fuel moisture of the H-layer (Figure 3-25). This applies to all categories of duff, from shallow strips under ponderosa pine forests to thick deposits of muck. The function of the surface fire is to ignite the duff, expose it to oxygen, and dry it out. Depending on fuel moisture, duff fire and surface fire may be linked in three ways. Above 120%, the moisture of extinction, duff refuses to burn regardless of the character of the surface fire. Below 30% or so, duff can burn independently of the surface fire, though most of this burning will take place after passage of the flaming front. Between these fuel moistures, however, the amount of duff consumed will vary with the intensity of the surface fire. The two tend to burn together, and there exists a correlation between the amount of duff consumed and the total consumption of 1, 10, and 100-hr TL fuels on the surface. In this case, the function of the surface fire is to dry out and reignite duff, a kind of transient ignition process. For peat or muck fires, this pattern may become complex. The initial wave of surface fire may dry out fuels beneath it, support a second wave of new surface fire, and so on until a level is reached at which the moisture of extinction persists (Sandberg, 1980).

In duff, moisture levels control burning levels. For the suppression of duff fires, heavy mixing with dirt and water (or in the case of muck fires, flooding) is the only means of extinguishment. From the standpoint of prescribed burning, the depth of the duff and the fuel moisture of its lower levels will dictate how long the fire will smolder and smoke and how much fuel will be

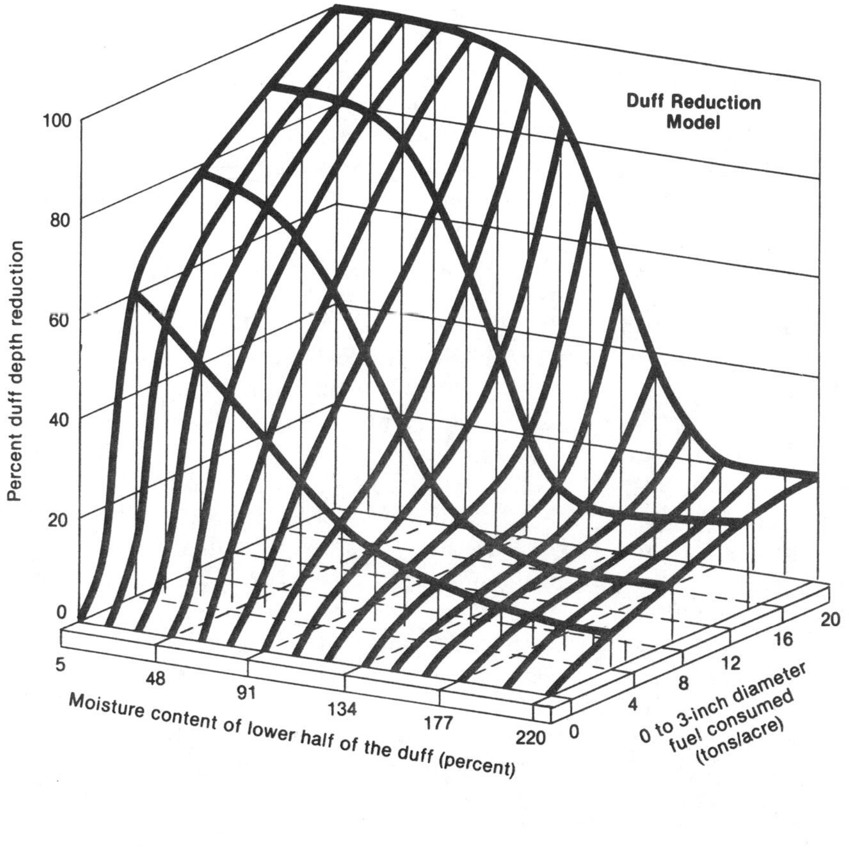


*Figure 3-25. Duff consumption by fire. The graph records duff consumption (%), as it varies according to moisture content (lower half of the duff) and to the consumption of woody surface fuels (under 1000-hr TL size). From Martin et al (1979).*

reduced. In most wildfires, duff insulates soil from the heat of combustion and protects it against undue erosion after the surface litter has been stripped away.

### *Crown*

Like duff fires, crown fires relate to surface fires in various ways. But unlike duff, crown fuels show a structure and chemistry that greatly influence the properties of a crown fire. To recapitulate, crown fires may be independent, propagating without contributions of gases or heat from the surface fire; active, spreading with some help from the surface fire; or passive, little more than an enlargement of the flaming zone of the surface fire. Which form of spread a crown fire shows will depend on certain structural properties of the fuel stratum (fuel load and bulk density) and on chemical properties of the particles (foliage chemistry and fuel moisture). For active or independent crown fires to develop, the crown fuel stratum must be dense, continuous, and live, and its fuel moisture low. This restricts crown fires, practically speaking, to early spring or fall; to certain tree species or fuel complexes such as jack pine or forests with a slash or dense brush understory; and to particular stages in fuel history, notably youth or early maturity. To these preconditions must be added the presence of an intense surface fire.

Crown fires produce large flaming fronts, and fuel consumption within the front is complete for fine (1-hr and 10-hr TL) fuels. The net effect, however, is an increase in fuel load by the destruction of large stands of trees. Live fuels instantaneously become dead fuels. The probability of reburn following reburn until the fuel complex has little fuel left is a real one. Especially as reproduction carpets the forest floor with a new stratum of surface fuels, a cycle of reburning is likely. The parameters of this cycle are not known. There is a gradual decomposition of the fuel complex as fuels break and fall, and as fire sweeps across the surface. But for the most part such sites, with intensive human intervention, gradually recompose. Fire control breaks the fire cycle; fuel reduction through logging and fuel rearrangement through felling and crushing accelerate decomposition; replanting and reseeding speed up the succession of large dead fuels to small live fuels.

## *REFERENCES*

An excellent starting point for fuel moisture is Schroeder and Buck (1970), *Fire Weather.* Simpler discussions are given in National Wildfire Coordinating Courses, especially S390, Fire Behavior. Detailed mathematical analysis is available in Fosberg. For the incorporation of FMC within fuel models, see Deeming et al. —which describes the application of fuel moisture models for fire danger rating. For a concise overview of live fuel moisture, see Burgan (1979), "Estimating Life Fuel Moisture for the 1978 National Fire Danger Rating System." For drought measurements, see Fosberg et al. (1981), "Moisture Content Calculations for 1000-hr Timelag Fuels"; Keetch and Byram (1968), "A Drought Index for Forest Fire Control"; Palmer (1965), "Meteorological Drought"; Haines et al. (1975), "An Assessment of Three Measures of Longterm Moisture Deficiency Before Critical Fire Periods."

For a superb overview of fuel models, complete with a capsule history of fuel classification schemas, see Anderson (1982), "Aids to Determine Fuel Models for Estimating Fire Behavior." For the NFDRS models, see Deeming and Brown (1975), "Fuel Models in the National Fire Danger Rating System," and Deeming et al., "National Fire Danger Rating System." For the NFFL models, see Albini (1976), "Estimating Wildfire Behavior and Effects," and Rothermel (1972), "A Mathematical Model for Fire Spread Predictions in Wildland Fuels." For gradient modeling as a technique, see Kessel (1976), "Gradient Modeling: A New Approach to Fire Modeling and Wilderness Resource Management," and (1979), *Gradient Modeling.* Fuel keys are described in Fahnestock (1970), "Two Keys for Appraising Forest Fire Fuels". The technique of the photo series is given in Fischer (1981), "Photo Guides for Appraising Forest Fire Fuels," and Ward and

Sandberg (1981), Maxwell and Ward, Blonski and Schramel (1981), Koski and Fischer (1979) provide actual, published series. The modeling of nonuniform surface fuelbeds is described in Frandsen and Andrews (1979), "Fire Behavior in Nonuniform Fuels"; of the dynamic slash model in Albini and Brown (1978), "Predicting Slash Depth for Fire Modeling"; of the palmetto–gallberry model in Hough and Albini (1978), "Predicting Fire Behavior in Palmetto–Gallberry Fuel Complexes"; of chamise chaparral in Rothermel and Philpot (1973), "Fire in Wildland Management: Predicting Changes in Chaparral Flammability."

Fuel inventory techniques are summarized in Brown (1974), "Handbook for Inventorying Downed Woody Material," and reviewed in Freeman (1981), "Inventory and Prediction of Forest Fuels." National programs for the classification, inventory, and appraisal of forest fuels, based on these techniques, are described in Bevins and Roussopoulos (1980), "National Fuels Inventory Library"; Radloff et al. (1982), "Coping with Uncertainty in Fuel Management Decision"; Roussopoulos (1980), "Improving Wildland Fuel Information for Widlfire Decisionmaking"; Hirsch et al. (1979), "Choosing an Activity Fuel Treatment for Southwest Ponderosa Pine," and (1981), "The Activity Fuel Appraisal Process: Instructions and Examples."

A summary of fire and fuel relationships is provided in Martin et al. (1979), "Effects of Fire on Fuels. A State-of-Knowledge Review." The digest synopsizes existing fuel models, fuel classification schemas, and fuel inventories as they were known at the time.

For biological controls on fuels, see the references in Chapter 5. For the relationship of fuel to combustion and fire behavior, consult the references in Chapters 1 and 2. For techniques and programs of fuels management, see Chapter 8.

Albini, Frank A., 1976, "Estimating Wildfire Behavior and Effects," U.S. Forest Service, General Technical Report INT-30.

Albini, Frank A. and James K. Brown, 1978, "Predicting Slash Depth for Fire Modeling," U.S. Forest Service, Research Paper INT-206.

Anderson, Hal E., 1982, "Aids to Determine Fuel Models for Estimating Fire Behavior," U.S. Forest Service, General Technical Report INT-122.

Barney, Richard J. et al., 1981, "Forest Floor Fuel Loads, Depths, and Bulk Densities in Four Interior Alaskan Cover Types," U.S. Forest Service, Research Note INT-304.

Bevins, Collin D. and Peter Roussopoulos, 1980, "National Fuels Inventory Library," pp. 146–150, in *Proceedings, Sixth Conference on Fire and Forest Meteorology;* American Meteorological Society and Society of American Foresters.

Blonski, Kenneth S. and John L. Schramel, 1981, "Photo Series for Quantifying Natural Forest Residues: Southern Cascades, Northern Sierra Nevada," U.S. Forest Service, General Technical Report PSW-56.

Brackebusch, Arthur P., 1975, "Gain and Loss of Moisture in Large Forest Fuels," U.S. Forest Service, Research Paper INT-173.

Brown, A.A. and Kenneth Davis, 1973, *Forest Fire: Control and Use,* 2nd ed. (New York: McGraw-Hill).

Brown, James K., 1966, "Forest Floor Fuels in Red and Jack Pine Stands," U.S. Forest Service, Research Note NC-9.

———, 1970, "Ratios of Surface Area to Volume for Common Fine Fuels," *Forest Science* 16(1): 101–105.

———, 1974, "Handbook for Inventorying Downed Woody Material," U.S. Forest Service, General Technical Report INT-16.

———, 1978, "Weight and Density of Crowns of Rocky Mountain Conifers," U.S. Forest Service, Research Paper INT-197.

———, 1981, "Bulk Densities of Nonuniform Surface Fuels and Their Application to Fire Modeling," *Forest Science* 27(4): 667–683.

Brown, James K. et al., 1977, "Handbook for Predicting Slash Weight of Western Conifers," U.S. Forest Service, General Technical Report INT-37.

———, 1982, "Handbook for Inventorying Surface Fuels and Biomass in the Interior West," U.S. Forest Service, General Technical Report, INT-129.

Brown, James K. and Michael A. Marsden, 1976, "Estimating Fuel Weights of Grasses, Forbs, and Small Woody Plants," U.S. Forest Service, Research Note INT-210.

Brown, James K. and Peter J. Roussopoulos, 1974, "Eliminating Biases in the Planar Intersect Method for Estimating Volumes of Small Fuels," *Forest Science* **20**: 350–356.

Brown, James K. and Thomas E. See, 1981, "Downed Dead Woody Fuel and Biomass in the Northern Rocky Mountains," U.S. Forest Service, General Technical Report INT-117.

Burgan, Robert E., 1979, "Estimating Live Fuel Moisture for the 1978 National Fire Danger Rating System," U.S. Forest Service, Research Paper INT-226.

Cooper, Robert W., 1976, "Tradeoffs Between Smoke from Wild and Prescribed Forest Fires," in National Academy of

Sciences, *Air Quality and Smoke from Urban and Forest Fires. Proceedings of International Symposium* (Washington D.C.: National Academy of Sciences).

Cramer, Owen (ed.), 1974, "Environmental Effects of Forest Residues Management in the Pacific Northwest. A State-of-Knowledge Compendium," U.S. Forest Service, General Technical Report PNW-24.

Deeming, John E. et al., 1972, "National Fire-Danger Rating System," U.S. Forest Service, Research Paper RM-84.

———, 1974, "National Fire-Danger Rating, U.S. Forest Service Research Paper RM-84 revised.

———, 1978, "The National Fire-Danger Rating System—1978," U.S. Forest Service, General Technical Report INT-39.

Deeming, John E. and James K. Brown, 1975, "Fuel Models in the National Fire Danger Rating System," *Journal of Forestry* **73**: 347–350.

Fahnestock, George R., 1970, "Two Keys for Appraising Forest Fire Fuels," U.S. Forest Service, Research Paper PNW-99.

Fischer, William C., 1981, "Photo Guides for Appraising Downed Woody Fuels in Montana Forests: How They Were Made," U.S. Forest Service, Research Note INT-299.

Fosberg, Michael A., 1970, "Drying Rates of Hardwood Below Fiber Saturation," *Forest Science* **16**: 57–63.

———, 1971, "Climatological Influenc on Dead Fuel Moisture: A Theoretical Analysis," *Forest Science* **17**: 64–72.

———, 1972, "Theory of Precipitation Effects on Dead Fuel Moisture," *Forest Science* **18**: 98–108.

———, 1977, "Forecasting the 10-Hour Timelag Fuel Moisture," U.S. Forest Service, Research Paper RM-187. U.S. Forest Service, Research Paper RM-84, rev.

Frandsen, William H. and Patricia L. Andrews, 1979, "Fire Behavior in Nonuniform Fuels," U.S. Forest Service, Research Paper INT-232; *Forest Science* **17**: 64–72.

Freeman, Duane R., 1981, "Inventory and Prediction of Forest Fuels," paper presented at "In-Place Resource Inventories: A National Workshop," University of Maine, 1981.

———, 1982, "Instructor's Guide for the Fuel Management Training Series: An Overview," U.S. Forest Service, Rocky Mountain Forest and Range Experiment Station.

Fosberg, Michael A. et al., 1981, "Moisture Content Calculations for 1000-hr Timelag Fuels," *Forest Science* **27**(1): 19–26.

Haines, Donald A. and William A. Main, 1978, "Variation of Six Measures of Fire Activity Associated with Droughts," pp. 5–7, in *Fifth National Conference on Fire and Forest Meteorology,* American Meteorological Society.

Haines, Donald A. et al., 1975, "An Assessment of Three Measures of Longterm Moisture Deficiency Before Critical Fire Periods," U.S. Forest Service, Research Paper NC-131.

Harrington, Michael G., 1981, "Preliminary Burning Prescriptions for Ponderosa Pine Fuel Reductions in Southeastern Arizona," U.S. Forest Service, Research Note RM-402.

Hirsch, Stanley N. et al., 1979, "Choosing an Activity Fuel Treatment for Southwest Ponderosa Pine," U.S. Forest Service, General Technical Report RM-67.

Hirsch, Stanley N. et al., 1981, "The Activity Fuel Appraisal Process: Instructions and Examples," U.S. Forest Service, General Technical Report RM-83.

Hornby, Lloyd G., 1936, "Fire Control Planning in the Nothern Rocky Mountain Region (Missoula, Montana: U.S. Forest Service).

Hough, W.A. and F.A. Albini, 1978, "Predicting Fire Behavior in Palmetto–Gallberry Fuel Complexes," U.S. Forest Service, Research Paper SE-174.

Howard, E.A., III, 1978, "A Simple Model for Estimating the Moisture Content of Living Vegetation as Potential Wildland Fuel," pp. 20–23, in *Fifth National Conference on Fire and Forest Meteorology,* American Meteorological Society.

Keetch, John J. and George M. Byram, 1968, "A Drought Index for Forest Fire Control," U.S. Forest Service, Research Paper SE-38.

Kessel, Stephen R., 1976, "Gradient Modeling: A New Approach to Fire Modeling and Wilderness Resource Management," *Environmental Management* **1**(1):39–48.

———, 1979, *Gradient Modeling. Resource and Fire Management* (New York: Springer-Verlag).

Kessell, Stephen R. et al., 1978, "Analysis and Application of Forest Fuels Data," *Environmental Management,* **2**(4): 347–363.

Koski, Wayne H. and William C. Fischer, 1979, "Photo Series for Appraising Thinning Slash in North Idaho. Western Hemlock, Grand Fir, and Western Redcedar Timber Types," U.S. Forest Service, General Technical Report INT-46.

Loomis, Robert M., 1977, "Jack Pine and Aspen Forest Floors in Northeastern Minnesota," U.S. Forest Service, Research Note NC-222.

Loomis, Robert M. and Richard W. Blank, 1981, "Summer Moisture Content of Some Northern Lower Michigan Understory Plants," U.S. Forest Service, Research Note NC-263

Loomis, Robert M. and William A. Main, 1980, "Comparing Jack Pine Slash and Forest Floor Moisture Contents and National Fire Danger Rating System Predictions," U.S. Forest Service, Research Paper NC-189.

Martin, Robert et al., 1979, "Effects of Fire on Fuels. A State-of-Knowledge Review," U.S. Forest Service, General Technical Report WO-13.

———, 1981, "Average Biomass of Four Northwest Shrubs by Fuel Size Class and Crown Cover," U.S. Forest Service, Research Note PNW-374.

Maxwell, Wayne G. and Franklin R. Ward, 1976a, "Photo Series for Quantifying Forest Residues in the: Ponderosa Pine Type, Ponderosa Pine and Associated Special Type, Lodgepole Pine Type," U.S. Forest Service, General Technical Report, PNW-52.

———, 1976b, "Photo Series for Quantifying Forest Residues in the: Coastal Douglas-fir-Hemlock Type, Coastal Douglas-fir-Hardwood Type," U.S. Forest Service, General Technical Report PNW-51.

———, 1979, "Photo Series for Quantifying Forest Residues in the: Sierra Mixed Conifer Type, Sierra True Fir Type," U.S. Forest Service, General Technical Report PNW-95.

———, 1980, "Photo Series for Quantifying Natural Forest Residues in Common Vegetation Types of the Pacific Northwest," U.S. Forest Service, General Technical Report PNW-105.

McClure, Joe P. et al., 1981, "Biomass in Southeastern Forests," U.S. Forest Service, Research Paper SE-227.

McCreight, Richard W., 1981, "Microwave Ovens for Drying Live Wildland Fuels: an Assessment," U.S. Forest Service, Research Note PSW-349.

Minnich, Richard A., 1983, "Fire Mosaics in Southern California and Northern Baja California," *Science* **219**: 1287–1294.

Mooney, H.A. and E. Eugene Conrad (technical coordinators),1977, *Proceedings of the Symposium on the Environmental Consequences of Fire and Fuel Management in Mediterranean Ecosystems,* U.S. Forest Service, General Technical Report WO-3.

Mooney, H.A. et al. (technical coordinators), 1981, *Proceedings of the Conference. Fire Regimes and Ecosystem Properties,* U.S. Forest Service, General Technical Report WO-26.

National Wildfire Coordinating Group (NWCG), "S390, Fire Behavior.'

Norum, Rodney A., 1977, "Preliminary Guidelines for Prescribed Burning Under Standing Timber in Western Larch/Douglas-fir Forests," U.S. Forest Service, Research Note INT-229.

Norum, Rodney A. and William C. Fischer, 1980, Determining the Moisture Content of Some Dead Forest Fuels Using a Microwave Oven," U.S. Forest Service, Research Note INT-227.

Olson, Craig M. and Robert E. Martin, 1981, "Estimating Biomass of Shrubs and Forbs in Central Washington Douglas-fir Stands," U.S. Forest Service, Research Note PNW-380.

Palmer, W.C., 1965, "Meteorological Drought," U.S. Weather Bureau Research Paper 45.

Paysen, Timothy E. et al., 1980, "A Vegetation Classification System Applied to Southern California," U.S. Forest Service, General Technical Report PSW-45.

Phillips, Douglas R., 1981, "Predicted Total-Tree Biomass of Understory Hardwoods," U.S. Forest Service, Research Paper SE-223.

Philpot, Charles W., 1971, "The Changing Role of Fire on Chaparral Lands," pp. 131–150, in Murray Rosenthal (ed.), *Symposium on Living with the Chaparral. Proceedings* (San Francisco Sierra Club).

———, 1977, "Vegetation Features as Determinants of Fire Frequency and Intensity," pp. 12–16, in Mooney and Conrad (technical coordinators), *Proceedings of the Symposium on the Environmental Consequences of Fire and Fuel Management in Mediterranean Ecosystems,* U.S. Forest Service, General Technical Report W0–3.

Puckett, John V. et al., 1979 rev., "User's Guide to Debris Prediction and Hazard Appraisal," U.S. Forest Service, Northern Region.

Radloff, David L., 1980, "Coping with Uncertainty in Fuel Management Decisions," in *Proceedings, Sixth Conference on Fire and Forest Meteorology,* American Meteorological Society and Society of American Foresters.

Radloff, David L. et al., 1982, "User's Guide to the National Fuel Appraisal Process," U.S. Forest Service, Rocky Mountain Forest and Range Experiment Station.

Rothermel, Richard C., 1972, "A Mathematical Model for Fire Spread Predictions in Wildland Fuels," U.S. Forest Service, Research Paper INT-115.

Rothermel, Richard C. and Charles W. Philpot, 1973, "Fire in Wildland Management: Predicting Changes in Chaparral Flammability," *Journal of Forestry* **71**(10): 640–643.

Roussopoulos, Peter J., 1980, "Improving Wildland Fuel Information for Wildfire Decisionmaking," pp. 138–142, in *Proceedings, Sixth Conference on Fire and Forest Meteorology,* American Meteorological Society and Society of American Foresters.

Running, Steven W., 1978, "A Process Oriented Model for Live Fuel Moisture," pp. 24–28, in *Fifth National Conference on Fire and Forest Meteorology* American Meteorological Society.

Sandberg, David, 1980, "Duff Reduction by Prescribed Underburning in Douglas-fir," U.S. Forest Service, Research Paper PNW-272.

Sandberg, David V. and Franklin R. Ward, 1981, "Predictions of Fire Behavior and Resistance to Control for Use with Photo Series for the Douglas-fir—Hemlock Type and the Coastal Douglas-fir—Hardwood Type," U.S. Forest Service, General Technical Report PNW-116.

Schroeder, Mark J. and Charles C. Buck, 1970, *Fire Weather...A Guide for Application of Meteorological Information to Forest Fire Control Operations,* U.S. Forest Service, Agriculture Handbook 360.

Southern Forest Fire Laboratory Personnel, 1976, "Southern Forestry Smoke Management Guidebook," U.S. Forest Service, General Technical Report SE-10.

Storey, T.G., 1965, "Estimating the Fuel Moisture Content of Indicator Sticks from Weather Variables," U.S. Forest Service, Research Paper PSW-26.

U.S. Forest Service, 1969, "Intermediate Fire Behavior," TT-81.

Van Wagner, C.E., 1977, "Conditions for the Start and Spread of Crown Fire," *Canadian Journal of Forest Research* **7**: 23–34.

Ward, Franklin R. and David V. Sandberg, 1981a, "Predictions of Fire Behavior and Resistance to Control for Use with Photo

Series for the Sierra Mixed Conifer Type and the Sierra True Fir Type," U.S. Forest Service, General Technical Report PNW-114.

———, 1981b, "Predictions of Fire Behavior and Resistance to Control. For Use with Photo Series for the Ponderosa Pine Type, Ponderosa Pine and Associated Species Type, and Lodgepole Pine Type," U.S. Forest Service, General Technical Report PNW-115.

Wright, Henry A. and Arthur W. Bailey, 1982, *Fire Ecology. United States and Canada* (New York: Wiley-Interscience).

Zimmerman, G. Thomas, 1982, "Preliminary Guidelines for Broadcast Burning Lodgepole Pine Slash in Colorado," *Fire Management Notes* **43**(1): 17–22.

*Chapter Four*

# Fire Weather

Air is one of the primary components of a fire environment. The interaction between a fire and the atmosphere is described by fire behavior, but the study of the important atmospheric properties and meteorological processes that affect fires comprises the field of fire weather. The properties of an air mass affect the character of the fuel-oxygen mixing process. The moisture content of an air mass is largely responsible for fuel moisture contents. The wind field powerfully influences fire behavior and shape. The variability in fire danger is principally the product of weather changes. Its fire climate is as fundamental a description of a region as a fuel model is of a biota, or a fire regime of an ecosystem. The unpredictability of fire behavior is, in good measure, the product of the unpredictability of weather. No component of the fire environment is so variable as weather and none so important for real-time fire behavior forecasting, short-term fire planning, and the long-term reconciliation of fire administrations with fire regimes.

The important properties of an air mass are its temperature, pressure, velocity, and moisture content. The important atmospheric processes are those that describe the exchange of heat and moisture, the movement of air from regions of high temperature to regions of low temperature, or, since heat and density are related, the movement of air from areas of high pressure to those of low pressure. The question of scale is important. Meteorology may be considered on a macro scale, the circulation of air around the globe; on a meso scale, the movement of particular air masses and large scale eddies; on a micro scale, where local heating and cooling differentials, site-specific winds, surface inversions, and microclimates are the objects of interest. The actual weather at a particular locale reflects the operation of the atmosphere on all these scales, though the behavior of individual fires tends, more often than not, to be site-specific. For a particular site, local conditions—or the interplay of micro with meso-scale processes—generally decides fire weather.

The concept of fire weather also suggests ways by which fires may be characterized. Geographically, one can speak of fire climates, regions showing a similar pattern of fire behavior and occurrence. Temporally, one can refer to fire periods, regular rhythms of fire starts and behavior. On a diurnal scale, this apparent periodicity leads to the concept of a burning period; on the scale of a week or so, to synoptic patterns associated with critical fire danger; on an annual scale, to the concept of a fire season; on still longer secular scales, to the concepts of fire climates and fire regimes. Just as a fuel model reduces an ecosystem to certain aspects of its physical chemistry and geometry, and much as a fire type or fire regime identifies a biota according to its biological reponse to fire, so the fire climate concept selects and organizes those meteorological features relevant to fire occurrence and behavior.

Assessing the probability of fire starts and the probable burning characteristics of the fires that result is the province of fire danger rating. The extent of the fire season, the distribution of critical fire periods within that season, and the likely occurrence of suitable periods for prescribed fire—all are fundamental for fire management planning and operations. No administrative unit is without some source for meteorological information, some procedure to forecast meteorological effects, or some plans to incorporate long-term fire climate patterns.

For most of the United States, large wildfires occur during special meteorological circumstances known as critical fire periods. The distribution of these periods has traditionally defined the fire season. But the fire season is not strictly a climatological phenomenon. It is reshaped by the human use of fire. Through ignition and fuels management, the natural period of burning can be extended, often to accomodate prescribed fire. To burn under circumstances that support wildfire would be treacherous, and most precribed burns occur before or after the wildfire season. Similarly, through suppression it is possible to constrict the natural fire season. Deliberation replaces chance; controlled fire, wildfire; a regular cycle of fire occurrences and behavior traits, the episodic distribution of wildfires and fire weather attributes.

## *4.1 ATMOSPHERIC STABILITY*

The Earth's atmosphere is much broader than it is deep. A vertical profile of the atmosphere shows a logarithmic decrease in density with altitude and a stratification of the atmosphere into more or less definable layers. Almost all weather occurs within the lowest layer, the troposphere, which contains nearly three-fourths of the atmosphere by weight and virtually all of its water vapor and carbon dioxide. Nearly half of the atmosphere, in fact, falls within 18,500 feet (5630 m) of the surface. The troposphere varies in depth between five miles (8 km) at the poles and about 10 miles (16 km) at the equator. Like the atmosphere at large, it shows its own internal structure: temperature and pressure decrease with height, and regular patterns of horizontal and vertical motion develop between different portions of the global air envelope. The stratification of the atmosphere is well-characterized by its vertical temperature distribution, and the average temperatures define the United States Standard Atmosphere (Figure 4-1).

Because the atmosphere is shallow relative to its extent, most motion within it is horizontal rather than vertical, controlled by *advection* rather than by *convection.* On a global scale, the horizontal flow of air by volume is several hundred times greater than the vertical flow. The principles that govern air flow apply to local pockets of air as well as to the global atmosphere. Everywhere air flows from regions of heat to regions of cold. But to this simple pattern there are many modifications. Some are due to the rotation of the Earth; some, to boundaries that develop between air masses; and others, to friction—in particular, to the thermal turbulence wrought by the heating of the Earth's surface and the mechanical turbulence occasioned by the Earth's irregular relief. One result is to homogenize the troposphere, mixing and equalizing its properties. But another outcome is to vastly complicate air flow by introducing many levels of turbulence. On the scale of most wildland fires, this turbulence, broadly conceived, is often more decisive than the general flow pattern.

Similar considerations apply to convection. Within the troposphere convection can operate on many scales: on a macro scale, with convective cells moving air between the equator and the poles and between continents and oceans; on a meso scale, with the lifting and subsidence of layers of air; on a micro scale, with the lifting and falling of air parcels or bubbles. The energy for

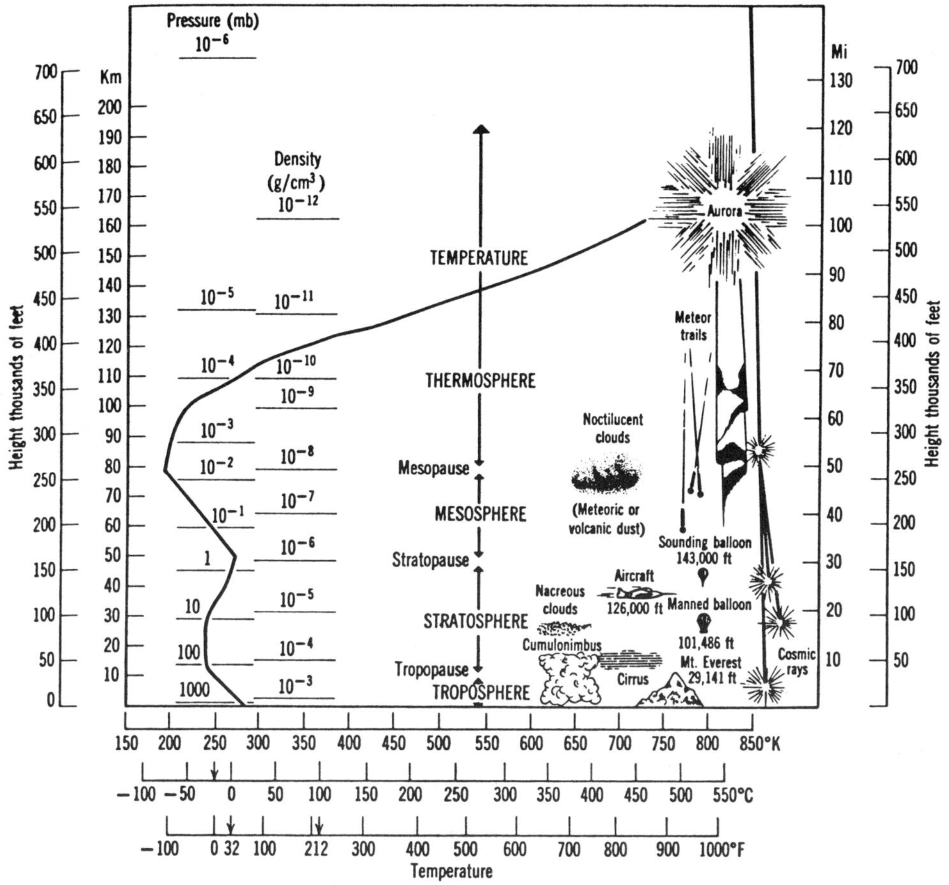


*Figure 4-1. Standard atmosphere. From Cole (1975).*

convection comes from solar heating, and so the vigor of convection shows diurnal and seasonal changes and differs according to geographic terrain and cover. How surface heating translates into convection depends, too, on the character of the air mass.

Consequently, convection is commonly discussed in terms of *atmospheric stability,* or the resistance offered by the air mass to vertical motion, of *lifting mechanisms* that initiate vertical movement, and of *adiabatic processes* that govern the expansion and contraction of displaced air. An atmosphere is considered *stable* when a parcel of air displaced up or down within it will return to its original elevation. It is *unstable* when a parcel will continue to move in the direction of displacement. And it is considered *neutrally stable* if the displaced parcel remains at its new vertical location, neither rising nor falling further. Convection requires that an atmosphere be unstable.

To illustrate, consider the effect of lifting parcels of air within a given air mass. The parcel of air concept is useful because air is a poor conductor of heat; air that is locally heated will tend to behave as a unit rather than quickly dissipate that heat to its surroundings. If the temperature of the displaced air is higher than the temperature of its newly surrounding air, then that parcel will rise, and it will continue to rise until its temperature equals the temperature of its new surroundings.

As it rises the parcel expands and cools. If the temperature of a displaced parcel is less than the temperature of the surrounding air, then the air parcel will descend back to its former position. If, after being displaced, the parcel has the same temperature as the surrounding air, then it will remain at its new location. In terms of stability, the atmosphere which allowed the displaced parcel to continue to rise was unstable, that which brought the displaced parcel back to its original position was stable, and that which allowed the displaced parcel to remain at its new location was neutrally stable. A number of mechanisms cause displacement, but once set in motion the parcel obeys *adiabatic processes* that determine its subsequent heating and cooling.

### *Adiabatic Lapse Rates*

Adiabatic processes involve changes in the state variables of an object—its temperature, pressure, and volume—during which no other energy or mass enters or leaves the system. During compression, the temperature and pressure of the system increase while its volume decreases; during expansion, temperature and pressure decrease while volume increases. Under true adiabatic conditions the total energy and mass of the system remain unchanged, and the process is reversible. If energy or mass leave the system, however, a pseudo-adiabatic process results (see Figure 4-2).

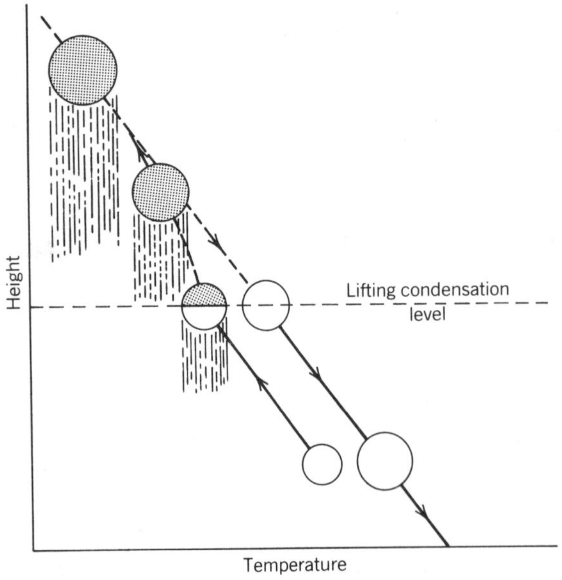


*Figure 4-2. A pseudo-adiabatic process. A parcel of air is lifted according to the dry adiabatic rate until it reaches saturation, from which point it rises according to the wet adiabatic rate. Precipitation means that energy and mass have been lost from the system, so the descent process does not simply reverse the ascent. The parcel returns warmer and drier than it began. From Cole (1975).*

When adiabatic processes are applied to layers or parcels of air, they result in regular, predictable changes. Lifting a parcel of air causes it to expand in volume and to diminish in pressure and temperature; depressing an air parcel downward causes a compression of volume and an increase in temperature and pressure. The changes in air temperature adequately describe the process. When the parcel is not saturated with water vapor, it shows a temperature change of 5.5° F/1,000ft (10°C/1000m); this is known as the *dry adiabatic lapse rate.* If the parcel is saturated with water vapor, however, the change in temperature approximates 3.0°F/1,000 ft (3–8°C/1000 m), known as the *moist adiabatic lapse rate.* The difference between the dry and moist rate results from the latent heat of water. It takes energy to convert water from a liquid state to a vapor; the water vapor stores this energy (as latent heat), then releases it upon condensation. The liberated heat warms the parcel of air, reducing its rate of cooling. The moist adiabatic lapse rate will differ according to the amount of water vapor present. At higher temperatures air holds more water; at saturation, then, more heat will be released and the lapse rate will drop.

Progressive cooling with increased elevation is a characteristic of the troposphere in general. A general value for this decrease, known as the *environmental lapse rate,* is 3.5°F/1,000 ft for latitude 45°. Like the temperature profile of the Standard Atmosphere, this rate is an average. At any given time the actual temperature structure of the atmosphere can be expressed as a *prevailing lapse rate.* Typically, the prevailing lapse rate will show considerable variability according to local circumstances and the stratification of the air mass. Its numbers are empirical, derived by atmospheric sounding.

Adiabatic processes characterize a wide range of meteorological phenomena, many of them significant to fire behavior. They apply to the development of thunderstorms, important sources of rain, wind, and lightning; to the establishment of convection columns, a mixing process fundamental to combustion; to vortical developments such as firewhirls and dust devils; to smoke dispersion, the venting and mixing of combustion products; to local wind systems, the characteristics of which are controlled by convection. Fundamentally, they provide a means of measuring atmospheric stability.

To determine the likelihood of vertical motion, compare the prevailing lapse rate with the adiabatic lapse rates. For a given parcel of air which is elevated or depressed to a new level, the dry lapse rate will give its new temperature. Comparing that new temperature with the temperature of the surrounding air mass determines whether the parcel of air will continue in motion, return to its former postion, or remain in its new location. Accordingly, the atmosphere may be considered stable, unstable, or neutrally stable. The respective prevailing lapse rates will be lesser, greater, or equal to the adiabatic lapse rates.

*Dry Adiabatic Lapse Rate.* Consider, first, the case with dry (unsaturated) air (Figure 4-3). In case I, the prevailing lapse rate is set at the environmental lapse rate; this makes it less than the dry adiabatic lapse rate. If a parcel of air is displaced upward under these conditions, it will cool faster and become more dense than the surrounding air.

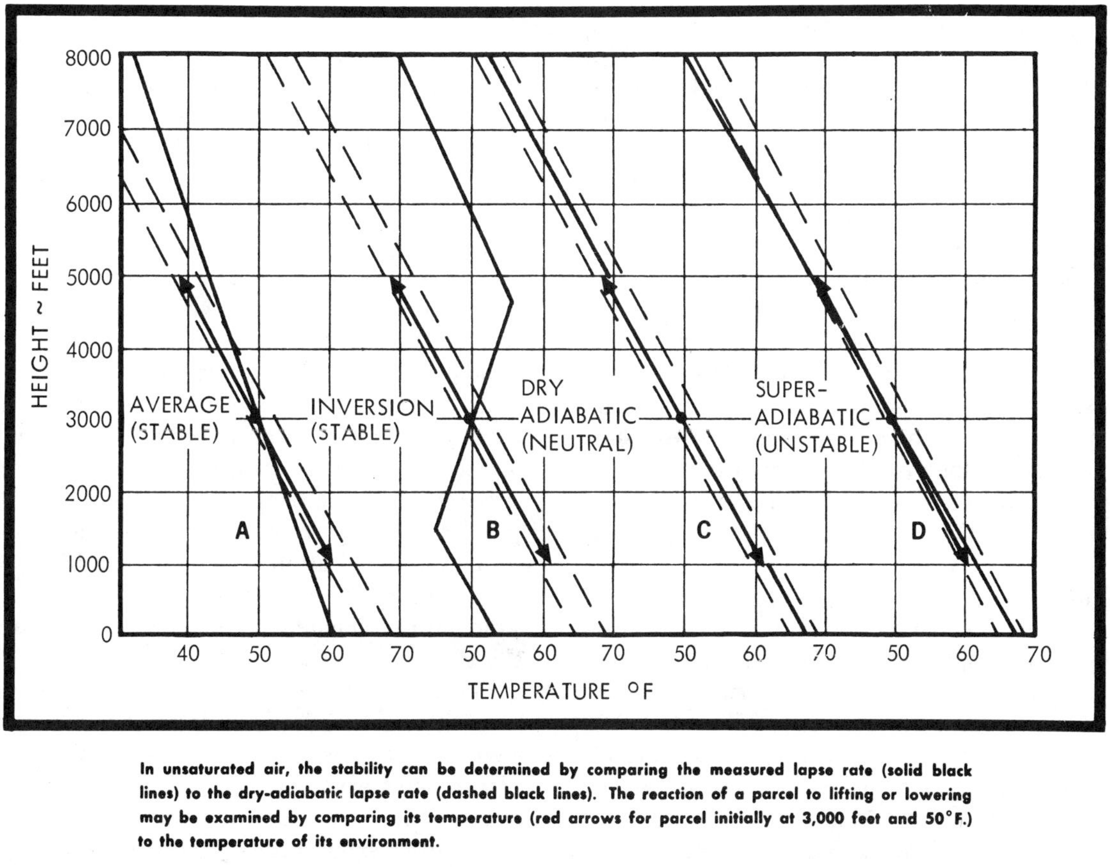
8000
7000
6000
5000
4000
3000
2000
1000
0
HEIGHT ~ FEET
AVERAGE (STABLE)
INVERSION (STABLE)
DRY ADIABATIC (NEUTRAL)
SUPER-ADIABATIC (UNSTABLE)
A
B
C
D
40 50 60 70 50 60 70 50 60 70 50 60 70
TEMPERATURE °F
In unsaturated air, the stability can be determined by comparing the measured lapse rate (solid black lines) to the dry-adiabatic lapse rate (dashed black lines). The reaction of a parcel to lifting or lowering may be examined by comparing its temperature (red arrows for parcel initially at 3,000 feet and 50°F.) to the temperature of its environment.

*Figure 4-3. Adiabatic lapse rates: four cases. The solid black line gives the measured environmental lapse rate; the dashed lines, the dry adiabatic lapse rate. Comparing the rates (or slopes of the lines) determines whether the atmosphere is unstable, neutral, or stable. From Schroeder and Buck (1970).*

When the lifting force is removed, the parcel will descend to its original location. Conversely, if the parcel were forced downward, it would warm faster and become more bouyant than the surrounding air; when the compressive force was withdrawn, the parcel would rise to its former height. The atmosphere is stable.

In case II, the prevailing lapse rate abruptly changes with height. In this instance, for a stratum of several thousand feet, the lapse rate reverses. The air warms, a phenomenon known as an *inversion.* Within the inversion, if a parcel of air were disturbed upward, it would quickly become much cooler than the surrounding air; if displaced downward, it would just as quickly become much warmer than its environs. In both instances, the tendency to stability would be very strong. By the same token, it is difficult to move a parcel of air through an inversion. Normally, the prevailing lapse rate and the adiabatic lapse rate proceed in the same direction but at different speeds, and the question of atmospheric stability is one of their relative ratio. With inversions, however, the two lapse rates move in opposite directions, any motion by a parcel of air only increases the difference between it and the surrounding air, and the air mass shows great stability.

In case III, the prevailing and adiabatic lapse rates are identical. Any movement of a parcel of air up or down in the atmosphere will put its new temperature at the same value as the surrounding air. The air mass is neutrally stable. Neutral stability does not long persist. Its condition is relative, and newly advected air along or above the surface can rapidly change the prevailing lapse rate.

Only when the prevailing lapse rate is greater than the dry adiabatic lapse rate, as seen in case IV, is the atmosphere unstable. A lifted parcel of air will cool less rapidly than its surroundings, will show greater buoyancy, and will continue to rise. The difference between the temperature of the parcel and that of its surroundings will increase as the parcel continues to rise. At a certain point, known as the *level of free convection,* the rise of the parcel no longer depends on an external force but results simply from the relative buoyancy of the parcel. A prevailing lapse rate of this type is called *superadiabatic.* Because it encourages vertical movement, however, superadiabatic lapse rates lead to considerable mixing; temperature and relative humidity tend to become more homogenized. Superadiabatic lapse rates are ideal for the development of cumulonimbus clouds, convective columns above a fire, and associated vortex formation such as whirlwinds. When the prevailing lapse rates are exceptionally high (19°F/1000 ft), a range of values known as *autoconvective lapse rates,* equilibrium is impossible. An initial displacement is no longer necessary to set a parcel of air in upward motion: it accelerates upward automatically.

*Moist Adiabatic Lapse Rate.* Consider, next, the case for saturated air. A parcel of air contains a certain amount of water vapor, a quantity known as its *absolute humidity.* The amount of water vapor a unit of air can contain will vary according to its temperature—high temperatures allowing greater amounts of vapor. Usually this vapor is not measured in terms of its partial pressure, but by comparison to the amount of vapor needed for saturation. For a given temperature, a unit of air can only contain so much water vapor until saturation occurs, the vapor condenses into liquid, and clouds form, often with precipitation. *Relative humidity* describes the ratio of the amount of water vapor present in the unit to the potential amount that the unit could contain at a given temperature. *Dew point* gives the temperature at which, for the amount of water vapor present, saturation would occur. Thus saturation can result from increasing the quantity of water vapor or decreasing the temperature. The dew point shows a lapse rate of 1°F/1000 ft.

Elevating a parcel of air cools it; its relative humidity increases; and at the point of saturation, it will form clouds. That the dew point also shows a lapse rate means that the level of saturation can be reached even sooner; the temperature of saturation drops even as the parcel of air cools. Up to the point of saturation, the parcel of air obeys the dry adiabatic lapse rate. But after saturation, it cools at the moist adiabatic lapse rate. The moist lapse rate will show a range of values because the amount of heat released during condensation will vary with the absolute amount of water vapor in the system; this depends on the temperature of the parcel. Different parcels of air will have different heat capabilities. The elevation at which, for a given air mass, saturation will occur is known as the *lifting level of condensation (LLC).*

A rising parcel of air will not automatically stop rising once saturation occurs. Only its rate of cooling varies. For atmospheres in which condensation occurs, there is not one lapse rate but two rates against which the prevailing lapse rate must be measured. It is possible—indeed, common—for an atmosphere to be stable with respect to the dry lapse rate but unstable with respect to the moist lapse rate. In this case if a parcel of air were to be lifted by some means to its condensation level, then free convection would occur and a thunderhead would result. This occurs when the prevailing lapse rate has values between the dry and moist adiabatic lapse rates. Such an atmosphere is considered *conditionally unstable.* It is stable as long as the parcel does not become saturated, and it becomes unstable once saturation is reached.

Free convection does not continue indefinitely, however. The saturated air loses moisture to precipitation, turbulence entrains colder ambient air, the adiabatic lapse rate of the parcel gradually changes from moist to dry, and the parcel eventually comes into equilibrium with the air around it. Typically, for a thunderstorm, this occurs at the tropopause, along the crest of the troposphere. To determine atmospheric stability, therefore, it is necessary to know the prevailing lapse rate, the individual temperature of the displaced air, and its water vapor content (usually expressed as a dew point; Figure 4-4).

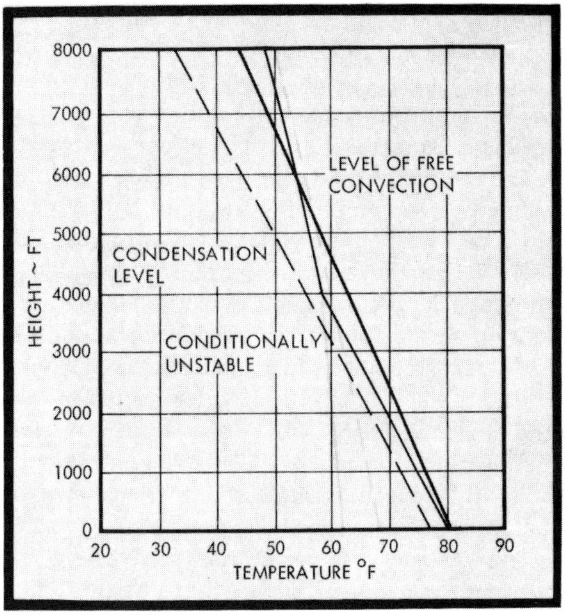


*Figure 4-4. Conditional instability. Whether a parcel of air reaches the level of condensation depends on how the dry lapse rate compares with the environmental lapse rate. Whether, after saturation, it reaches the level of free convection depends on how the wet lapse rate compares with the prevailing lapse rate. Modified from Schroeder and Buck (1970).*

*Parcels and Layers.* The atmosphere does not consist simply of aggregates of air parcels. It contains strata, or layers, of air, and stability determinations must compare the temperatures of the different layers which make up the air mass as a whole, not merely the temperatures of quasi-isolated parcels. This stratification experiences regular diurnal and seasonal changes in addition to changes brought about by advection. Like parcels of air, layers compress and warm during descent, extend and cool during uplift. But although a parcel of air may be considered a single unit that responds uniformly to displacement, a layer of air shows differences between its top and bottom. Adiabatic process depend on the distance traveled, and the tops and bottoms of a layer will warm or cool differentially. The distance between top and bottom, and thereby the magnitude of the temperature difference, will vary according to whether the layer is compressed or extended. Compression shortens the distance between layer top and layer bottom, reducing the temperature difference within the layer; lifting spreads top and bottom further apart, increasing their temperature differential. This temperature difference will affect the atmospheric stability of the layer.

During uplift, the top of a layer cools more than its bottom. This places warmer air under cooler air, an unstable condition. Further lifting means even less stability. The situation worsens if the bottom of a layer contains more moisture, condenses, and proceeds at the moist lapse rate. In this case the bottom becomes progressively warmer than the top. During descent (subsidence) the opposite occurs. The top of the layer warms more than the bottom, enhancing the stability of the layer as a whole. The more the layer descends, the more stable it becomes. Expressed graphically, the descending layer will appear as an inversion. In brief, lifting encourages instability; subsidence, stability. Advection can lead to lifting and instability when, as with a low pressure system, air converges and is forced to rise. Conversely, advection promotes subsidence and stability when, in the form of a high pressure system, it leads to divergence (Figure 4-5).

The effects of ascending and descending layers of air on fire behavior are complex. The instability of rising air promotes convective development, but whether convection takes the form of a smoke column above a fire or a thunderhead that rains on a fire depends on the moisture content of the air mass. Conversely, the stability of a descending layer weakens the prospect of convective development. But by pushing down air from the upper atmosphere the process can cause temperatures to rise, humidities to plummet, and strong winds to develop.

### *Mechanisms of Displacement*

Once past the level of free convection a parcel of air will continue to ascend because of its buoyancy. But to get a parcel to that level requires force, some mode of displacement. Elevating processes include thermal lifting, orographic lifting, frontal lifting, turbulent mixing, and convergence; lowering processes usually result from divergence, a condition which can lead to subsidence (Figures 4–5 and 4–6). *Thermal lifting,* or convection, results from surface heating. The Earth's surface warms the surrounding air by conduction. With the proper triggering mechanism this warm, less dense air near the surface rises. This leaves a partial vacuum, cooler air rushes into the surface, and a convective cell may develop. An important variant is thermal lifting induced by a fire. In this case the lifting is not important as a prelude to cloud formation but as a means of developing a convective column. *Orographic lifting* refers to the elevation given a mass of air as it passes over mountainous country. The flow of air has its origin in general circulation patterns, so it persists despite the resistance offered by the terrain and is forcibly lifted. Whether it continues to rise, remains at its new elevation, or returns to its former height after crossing the mountains depends on the prevailing lapse rate. A similiar situation characterizes *frontal lifting.* The air mass is displaced upward by the interaction of two air masses. The colder air mass forces the warmer air mass abruptly upward in the case of a cold front, whereas warmer air gradually overrides colder air in the case of a warm front. Large-scale *turbulence* causes some vertical movement through mixing. The *convergence* that characterizes the airflow around an area of low pressure, behaves like a large, slow vortex. Air converges, rises, and induces instability.

Most large-scale lowering, or subsidence, follows from horizontal *divergence.* The process is associated with high pressure cells, much as convergence is associated with low pressure cells. The two processes are, in fact, coupled: low pressure cells result in an upward transport of air and high pressure cells return that air as part of a grand system of atmospheric recycling. With its origin in the upper troposphere, the subsiding air is exceedingly dry, and as it descends through the troposphere it warms adiabatically. The lower boundary of the layer marks an inversion, trapping the water vapor, smoke, pollution, and even clouds of the lower levels of the troposphere beneath it. The interaction of a subsiding layer with surface layer below it can be maddeningly complex and unpredictable.

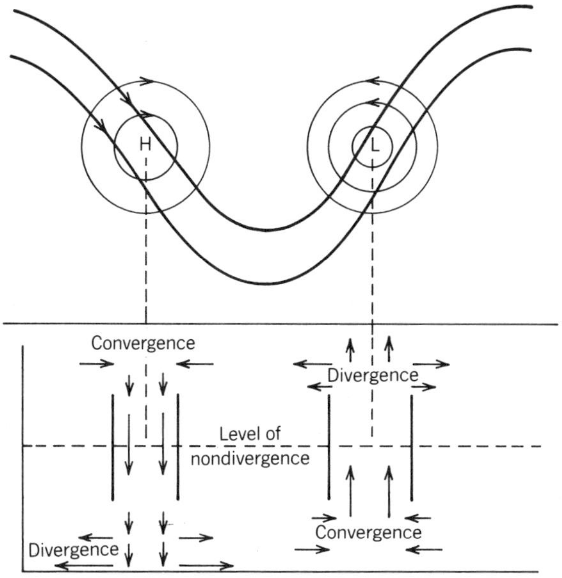


*Figures 4-5 and 4-6. Lifting processes. The upper diagram shows the downward flow (divergence) associated with high pressure cells and the upward flow (convergence) associated with low pressure cells. The lower diagrams show more local forms of lifting. From Cole (1975) and Schroeder and Buck (1970).*

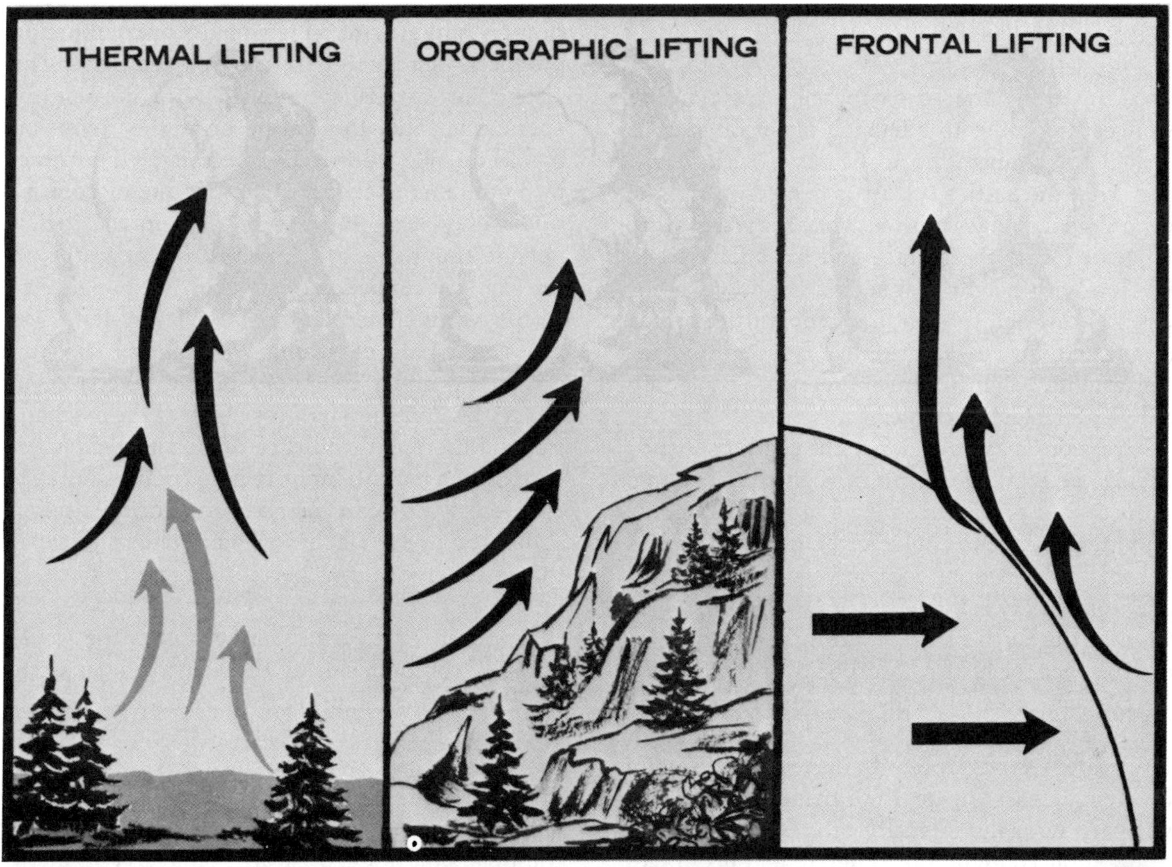

A number of mechanisms may transfer the properties of the subsiding air, if not the air itself, to the surface. Primarily this means turbulence. Thermal or mechanical turbulence within the surface layer—induced by strong surface heating or by mountainous relief—may eat away at the inversion boundary and transfer some subsiding air to the surface. Descent will be irregular, and the lower layer boundary may show a pattern of relief resembling inverted mountains. In special locations, mountain waves (perhaps in conjunction with foehn winds) can bring subsiding air from great heights down the lee side of mountains. Fire management can be difficult under these conditions. The transfer of subsiding air can lead to severe fire weather. Equally, the presence of an inversion may make smoke dispersion almost impossible, seriously constricting the prospects for prescribed fire.

### *Inversions*

Since, within the troposphere, temperature changes according to adiabatic lapse rates, it is normal for temperature to decrease with elevation. That the Earth's surface is heated during the day only encourages this tendency for warm air to be located near the surface and for cooler air to develop with height. Yet such a condition is inherently unstable. When this temperature profile is reversed, an *inversion* results. Temperature increases with height, warm air overrides cooler air, and the air mass is stable.

Inversions may be regional, the product of subsidence, or local, the result of nighttime cooling. In the case of subsidence a layer of warm air gradually descends onto cooler air beneath it (Figures 4–7 and 4–8). Among local inversions, the most common variety is the *surface* or *evening inversion.* In this case the Earth's surface during the night radiates to space the heat it absorbed over the day. As the surface cools, it chills the adjacent air; the chilled air, being denser than the air above it, flows down topographic drainages to collect in low-lying areas. The depth of these cold air lakes will vary with the intensity of cooling and with topographic configuration. High relief encourages the flow of cool air, and deep basins, the dense ponding of this air. The intensity of evening cooling will vary according to the presence of clouds and water vapor, which intercept long-wave radiation from the surface and reradiate some of it back; of winds, which may cause mixing with warmer air aloft; and of unstable conditions, again because of mixing. Conversely, inversions are most easily broken up where relief is not pronounced, where solar radiation (morning heating) is intense, and when unstable conditions favor mixing.

The consequence of evening inversions for fire behavior, like the effects of subsidence, are mixed (Figure 4-9). Obviously, the removal of solar heating and the establishment of a stable atmosphere near the surface diminish some of the factors encouraging fire spread and burning intensity. Under normal circumstances fire activity slows at night. But inversions trap combustion products—smoke and even pyrolyzed gases. The smoke can interfere seriously with fire control operations, and the gases, once the inversion breaks up, may contribute to accelerated burning. In mountainous country, the pooling of cold air into valleys can improve burning in the region outside the inversion. A *thermal belt* develops on the slopes just above an inversion, a zone of relatively warmer temperatures. And low-level *evening jets*, bands of strong winds—may develop across the upper boundary of the inversion. In effect, the inversion reduces much of the mechanical and thermal turbulence that characterizes the surface of a mountainous region; frictional drag is reduced; winds can increase in velocity, ripping across the upper reaches of mountains and causing an otherwise unexpected increase in burning rates.

Another local form of inversion is the *marine inversion,* common during the warm season along the Pacific coast. During the day cool air collects along the ocean surface at a depth from a few hundred to a few thousand feet. Advection, driven by the temperature difference between land and ocean, then brings the cool air inland. If the inversion is deep, stratus clouds form along

**A nighttime inversion in a valley effectively shields the downslope and downvalley winds from the general wind flow above.**

*Figures 4-7 and 4-8. Inversions. Upper diagram shows an evening inversion. Note the thermal belt which occupies the middle range of mountains. Lower diagram shows an upper-air inversion due to subsidence. From Schroeder and Buck (1970).*

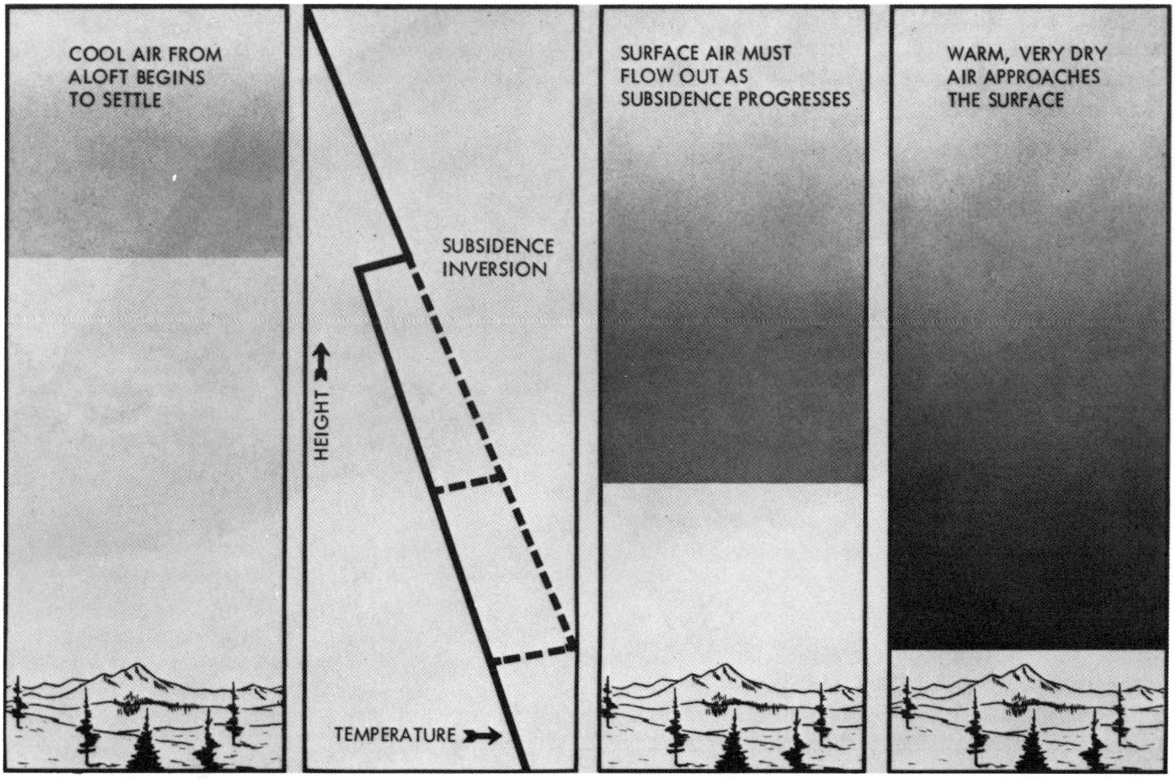

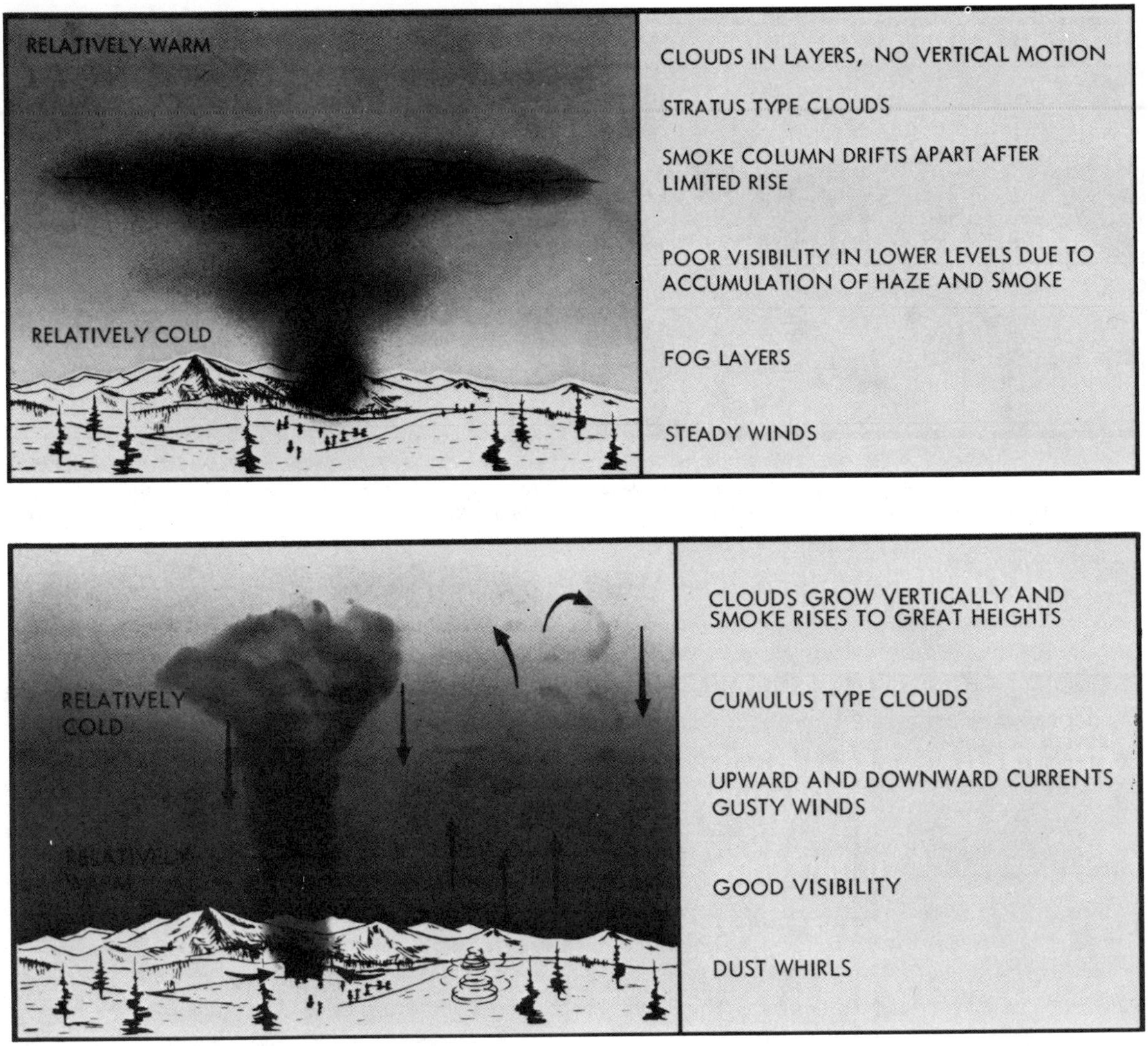


*Figure 4-9. Visible indicators of stability (upper diagram) and instability (lower). From Schroeder and Buck (1970).*

the upper boundary; if the inversion is shallow, fog results. The penetration inland by marine air is generally limited by the mountain ranges that fringe the coast, but the interaction of cool marine air and other winds, often from the interior, can create complex, frequently sporadic wind fields, with a fire responding alternately to one wind or the other. This pattern of advection, by which marine air invades the coast, may occur even underneath a subsiding layer of air.

## 4.2 WINDS

No component of a fire environment is more variable than weather, and no aspect of weather is more unpredictable than wind. Wind is a vector quantity, described in terms of its speed and direction. Different strata in the atmosphere will show different wind vectors, and even within a single stratum wind speeds will display a profile with slower speeds resulting from friction along the layer boundaries. Moreover, winds may be categorized according to a simple hierarchy of scale. Primary winds are those related to the general circulation of air around the globe; second order winds, to synoptic conditions such as fronts and pressure cells; third order winds, to local conditions responsible for assorted small-scale convective systems. Most second order systems are advective, the product of broadscale pressure differentials on the size of the synoptic charts used for most weather maps. Third order systems are convective, resulting from local differentials in heating and cooling on a scale too fine to appear on most meteorological charts, but of significance at the scale of wildland fires.

Whether advective or convective, wind flow is far from simple. The troposphere is commonly stratified, with individual layers characterized by changes in wind speed and even wind direction as well as by changes in temperature and pressure. Where these changes are abrupt, a zone of *wind shear* can be identified. Within a layer, flow may be either laminar or turbulent. *Laminar flow* shows a little vertical mixing, as though the flow consisted of an infinite number of fine planes, each sliding over its neighbors. *Turbulent flow,* by contrast, involves considerable vertical mixing as well as forward velocity, as though the wind were a series of tumbling eddies.

Turbulence results from friction—either the internal friction due to flow, measured as viscosity, or the friction that results from contact with boundaries having different properties. Viscosity varies primarily with the speed and density of the flowing air; boundary friction, with the roughness of the surfaces in contact with each other. When the boundary is the Earth's surface, the degree of roughness may be very large indeed. Trees, structures, and mountainous topography all can produce mechanical resistance to flow, and thereby turbulence. Similar effects can result from strong thermal processes. If the ground warms up sufficiently, convective flow rises up, enhancing the "roughness" of the surface. How greatly thermal or mechanical obstructions modify wind flow varies, though a rule of thumb puts the range of influence at 8–10 times the height of the obstacle. The principal outcome is eddy formation—a form of turbulence that translates into gustiness.

Not all of the troposphere is affected by surface friction, and general winds are distinguished between *surface winds* and *winds aloft* (or *free winds)*. The winds aloft are steadier in both speed and direction and, because they exist outside the friction layer, or layer of mixing, they are faster than suface winds. The depth of the friction layer will vary with wind velocity and the roughness of the terrain, as determined by topographic relief and intensity of surface heating. Fast winds are invariably turbulent. The effect of winds aloft on fire behavior is indirect, but significant. They help determine the possibility of convective column formation, of long distance spotting by firebrands, and of smoke dispersion. Most fire weather forecasts will include both free and surface winds, though the actual properties of the surface wind will reflect complex interactions between it and the free winds. Surface winds will vary greatly, too, according to vegetative cover. Most forecasted wind velocities apply to the tops of the vegetative crown, whereas the significant velocity for fire behavior forecasts is the velocity at midflame height. Procedures exist by which to reduce, for various vegetative covers, the general surface windspeed to midflame windspeed (Figures 4–10 and 4–11; Albini and Baughman, 1979).

### *General Winds*

The primary circulation of the Earth follows from unequal temperatures between the poles and the equator, and between the oceans and the continents. For the temperate zone, which includes the United States, this means that prevailing winds come from the west. But the exchange between pole and equator is not simple; it works episodically through such means as the migration of large-scale eddies and the movement of large air masses as more or less distinct bodies. Such events may last from 3 to 14 days before the transfer of energy is complete—the meso-scale of meteorology recorded on the synoptic charts typical of most weather maps. Within this range are found the *general winds.* Their flow is horizontal. *Gradient winds* flow from areas of high pressure to zones of low pressure, while *frontal winds* flow according to the interaction of air masses along a boundary. In special circumstances gradient winds interact with mountainous topography to produce a distinctive category of high-intensity wind fields known as *foehn winds.* In other circumstances, most commonly on level terrain or on mountainous terrain made level by evening inversions, a high velocity river of wind, *a low level jet,* can result. Such jets may be associated with either strong frontal or gradient wind systems.

*Frontal Winds.* From their source regions air masses acquire certain properties: temperature, pressure, and moisture. When air masses move, a dynamic boundary known as a *front* develops along their advancing edge. Air masses are characterized as cold or warm depending on their temperatures relative to the air into which they move. If the advancing air mass is colder than the air mass it replaces, the boundary becomes a cold front; if warmer, then a warm front. If the two air masses are stalled, with neither advancing on the other, then the front is stationary. The varieties of interaction are infinite. For fire behavior purposes, however, one form of frontal passage is particularly significant for the wind field it produces: this is the cold front.

Cold fronts can affect fire behavior in a number of ways. By lifting air masses ahead of it, a front can release the instability of a conditionally unstable atmosphere. If sufficient moisture is present, precipitation and even thunderstorms can result. A row of thunderstorms will produce erratic winds of their own. Ahead of the front, winds will pick up in intensity, blowing parallel to the front from the south or southwest. As the front passes, there occurs an abrupt change in wind speed and direction. The wind shifts to the north or northwest, becoming gusty during passage, then blows strong and steady when passage is completed. This change in wind field comes even in the absence of precipitation along the

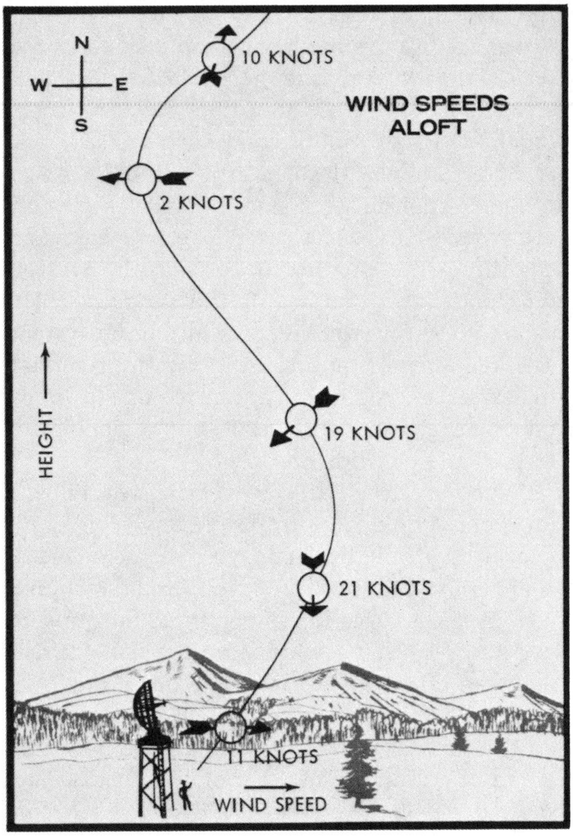


*Figures 4-10 and 4-11. Surface winds. Surface wind flow will differ from the free wind regime. The air mass may be stratified (above diagram), with different wind speeds and directions for each strata. Topography will interact with surface flow affecting speed, direction, and turbulence (opposite diagrams). And vegetative cover will diminish, often in complex ways, the effective winds at midflame height (see Figure 2-12). From Schroeder and Buck (1970).*

front. In the eastern United States the passage of "dry" cold fronts in the spring has been responsible for many of the larger regional fires—the prefrontal winds drive the fire into a long ellipse with its major axis roughly southwest to northwest, and the postfrontal winds transform the long eastern flank into a series of multiple heads advancing east or southeast. In the western United

Large roll eddies are typical to the lee of bluffs or canyon rims. An upslope wind may be observed at the surface on the lee side.

Ridgetop saddles and mountain passes form important channels for general wind flow. The flow converges and the wind speed increases in the passes. Horizonal and vertical eddies form on the lee side of saddles.

*Figure 4-11*

States similar fronts during the late summer fire season have resulted in large fire complexes, especially when a series of such fronts passes through in quick succession.

*Gradient Winds.* Gradient winds flow from regions of high pressure to those of low pressure. The flow is complicated, however, because of effects due to the Coriolis force and surface friction, with the result that winds do not flow directly between zones of unequal pressure but with a vortical motion. For the northern hemisphere the wind field around a low pressure cell (or *cyclone*), a zone of convergence, is counterclockwise. Around a high pressure cell (or *anticyclone*), a zone of divergence, it is clockwise. Some cells are semipermanent in nature, forming over oceans or continents according to the seasons; others freely migrate in belts across the temperate zones (Figures 4–12 and 4–13).

Collectively, the distribution of cells determines the normal field of general winds, and it is this flow that is recorded daily on synoptic weather maps. The gradient winds, then, set the prevailing winds for the lands of the United States, the westerlies, but they may also interact in important ways with gross topographic features such as mountains and plains to produce subsidiary groups of winds, such as low-level jets and foehns. These winds are quasi-general in that their energy derives from large-scale pressure gradients, but local in that they occur at particular sites because of topographic features.

*Low-Level Jets.* It sometimes happens that the wind field takes on an organized flow around a core of high-velocity winds. When such a flow occurs at or within a few thousand feet of the surface of the Earth, it is known as a low-level jet. The effects are local, and the air current should not be confused with the high-level jet streams that are a part of primary and second order circulation patterns. The cause for such jets is not well understood, and a variety of mechanisms may be possible. "Jet" is used as a descriptive term, not a

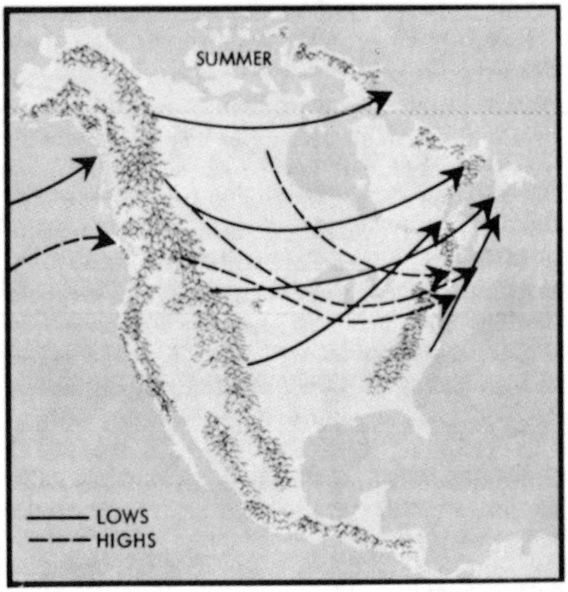


*Figure 4-12. Preferred summer tracks of migratory Lows and Highs. From Schroeder and Buck (1970).*

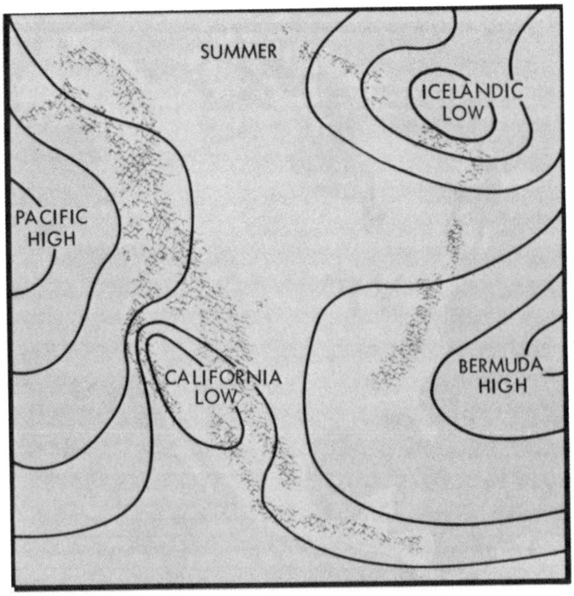


*Figure 4-13. Average July sea-level pressure pattern, showing the distribution of semipermanent seasonal Highs and Lows. From Schroeder and Buck (1970).*

genetic one; it indicates a particular wind pattern, without regard to its cause. But common to all of its occurrences is a relatively smooth boundary surface.

Low-level jets occur frequently on the Great Plains. Essential to the mechanism of formation is a layered atmosphere and the smooth relief of the plains. Most jets appear during the evening, when thermal turbulence no longer roughens the lower boundary. Similar jets have been reported within the marine inversion that develops over San Francisco Bay. And in mountainous areas, where strong evening inversions appear, low-level jets often characterize the wind regime above the inversion (Baughman, 1981). In all these cases, the inversion creates a favorable surface, with little roughness, for the winds move across (Figure 4-14).

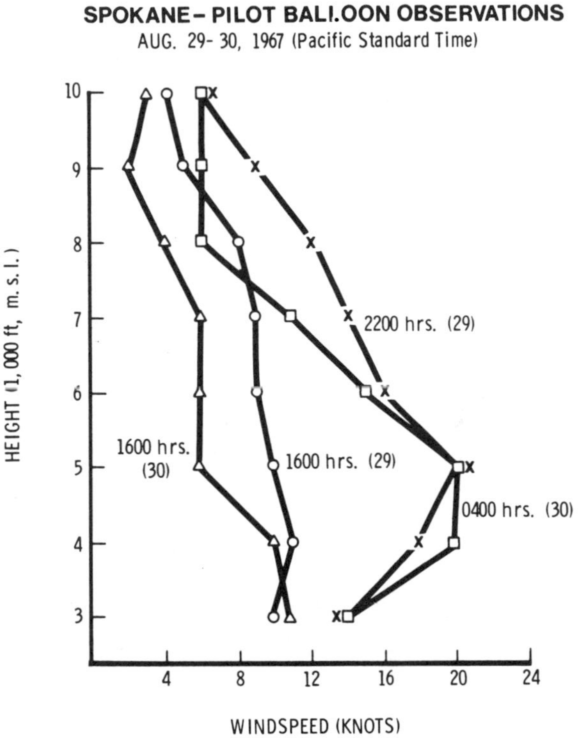


*Figure 4-14. The Sundance fire low-level jet wind, an example of strong evening wind development above an inversion. From Baughman (1981).*

*Foehn Winds.* Foehn winds belong to a worldwide category of strong mountain downslope winds (SMDW) that are known locally by an assortment of names and for a variety of somewhat different properties. Characterized by high velocities and warm temperatures foehn winds in the United States radiate outward from the Great Basin (Figure 4-15). East of the Rocky Mountains they are known as the *chinook,* and west of the Sierra–Cascade Range, as the *east wind, north wind, mono,* or *Santa Ana.* Though often considered a local wind, because its effects are localized, a foehn wind actually results from the interaction of mountains with meso-scale conditions.

The original theory of foehns characterized winds that occurred seasonally in the European Alps. The winds resulted from air that was lifted across the mountains, drained of its moisture, and then warmed adiabatically in its descent down the leeward slope. Some foehn winds in the United States do result from this mechanism (Figure 4–3). A high pressure cell develops in such a way that it extends across a mountain chain. This may occur after the passage of a cold front, or it

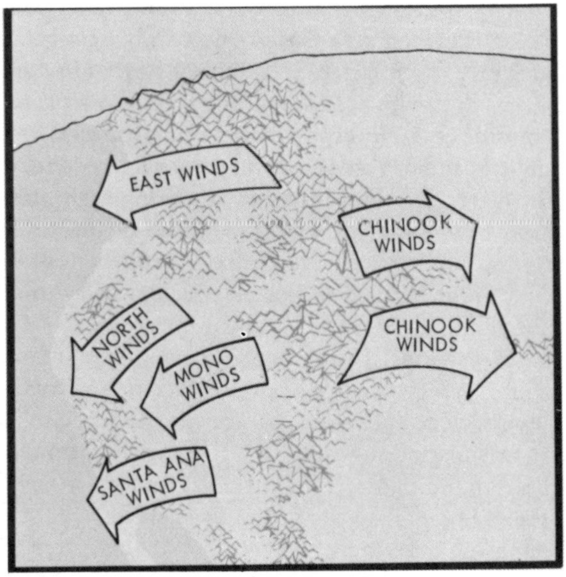


*Figure 4-15. Geography of foehn winds in the United States. From Schroeder and Buck (1970).*

may represent a bulge of a semipermanent feature like the Pacific High. Gradient winds sweep over the mountains and then warm dramatically as they plunge down the lee side. But other mechanisms are also common that rely on the topography of the Great Basin as a source of air, on strong pressure gradients to encourage flow, and on mountain waves as a means of actual transport.

In this case, foehns require the presence, and persistence, of a cold, dry high-pressure system in the Great Basin—an atmospheric no less than hydrographic basin that traps air between the Rocky Mountains and the Sierra–Cascade range. The mountain border stops the divergent flow that would otherwise spread out along the landscape; then if atmospheric pressure continues to build, the subsiding air aloft splashes across the ranges as a cascade of wind. If a strong pressure gradient develops on the opposite side of the mountains, then conditions may favor the violent release of air through mountain passes. The air rushes from an inland area of high pressure to an offshore region of low pressure. The flow is a veritable avalanche of air. Surface wind speeds may reach 40–60 mph (64–97 km/hr), or even higher. Because they come as waves, the winds may cease abruptly, or fluctuate as ridges of descending air succeed troughs. Because they reflect regional conditions, foehns may persist for as long as three days or more. When less stagnant Highs develop in the region—that is, when migrating high pressure cells pass through the Great Basin—mild foehn winds may develop for shorter periods.

Though warm, and often much warmer than the air it invades, a foehn can drive away local surface winds. Unlike katabatic winds, such as those which develop over glaciers, adiabatic winds such as foehns are not necessarily denser than the air they displace. How foehns can replace more buoyant air is the consequence of either a favorable pressure gradient or the phenomenon of mountain waves. A strong pressure gradient may lead to a partial vacuum on the lee side of the mountain range which the foehn fills. The amplitude and length of the mountain wave vary according to the momentum of the flow and the stability of the layer in which the wave occurs. If the waves are short, then the foehn may override local winds; if long, the foehn may actually surface, scour the topography, lift unexpectedly, and then surface again. Only when the system weakens do local or other general winds dominate. Flow conforms closely to topography, seeking out mountain passes and saddles, and depending on the character of the wave, a wind shadow may develop on the lee side of the range. This shadow may shelter local winds (Figures 4–16 and 4–17).

Foehn winds present extremely serious fire hazards. Because the source is drawn from the upper atmosphere, the relative humidity of foehns is low, 5% or less. Because of their long descent down the mountains, foehns heat adiabatically until they are much warmer then the air they replace. Because of the concentrated buildup of pressure and damming by the mountains, the pressure gradient that drives them will be great and foehns can reach fantastic surface velocities. Most of the largest fires of historic record in the Pacific Northwest and Southern California have occurred under foehn conditions—the east wind and the Santa Ana, respectively.

### *Local Winds*

Fires are local rather than regional events, and they respond most closely to local rather than to general winds or to general winds shaped by local circumstances. Even in the case of foehn winds, local topography directs the actual outrush of air, and a fire must react to a complicated interplay of foehn and local wind fields. The accuracy of most fire behavior forecasts will depend on the topography of the site. Most obey general principles that govern convective heating and the exchange of air from warm to cool regions. All require site-specific data. Knowledge of local wind systems is fundamental to fireline operations, both for the suppression of wildfire and for the execution of prescribed fire. General winds will be given in synoptic weather forecasts, but the prediction of

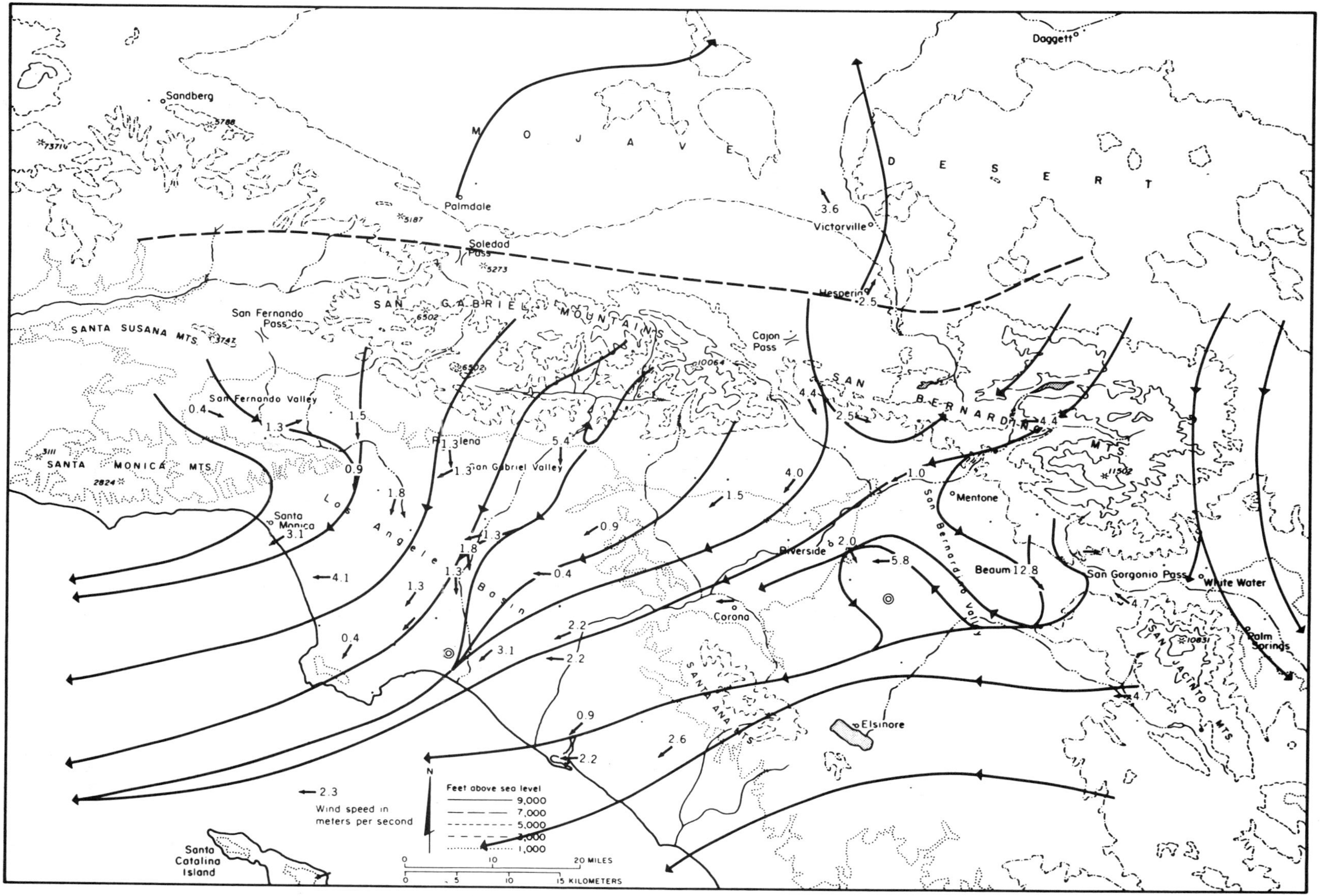


*Figures 4-16 and 4-17. Interaction of Santa Ana winds and marine air flow. The map on this page shows mature Santa Ana flow (0100 hrs) dominating local winds, which are blowing offshore anyway. The map on the next page shows the reestablishment of onshore marine air (1600 hrs). The main penetration occurs behind the wind shadow created by the San Gabriel Mountains, but flow is turbulent and sporadic. From Fosberg et al. (1966).*

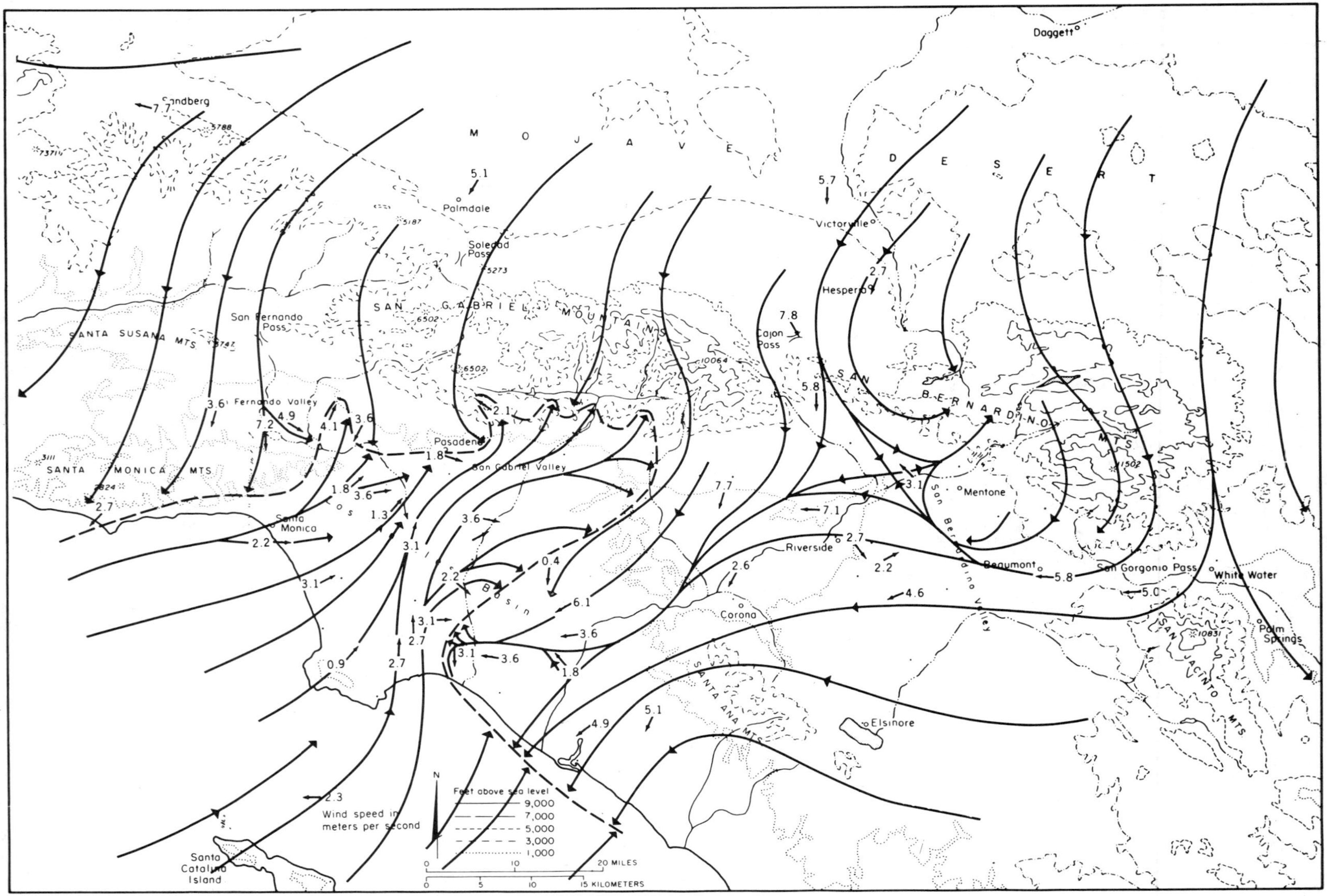
MOJAVE DESERT
Daggett
Sandberg
Palmdale
Soledad Pass
Victorville
Hesperia
Cajon Pass
SAN GABRIEL MOUNTAINS
SAN BERNARDINO MTS
SANTA SUSANA MTS
SANTA MONICA MTS
San Fernando Pass
Fernando Valley
Pasadena
San Gabriel Valley
Santa Monica
Los Angeles Basin
Mentone
San Bernardino Valley
Riverside
Beaumont
San Gorgonio Pass
White Water
Palm Springs
SAN JACINTO MTS
Corona
SANTA ANA MTS
Elsinore
Santa Catalina Island
Wind speed in meters per second
Feet above sea level
9,000
7,000
5,000
3,000
1,000
N
0 10 20 MILES
0 5 10 15 KILOMETERS

*Figure 4-17*

local wind regimes will depend, as often as not, on fire behavior officers or fire weather meteorologists at the scene.

Local winds are convective winds, the product of differential heating and cooling. Surfaces irradiated by the sun heat adjacent air through conductive transfer. This air expands upon heating, rising buoyantly because of the pressure difference between it and the surrounding air. Conversely, surfaces that are cooling drain the air around them of their heat; this air, now denser than its surroundings, descends. The temperature differential may arise because of differences in the amount of solar radiation (insolation) received by a surface or because of differences in the capacity of various surfaces to absorb that radiation. Insolation will vary with diurnal and seasonal changes, with the orientation of a surface relative to the sun (aspect), or with the presence of clouds, particulates, and water vapor in the atmosphere—all of which have the effect of intercepting the original radiation. Moisture of any kind—in the atmosphere, on the surface, or in vegetation—has a moderating effect, acting as a kind of buffer to reduce extreme temperature differences. Bare surfaces heat faster than vegetated ones, lands faster than oceans, arid climates faster than humid.

Convection may be organized in various ways. If the temperature differentials result from a shoreline separating land and sea, convection may take the form of a cell, rising over the zone of heating and descending over the zone of cooling. If the differentials occur in mountainous country, convective flow will tend to follow topographic contours. During the day winds will rise upslope and upvalley and, converging along the summits, ascend as a kind of plume. At night, air flow will reverse, spilling downslope and downcanyon much like water runoff, ultimately to pool in deep valleys. In flatter terrain, intensely heated air along the surface may escape as rising parcels or bubbles, as though the atmosphere were slowly boiling. If superadiabatic lapse rates are present, dust devils or whirlwinds may develop—a yet more abrupt means of restoring the vertical imbalance between warm surface air and cooler air aloft (Figure 4–18).

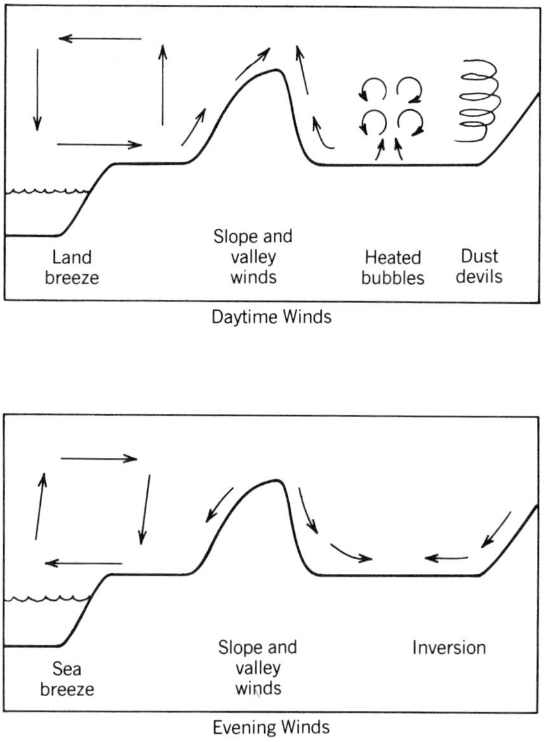


*Figure 4-18. Local convective winds.*

*Land and Sea Breezes.* The convective cell that may develop along a shoreline experiences a diurnal cycle. Land surfaces heat and cool more rapidly than do ocean surfaces, and hence are warmer in the daytime and cooler at night. During the day the air over land rises, and cooler air from the ocean moves in to fill the partial vacuum created by convection. The ascending air, in turn, expands, cools, and eventually sinks back over the ocean, completing a gigantic cell. At night this process reverses. The ocean is warmer than the land, air above it rises, and cooler air from the land flows out to take its place. The daytime breezes are referred to as *onshore winds,* and the nighttime, as *offshore winds.*

This pattern, however, interacts in complex ways with other, larger wind fields. When the general wind flows in the same direction as the local wind, it strengthens and masks the local wind regime. When both general and local winds flow in opposite directions, however, the general wind may block the local flow entirely, stall it temporarily, skim across the top of it—or a mixture of all three. It is common, for example, for onshore local winds to be momentarily halted. This event allows the onshore winds to pile up, ultimately to invade the land more as a small-scale cold front than as a breeze. Such fronts can be found on both the east and west coasts of the United States. Similar effects can result from the interaction of local winds with topography. Mountain ranges bordering the shoreline can function much like moderate offshore winds, allowing the onshore winds to build up in intensity and then punch through passes or saddles, often with considerable velocity.

This occurs along the western United States. In California the onshore winds may reach speeds of 10–15 mph (16–24 km/hr). At the same time, onshore winds that strike mountains may be assisted up the mountain flanks by local upslope and upvalley winds, bringing moist marine air to much greater heights than the depth of the onshore wind (1200–1500 ft; 366–457 m) would suggest. Even more complicated patterns between convective onshore winds and general winds may occur in mountains where strong downslope mountain winds make an appearance. Such winds may drive off the onshore winds entirely, or, being frequently much warmer than the marine air, they may override the shallow invasion of onshore air, or, all too often, they may interact in kaleidoscopic patterns, creating wind fields of constant change and great hazard.

In the eastern United States, land and ocean temperature differentials are greater in the spring and early summer. Land and sea breezes may thus influence the spring fire season, though their effect on the fall fire season is much less. Often during fire season the general winds obscure the local winds by generating strong onshore breezes of their own. But when a general offshore wind opposes a local onshore wind, a sea breeze front develops which can travel at 3–4 mph (5–6.4 km/hr), and which may advance and retreat many times across a stretch of coast before breaking up. The effect on a fire is akin to having a series of cold fronts pass across it, with sudden changes in wind speed and direction. The opposing winds, moreover, create a pattern of convergence for which strong vertical development is a predictable outcome. The combination of small cold fronts, turbulent vertical motions, and the high temperatures that are necessary to power the convective cell—all are conducive to fire spread and intensity, and all are likely to occur during the spring fire season.

A somewhat different mechanism operates along the western United States. Unlike the ocean temperatures of the east coast, which derive from the warm waters of the Gulf Stream, the temperatures of the west coast are set by the cold California current. The highest temperature differentials thus occur during the summer fire season. The onshore winds are stronger and, flowing from west to east, they are more often supported by general winds than opposed to them. Marine air may enter the coastline in thick surges. Where conditions favor turbulence, the general surge may break down into a series of advancing, smaller leaps inland. The spread of the onshore breeze can reach 10–15 mph (16–24 km/hr), and it may advance as much as 30–50 miles (48–64 km) from the shoreline.

Although the general land–sea convective winds combine to strengthen the invasion by marine air, mountainous topography complicates the scene by introducing other wind systems, both local and general. The onshore winds must interact with slope and valley winds created in the mountains, often of unusual properties and irregular occurrence, and they must confront occasional strong downslope mountain winds, primarily foehn winds that spill across the mountain passes in late fall and early winter. In such a situation relative humidity plummets, and fuel which was—only hours previously—unavailable for combustion because of its high fuel moisture suddenly becomes available. A fire caught between two such opposing wind fields will behave erratically and somewhat violently.

*Slope and Valley Winds.* A similar diurnal cycle characterizes the wind field that develops along slopes and valleys. During the day solar heating causes air to rise along slopes. Though the flow is shallow, it can function as a broad channel for collecting and directing the winds that pass through it; ravines or natural chimneys accentuate this trend towards organization; as the volume of air flow increases, so does the depth of the upslope winds. Powered by solar radiation, winds may reach 10–15 mph (16–24 km/hr) by midafternoon; flow is turbulent. At night the surface cools and the entire process is reversed. A shallow layer of cool air flows downslope, following drainage configurations. Driven by gravity, the flow is laminar, with wind speeds of 2–5 mph (3–8 km/hr). Strong upslope winds in mountainous topography can give rise to convective flow upward into the atmosphere; intense downslope winds, to inversions.

An analogous, and not unrelated pattern, develops for valleys. In this case the entire basin must be warmed, not merely a slope. But valleys offer several advantages for solar heating. They are shielded to some extent from general winds that might mix and disperse warmed air in the basin. By volume they contain less air and more land than a plain, and it is the land surface that offers proportionately more area for solar heating. The air in the basin warms; and when the temperature differential becomes great enough, upvalley winds result. At night the process reverses. The same features that allowed for a buildup of warm air support a buildup of cold. When upvalley and downvalley winds are well-developed, they dictate the wind field of the slopes.

Valley and slope winds interact not only with each other, but with the general winds as well. When the general winds consist of cold, dense air they may well dominate local circulations. If the general winds are warmer and lighter, however, they may assist daily upward flow, or ride over the inversions of dense air that form at night. The combinations are legion. A pattern of stiff downslope winds, for example, routinely emerges in the afternoon along the eastern slopes of the Pacific Coast Ranges, when slope and valley winds combine with onshore winds. Similarly, strong downslope winds can develop at night along the lee side of mountains in the form of mountain waves or foehn winds. Though regional conditions establish general parameters that govern the range of potential behaviors for a fire, the fire responds to particular events at a specific site. Its outcome is probabilistic, and its environment ultimately local. On large fires, fire behavior officers (whose job it is to forecast fire behavior) are drawn from the ranks of those who have not only received formal training but who are knowledgeable about local idiosyncracies. Most of those pecularities will involve local wind systems.

*Thunderheads.* For a thunderstorm to develop it is necessary, first, to have an air mass that contains ample moisture in a state of conditional instability and, second, to have some mechanism by which to release this instability. The instability must be deep enough that a level of free convection can be attained, and the triggering device—some form of lifting—must be powerful enough to raise a parcel or layer of air to this elevation. For the potential energy inherent in the atmosphere to be released, that is, a certain amount of energy must be applied to it. When the energy liberated is greater than the amount invested, a thunderstorm can result.

The particular mechanics of cumulus cloud formation are not completely understood, and for purposes of analyzing the wind fields of a thunderhead are not essential for understanding. Nor is it necessary to catalog the many varieties of thunderstorms, a taxonomy that relates thunderstorms to the method of lifting. One special type of storm, however, is worth singling out. This is the *dry thunderstorm,* characterized by a high cloud base and the evaporation of precipitation before it reaches the ground. Dry thunderstorms bring the winds and lightning of a thunderstorm without its moisture; they contribute to fire ignition and spread without assisting in fire suppression. In the Southwest such storms develop when moist air from either the Gulf of Mexico or the subtropical Pacific enters the region at high elevations (10,000–18,000 feet; 3048–5486 m). When this air is lifted across the mountains—a displacement supported by strong convective cells over heated desert surfaces—high-level thunderstorms can result. In other areas of the far West, the presence of a cold low pressure cell aloft produces similar effects. In this case a closed low pressure system becomes separated from the main storm tracks, wanders randomly at upper elevations, and touches off dry thunderstorms as the cold air it contains interacts with the surrounding air.

A thunderstorm can be analyzed as a thermodynamic cycle. It begins with the uplift and inflow of air in a convective pattern. It matures when the ascending air reaches equilibrium, and a return flow of descending air sets in. It ends with general dissipation: the air of the cell mixes with the surrounding air, and no further energy is introduced into the system (Figure 4–19). Each thunderstorm is, in turn, an amalgamation of individual convective cells—each of which may be in a different stage of development and each of which, in the course of its own life cycle, may initiate other cells. In brief, the whole process begins with an investment of energy in the form of lifting. Above the lifting level of condensation the energy stored by water vapor as latent heat is released to power the system. Free convection and precipitation remove this potential energy from the system.

With or without surface precipitation the wind field associated with a thunderstorm can greatly influence fire behavior. If lifting is by thermal convection, then the updrafts that lift air to the condensation level may be felt on the surface as part of normal diurnal upslope and upvalley winds. Updrafts may accelerate the convective flow above a fire. The downdrafts may be localized but intense. The first gust strikes the surface in a surge. If the terrain is flat, then a dome of cold, descending air builds up and the air mass rushes out much in the manner of a fast-moving cold front. A sharp change in wind speed and direction marks its passage. Winds may gust up to 60 mph (97 km/hr). If the terrain is rugged, then the descending air will follow the existing drainage configuration, at great speeds and with turbulence. The surface flow of air from the thunderstorms is always outward from the cell. Regardless of whether a fire is to the front, to the side, or to the rear of a passing thunderstorm, the effect of the thunderstorm's winds will be to drive the fire away from the storm center. Their speed and gustiness make such winds treacherous. More than once thunderheads have caused fires to break out of control, have led to spotting, and have threatened the safety of crews working control lines.

## *4.3 FIRE DANGER RATING*

As natural events, wildland fires are episodic, not periodic. They represent the compounding of various meterological, physical, and biological processes. Neither for fire, nor for the elements that come together to make the fire environment, does history proceed according to regular cycles. Like floods, fires are better described by a frequency–intensity distribution than by a simple frequency-time distribution. Yet fuels and weather do follow daily and seasonal trends, the product of solar heating; fires tend to concentrate within certain portions of the day and year; and an apparent periodicity is evident.

These apparent periods have given rise to a series of concepts by which to characterize the distribution of fires or of fire behavior traits: the burning period, on the scale of a day; the critical fire period, on a synoptic scale; the fire season, on an annual scale; the fire climate, on the scale of decades. These periods are statistical averages not cycles, strictly speaking. They simply describe the likelihood of fire occurrence and behavior. Since the greatest variable in the construction of these periods is weather, fire periodicity—and the means to assess fire danger—may be considered under the general topic of fire weather.

The apparent periodicity of fire both simplifies and complicates the planning and administration

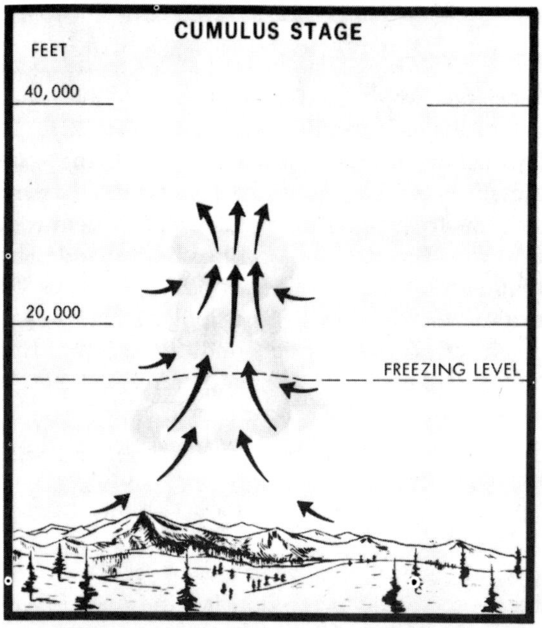


**The cumulus stage of a thunderstorm cell is characterized by a strong updraft, which is fed by converging air at all levels up to the updraft maximum. Rain does not occur in this stage.**

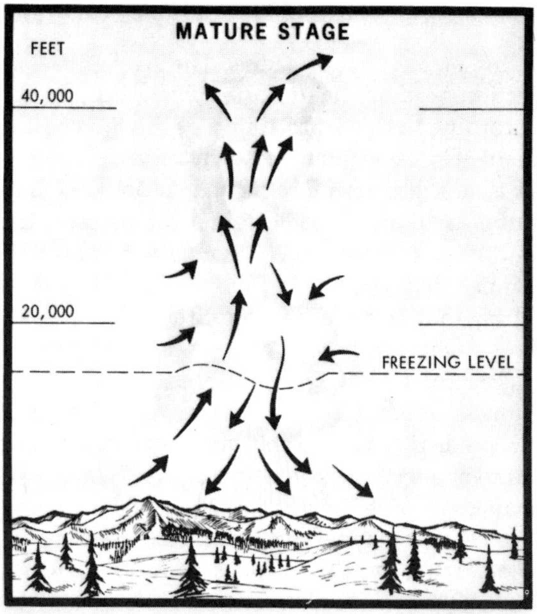


**The mature stage, the most active portion of the thunderstorm cycle, begins when rain starts falling out of the base of the cloud. The frictional drag exerted by the rain or other precipitation initiates a downdraft. There is a downdraft in part of the cell and an updraft in the remainder. The updraft is warmer, and the downdraft is colder, than the air surrounding the cell.**

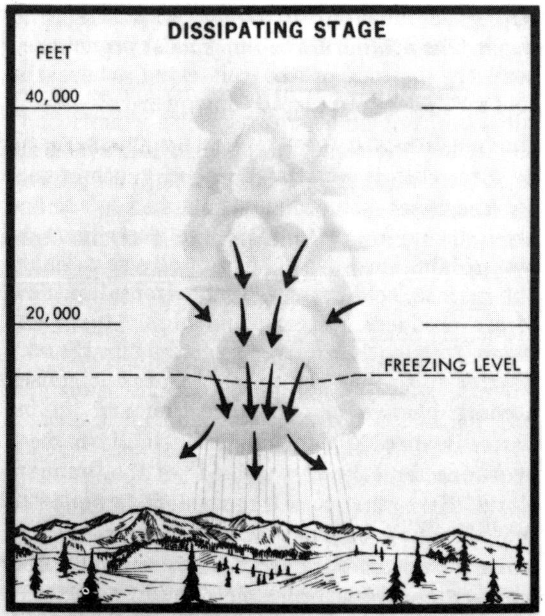


**The downdraft spreads over the entire cell, and the updraft disappears in the dissipating stage. Light rain falls from the cloud. Gradually the downdraft weakens, rain ends, and the cloud begins to evaporate.**

*Figure 4-19. Thunderstorm development. Updrafts dominate during the cumulus stage; strong downdrafts dominate the surface wind flow during the mature stage; gentle downdrafts characterize the dissipation stage. A young thunderhead passing over a fire may accelerate convective flow upward, while a mature thunderhead will drive the fire perimeter outward. From Schroeder and Buck (1970).*

of fire management systems. The level of fire protection need not be uniformly high, but can vary according to the flammability of the fire environment. Fires burn more readily during some parts of a day than during other parts; on some days, rather than on other days; in some seasons, than in others. As a guide to short-term planning, fire danger rating systems have been developed. These indices replace general forecasts based on fire seasons with forecasts specific to particular places for specific periods of time. Nor is the human response to fire periodicity restricted to the prediction of such episodes. People can and do intervene in the distribution of fire. Natural fires may be suppressed and anthropogenic fires set. Fire seasons may be compressed or lengthened. In general, the outcome of this intervention is to make fire occurrence more periodic rather than less so, to reduce much of the element of chance that normally dictates fire starts and determines fire behavior.

### *Concepts of Fire Periodicity*

*Burning Period.* The concept of a burning period relates the probability of fire ignition and fire behavior to a diurnal cycle of temperature and relative humidity. As a day progresses, temperature rises, relative humidity (and fuel moisture) decrease, and convective winds pick up in intensity. It is more likely that a firebrand will lead to ignition and that an initiating fire, or a steady-state fire, will escalate in intensity. On the average, this transformation begins around 1000 hrs in the morning, with the breakup of the evening inversion, and it wanes about 1800 hrs in the afternoon, with the approach of evening. The time between 1000 and 1800 hrs is traditionally known as the *burning period,* and almost from the start of organized fire protection the practical objective of fire control has been to contain a fire before the onset of the next daily burning period. Control by 1000 hrs remains a useful rule of thumb. The concept applies, however, to local, not to synoptic weather. Its actual timing and extent—even its existence—will vary with local peculiarities.

*Critical Fire Periods.* High fire danger tends to be associated with certain synoptic conditions, meteorological events that extend from 2 to 10 days, that appear episodically, and that are known as *critical fire periods.* The typical pattern varies with geographic region, and it defines a regional rather than a local burning period. The frequency of critical fire periods also varies by region: some areas experience long periods during which critical synoptic conditions come frequently; others know them only rarely. The great majority of wildfires and the vast proportion of burned acreages will occur during critical burning periods (Schroeder et al., 1964). The problems of coping with large wildfires reduce, in most cases, to the problem of forecasting the episodic appearance of critical fire periods and of providing greatly increased protection for the duration of these events.

Synoptic conditions that produce critical fire hazards can be divided into *surface types* and *upper-air patterns.* The surface types all relate to high pressure cells (Figure 4–20). The general process involves regional subsidence, an unstable atmosphere, stiff winds, and a blockage of the normal flow of humid (often marine) air. In the western United States, this blockage results from offshore winds; in the east, from stable high pressure that interferes with the general circulation of moist air from the Atlantic Ocean and the Gulf of Mexico. The consequences are a drop in fuel moisture, perhaps even drought, a prevailing lapse rate favorable to surface convection, and a synoptic condition that persists stubbornly.

High pressure cells are known by their source regions, of which five exist for North America. Seasonally, such cells tend to be semipermanent features—though they do show a tendency to migrate, and even when stable they will often spill into areas beyond their normal range. If a cell does break loose to wander, the source region builds a new replacement. When these Highs reach beyond their source region, they bring fire danger with them. When the Pacific High, for example, extends a ridge into the Pacific slope region, it can lead to foehn winds (east or Santa Ana); when it

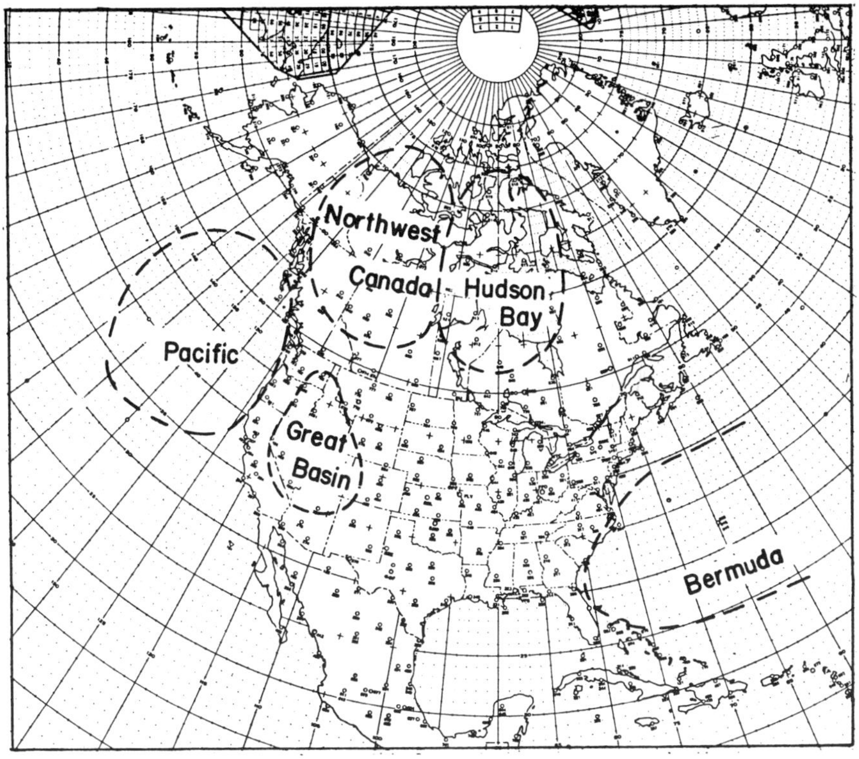


*Figure 4-20. Critical fire periods: surface types. (A). Locations and names of principal high pressure cells*

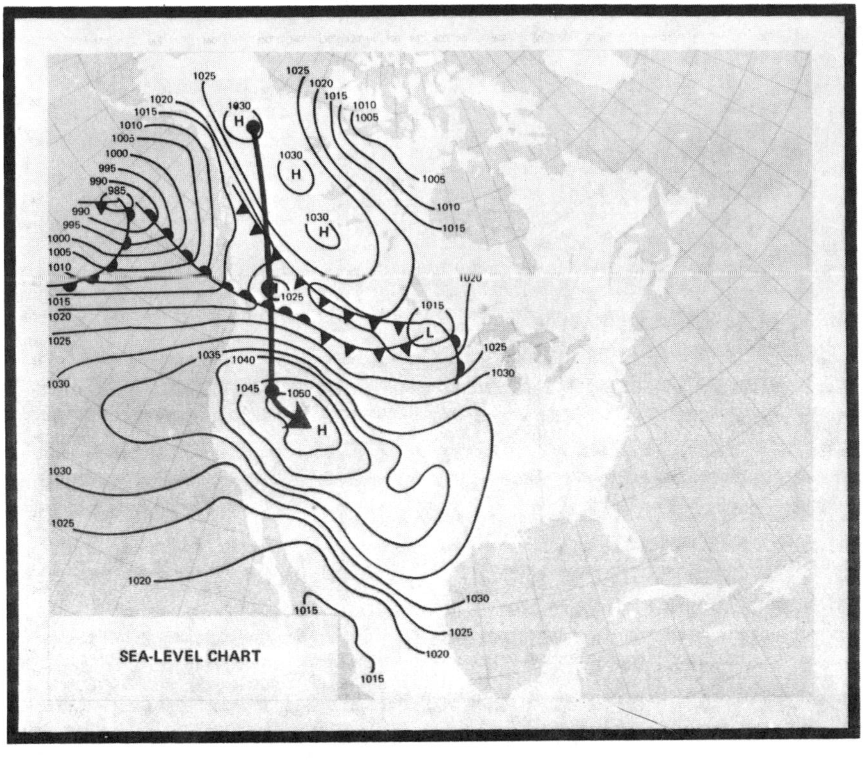


*Figure 4-20 (B). Extension of Pacific High inland; dashed lines show past positions of front.* *(Continued on next page)*

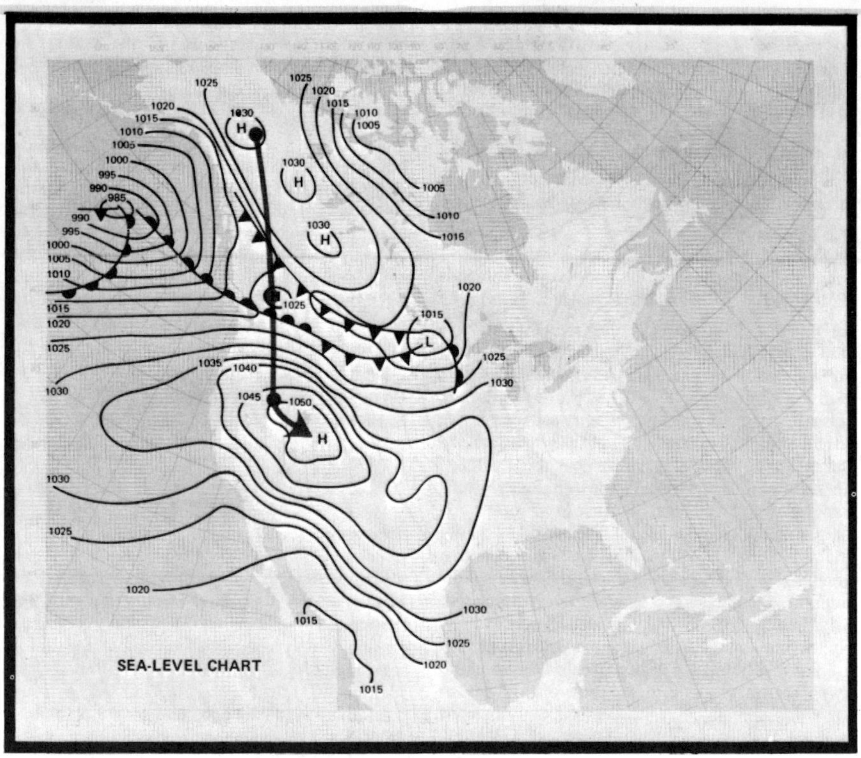


*Figure 4-20 (C) Great Basin High, tracked by arrow. Coupled with an offshore Low, foehn winds are likely.*

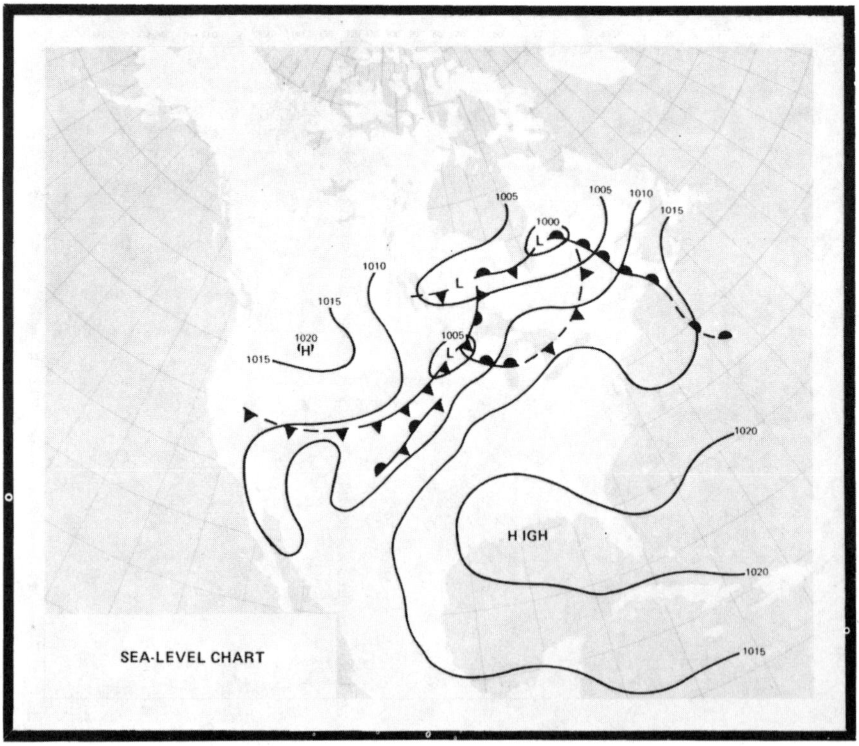


*Figure 4-20 (D) Enlargement inland of Bermuda High, cutting off moisture from the Gulf.*

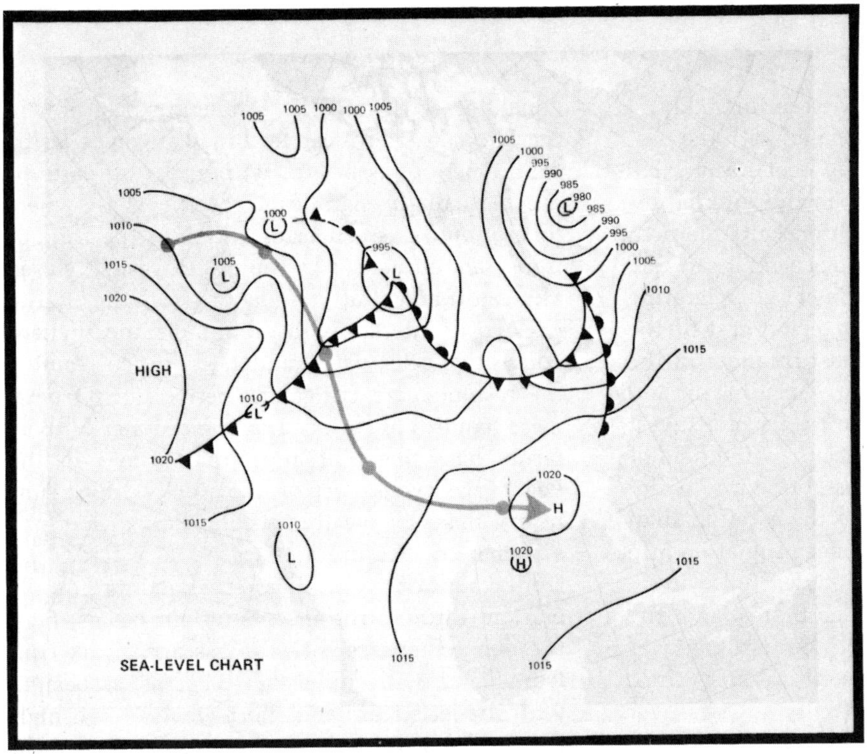


*Figure 4-20 (E) Pacific High tracking across United States. Usually the western or northwestern portion of the High produces the most critical fire danger.*

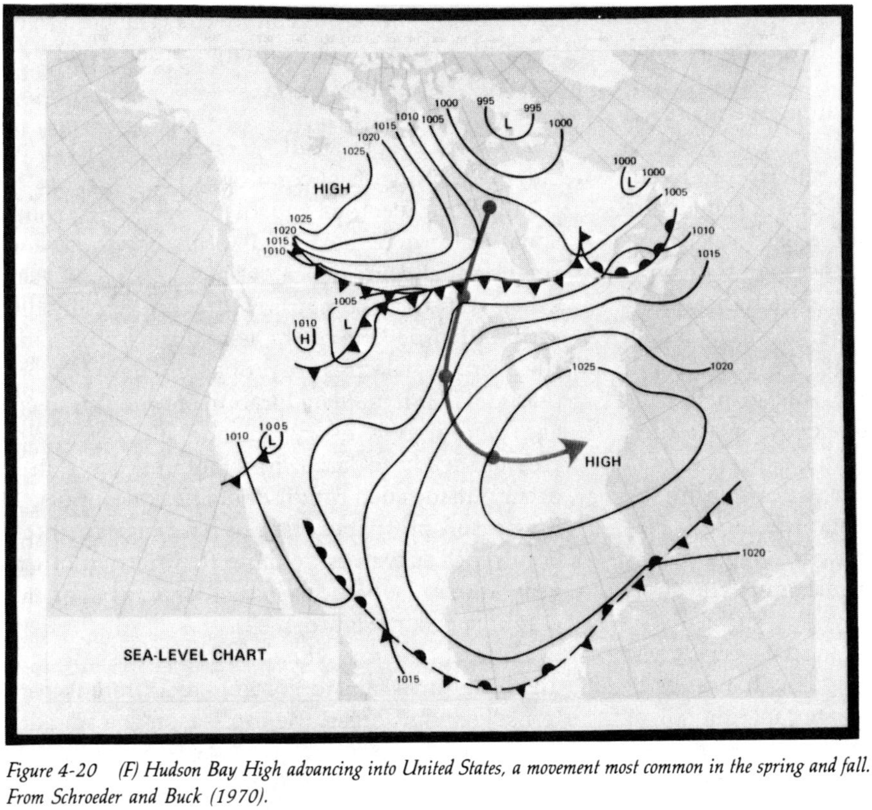


*Figure 4-20 (F) Hudson Bay High advancing into United States, a movement most common in the spring and fall. From Schroeder and Buck (1970).*

wanders east like a gigantic, descending eddy, it generates the wind fields associated with the passage of a dry cold front. The Great Basin High creates excellent fire weather in the intermountain region; if a strong enough pressure gradient exists with a Low off the Pacific coast, it can lead to foehn winds. The Hudson Bay High, migrating southward, can influence any region east of the Rocky Mountains. Usually most frequent in the spring and fall, it brings warm, subsiding air and especially critical conditions to its northwest flank. Its passage, too, generates the wind field of a cold front. By a similar process, but with a wider range for movement, the Northwest Canadian High also moves south—at times stagnating into a Great Basin High; at other times following a storm track though the Lake States and the Northeast. The north and northwestern flanks of the cell are generally the most critical areas since descending air has no opportunity to pass over large bodies of water and acquire moisture. Strong, dry winds mark the boundary of the High. Moreover, a High may extend its influence by a process of enlargement rather than by wandering. Like the Pacific High, the Bermuda High—normally a resident off the southeastern coast of the United States—induces drought and critical fire weather simply by broadening itself inland. In this case it reaches across the southern Gulf plains and shuts down the normal flow of moisture pumping north from the Gulf of Mexico.

Among upper-air patterns there are four typical arrangements: meridional, in which the upper-air chart shows long amplitude; zonal, in which the amplitude is shorter but the wavelength longer; short wave trains, in which small scale waves pass through the general belt of westerlies; a blocking High, in which the existence of a closed High north of a Low causes flow within the belt of westerlies to pass around the region (Figure 4–21). The blocking High encourages subsidence and can lead to serious drought. The short wave train is accompanied by dry, surface cold fronts, magnifying the effects (and especially the winds) typical of a surface High. Such a series of troughs can exist within a larger meridional or zonal flow. By contrast, a stable upper-air flow in meridional patterns can lead to heat waves. Critical fire periods associated with upper-air patterns are particularly significant for the West where lofty mountains can bring fuel into the zone of upper-air influence. In the Southwest, a closed High aloft may draft in moisture from the Gulf of Mexico to interact with mountains and thermal convective cells; dry thunderstorms can result.

To summarize, critical fire periods stem from a mere handful of surface types and upper-air patterns that appear at appropriate times in the vegetative cycle. East of the Rocky Mountains nearly all severe weather occurs along the periphery of high pressure areas, especially across the frontal zone of dry air masses. Along the Cordillera and through the intermountain basins, high fire danger appears with the passage of dry cold fronts and with upper-air patterns associated with the jet stream and short-wave troughs. For the far West critical weather results from heat waves, the product of subsidence. And everywhere that foehn winds are possible, they present enormously dangerous conditions. The pattern of its critical fire periods will be recorded vividly in a region's fire history.

*Fire Season.* Proper weather conditions must coincide with proper fuel conditions for large fires to result. Both weather and fuel relate to the seasonal pattern of solar radiation, however, and their annual coincidence defines a *fire season.* Another statistical composite, the fire season describes that portion of the year during which, on the average, wildfires are probable. The actual fire season will fluctuate in duration around this average period. Not every episode of critical fire weather leads to large fires, but few large fires occur outside such conditions. The concept of the fire season prescribes the normal range for their occurrence as well as defining that portion of the year during which the vast majority of all wildfires occur (Figure 4–22).

Different regions will experience fire seasons at different times and according to different patterns. For some areas, the fire season is a singular

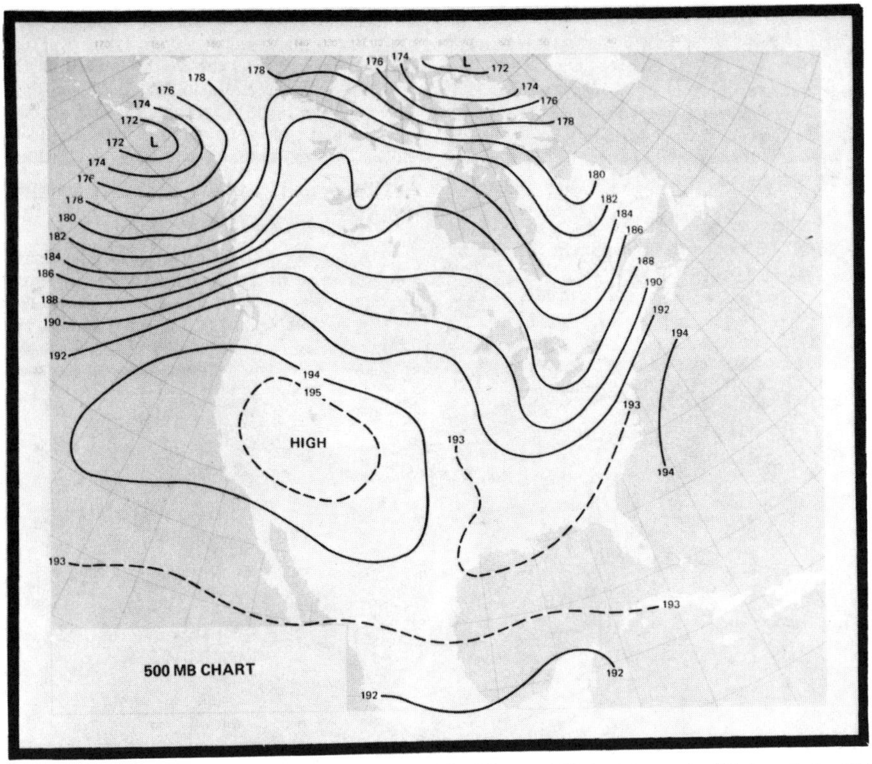


*Figure 4-21. Critical fire periods: upper-air patterns. (A) Subtropical High aloft over the West, producing high temperatures, low humidities, and unstable air near the surface.*

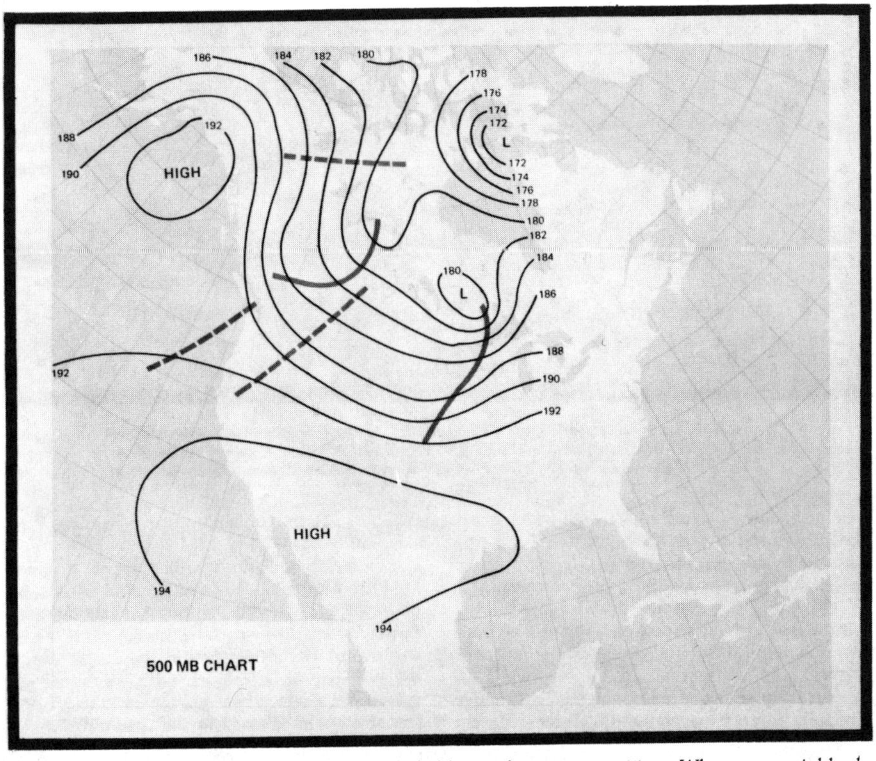


*Figure 4-21. (B) Short-wave troughs aloft, with dashed lines indicating past positions. When accompanied by dry surface cold fronts, high fire danger results.*

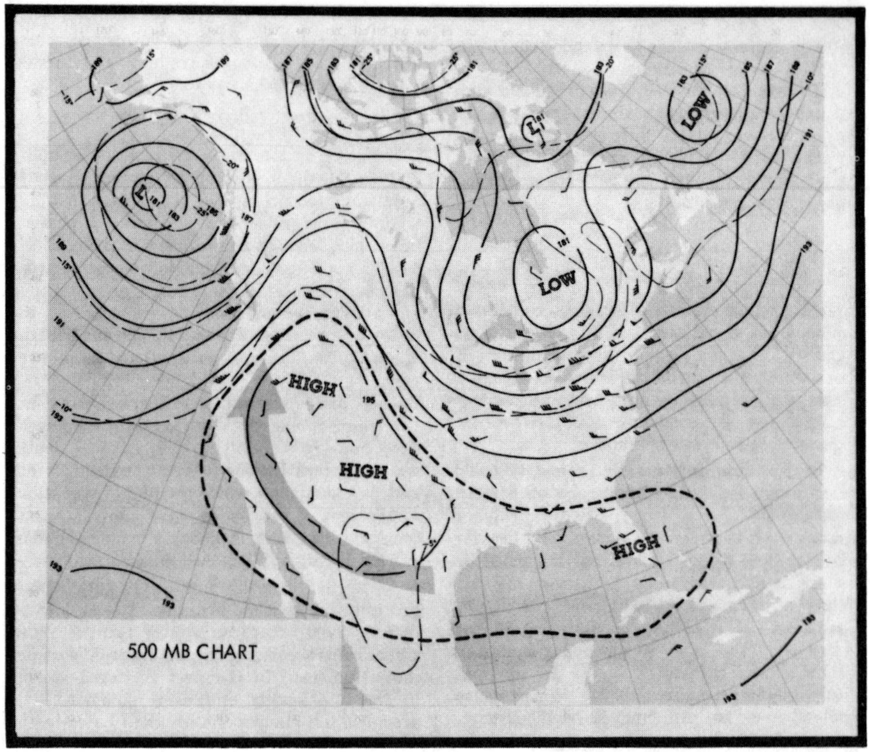


*Figure 4-21* *(C) Large High over Gulf of Mexico brings moist air into the Southwest which is then lifted by thermal and orographic means.*

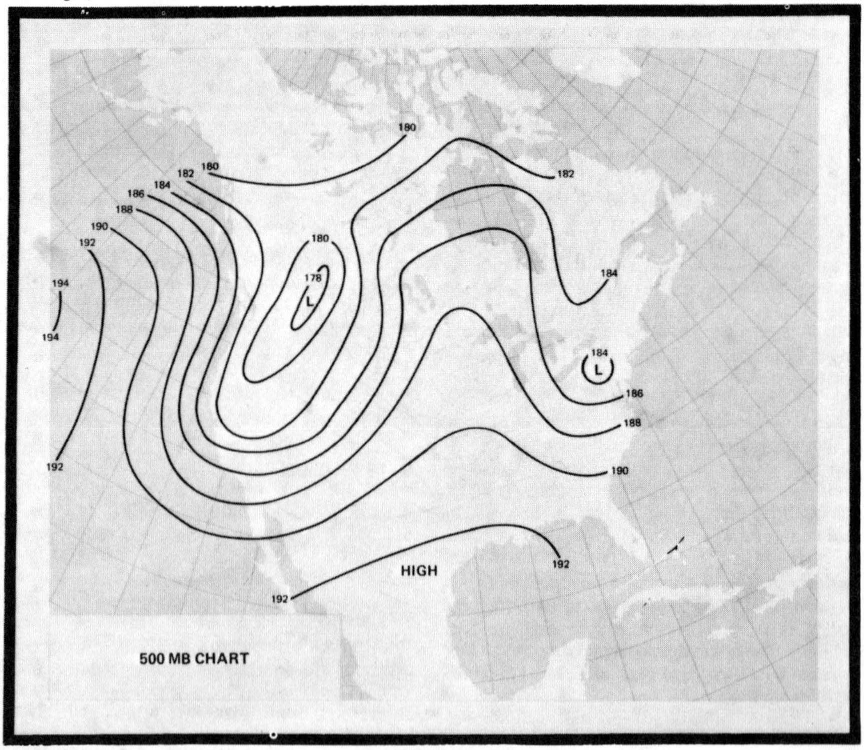


*Figure 4-21* *(D) Southwesterly flow alof, with peaks in fire danger occurring as short-wave troughs and surface cold fronts pass over region. From Schroeder and Buck (1970).*

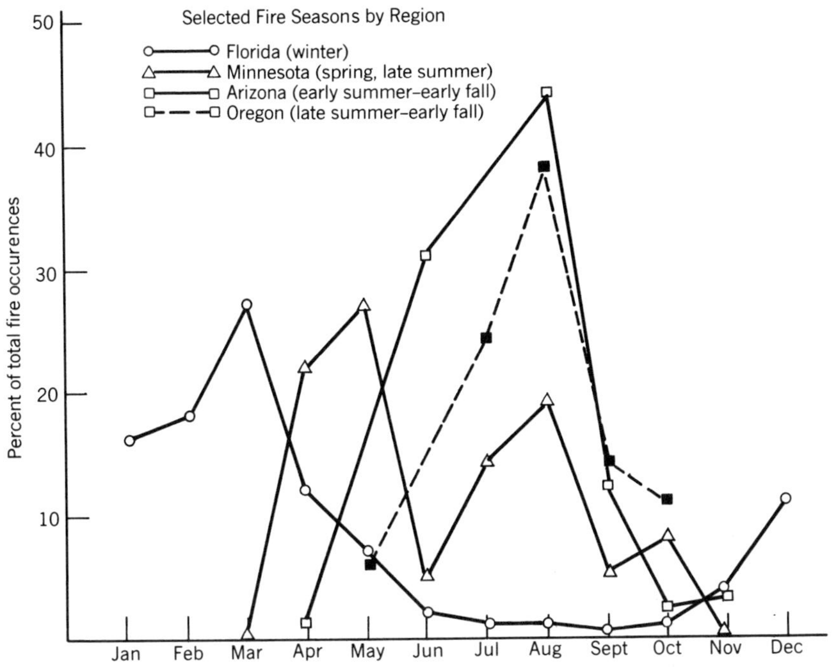


*Figure 4-22. Fire seasons: wildfire. Regional occurrence of wildfires as a monthly percentage. Statistics from Cooperative Fire Protection, U.S. Forest Service.*

phenomenon, a short if occasionally brilliant annual event. Such a pattern characterizes the northern conifer regions, with their brief summers, or tropical areas, with their annual dry periods. For some areas, the fire season is bimodal. It appears in the spring and the fall, tied to the cycle of grass curing, and it subsides during the summer rainy season, when fuel moisture reaches a peak. Such a pattern typifies broadleaf deciduous forests and southern coniferous forests for which the primary surface fuel is grass. For still other areas, the fire season shows a sequence of patterns—an annual event of moderate proportions during a hot dry summer and an episodic event during outbreaks of foehn winds in the late fall or early spring. Such a condition characterizes many mountain regions with Mediterranean climates, such as Southern California, and it makes the fire season virtually a year-round affair.

Traditionally, the concept of a fire season applied to the period of wildfire occurence. Yet fire management has greatly expanded the annual appearance of fire. The fuel cycle is modified, the ignition pattern is distorted by the addition of accidental or deliberate fire, and the fire season, as actually practiced, is significantly reorganized (Figure 4–23). Curiously, the effect of such change is to increase the periodic nature of fire. To the extent that prescribed fire supplements, or even replaces, wildfire, the stochastic nature of wildland fire gives way to more predictable periods of applied fire. Much of the element of chance is removed. Critical fire periods may or may not come; if they appear, they may or may not coincide with proper ignitions. But periods of prescribed fire—especially when combined with fuel preparations—are vastly more predictable, more regular, and more periodic. Commonly, the prime periods for prescribed fire are either before or after the wildfire season, lengthening and smoothing out the annual fire load. With more land burned annually from prescribed fire than from wildfires in many regions, the effect on the shape of the fire season is pronounced. Computer-assisted programs of fire climate analysis such as the Firefamily group produced by the U.S. Forest Service, will likely enhance this trend (Main et al., 1982).

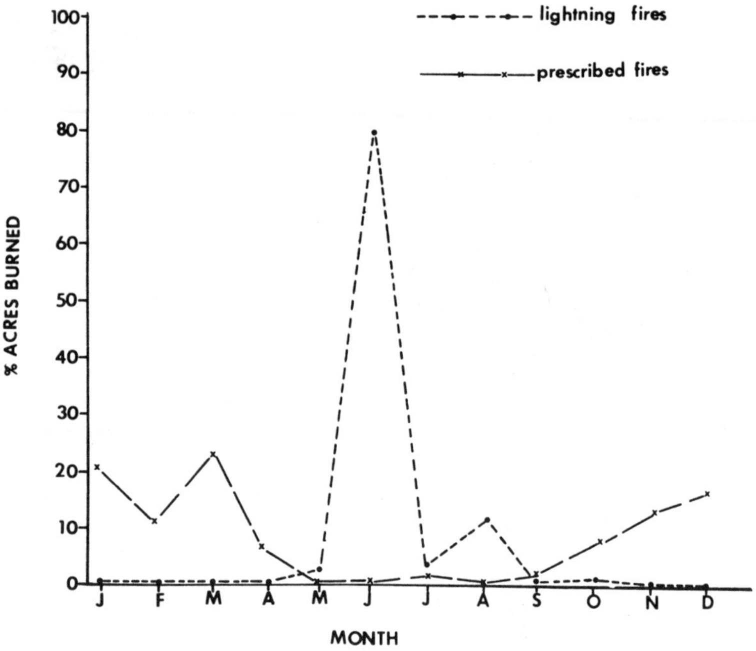


*Figure 4-23. Fire seasons: effect of prescribed burning. Note how prescribed fire extends the fire season. Example from Everglades National Park. From Taylor (1981).*

*Fire Climates.* An averaging of fire seasons defines a *fire climate.* The concept refers to a recurring pattern of meteorological conditions, the long-term distribution of fire seasons and critical fire periods (Figure 4–24). Each fire climate region shows a characteristic fire season, a distinctive mode for the appearance of critical fire periods, and, usually, a particular problem fire. For purposes of planning, the fire climate concept is considered in terms of an "average worst" fire season or average worst critical fire period—typically the third heaviest fire season or critical period of a 10-year average. Staffing for an average season or an average fire is not a useful management strategy: it is the episodic large fire and the occasional bad fire season that must shape plans. Forecasting these events is only possible in broad probabilistic terms based on historical statistics. The fire climate concept identifies place, while various fire danger rating schemes identify time.

The concept of a fire climate region is not synonymous with the concept of a fire regime. A fire climate describes a recurring pattern of weather; a fire regime, recurring patterns of ignition and fuel. The fire regime assumes a fire climate, but it amalgamates weather with other components of a fire environment. Under natural conditions (for which lightning is the sole form of ignition and within which only natural processes set fuel loadings) the fire regime is determined by the fire climate. But such conditions rarely exist. Almost everywhere humans intervene to modify fuels, to introduce new ignition sources and remove traditional ones, and to alter the timing of ignition. By definition, prescribed fire occurs outside the critical fire periods that lead to most fire effects. The processes that shape a fire climate provide the occasion for, and in some cases the ignition for, fire regimes, but regimes are biological and cultural systems that integrate processes far beyond their meteorological components. These processes have extended across the landscape of America for hundreds, perhaps thousands of years. To the extent that knowledge of natural fire regimes is desirable, that information today resides principally in fire climate data. Fire weather is the component of a fire environment over which humans have the least control and to which must be at-

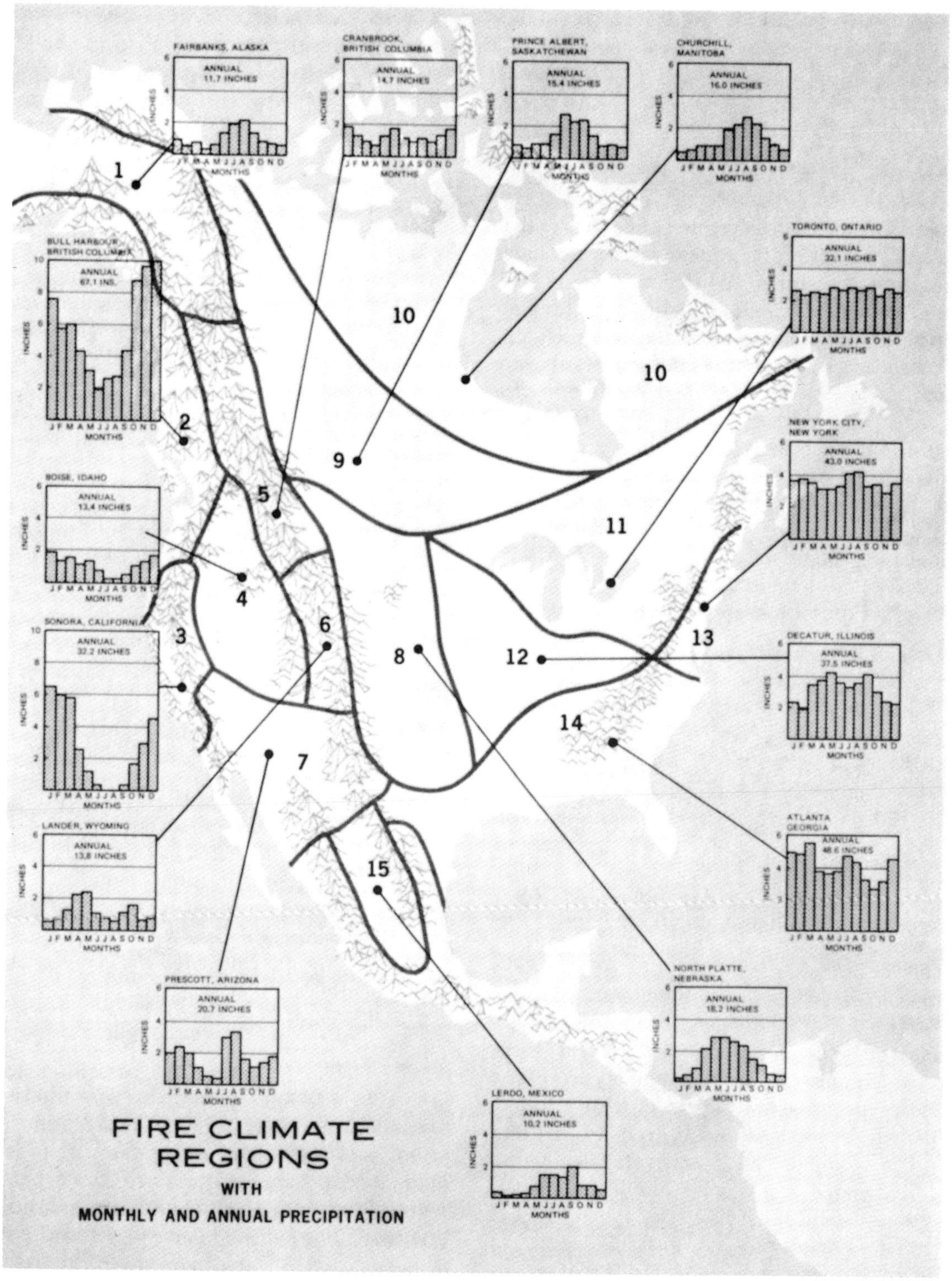


*Figure 4-24. Fire climate regions of North America. The associated bargraphs give the monthly and annual precipitation for a representative station. From Schroeder and Buck (1970).*

tributed some of the extraordinary variability of fire behavior, of critical fire periods, and of fire seasons.

### *National Fire Danger Rating System*

The need for a fire danger rating system has both stimulated and integrated much of the wildland fire research programs conducted in the United States. Typically, the first research program for a region newly experiencing fire protection is to devise some method for the prediction of fire behavior and fire frequency, that is, an assessment of fire danger. As an object of scientific interrogation, fire danger rating has integrated research on the physics and meteorology of fire behavior, on the properties of lightning, on the character of fuel complexes and the processes that determine fuel moisture, and on the operation of fire control systems. The primary outputs of a rating system are the probability of ignition and the likely behavior of an initiating fire; its principal inputs, fuels and weather, as integrated by fuel moisture.

Fire danger rating has proved to be a useful instrument of administration. It provides a means of comparing fire seasons experienced by a particular region and of comparing the fire load of different regions, even those which exhibit distinctive fuels and weather. Because it relies on local data, the calculation of fire danger can transform general fire periods, like a fire season, into meaningful daily statements about the condition of fuels, weather, and sources of ignition. It renders into specific probabilities what is otherwise a statistical average. Where special funds exist for presuppression activities, fire danger rating serves as a guide for their allocation. And not the least among its administrative assets, a widely accepted method of fire danger rating can provide a common language among different agencies, a vital medium of communication and comparison. In the United States, the National Fire Danger Rating System (NFDRS) developed by the U.S. Forest Service has functioned in just this way, assisted by a computer network (Affirms) to process fire danger data and exchange administrative information.

*Synoptic History.* From the beginning of systematic fire protection in the United States, the architects of the program recognized the value of an index for assessing and predicting fire danger. It would be a kind of liability rating, and it could be calculated through the techniques of statistical correlation used in experimental silviculture. Even prior to 1920 researchers and administrators began correlating fires to "climate," and by the mid-1920s an organized search was underway by the U.S. Forest Service to find appropriate measurements. In the Pacific Northwest this inquiry aimed at developing a simple measurement that could be understood by loggers and that could provide a standard for forest closure during periods when the fire potential became extreme. Researchers soon settled on relative humidity, an easily determined property that correlated directly to fuel moisture. In the normally humid Northwest, moreover, a drop in relative humidity below 35% or so was quickly reflected in fuel moisture among the finer fuels.

For the Northern Rockies, however, a multiple index in the form of a meter became the object of investigation. The basic concepts and data appeared between 1928 and 1932, and the Gisborne meter (named after its inventor, H.T. Gisborne), was soon ready for the 1934 fire season—the worst in the region since 1910. Soon afterward the Forest Service promulgated its famous 10 AM Policy as a guide to suppression standards and authorized the use of a special emergency fund for presupppression activities, with expenditures to be justified by the calculated fire danger rating. Subsequently, a major objective of fire research by the U.S. Forest Service was to refine the meter and extend it, with appropriate modifications, to other regions. By the mid-1950s some 8 to 11 meters were in existence (the exact number depending on how many revisions added up to a new meter), and a version was even adapted for the interior of Alaska.

But if the Gisborne meter and its progeny solved some problems, they created others. The meter was a kind of slide rule, integrating the proportionate effects of assorted variables into a single index. The environmental variables, however, were factors similar to those measured in silviculture, not components directly relevant to fire behavior. The meters were highly specific, too. Although they shared a general form, they were founded on empirical statistics derived from local climate records. They could be used to compare seasons experienced by a particular region, but they could not compare regions. Even within a single region several meters were liable to exist for different fuels—one for grass, one for brush, and another for timber. Meters proliferated for regions, agencies, and fuel types. Exploiting new research on fire physics, the Forest Service decided in 1958 to devise a truly national fire danger rating system.

When work began the following year, plans called for a rating system based on four phases—ignition, risk, fuel energy, and rate of spread. By 1961 the spread phase was ready for field testing. Rate of spread was conceived as a composite of three indices—two related surface spread to fuel type (closed timber stand or open grass) and a third incorporated a droughtlike element to account for cumulative drying among the larger forest fuels. Generally speaking, fire danger was expressed for a particular locale as a burning index composed of two indices: a spread index, largely set by fine fuel moisture and wind, and a buildup index, primarily the product of precipitation deficits. By the mid-1960s most agencies used a form of the spread phase. The remaining phases did not yet exist.

In 1965 the entire program was reviewed, and in 1968 the project was reorganized. It was determined that a new system would be brought out as a whole package, but that it would be designed in such a way that it could incorporate further information without major overhauls. This meant a carefully conceived systems design, and it required that physical relationships be based on analytical, fundamental principles involved in heat and moisture transfer, not merely on empirical correlations to general climatic variables. The NFDRS appeared in 1972, and it relied on the Rothermel fire model to integrate weather and fuel data into projected fire behavior (Deeming et al., 1972). Amendments in 1974 and again in 1978 refined some of the relationships, added new fuel models, and incorporated a better measurement of drought (1000-hr TL). But the fundamental structure of the system remained intact. The Affirms system provided a national time-sharing computer network for the automatic calculation of fire danger from local weather and fuel data (Helfman et al., 1980). By 1977 the NFDRS was used by all federal agencies and 35 state agencies, and ratings were being calculated at nearly 2000 stations. It is anticipated that the system will continue to expand as an administrative apparatus, and that future modifications will, like those of the 1970s, result in refinements rather than in major reforms in the conception, philosophy, purpose, or structure.

*Philosophy of the NFDRS.* The NFDRS is a mathematical model composed of many submodels, and like any such device it is only applicable within certain conditions. The boundaries of the NFDRS were expressed in the form of a philosophy, or set of principles, when the system was released in 1972. The conditions they establish apply to both the input and the output of the system.

The NFDRS addresses only the potential for a steady-state fire. It will not predict behavior for either the upper ranges of fire intensity, where the fire propagates by other than steady-state means, or the lower ranges, in the realm of an initiating fire. The NFDRS considers fire control in terms of fire containment, not extinguishment, so it forecasts only those properties that pertain to the flaming front: rate of spread, rate of energy release, and flame length. Nor is the system designed for all-purpose fire behavior forecasting. Rather it is a technique for short range planning; it estimates the likelihood of a fire and sets upper limits for fire behavior. It predicts some very specific attributes for a fire that occurs under the specific fuel and weather conditions entered into it. These attributes are expressed as a set of fire behavior indices. Though originally designed to operate across a scale of 0–100, to improve sensitivity most of the indices are now open-ended.

Perhaps its principal constraint is that the NFDRS assesses the average worst condition for a potential fire. Fire weather is measured under the average worst conditions: at the warmest time of the day (1300 hrs), in open surroundings, on southerly exposures, with elevated fuel sticks. Clearly some portions of the area being sampled will experience conditions even more favorable for fire, but far more will have less severe conditions. Fire danger rating will inherently tend to overpredict fire frequency and fire behavior. But in much the same way that engineers design a bridge to withstand a flood of a certain size and frequency (e.g., a 50-year flood), so fire planners must design protection systems to meet a certain fire load; and for this purpose the average worst concept has, over the years, proved useful (Figure 4–25).

*Structure.* The NFDRS combines the probability of ignition with the probable behavior of resulting fires. For input the system requires meteorological observations and the selection of appropriate models for fuel, climate, and slope. For output the system predicts ignition, spread, and energy release components, organized into several indices. The risk of ignition components, summarized as an *occurrence* index (OI), are somewhat subjective and in some degree the result of statistical analysis of past fire history. The fire behavior components, expressed as a *burning index* (BI), have surer grounding in fundamental fire physics and are derived from objective environmental data. The burning and occurrence index combine to give the *fire load index* (FLI) (Figure 4–26). There exist a number of means by which to calculate the indices. They can be done manually through published tables and monographs, by programmable hand calculators with appropriate instruction chips, or automatically through the Affirms computer network.

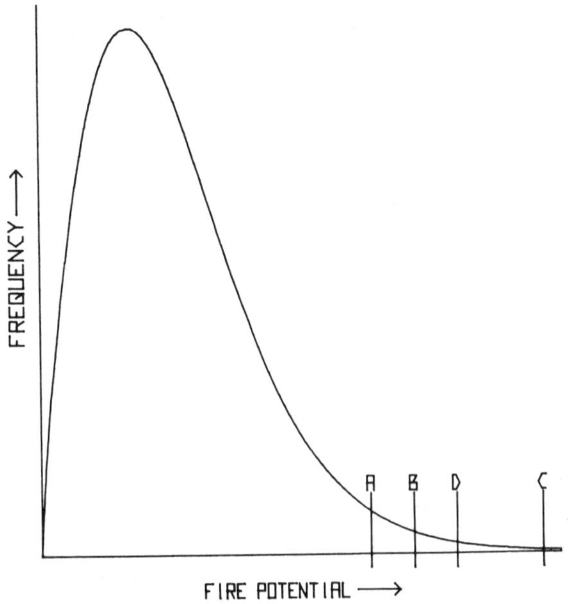


*Figure 4-25. Unit area fire potential. Theoretical frequency distribution for potential fires started randomly over a 24-hr period. Since fire danger data is collected under the worst conditions for the most severe sites, they will record fire potential in the range of A–C. The "average worst" of these potentials (D) is what the NFDRS predicts. Obviously, most fires will start or burn under less severe conditions. From Deeming et al. (1978).*

Each set of components, in turn, integrates others. Fire occurrence depends on another set of factors, including an *ignition component* (IC), and estimations of both lightning risk (LR) and man-caused risk (MCR). The ignition component measures the susceptibility of fine fuels, both live and dead, and it is expressed as the probability that a firebrand will produce a fire requiring suppression. Since fire spread consists of an infinite number of smaller ignitions, the IC is fundamental for the calculation of the occurrence index and useful in interpreting the burning index. The risk factor sets the probability that either lightning or human activity will result in ignition.

The burning index combines a *spread component* (SC) and an *energy release component* (ERC). The spread component integrates windspeed and slope with the properties of the fine fuels, and it is calculated from the equations of the Rothermel fire model. The energy release component describes the general rate of combustion (reaction intensity) and it includes the contribution of the 1000-hr TL category fuels. Rate of spread and reaction intensity combine to give the fireline intensity, and fireline intensity, the flame length.

*Interpretation.* Part of the strength of the indices is that they are synoptic, integrating many lesser

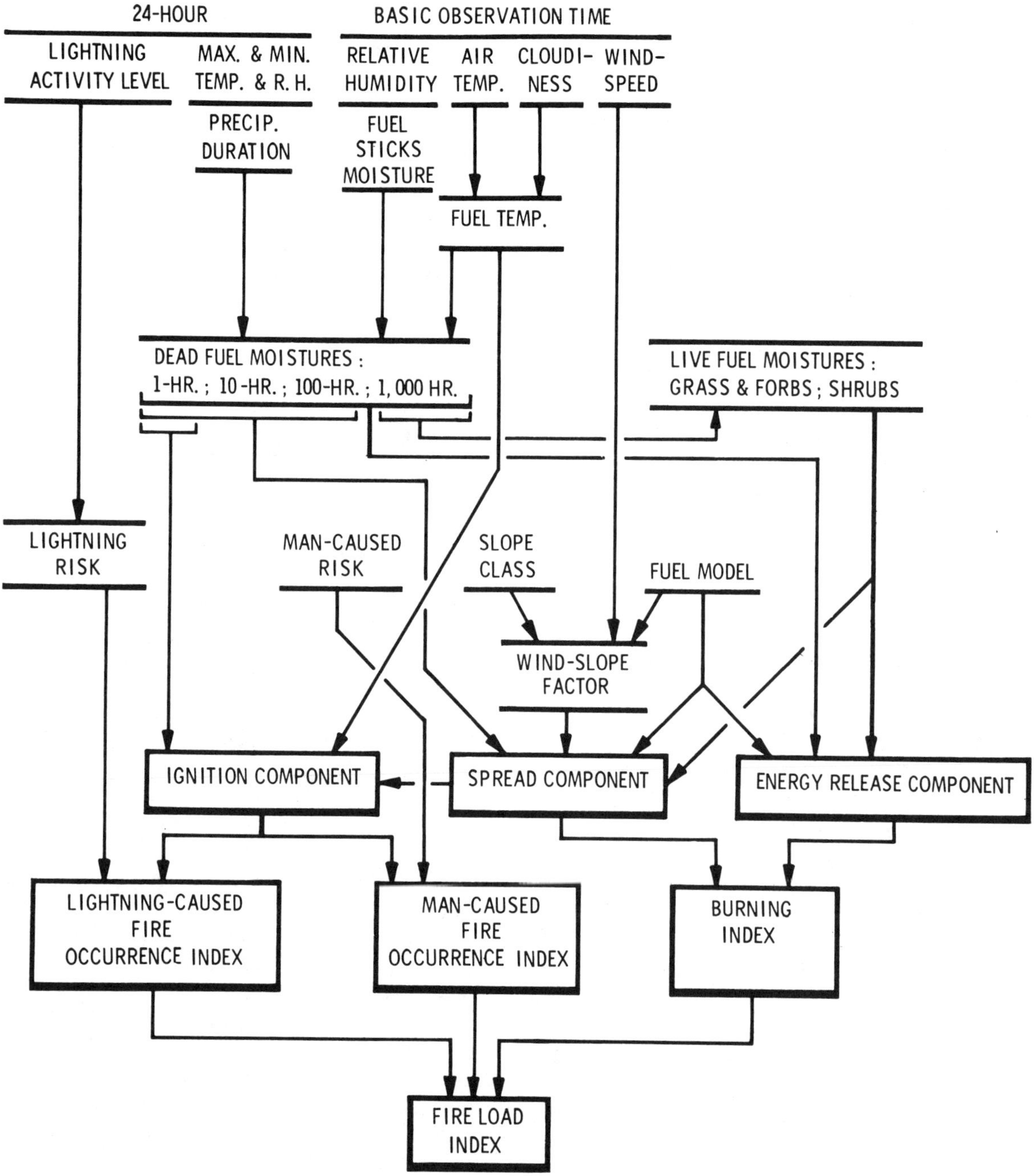


*Figure 4-26. National Fire-Danger Rating System, 1978. The structure builds on basic observations (top) which it combines with fuel moistures by size class to get an assessment of combustion. Wind, slope, and fuel model then describe fire behavior. To estimate the fire load, risk factors are also entered. From Deeming et al. (1978).*

components. But this is also a weakness. Each synthesis compounds one level of uncertainty with another, and as the index becomes more general it loses some of the specific information that may be of special interest to a particular fire program at a particular time during the fire season. For this reason, the meaning of the fire load index is left to local interpretation. Typically, when describing the fire danger for an area, all of the fire indices will be cited—each delineating a particular aspect of the total fire problem.

The fire behavior components are scaled such that they translate directly into measurements significant to the problem of containing a flaming front. The spread component is numerically equal to the theoretical rate of spread in feet per minute. An SC of 12 predicts a forward rate of spread of 12 ft/min (3.7 m/min). The energy release component describes the available energy in Btu/$ft_2$/min (2.7 kcal/$m_2$/min) divided by 25. An ERC of 100 thus translates into a reaction intensity of 2500 Btu/$ft_2$/min (6780 kcal/$m_2$/min). The burning index equals the predicted flame length times 10. A BI of 29 thus predicts a flame length of 2.9 ft (88 cm). The BI correlates directly with the difficulty of control. A BI over 40 indicates that the average worst fire would be beyond control by manual methods; over 80, beyond the control of any means; and over 100, beyond the range of steady-state fires altogether (see Figures 2–5 and 4–27).

More difficult to interpret, because they are more subjective in their source, are the occurrence indices. The OI forecasts the number of fires per million acres, with each unit of the index equal to 0.1 fires. Its interpretation is much like that of weather predictions—a description of what, on the average, would happen given similar conditions. Often the risk component is simply ignored in any quantitative sense; the fire load index becomes, in effect, synonymous with the burning index. For purposes of public information and administrative decisions, fire danger is commonly expressed in terms of five qualitative levels: low, medium, high, very high, and extreme. To correlate this scale with the BI or FLI requires a com-

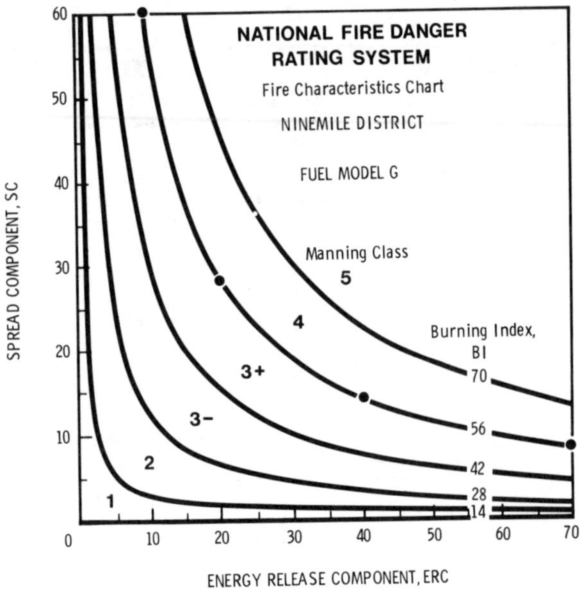


*Figure 4-27. Interpreting the burning index. Since a given burning index integrates the spread rate (SC) and energy release component (ERC), the BI can be expressed as a fire characteristics curve. A curve exists for each BI; plotting a fire along the curve gives useful information about its peculiar behavioral traits. In this example, ranges of BIs are correlated with manning classes, a presuppression guide. From Andrews and Rothermel (1982).*

parison with past records and fire history. A period of five years is probably minimal for any meaningful association. Based on these levels it is possible to draw up guides for manning and dispatching that prescribe, for various levels of fire danger, certain automatic responses: the activation of designated prevention programs, or even closure of the land to any human activity; the buildup of suppression resources and their automatic dispatch, both by type and amount; an increase in detection efforts.

For all this sophistication, the NFDRS does have limitations. Some are internal to the model, the product of yet poorly understood relationships or of simplifications of data. Some are inherent in the fundamental complexity of natural systems; the variability of the environment remains greater than the limits of the models used to describe them. These are amenable to further refinement and research, and the NFDRS was designed to accommodate future revisions of this sort. Its fundamental difficulty is that the NFDRS uses only one observation to characterize an entire diurnal burning cycle and that cycle the one typical of the Northern Rockies. It does not yet accurately predict fires that start outside peak burning times, or fires that burn within other diurnal burning cycles. Other problems result from improper use of the model: poor field data, inappropriate selection of fuel, slope, or climate models, or the application of the forecasted indices beyond their intended scope and purpose. The NFDRS indices do not constitute a general purpose fire behavior model.

### *Weather Services*

Most land and fire management agencies maintain permanent weather stations for use in fire danger rating, and nearly all of these stations form part of the national meteorological reporting network used by the National Weather Service (NWS). For the climatological purposes, the important observations are made at 0800 and 1700 hrs. For fire danger rating, observations are made at 1300 hrs. All of these observations are permanently recorded; the fire weather data is automatically accessioned by the Affirms system into a National Fire Weather Data Library maintained by the U.S. Forest Service (Furman and Brink, 1976). Out of this depository comes the raw information for the various computer-assisted programs that sketch fire climates and translate the FLI into administrative manning and action guides. This network of permanent stations is serviced by hand, but for more inaccessible areas there exists the Remote Automatic Weather Station (RAWS), frequently outfitted with lightning recording devices in addition to normal weather instruments (Figure 4–28). Communications are often achieved by satellite telemetry. Temporary weather stations are common at the sites of large fires, both wild and prescribed. A special hand-kit of weather instruments that can be carried on a belt is widely employed, and is suitable for fireline use.

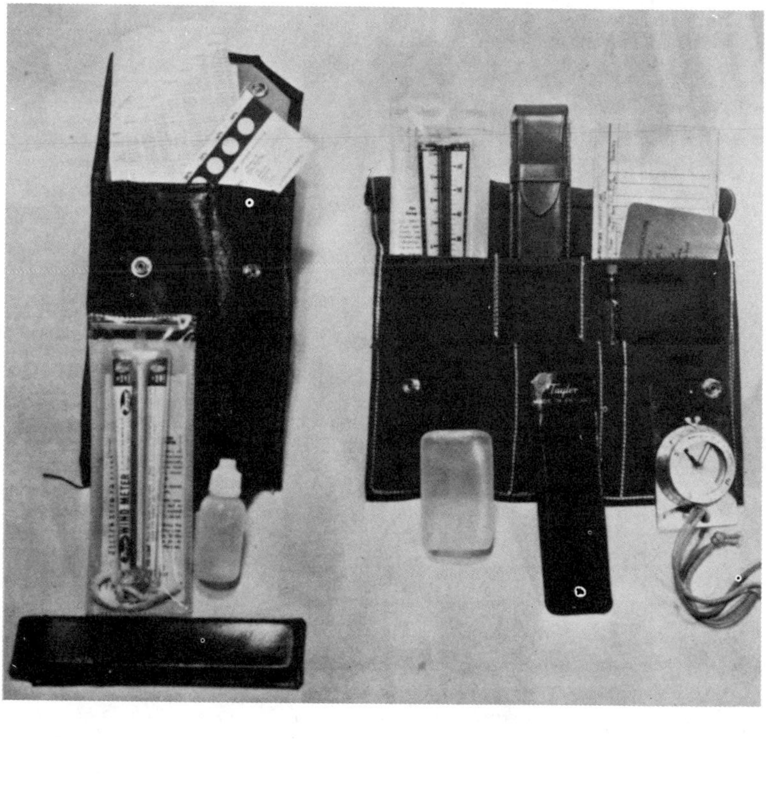

*Figure 4-28. Portable weather kit, suitable for attachment to a belt. Courtesy U.S. Forest Service.*

Out of these basic observations come a variety of weather and fire behavior forecasts. The 1300-hr observations enter into the calculations of the NFDRS. The NWS sponsors a Fire Weather Service that routinely issues general fire weather forecasts during fire season; this service can be tapped by teletype or the Affirms network. When more specific forecasts are needed, the Fire Weather Service will issue spot weather forecasts or, on request for large wildland fires, will dispatch a special mobile weather unit and meteorologists to the scene (Figure 4–29). The meteorologists will take observations and make specific fire weather forecasts. Where regional fire needs are large, other arrangements have been tried. In Alaska, fire agencies have contracted with a private firm for fire weather forecasts; and in the South, where prescribed burning creates a large demand, a Forestry Weather Information Service (FWIS) has been established by federal and state forestry agencies to amplify the normal flow of agricultural and fire weather forecasts (Paul, 1981).

*Figure 4-29. Mobile weather unit of the Fire Weather Service (NWS). Units are dispatched, on request, to large fires. Courtesy U.S. Forest Service.*

## 4.4 SELECTED FIRE CLIMATES: METEOROLOGICAL CHARACTERISTICS

Of the 15 fire climate regions that characterize North America, 12 exist in the United States (Figure 4–24). Sketches of six of these regions follow. With regard to fire, each region shows a general climate that determines the nature of the fire season and each experiences distinctive mechanisms for creating critical fire periods within the larger seasons. Nearly all fires of historic significance have occurred during these critical fire periods.

### *Southern California*

The general climate is Mediterranean, with short wet winters and long dry summers. Annual precipitation is light, from 10 to 20 in. (25–51 cm) in lower elevations and several times greater in the mountains. Droughts are frequent. Unusually wet winters, however, produce a superabundance of grass and can increase the area where summer fires are possible. Marine air locally influences the coastal area, but mountains restrict its penetration inland. Dry thunderstorms can occur during the summer. Rugged mountains make topography an important variable in fire behavior.

A variety of critical fire periods occurs, and collectively they have the effect of making an otherwise lengthy summer fire season into a yearlong affair. One mechanism is subsidence, the persistence of a ridge of high pressure or a closed High over the region. Temperatures rise, humidities drop, and the air mass becomes unstable. Another mechanism is the foehn wind—the Santa Ana—produced by a stagnating Great Basin High or an extension of the Pacific High across the Transverse Range. Most large fires driven by Santa Anas occur in November and December, with a secondary concentration in March. Nearly all fires and fire complexes of historic size have broken out during Santa Ana winds.

### *Southwest*

Its mountainous topography means that the Southwest has not one climate but several. The lower reaches are desert, with an annual precipitation of 5–10 in. (13–25 cm); high elevation mountains show 30–40 in (76–102 cm). Rainfall is concentrated into two periods: a summer "monsoon," reaching a peak in August, and a short winter season. The fire season is defined by the summer rainy period and the annual cycle of grass, the principal fine fuel. The pattern is bimodal—one peak coming before the onset of the monsoon and new grass growth (May and June), and the other after the rains have abated and the grass has cured (September and October). Early monsoon thunderstorms often feature dry lightning. Droughts are frequent. Heavy winter rains can bring grass, and summer fires, to areas that are otherwise desert.

Critical periods can result from an intensification of the normally dry climate and from certain characteristics of the summer monsoon. Both relate to upper-air patterns. Prolonged high fire danger results when a broadscale pattern aloft establishes a southwesterly flow; as short-wave troughs pass through this band, windspeed accelerates locally. The monsoon pattern is complex, but generally results when a subtopical High aloft positions itself north of the region and brings in southeasterly flow from the Gulf of Mexico. The initial layers of air are shallow, interact with the thermal convection over the desert surfaces and high mountains, and produce local thunderstorms, often dry thunderstorms. Historic problem fire complexes have followed large lightning busts in the spring or early summer. As the monsoon pattern intensifies, more moisture pumps into the region and lightning-caused fires, though abundant, burn in fuels too wet for rapid spread.

### *Pacific Northwest*

Maritime influences dominate the climate, characterized by heavy rainfall and heavy fuel loads. Rainfall is light during the summer months, which define the fire season. Lightning-caused fires increase away from the coast.

Virtually all historic fires have occurred when strong offshore winds drove off marine air. The mechanisms of the east wind, a foehn wind, are analogous to those typical of Southern California. In one form, high pressure (Northwest Canadian High) builds up east of the Cascades with low pressure off the Pacific coast; the resulting gradient flow warms adiabatically as it falls from the mountains and pushes marine air out to sea. A second form follows the passage of a cold front. In this case the Pacific High bulges inland, a northeasterly flow across the Cascades and Coast Ranges results, adiabatic processes warm and dry the offshore flow of air. Historically, large fire complexes have arrived in August or September under the influence of an east wind.

### *Northern Rockies*

Like the Southwest, the mountains of the Northern Rockies present a varied climate. Precipitation in the valleys ranges from 10 to 20 in. (25–51 cm), and in the mountains, 40–60 in. (102–152 cm). Most of the annual precipitation is concentrated in winter snows and spring rains. By June rainfall slows, drying occurs, and the fire season begins a two to three month tour.

Again like the Southwest, critical fire periods result more from upper-air patterns than from surface weather. Three situations are especially hazardous. One coincides with large-scale subsidence, accompanied by low surface humidity. Another features a succession of short-wave troughs passing through a general zonal flow from northwest to southeast—each frontal passage bringing strong winds and a wind shift. And, finally, the same closed High that sweeps moisture into the Southwest can, under the proper circumstances, carry that moisture into the Northern Rockies. Dry lightning and thunderheads with strong downdrafts result, increasing ignition and spread without increasing fuel moisture. Historically, the greatest fire loads have occurred in August from the combination of dry lightning storms succeeded by the passage of dry short-wave troughs.

### *South*

Dominated by the Gulf of Mexico, a great reservoir of heat and moisture, the southern climate is humid and generally warm. Annual precipitation averages 40–60 in (102–152 cm), with locally higher amounts. The fire season, however, follows from the region's principal fuels, grass and fine shrubs, that can dry out quickly. The fire season tends to be bimodal, concentrating in the spring and fall. But fires can occur throughout the year and most prescribed burning is conducted during the winter.

Critical fire periods result when the normal northward flow of Gulf air is blocked. Two mechanisms typically account for this situation. One relies on migrating Highs that pass through the region in the form of cold fronts. The Pacific High is the most common type; when its passage in the form of a dry cold front coincides with the life cycle of grass, as frequently happens during the spring, the outcome is ideal for the promotion of large fires. Strong gusty winds precede the front, a wind shift marks its passage, and stiff winds and dry unstable air suitable for convective column development follow in its wake. Regional fire loads, however, are associated with droughts, and the normal drought mechanism requires an inland extension of the Bermuda High during summer and fall. Large numbers of fires can result, with large cumulative acreages burned. Regional fire problems may intensify if the High is accompanied by the passage of short-wave troughs along its margins. Historically, most large individual fires have occurred in the spring, although most fire complexes have come in the fall.

### *Lake States*

The Lake States feature long cold winters and variable summers, the product of mixed air masses entering the region. Annual precipitation exceeds 30 in. (76 cm), slightly heavier in the summer than in the winter. Relative humidity and fuel moisture are similarly moderate to high. Consequently, the fire season follows the vegetative cycle for grass and deciduous trees, with bimodal peaks during the spring and fall.

Critical periods result when certain synoptic patterns coincide with the primary fire season. Fundamentally, this means the movement of Highs through the region. The Pacific High accounts for the preponderance of serious episodes, although the Hudson Bay and Northwest Canadian Highs are also influential. If the front preceeding the High is dry, large fires may result. This mechanism accounts for the pattern of large fires during the spring (April). But high fire danger also exists along the rearward portion of the air mass, once it has passed through the Lake States proper. During its travel and subsidence, the air mass has been warmed and dried, and it blocks the normal flow of moisture from the Gulf of Mexico. Local drought results. This pattern typifies the autumn months, and most of the large fire complexes of historic record—some of enormous extent—have occurred under these conditions. The size of these complexes, however, was as much a product of extensive landclearing and abundant slash firing as of meteorological circumstances.

### *Interior Alaska*

With its continental climate, the Alaskan interior experiences long dry winters and brief warm summers. The Alaska Range normally shuts off the flow of marine air from the south. The climate is arid, with 10–15 in. (25–38 cm) annual precipitation. Combined with the lengthy summer days of the higher latitudes, however, fuel moisture can be low. The summer fire season is brief, but potentially explosive. Perhaps the key is the pattern of summer thunderstorms, occasionally pounding the taiga and tundra with dry lightning. Moderately moist air is drafted out of the southwest into the topographic corridor roughly traced by the Yukon River. The historic record suggests a distribution of large fire complexes much like that characteristic of the Northern Rockies. In the heavy fire years, the number of fires increases briskly through the summer, reaching a peak in August.

## *REFERENCES*

An excellent summary of meteorological considerations for fire behavior, profusely illustrated, is contained in Schroeder and Buck (1970), *Fire Weather.* Useful overviews of advances in fire weather study since its publication can be found in the biennial conferences jointly sponsored by the American Meteorological Society and the Society of American Foresters, of which the proceedings for those since 1976 have been published: Baker and Fosberg (1976), "Proceedings of the Fourth National Conference on Fire and Forest Meteorology"; Fosberg (1978), "Fifth National Conference on Fire and Forest Meteorology"; American Meteorological Society and Society of American Foresters (1980), "Sixth National Conference on Fire and Forest Meteorology." For a global context, see Reifsnyder (1978), 'Symposium on Forest Meteorology." A general description of fire weather as an influence on fire behavior is given in U.S. Forest Service, 1969, "Intermediate Fire Behavior." Also recommended are numerous fire behavior case studies, such as those mentioned in Chapter 1, which include analyses of fire weather.

The NFDRS is ably described in Deeming et al. For the fuel moisture components and fuel models used, consult the references in Chapter 3. Descriptions of the diurnal burning period can be found in the NWCG fire training courses S190, S390, and S590.

General sources of information about weather services and weather data processing capabilities are Main et al (1982), "FIREFAMILY: Fire Planning with Historic Weather Data"; Furman and Brink (1976), "The National Fire Weather Data Library: What It Is and How To Use It"; Bradshaw and Fischer (1981), "Computer System for Scheduling Fire Use"; Johnson and Main (1978), "A Climatology of Prescribed Burning." The Affirms system is described in Helfman et al. (1980), "User's Guide to AFFIRMS"; the FWIS network, in Paul (1981), "A Real-Time Weather System for Forestry." For the operation of manual weather stations, see Fischer and Hardy (1976), *Fire Weather Observer's Handbook,* and for RAWS, see Warren and Vance (1981), "Remote Automatic Weather Station for Resource and Fire Management Agencies."

Albini, F.A. and R.G. Baughman, 1979, "Estimating Windspeeds for Predicting Wildland Fire Behavior," U.S. Forest Service, Research Paper INT-221.

American Meteorological Society and Society of American Foresters, 1980, "Proceedings. Sixth National Conference on Fire and Forest Meteorology."

Andrews, Patricia L. and Richard C. Rothermel, 1982, "Charts for Interpreting Wildland Fire Behavior Characteristics," U.S. Forest Service, Research Note INT-131.

Baker, Douglas H. and Michael A. Fosberg (technical coordinators), 1976, "Proceedings of the Fourth National Conference on Fire and Forest Meteorology," U.S. Forest Service, General Technical Report RM-32.

Bancroft, Larry, 1976, "Natural Fire in the Everglades " pp. 47–60, in U.S. Forest Service R-8), "Proceedings. Fire By Prescription Symposium," U.S. Forest Service, Southern Region.

Baughman, Robert G., 1981, "Why Windspeeds Increase on High Mountain Slopes at Night," U.S. Forest Service, Research Paper INT-276.

Bradshaw, Larry S. and William C. Fischer, 1981a, "A Computer System for Scheduling Fire Use. Part I: The System," U.S. Forest Service, General Technical Report INT-91.

———, 1981b, "A Computer System for Scheduling Fire Use. Part II: Computer Terminal Operator's Manual," U.S. Forest Service, General Technical Report INT-100.

Brotak, Edward A., 1978, "Low-level Wind and Temperature Profiles Associated with Major Wildland Fires," in Fosberg (ed.), "Fifth National Conference on Fire and Forest Meteorology," American Meteorological Society and Society of American Foresters.

Brotak, Edward A. and William E. Reinsnyder, 1976, "Synoptic Study of the Meteorological Conditions Associated with Extreme Wildland Fire Behavior," in Baker and Fosberg (eds.), "Proceedings of the Fourth National Conference on Fire and Forest Meteorology" U.S. Forest Service, General Technical Report RM-32.

Brown, A.A. and Kenneth P. Davis, 1973, *Forest Fire: Control and Use,* 2nd ed. (New York: McGraw-Hill).

Burgan, Robert E., 1979, "Fire Danger/Fire Behavior Computations with the Texas Instruments TI-59 Calculator: User's Manual," U.S. Forest Service, General Technical Report INT-61.

Burgan, Robert E. et al., 1977, "Manually Calculating Fire-Danger Ratings—1978 National Fire Danger Rating System," U.S. Forest Service, General Technical Report INT-40.

Byram, George M., 1954, "Atmospheric Conditions Related to Blow-up Fires," U.S. Forest Service, Southeastern Forest Experiment Station Paper No. 35.

Chandler, Craig C., 1976, "Meteorological Needs of Fire Danger and Fire Behavior," in Baker and Fosberg, "Proceedings of the Fourth National Conference on Fire and Forest Meteorology," U.S. Forest Service, General Technical Report RM-32.

Cole, Franklyn, 1975, *Introduction to Meteorology,* 2nd ed. (New York: Wiley).

Deeming, John E. et al., 1972, "National Fire-Danger Rating System," U.S. Forest Service, Research Paper RM-84.

———, 1974, "National Fire-Danger Rating System," U.S. Forest Service, Research Paper RM-84 revised.

———, 1978, "The National Fire-Danger Rating System—1978," U.S. Forest Service, General Technical Report INT-39.

Fawcett, Edwin B., 1976, "Current Capabilities in Prediction at the National Weather Service's National Meteorological Center," in Baker and Fosberg (technical coordinators), "Proceedings of the Fourth National Conference on Fire and Forest Meteorology," U.S. Forest Service, General Technical Report RM-32.

Fischer, William C. and Charles E. Hardy, 1976, *Fire Weather Observer's Handbook*, U.S. Forest Service, Agriculture Handbook No. 494.

Fosberg, Michael A. (chairman), 1978, "Fifth National Conference on Fire and Forest Meteorology," American Meteorological Society and Society of American Foresters.

Fosberg, Michael A. et al., 1966, "Some Characteristics of the Three-Dimensional Structure of Santa Ana Winds," U.S. Forest Service, Research Paper PSW-30.

Furman, William R. and Glen E. Brink, 1976, "The National Fire Weather Data Library: What It Is and How To Use It," U.S. Forest Service, General Technical Report RM-19-FR8.

Furman, William and Robert S. Helfman, 1973, "A Computer Program for Processing Historic Fire Weather Data for the National Fire-Danger Rating System," U.S. Forest Service, Research Note RM-234.

Helfman, Robert S. et al (compilers), 1980, "User's Guide to AFFIRMS: Time-Sharing Computerized Processing for Fire Danger Rating," U.S. Forest Service, General Technical Report INT-82.

Johnson, Von J. and William A. Main, 1978, "A Climatology of Prescribed Burning," in Fosberg, "Fifth National Conference on Fire and Forest Meteorology," American Meteorological Society and Society of American Foresters.

Krumm, William, 1959, "Fire Weather," in Kenneth P. Davis (ed.), *Forest Fire: Control and Use* (New York: McGraw-Hill).

Main, William A. et al., (1982), "FIREFAMILY: Fire Planning with Historic Weather Data," U.S. Forest Service, General Technical Report NC-73.

Marlatt, W.E. et al., 1980, "The Shape of Air Temperature Inversion Surfaces of Mountain Valleys," pp. 173–178, in "Proceedings. Sixth National Conference on Fire and Forest Meteorology," American Meteorological Society and Society of American Foresters.

McCutchan, Morris H., 1977, "Climatic Features as a Fire Determinant," in H.A. Mooney and C. Eugene Conrad (coordinators), "Proceedings of the Symposium on the Environmental Consequences of Fire and Fuel Management in Mediterranean Ecosystems," U.S. Forest Service, General Technical Report WO-3.

National Wildfire Coordinating Group, n.d., S190, "Introduction to Fire Behavior"; S390, "Fire Behavior"; S590, "Fire Behavior Officer."

Paul, James T., 1981, "A Real-Time Weather System for Forestry," *Bulletin of the American Meteorological Society* **62**(10): 1466–1472.

Reifsnyder, William E. (director), 1978, "Symposium on Forest Meteorology," World Meteorological Organization Publication No. 527, Canadian Forestry Service.

Schroeder, Mark J. and Charles C. Buck, 1970, *Fire Weather...A Guide for Application of Meteorological Information to Forest Fire Control Operations,* U.S. Forest Service, Agriculture Handbook No. 360.

Schroeder, Mark J. et al., 1964, "Synoptic Weather Types Associated With Critical Fire Weather," U.S. Forest Service and U.S. Weather Bureau.

Sommers, William T., 1978, "On Forecasting Strong Mountain Downslope Winds," in Fosberg, "Fifth National Conference on Fire and Forest Meteorology," American Meteorological Society and Society of American Foresters.

———, 1981, "Waves on a Marine Inversion Undergoing Mountain Leeside Wind Shear," *Journal of Applied Meteorology* **20** (6): 226–236.

Taylor, Dale L., 1981, "Fire History and Fire Records for Everglades National Park, 1948–1979," South Florida Research Center, Report T-619 (Homestead, Fla.: National Park Service).

Warren, John R. and Dale L. Vance, 1981, "Remote Automatic Weather Station for Resource and Fire Management Agencies," U.S. Forest Service, General Technical Report INT-116.

U.S. Forest Service, 1969, "Intermediate Fire Behavior," TT-81.

Waters, Marshall P., III, 1976, "An Application of Geosynchronous Meteorological Satellite Data in Fire Danger Assessment," in Baker and Fosberg, "Proceedings of the Fourth National Conference on Fire and Forest Meteorology," U.S. Forest Service, General Technical Report RM-32.

*Part Two*

# The Fire Regime

## *Chapter Five*

# *Fire and Life*

Combustion requires only that fuels have certain physical-chemical properties. But an ecosystem is more than a fuel complex, and a fire does more than simply rearrange the fuel elements and flammability of the system. Fire is a profound biological event. The chemistry of combustion derives from the chemistry of organisms, and the geography of fire is synonymous with the range of organic matter. Although the combustibility of matter may depend on the chemical properties of its fuel elements and on their physical arrangement, that matter consists of biochemical assemblages shaped under genetic direction and organized through the shared dynamics of an ecosystem. Although fire may be measured by certain physical manifestations, its effects are described or reconstituted as organic entities or ensembles of such entities.

The relationship between fire and biota is, in fundamental ways, reciprocal. How a fire burns is clearly related to the chemical composition of its fuels and how these fuels are physically arranged. In wildland environments these properties are determined by biological considerations—the sum of constituent organisms and the processes of their interaction which make up an ecosystem. Yet the presence of fire can become a selective force in the evolution of those constituent organisms, favoring some traits over others; a shaper of ecological systems, favoring certain structural arrangements over others; and an ecological process, promoting certain cycles of nutrients and particular pathways of energy to the partial exclusion of others. Over long periods of time it seems likely that certain organisms would achieve special fire-related properties, that certain biotas would show special dependence on the presence of fire, and that certain processes would become specially prominent as mechanisms to cope with fire.

This has apparently happened. Fire and life show a fundamental kind of interdependence. The Earth is not only the great water planet but the great fire planet. What binds fire and water together is life. The study of how fire and life interact is the province of fire ecology; its informing idea, the concept of the fire regime; its administrative objective, the incorporation of fire management within the context of land management.

## *5.1 LIGHTNING*

The study of fire as a biological phenomenon may properly begin with lightning. Not only has lightning been the predominant source of ignition over evolutionary time, but it may have helped catalyze a primordial swamp of chemicals into the proto-organic compounds essential for living systems. As an electrical phenomenon, lightning continues to influence biological systems, and as a source of fire, it commands considerable attention from fire managers. Many of the most troublesome fires of the 1960s and 1970s began from lightning, both in areas with a long history of lightning fire and in areas for which lightning fires were rare. As other sources of ignition are identified and controlled, lightning will persist. It will provide a background count of natural fires, an ignition source whose relative influence may increase over time.

The mere fact of lightning fire and its continuance through geologic time, moreover, sustains the contemporary conviction that wildland fire can be a natural process—one that in systems managed as wilderness ought to be preserved, and one that in other wildland areas ought to be used in surrogate forms, as prescribed fire. To a remarkable extent reforms in fire policy during the 1970s related to the practical and philosophical problems posed by lightning fire.

### *Physical Properties of Lightning*

As a physical event lightning maintains the electrical equilibrium of the Earth. Between the ionosphere, whose electrical charges are balanced, and the Earth, which has an excess of charge, there exists a strong voltage potential. The intervening atmosphere acts as an insulator, and the three layers together resemble that primitive spark machine, a Leyden jar. But the atmosphere is flawed as an insulator; some electricity leaks through it along the gradient, and when this leakage concentrates, a sudden discharge will restore the former equilibrium.

The common mechanism for concentrating electrical potential is a thunderstorm, and its method of discharge, a lightning bolt. The discharge may occur between any locally significant potential: within a cloud, between clouds, between cloud and Earth. Without some process of restoration, the Earth would lose its entire charge within an hour. Thunderstorms are not the sole mechanism for returning that charge, but they are the most spectacular; they should be considered as an electromagnetic as well as a thermodynamic necessity. The Earth experiences perhaps 1,800 storms per hour, or 44,000 per day. Collectively, these storms produce 100 cloud-to-ground discharges per second, or better than 8 million per day globally. About 75% of the energy of a bolt is lost to heat.

In fair weather, the electrical gradient between atmosphere (positive charge) and Earth (negative charge) is about 30 v/ft (9 v/m). With the cumulus cloud development that leads to a thunderhead, however, this simple pattern alters. The upper reaches of the cloud become positively charged, the lower reaches negatively charged, and the ground directly beneath the cloud experiences a charge reversal induced by the distribution of charges in the cloud (see Figure 5–1). Most lightning discharges occur within or between clouds, a process that is stimulated by the onset of rain since the rain carries a positive charge from the upper reaches of the cloud down to and through the cloud base. During this time—the mature phase of thunderstorm development—the cloud has reached its greatest height, has altered the electrical potential over as great a distance as possible, and begins maximum lightning discharge.

The mechanisms by which the electrical potential between cloud and earth overcome the resistance of the atmosphere generally initiate in the cloud and proceed to the Earth in two stages. In the first stage, the cloud sends out a probing feeler, a *leader stroke.* The leader proceeds by a series of brachiated, discrete steps. The stepped leader probes at a velocity of about 33–164 ft/sec (10–50m/sec), and makes a total trip in about 20 msec. Shortly before reaching the ground, the stepped leader encounters a similar probe emanating from an elevated object on the surface. Upon meeting they create a complete conduit between cloud and earth. Now the second stage begins, the *return stroke.* This process creates the actual lightning bolt or flash. Though the speed is

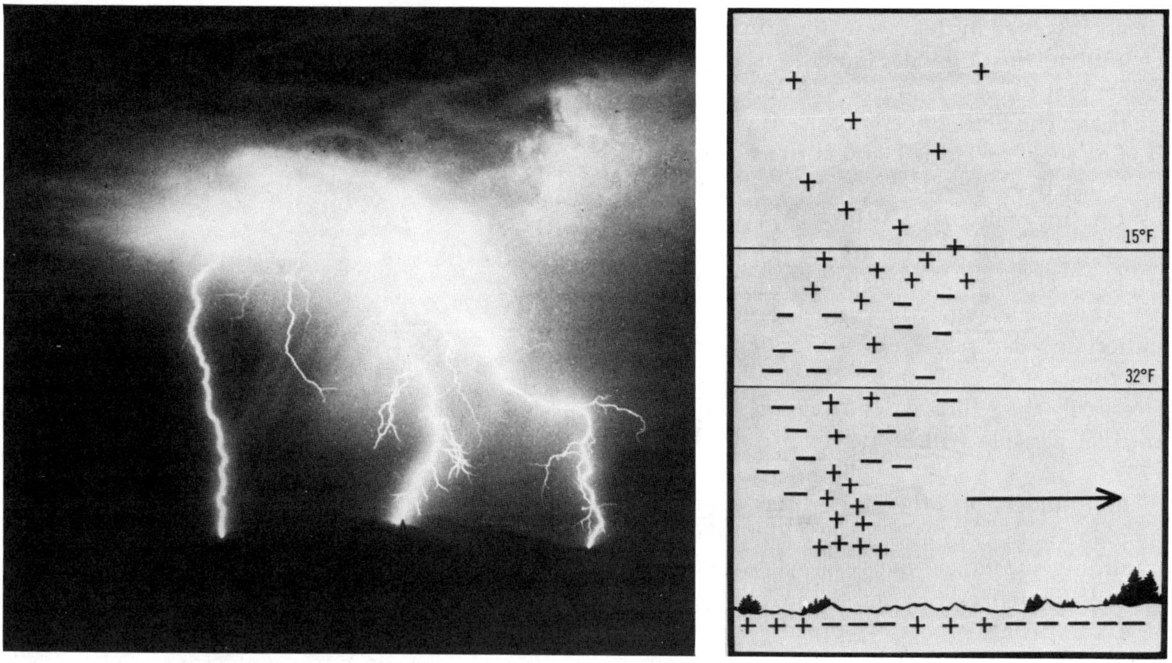


*Figure 5-1. Electrical charge distribution in a thunderhead. From Schroeder and Buck (1970), photo courtesy U.S. Forest Service.*

such that the human eye cannot resolve it, the return stroke consists of a series of strokes or darts, about four darts on the average. The flow of current is continuous, but the reduced charge between darts lasts about 40–50 msec.

Depending on the character of this continuing current, two types of bolts are distinguished: the *cold stroke,* whose main return stroke is of intense current but of short duration, and the *hot stroke,* involving currents of lesser voltage but longer duration, a phenomenon known as *long-continuing current* (LCC; Figure 5–2). As much as half of the total electrical charge may consist of LCC. The cold stroke generally has mechanical or explosive effects, the hot stroke is more prone to start fires. About 20% of lightning bolts in the Cordillera of the American West display LCC characteristics.

Whether ignition results or not depends on the character of the bolt and the character of the fuels, if any, that it strikes. Hot strokes are superior to cold strokes, fine fuels better than coarse, and dry fuel more suitable than wet. The actual

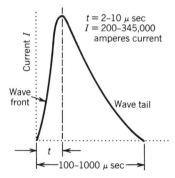

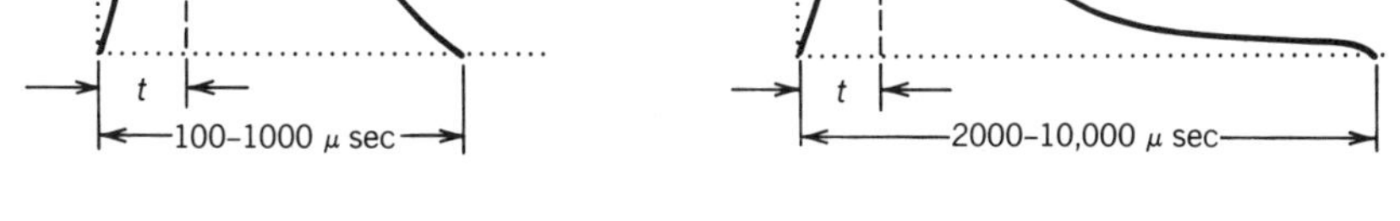

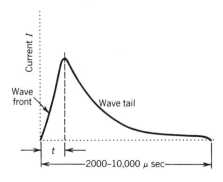


*Figure 5-2. Physical properties of a lightning bolt. "Cold lightning" properties are given in the left graph, and "hot lightning," in the right. From Viemeister (1972).*

process involved in this electrostatic match is not yet known. But a good candidate is the debris shower that a bolt generates as it furrows through tree bark. The bolt converts coarse fuel, the tree, into fine fuels, an ejected cloud of wood particles and extractives; these materials ignite through flashover into a fireball. The fireball in turn ignites fine fuels at the base of the tree or resin along the torn bole. The fireball burns only briefly, like the flaring of a match when first struck. For persistent ignition to occur the heat of the fireball must exceed the heat of ignition for the available fuels. Necessarily these fuels will be fine fuels, and their principal variable will be fuel moisture (Taylor, 1973).

The thunderstorm thus introduces a paradox. It supplies both an ignition source (lightning) and a retarding substance (rain). The distribution of lightning fire will correspond to the ratio between these competing processes, broadly conceived, not simply to the frequency of thunderstorms. Some lightning will strike outside the zone of precipitation, some precipitation will evaporate before striking the ground, some fuels will be shielded from rain (the base of a tree), and some rain will be ineffectual. Fuel moisture is more responsive to the duration of precipitation or high relative humidity than to its amount. Most of the moisture from a sudden downpour will be quickly shed off the fuel surfaces, and fine fuels can, in any event, soon dry out. Once ignited in a sheltered site a fire may spread into surrounding fuels in short order afterwards.

*Figure 5-3. Lightning effects: fire. Note the spiralling of the bolt around the bole. The gouge—often widened by the ripping of wood chunks—provides a favorable, somewhat protected niche for fire to burn in. From U.S. Forest Service.*

## *Lightning and Life*

The biological consequences of lightning are widespread. Fire may be its most prominent effect, but it is by no means isolated. Lightning has been identified as a predator on old and decadent trees, inducing both structural and physiological trauma; as a precursor to subsequent damage by insects, diseases, and wind; and as a minor fixer of atmospheric nitrogen. The bolt travels through the cambium, where water provides a conductor (Figure 5–3). For individual trees, the amount of structural damage that follows a bolt can vary widely. Where the bark is rough, the path of the bolt follows the grain of the bole, gouging and skipping in the general form of a spiral. The path forms a furrow with a strip of crushed inner bark along its axis. When the bark is smoother, the furrow is replaced by a series of tears and slabs. In some cases, the tree may simply disintegrate.

But this de facto girdling does not normally kill the tree. A more likely source of physiological trauma lies in the root system, where the path of the bolt provides a point of entry for insects and diseases. Yet many trees in areas of heavy lightning bombardment survive—this despite the fact that trees are struck preferentially. Of the trees hit, 96% are level with or above the group in which they occur, and 84% of these are dominant species within that group. Obviously, trees in such environments have evolved some adaptations to lightning and its associated effects.

Nor are the effects limited to isolated trees. Electrocution commonly affects organisms within a radius around the actual path of the bolt. A struck tree may become an epicenter for insect and disease infestations. Once begun, a small clearing may broaden under the influence of wind. Interestingly, many of these effects show up best in tropical forests or where temperate forests have been converted to citrus orchards or farms—all sites where fire cannot camouflage the impact of lightning alone (Figure 5–4).

But it is as an ignition source that lightning has its broadest biological consequences. A rough estimate suggests that 1% of all lightning discharges start fires, accounting for about 50,000 fires per year. These fires are not randomly distributed. Globally, the heaviest lightning activity occurs in the tropics, where natural fire is rare; the lightest lightning load, in the upper latitudes of the boreal forest, where natural fires, though infrequent, can reach conflagration size (Figures 5–5 and 5–6). In the summer of 1957, for example, some 5 million acres burned in the interior of Alaska, largely due to lightning. In the temperate United States the heaviest eruption of lightning fire happened during a 10-day period in June, 1940, when 1488 lightning fires broke out in the Northern Rockies. This is the largest known concentration by a factor of two. But from 1960 to 1971 the Northern Rockies and the Southwest regions of the U.S. Forest Service witnessed six separate 10-day outbreaks of 511 to 799 lightning fires each. These busts must be superimposed over a long-term background count of natural fires, a climatic constant. Between 1940 and 1975 a total of 59,518 lightning fires occurred on the national forests of the Southwest, 79,131 on the forests of the Rocky Mountains, and 88,680 on the forests of California and the Pacific Northwest. For the United States as a whole lightning

*Figure 5-4. Lightning effects: electrocution. Deadened area caused by a lightning strike in a mangrove swamp, south Florida. From Wade et al. (1980).*

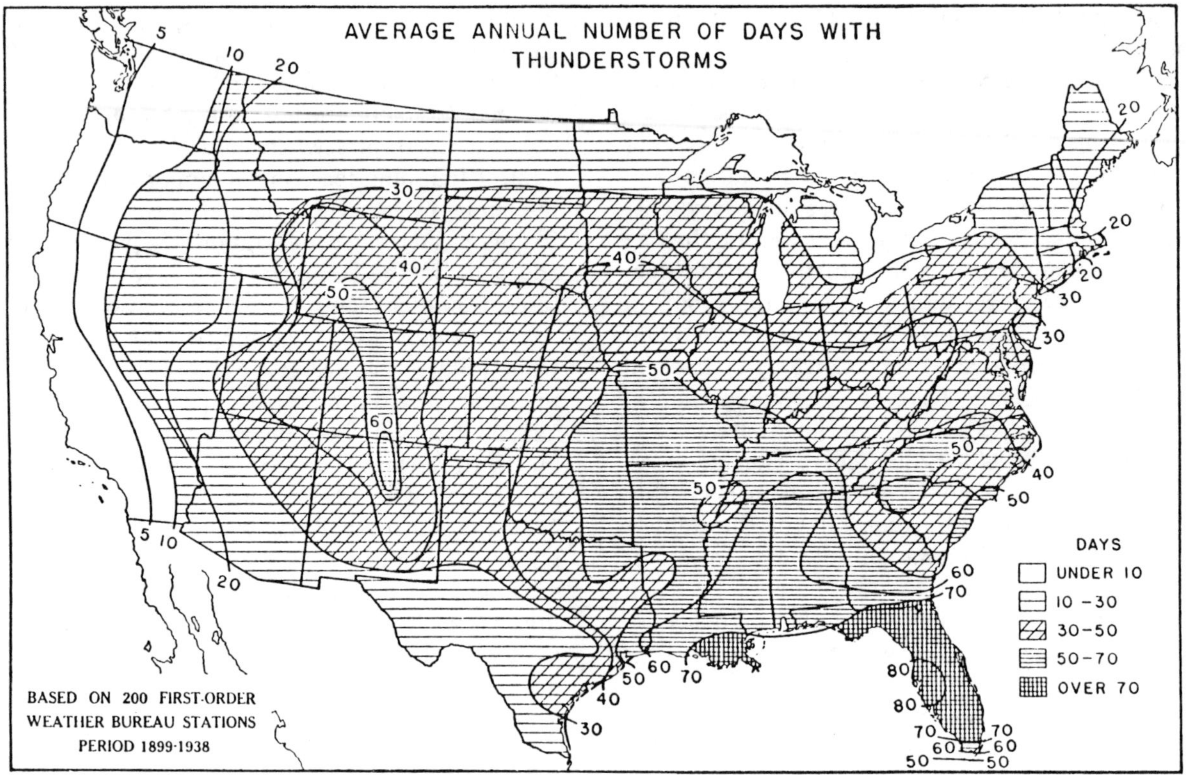


*Figure 5-5. Thunderstorm days. From U.S. Department of Agriculture (1941).*

is responsible for about 10% of all fires. Locally, however, lightning may account for over 90% of all fire occurrences and virtually all problem fires (Barrows, 1978; Barrows et al., 1977).

The evolutionary significance of this process may be large. Lightning predates the existence of all terrestrial biotas, and lightning fire must have exerted some selective as well as ecological pressures. Fossil lightning strikes of fused sand, called *fulgurites,* and fossil fire residue in coalbeds, *fusain,* testify to the longevity of lightning fire over evolutionary time. Char from ancient fires can be found in all the coal-bearing strata of the geologic record. Adaptation to lightning and its effects—fire among them—has gone on for most of the Phranerozoic Era. Only with the advent of mankind did the influence of natural fire apparently recede. Since at least the time of *Homo erectus* humans have manufactured fire, used fire, and controlled fire. From this point on the ubiquity of anthropogenic fire would begin to swamp the presence of lightning fire as an ecological process.

Yet the reality of natural fire is incontestable. As a manifestation of climate, lightning fire is the basis for fire ecology, fire behavior, and the possession of fire by mankind. There are reports of fires starting from other ignition sources in nature —from branches rubbing together, stones accidentally striking against each other, volcanic discharges, and even spontaneous combustion (in caves). But such sources, if they exist, are too intermittent to account for the widespread adaptations to fire by natural communities or for the universal capture of fire by humans. In the folklore of most peoples fire and lightning are intimately connected. So they are in nature, and no other source of ignition is really necessary. Lightning and lightning fire are abundant. The evolu-

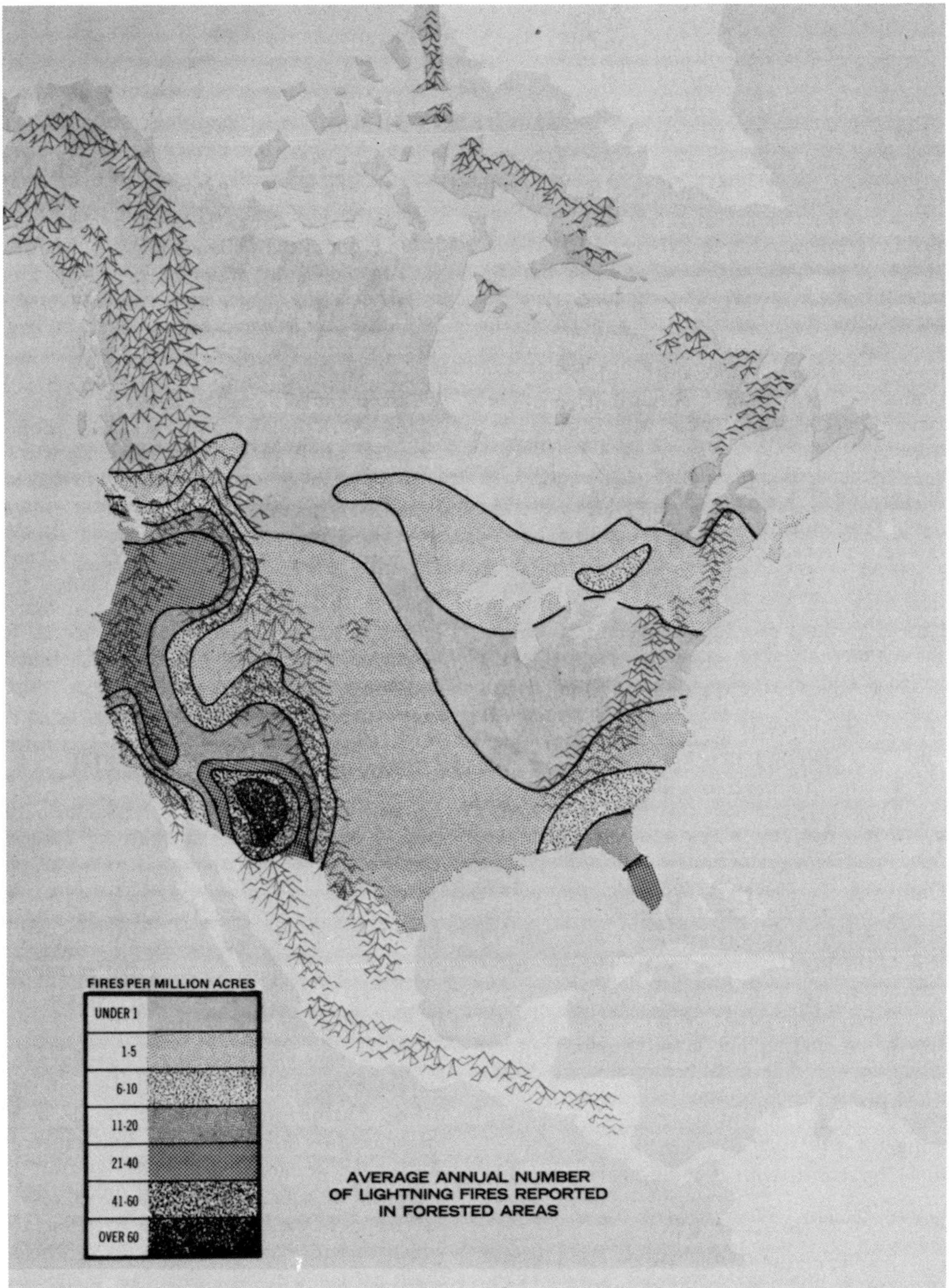


**Most lightning fires occur in the mountainous West and the Southwest. More thunderstorms occur in the Southeast but start fewer fires because of the accompanying rain.**

*Figure 5-6.* *Geography of lightning fires. Note the contrast to the geography of thunderstorms. Thunderstorms are distributed with moisture from the Gulf of Mexico; lightning fires, with other sources (western U.S.), or with peripheral effects (Southwest), or in dry seasons (Florida). From Schroeder and Buck (1970).*

tionary reality of lightning is an inescapable fact; lightning fire provides a scientific foundation for the theoretical assumptions underscoring fire ecology, and natural fire has become the philosopher's stone for nearly all contemporary thinking about the objectives of wildland fire management.

## 5.2 *EARTH, AIR, WATER, AND FIRE*

The heat released by a fire leads to thermal degradation, while the oxidation reactions that constitute the fire proper lead to chemical degradation. Many of the products of pyrolysis will be consumed as fuel by the flaming front, and in the process they will be further transformed chemically and even transported physically out of the reaction site. Of the oxidation products, some will become mobile and leave the system by vaporization or leaching; others will acquire new forms, to be incorporated into new organisms or new biochemical compounds; still others will acquire new niches, leaving the layer of surface litter to enter the soil profile or components of the living biota.

Among the constituents that will leave the system are organic nitrogen, by volatilization, and phosphate and sulfate, by leaching or erosion. Among those constituents that remain there will be carbon, in the form of charcoal; ash, rich in basic cations such as potassium, calcium, and magnesium; inorganic nitrogen, in the states of ammonia and nitrate. The release of soluble elements, like cations, will alter soil pH. The flaming front will kill microorganisms throughout the litter it consumes, and, insofar as it penetrates the soil profile, it will pursue them into mineral soil.

Stripping off the surface layer of vegetation will induce physical as well as biological changes. Since it functions as a protective cover, removal of duff will encourage erosion by wind and water; as a source of shade and insulation, its removal (and replacement by blackened ash) will increase soil temperature, thereby stimulating chemical and biological activities. Although fire itself resides in the biota, feeding on organic matter for fuel, its effects filter outward into the abiotic components of the ecosystem as well. Primarily this means soil, air, and water.

### *Litter*

Not all components of a biota are affected equally by a fire. Fires follow the distribution of fine fuels, and the fine fuels tend to gather on the forest floor. In virtually all cases this surface layer of fuels includes an array of dead and downed debris known as *litter* and *duff.* The litter layer is usually the most mobile component of the forest biota, the site of the most intensive modes of decomposition, and the principal locus for fire. In grass and brush, fire travels primarily through the standing stems, though there is generally some litter accumulation which helps sustain its passage. In forests, the litter is the primary medium of propagation, and a forest fire, properly speaking, is really a litter fire. Even when it flares into the crowns, a sustaining surface fire in litter is required for the crown fire to propagate for any distance or be reinitiated with any regularity.

The reason is that wildland fire is part of a dynamic equilibrium between the production and decomposition of biomass. This biomass represents stored chemical energy; its properties are both physiochemical and biochemical. Where it is stored in the ecosystem, and in what amounts, determines in large measure the character of the fire regime as a whole. In general that biomass which is most available as fuel—that is, that which is dead and accessible—is the litter. Deposited by needle cast, leaf fall, and windfall the litter merges, by stages of decomposition, into duff and the duff into mineral soil. An excess of biomass, litter not only makes fire possible but biologically necessary.

Without fire many ecosystems would continue to accumulate litter in a kind of biochemical reservoir, a holding tank of nutrients, some of which, like nitrogen and phosphorus, might be limiting for the entire system. Unless these materials are reintroduced into the system, its productivity will decline. Moreover, a thick veneer of litter can make otherwise suitable sites unusable for the reproduction of many organisms. For site maintenance and for the recycling of nutrients both, decay of the litter is essential. In many areas the only practical mode of decomposition is fire.

In a sense, wildland fire responds according to the amount of biological work, as manifest by litter, required of it. Fire's effects are correspondingly greater in forests than in grasslands, in temperate regions than in tropical. In warm, humid climates the production of litter is high, but so is the rate of its decomposition by organisms. In the extreme cases, such as tropical rain forests or areas infested with termites, the rates of production and decomposition may be equal. No litter builds up, and under natural conditions no fires occur. Fires can be introduced only by putting large amounts of fuel on the ground; this occurs with storms or with slashing by humans. Conversely, in cold, dry regions biomass is produced slowly, but it decomposes even more slowly; litter builds up, and fires, though rare, occur in the natural order of things.

The temperate regions of the United States lie between these two extremes, though almost everywhere the buildup of surface fuels by natural means exceeds their decomposition by purely biological means. Where the climate is warm and humid, as in the South, the buildup is relatively rapid, the outcome of high rates of production; fires tend to be frequent, low-intensity events more or less keeping pace with the production of fuels. Where the climate is colder, as in the Lake States, the buildup results from the slow rate of decomposition; fires tend to be episodic, high-intensity events. Where both growth and decay depend on biological processes, their periods tend to be identical, and they tend to correspond to the fire season. In the case of the South the fire season and growing season are relatively long; in the case of the Lake States, both are short.

Removal of the litter layer by fire alone is rarely complete. Where more or less total destruction occurs, the site will be found to be almost always marginal to the biota occupying it. Moreover, in nearly all such cases, the fires are anthropogenic in origin and are accompanied by other anthropogenic activities, such as logging, landclearing, farming, herding, or swamp draining. For this property of self-protection, certain physical attributes of fire and duff are responsible. Most of the heat generated by a fire rises. Even in intense chaparral fires, as little as 8% of the total heat flux is propagated downward. Since duff and mineral soil are poor conductors of heat, the heat of combustion diffuses downward through them for only small distances (see Figure 5–7). Most of the duff, moreover, is not immediately available for fuel. It retains moisture, especially in its lower layers, only drying out during exceptional periods of drought. Where it is thick and stratified, it tends to burn layer by layer in a series of slow reburns. As a fuel, duff resembles large diameter particles in its ability to acquire

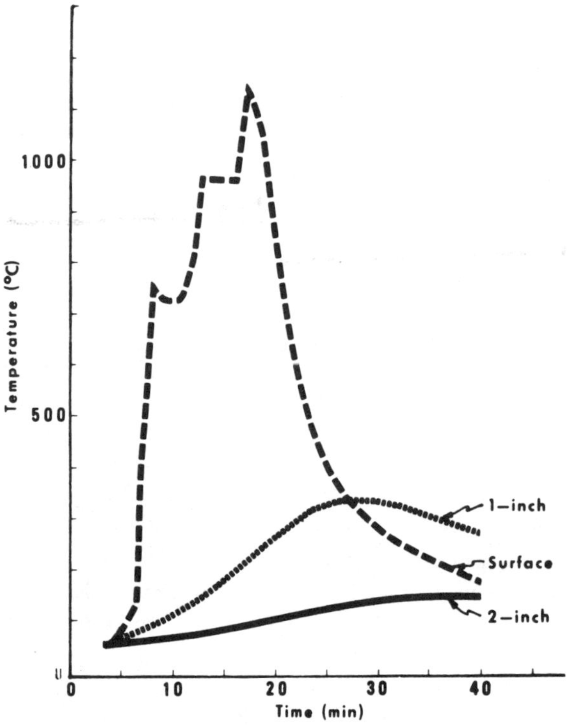


*Figure 5-7. Soil heat pulse. Note the effective insulating properties of soil. Most of the heat of combustion, moreover, is transferred upward by convection, not downward by conduction. From DeBano et al. (1977).*

and shed moisture and heat. Most combustion occurs as residual burning, with total duff consumption determined by fuel moisture (see Figue 3–25).

Between the fire and the litter-duff fuelbed there exists a reciprocity. The effect of the fire on the surface fuels will largely depend on fire intensity, but insofar as fire intensity depends on fuels the character of the duff and litter will influence fire intensity. If the typical pattern of surface fuels promoted holocausts that vaporized the soil, there would be no biota to reestablish those fuels. Where high-intensity fires are relatively common, they tend to be crown fires—further distancing the heat of combustion from the soil. The regime can perpetuate itself only to the extent that it does not irretrievably destroy the site. Its soil is a fundamental component of that site.

## *Soil*

The effects of fire on soil vary with the properties of the fuel, the soil, and the fire, especially the fire's frequency, intensity, and timing. The consequences are physical, biochemical, and biological. As a physical event fire may influence soil temperature, soil structure, and the ability of the soil to absorb and store water. All of these properties are related, and all depend on how thoroughly the duff and litter are burned. An exposed blackened surface will be warmer than an unexposed surface. Fire by itself rarely strips duff completely; almost always there remains a matlike veneer from the lowest (F) layers that protects mineral soil against raindrop impact and erosion by wind or water. The heat of the fire tends to harden this layer temporarily and to make the soil somewhat more impermeable. But the effect is quite variable, and when it occurs it is temporary; like most fire effects it moderates quickly with time.

A similar tendency is apparent with regard to water infiltration and storage. Where the fire is intense, infiltration rates and water storage capacity decrease temporarily. With a less intense burn, interception of rain and evapotranspiration will be reduced, and there will be a subsequent increase in the soil water available for storage. The exact outcome tends to be highly specific to the site and the character of the fire.

*Water Repellency.* Where soil is dry and coarse, and when a fire is hot, a special phenomenon known as *water repellency* may occur. The process depends on a class of organic chemicals that are hydrophobic, that will not bond to water. These chemicals collect in the lower layers of duff, a product of biological decomposition; there they may coat the grains of soil, clog interstitial pore spaces, and form a layer that is relatively impermeable to water. During a fire this process is accelerated and the impermeable layer is relocated. The chemistry of pyrolysis and combustion distills the active hydrophobes, increasing their quantity and improving their mobility. A temperature gradient develops downward into the soil, and the distillates follow it; at a subsurface zone along this gradient, they condense, bonding tightly to surrounding mineral particles. The result is a water repellent layer of variable thickness and depth (Figure 5–8).

The character of the layer is a function of the soil texture, the amount of biomass, and the intensity of the fire. Obviously, the latter two variables can be related: the biomass serves equally as a source of distillates and as fuel. Near the surface, where temperatures remain high, the chemicals continue to vaporize. The most intense water repellency results from soils that are heated between 350–400°F (176–204°C). Although evidences of fire-induced water repellency can be found in many environments, the most spectacular examples occur in the chaparral lands of Southern California. Erosion is a problem under the best of circumstances in this region. But the downward migration of the water-repellent layer worsens the tendency for erosion by enlarging the the mobile layer of soil open to saturation (DeBano, 1981).

*Erosion.* Like nearly all studies of fire effects, those pertaining to watersheds show contradictory results. General observations, however, sug-

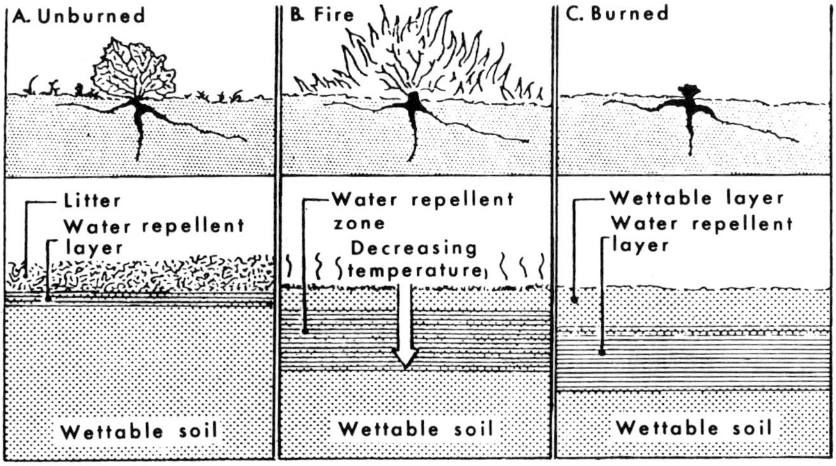


*Figure 5-8. Response of water-repellent soil to fire. (A) Before fire, hydrophobic substances accumulate in the litter layer and in the mineral soil immediately underlying it; (B) fire burns the vegetation and litter layer, causing hydrophobic substances to move downward along temperature gradients; (C) after fire, a water repellent layer is present below and parallel to the soil surface on the burned area. From DeBano (1981).*

gest that areas normally subject to erosion find their erosion rates accelerated temporarily by fire, and that areas which suffer little erosion under normal conditions show little increase as a function of fire alone. Again fire intensity, fire frequency, and soil properties are important variables, and to them must be added precipitation rates and topographic configurations such as slope. Even under unburned conditions steep slopes are subject to greater soil creep, mass wasting, and sheet and gully erosion than gentle slopes or level terrain. Where fire—especially fire in conjunction with some other activities—strips off the vegetative cover and prevents rapid regeneration, erosion on such slopes may be severe. Where ground cover can be restored quickly, however, erosion is soon contained.

Similarly, in areas where flooding is common as a result of heavy rains, a fire followed by heavy rain can result in large-scale, if temporary, erosion. But such a combination of events under natural conditions is rare. Perhaps the most prominent example is again Southern California, where large fires (of anthropogenic origin) burn in the late fall under Santa Ana winds and are then followed by a winter rainy season. Even without rains an acceleration in creep and dry ravel may be expected, but its duration is short. Most historic episodes of extreme erosion associated with fire resulted not simply from fire, but from fire in conjunction with other human activities.

*Soil Biology.* As a biochemical event, fire tends to increase the concentration and mobility of certain soluble elements, notably the ions of potassium, calcium, and magnesium; to decrease the presence of certain anions such as phosphate and sulfate; to decrease the quantity of organic nitrogen and to increase that of inorganic nitrogen; to raise the pH of the organic soil; and to leave a residue of carbon, in the form of charcoal. Depending on fire intensity and postfire events, these materials may be lost from the system, by erosion into wind or water or by leaching through the soil profile. They may be simply redeposited within the soil profile, in some cases leading to a nutrient reservoir on the surface and in others to a water-repellent layer in the soil profile. Or they may be, in part, absorbed into charcoal or seized by especially opportunistic plants. In general there tends to be a short-term transfer of phosphorus, potassium, calcium, magnesium, and nitrogen from the litter to the soil. What happens in the long term depends on what other events follow the fire and on the availability of organisms, notably plants, to utilize the sudden nutrient flush.

Several mechanisms are involved in these transformations. Among the elements that are converted from biological to mineral forms by volatilization are nitrogen, phosphorus, potassium, calcium, magnesium, and the trace metals copper, iron, magnanese, and zinc. The cations of calcium, magnesium, and potassium are oxidized and deposited as ash, in which form they react with atmospheric water and carbon dioxide to form bicarbonate salts. Bicarbonate salt exists naturally in the soil as a product of root respiration, but after a fire its quantity increases moderately and its solubility, dramatically. Both in soil solutions and in streams, the bicarbonate ion multiplies until it becomes the principal ion in the soil and the primary carrier of cations manufactured during the combustion process. The subsequent history of these cations—whether the newly constituted elements will be lost from the site or reincorporated into it—depends on the biota. Plants have means of capturing these elements, which are, from the plant point of view, macro- and micronutrients. Some of these mechanisms are latent, only released by a major disturbance such as a fire. The postfire history of these chemicals belongs in the larger realm of fire ecology (Wells et al., 1979).

*Nitrogen.* Especially significant is the nitrogen cycle. Organic nitrogen has a low temperature of volatilization, yet it is a critical, often limiting nutrient for most ecosystems. Nearly all evidence shows that large losses occur during a fire, but the significance of this loss is not obvious. Clearly most of the loss consists of nitrogen which was already used by plants or which was, as litter, unavailable; in either case, the loss would not affect the ability of the system to recover quickly. Moreover, if lost in one form, nitrogen may become more abundant in other forms. The concentration of ammonium and nitrate generally increase after a fire. Ammonium nitrogen is evidently a chemical byproduct of soil heating and of microbiological activity shortly after a fire. Nitrate nitrogen, by contrast, results indirectly from mineralization, decreased acidity, and increased ammonification (Figure 5–9).

Curiously, little of the nitrogen that is immediately produced comes from nitrifying bacteria. In

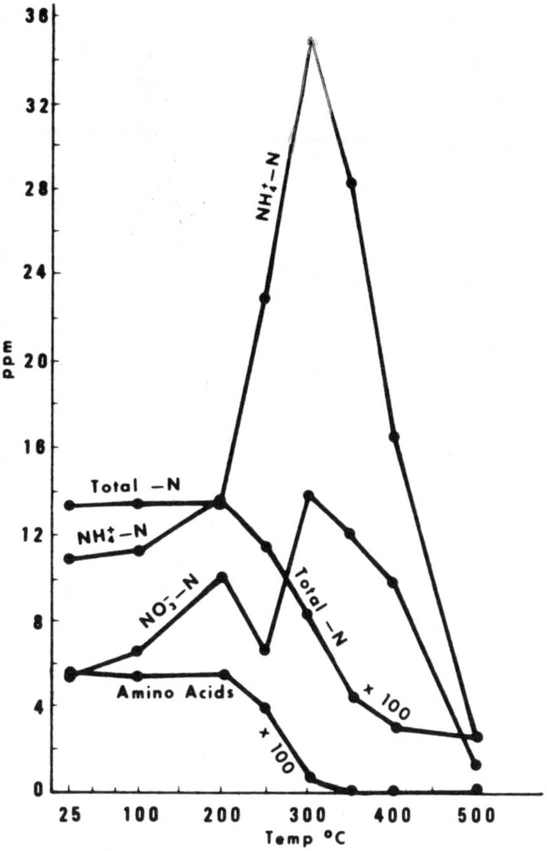


*Figure 5-9. Soil nitrogen changes. From DeBano et al. (1977).*

the initial stages, nonsymbiotic nitrogen fixation dominates over symbiotic processes. But where fires are common, nitrogen-fixing legumes are usually abundant, quickly seize burned sites, and begin the conversion of inorganic to organic nitrogen. Moreover, although there may be insufficient organic nitrogen available for the type of organisms typical of the community prior to the fire, there may be ample quantities available to those organisms who may occupy the site after the fire. Thus fire typically involves a transfer of nitrogen from litter to the air and soil regimes, a transformation of nitrogen from organic to inorganic forms, and a conversion from nitrogen-consuming to nitrogen-fixing plants (Figure 5–10).

In many areas where prescribed fire is used on an annual basis, there is evidently no loss in the available nitrogen. There are historical examples

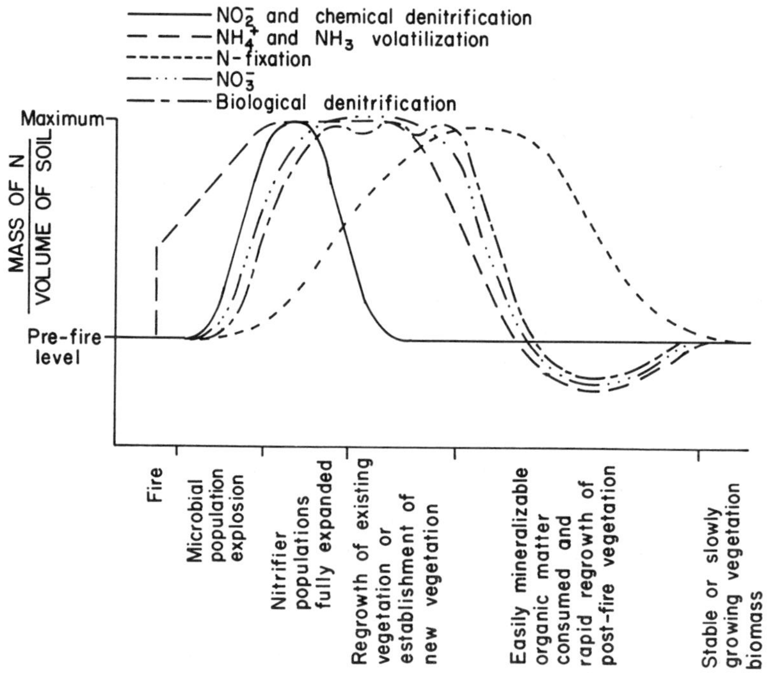


*Figure 5-10. Relative time course of hypothetical responses of organisms to fire, with an emphasis on the nitrogen cycle. From Woodmansee and Wallach (1981).*

in which existing quantities of nitrogen were virtually obliterated and the capacity of the system to reestablish its nitrogen cycle all but eliminated. But close inspection of the historical record invariably reveals that fire in these episodes was combined with other activities and that the combination of fire with ax, plow, or hoof was applied to marginal sites.

*Microorganisms.* As a biological event on the scale of organisms, a litter or soil fire has an immediate, usually lethal effect on populations of microorganisms and a more significant effect on their habitat. By altering soil temperature and by removing the litter cover, a fire changes the characteristics of the site as a habitat and as a microclimate. Microorganisms fare poorly—even worse when the fire burns with wet heat rather than with dry. Many are easily killed by the fire, but the instantaneous change of habitat is probably responsible for the largest decline in population. The effects are temporary, with recovery in a handful of years.

This lethal heat wave accounts, in part, for the relative poverty of nitrogen-fixing bacteria immediately following a fire. But elimination of microfauna and microflora also reduces the competition for the nutrients made newly available by the burn. In a sense, fire is in competition with biological agents as a decomposer. Nonsymbiotic nitrogen is soon generated, however; the increased soil temperature accelerates the subsequent decomposition of the remaining litter; a new vegetative regime is established, and with it comes the reestablishment of the microorganism population.

In some instances the destruction of a microorganism population must be counted as a benefit of fire in systems managed by humans. Some microorganisms are active parasites, and all are competitors for soil nutrients. Fire can have a sanitizing effect by eliminating, at least temporarily, certain pathogens. Conversely, it can release assorted latent mechanisms and promote existing mechanisms of reproduction that ensure the self-perpetuation of the biota. Prescribed burning for the control of pests, vermin, insects, and assorted diseases and for stimulation of selected organisms is an ancient fire practice, and one that is enshrined in the folklore of many peoples.

## *Water*

It is an interesting coincidence that the components of a watershed—topography, weather, and vegetative cover—are identical to the components of a fire environment. As natural phenomena fire and flood show remarkable parallels. They exhibit the same kind of intensity–frequency distribution. The modification of cover influences the intensity of both fires and floods. Weather is the supreme variable for both, and the one that most restricts human modification of flood and fire. But their interaction comes indirectly through their shared biota rather than by direct coupling; the effects of a fire are usually first assimilated into a biota, then made available to the hydrologic cycle; and the impact is moderated by the many intervening processes and the numerous self-regulating mechanisms that characterize an ecosystem. Reported results vary wildly, too, as a function of fire intensity and as a consequence of whether, due to heavy precipitation, flooding follows quickly after a fire. The outcome of a fire on a water regime may be physical, pertaining to the movement of water and sediment through the hydrologic cycle; chemical, describing the array of chemicals and nutrients released by the fire; or biological, relating to the changes in aquatic habitat as a result of the fire.

The hydrologic or geologic consequences of fire are perhaps the best researched. Yet the documented effects of fire tend to be ambiguous—the result of having to integrate the effects of change in one complex system, the biota, with effects in another system, the stream. In general, to the degree that it removes cover type and litter, a fire reduces the amount of rainfall intercepted and the rate at which it may infiltrate into the soil profile, while at the same time it increases the amount of snow that may accumulate and the rate at which snow will melt. The quantity of overland flow and of surface erosion are mutually related; how much results depends on the intensity of the fire, the steepness of the slope, the character of the soil, and the severity of rain. When combined with other activities such as logging or plowing, the rate of sediment and water runoff on mountain terrain may be high. By reducing evapotranspiration, fire may significantly enrich the amount of soil moisture in storage, but by lowering infiltration rates, it also depresses the quantity of new water being stored. Stored moisture, moreover, may lead in certain cases to mass movement. Where water-repellent soils have been created, the effects may be broad (Tiedemann et al., 1979).

The time required to return to prefire conditions varies widely. In some regions the former regime is quickly established; in others, perhaps 5–10 years may pass before restoration is more or less complete, assuming that no additional disturbances occur. Erosion problems associated with postfire runoff tend to be specific to certain fire regimes rather than inherent properties of all fires.

Of the soluble chemicals created by the fire or immediately following it, many are stored within the soil profile; a few are leached through the profile; and some are carried, as surface debris, by runoff. The increase in nitrogen evident in many onsite and offsite streams results most probably from organic detritus swept directly into the streams as part of their total sediment load. Cation quantities similarly increase, many bound to sediments and carried along with the general flux. The increase in runoff is responsible for entraining this debris, but that very increase dilutes their impact. Reported discharges vary considerably. But they have in common that, although large relative increases in chemical debris occur, the absolute amount involved is slight. This observation holds true whether the quantity is measured either as loss of nutrient capital from the site or as the introduction of a pollutant in the stream. Neither at the point of origin nor at the place of deposition is there significant deterioration of habitat. Especially when considered in terms relative to the increase of water, the nutrient flux tends to fall well within normal ranges of tolerance.

Partly for this reason, and partly perhaps because the question has been little studied, there is apparently little change in the aquatic habitat.

Postfire changes are less chemical in nature than physical—the product of increased water temperatures due to the removal of shading vegetation; of channel scour, due to the elimination of log barriers in the stream or to variable changes in the runoff ratios of sediment and water; or, where organic soils are involved, to the creation of new niches. Fires in deep organic soils not only release nutrients otherwise unavailable, they create a topography of basins that, in the wet season, furnishes feeding sites for birds and animals and, in the dry season, may become refugia for many aquatic species, source areas for future colonization. Even when fires occur in grass within marshes, changes in water chemistry are promptly diluted and disappear.

Increasingly pertinent are those modifications in stream regimes and water quality produced by fire suppression activities. Bulldozed lines on hillsides and low-lying permafrost, small dams constructed as water reservoirs, chemical retardants dumped near lakes and streams—all influence the chemistry and hydrology of runoff. Nearly all such means of fire control are, in some way, subject to restrictions, and in some sensitive areas are banned altogether. Similar reservations apply to those projects that attempt systematic rehabilitation after fire, especially when these programs involve further soil disturbance, the use of fertilizers, or the introduction of exotic plants, including the non-native grasses often sown in the name of erosion control.

### *Air*

The atmosphere shares with soil and hydrologic systems that it is a sink for certain products of wildland fire, that it relies on a similar fuel complex and fire regime, and that it dilutes and dampens the effect of a fire. But there similarities end. Air acts as a heat sink; soil and water systems do not. The chemical products that fire delivers to air are different from those deposited in soil and water systems. To the atmosphere, by and large, go the volatiles and extractives, or free radicals in a volatile state that can lead to vapors, aerosols, and particulates; to soil and water systems go solid residues such as charcoal, and quickly oxidized biochemicals, the macronutrients.

The chemical processes are different, too. For those chemicals destined for the atmosphere, the physical attributes of the fire itself are especially significant. The flaming zone of the fire creates a special chemical environment for subsequent reactions that yield a host of products, primary and secondary both, which never reach soils or streams. Fire tends to simplify the number and variety of chemicals delivered to soil and streams, but it tends to multiply and make more complex the compounds that are discharged into the atmosphere. How varied these compounds become and how great their quantity will depend to a large degree on the nature of the fire that creates them.

But whereas soil and water do not directly interact with fire, providing only boundaries and sinks for the burning process, the atmosphere is an active presence—an important component of the fire environment that helps determine how a fire burns, what heat and chemical byproducts a fire generates, and how these materials will be dispersed within the surrounding air mass. Because it actively enters into the combustion process, the air–fire relationship becomes far more complex than the soil–fire or water–fire relationships. Against this increased complexity, however, must be put the relative unimportance of vegetation as an intervening presence. As canopy, vegetation affects turbulence, thereby influencing surface mixing; it intercepts some smoke, acting as a scrubber; ultimately it reincorporates particulates back into the biota. Because smoke passes through the biota smoke acquires a biological agency. And because smoke passes into cultural systems it becomes a cultural phenomenon. Smoke is considered a pollutant in ways that other postfire residues left in soil or streams are not.

The outcome of fire on the atmosphere is known well in gross terms, and understood poorly in its particulars (Sandberg et al., 1979). As a general summary it may be said that fire ejects into the atmosphere an array of primary and secondary products of combustion—largely water, carbon dioxide, and matter aggregated as particulates; that these products enter the surrounding air mass as a smoke plume or column; that they are dispersed and mixed by atmospheric winds and quickly diluted; and that, at all stages in this sequence of events, the relationship between the fire and its products, on one hand, and the atmosphere and its processes, on the other, is symbiotic.

Everywhere in the sequence complexity arises from the great variability that is possible in the system—variability in the source, the chemistry of wildland fuels; variability in the processes that act on those sources, the dynamics of the fire environment; variability in its products, their chemistry and shape; and variability in the mechanisms that disperse these products through the atmosphere, processes that are ultimately stochastic. Smoke is a product of fire, fuels, and air. Fire and fuel are largely responsible for what is ejected into the air, and fire and air for how that ejecta is dispersed.

*Chemical Effects.* The detailed chemistry of combustion products was described in Chapter 1. Recall that within a forest biota some 14 categories of fuel with different chemical characteristics have been identified. Even for a single element such as a conifer tree, there may be large variation in the chemical constituency of duff, needles, twigs, bark, and heartwood. From fuel models may be determined how much of each kind of fuel is available for combustion, and from fire behavior models, how much actually burns. Wet fuel generates more particulate matter than does dry fuel, smoldering combustion more than flaming combustion, ground fires more than surface fires, and heading fires more than backing fires. Recall, too, that wildland fires are free-burning fires, that the burning process is far from efficient, and that the products of combustion are often far from complete in a chemical sense. At any point in the process some pyrolyzed, combusted, or pyrosynthetic chemicals will escape from the system, ready for subsequent reactions outside the fire proper. The ejection or escape of combustion byproducts or partial byproducts will depend on the character of the diffusion processes directing combustion.

*Chemistry of Smoke.* If combustion were total, the end products would be two volatiles, water vapor and carbon dioxide, and a minute residue of white ash, consisting of assorted noncombustible minerals. But some of the pyrolysates will escape as vapors, and others will cool and condense without flaming. Some will flame, but in the oxygen-poor environment of the fire only partially oxidize, generating assorted intermediate compounds. Others will enter into pyrosynthesis, welding hydrocarbon radicals into polynuclear aromatic hydrocarbons. Where compounds of low molecular weights result, they will often be lost as vapors to the atmosphere. Where compounds of higher molecular weight are common, condensation may form aerosols, collect on solids to create particulate matter, and combine with water vapor to yield white smoke.

Among other primary products—compounds generated directly from the burning process—are carbon monoxide, hydrocarbons, both unsaturated and aromatic (saturated), nitrogen oxides, sulfur oxides, and perhaps hundreds of other organic gases, including carboxylic acids, reactive aldehydes, and formaldehyde. Many of this last group are known to affect plants, and a few, human health. The nitrogen oxides, in general, require very hot temperatures, greater than occur in most prescribed burns; but they can result from lower-temperature combinations of hydrocarbon-free radicals. Their total quantity is probably low, and the significance of these compounds resides largely in their potential for secondary products. Sulfur oxides are rare, dependent on special fuels. From these compounds a host of secondary products can form; but of their chemistry little is known. Unless the compound is a health hazard, it has not been studied. Ozone can form; the aggregation of organic compounds into particulates can, as condensation nuclei, collect other noxious vapors; some trace quantities of hydrocarbons may, under special circumstances, lead to photochemical smog.

Few of these compounds, however, build up to any significant degree; all are quickly swamped by the surrounding air mass, converted into secondary products, or diluted into negligible traces. In the case of carbon monoxide, for example, concentrations as high as 200 ppm have been reported close to the flaming zone, but the level fell to less than 10 ppm within 100 ft (30 m) from the fire. No models exist to predict the emission factors for any of these compounds.

*Physics of Smoke.* The products of burning exist as vapors, liquid aerosols, and solid particulates. Certainly the most visible of emissions, particulates are a major cause for smoke formation. They scatter light and provide adsorption surfaces and condensation nuclei upon which other products of burning collect. Chemically, particulates are of two general varieties: those particles which represent organic debris entrained more or less intact within the convective plume, and those particles which have formed chemically from the bonding of assorted products generated by the fire. As much as 40–75% of this matter is organic, as measured by benzene solubility.

The physical impact of particulates on the atmosphere varies with their residence time in the air, their adsorptive capacity, their total quantity, and their ability to scatter light—all factors that follow from their size, and are generally properties of the fire rather than of its fuels. Physically, particulates tend to cluster around an average size regardless of source fuel: 0.1 micrometers, as measured by the numbers of particles, or 0.3 micrometers, as measured by their mass distribution. Approximately 91–95% of the particle masses are below 1 micrometer. Of the total, particles from 60 to 70% will exist as liquids. The particles tend to adsorb other products and to act as points of nucleation for further condensation. In one respect, then, they function as scrubbers, concentrating and cleansing the atmosphere of burning products (Southern Forest Fire Lab, 1976).

In another sense, however, they magnify the concentration of pollutants by reducing the effect of dilution, an important consideration where other toxic gases (e.g., sulfur dioxide) may be present in addition to those gases generated by the fire. Particulates whose diameters approximate the wavelengths for visible light (0.3–0.8 microns) cause the greatest light scattering, but this size of particle is also highly amenable to long-term suspension in the air. Particles of somewhat larger dimensions (from 2 to 3 microns) are likewise prone to long residence times in the atmosphere. Unlike the smaller particles, this size category does not affect visibility but it does affect health. Particles on this order are especially susceptible to respiration, penetrating deeply into the lungs with whatever toxins they may have acquired.

The effects of smoke on the atmosphere tend to be ephemeral and local. For a short period of time, and that usually confined to a particular season, the atmosphere is more heavily charged with particulate matter and various chemicals, collectively known as smoke. Over a large enough area, this condition will affect incoming solar radiation, and perhaps upset the local wind regimes. Extremely large historic fires have obscured the sky over relatively large areas, the "Dark Days" of chronicles. But no known meteorological disturbances resulted. Such overloads are not beyond the capability of the atmosphere to handle, and no known fire has exerted a global effect on climate comparable to those that have resulted from major volcanic eruptions. Where fire complexes persist over large regions, some local effects are likely.

*Biology of Smoke.* Eventually, however, this debris will collect back on the surface, there to be incorporated into biotic processes. In most fires, moreover, smoke passes through at least some portions of the biota before entering the free atmosphere. Thus smoke has the potential to become a biological agent as well as a meteorological event. Smoke seems to inhibit certain fungi, rusts, and algae; to stimulate (through propylene ethenol, a minor secondary product) the flowering of certain plants; and to retard the growth and spread of certain pathogens and parasites such as mistletoe. The details of the processes by which this occurs and the identification of the active agents in the smoke have not, for the most part, been worked out. But the association is evident, and smoke, as both stimulant and fumigant, has been used as an agricultural tool for centuries.

Not least among its biological effects, smoke can affect humans. Again little is known of the particulars, and the outcome depends on the character of the fire that, as a result, may produce active toxins. For the most part, however, wildland smoke seems to act mostly as an irritant on its own, and to show hazardous properties only when combined with other, industrial pollutants, in which case particulates from wildland smoke can collect toxins from other sources and deposit them in the lungs. Though, as a natural component of natural fires, smoke may be significant in the management of natural areas, it has also been implicated in several actual air pollution episodes.

For most regions the principal complaint against smoke is an aesthetic one, that it seriously impairs visibility. This is not an inconsiderable issue in areas such as many recreational sites, where scenery is a valued resource. Wildland smoke thus comes under the control of air quality regulations, and much of the future of prescribed fire may very well depend on what future research says about the innumerable products of wildland smoke, on how smoke is perceived as a contaminant of scenery, and in what way air quality criteria are interpreted. But regardless of constraints that might be imposed on it, smoke may be managed, much as fire behavior and fire ecology. For many areas, the chief limitation on prescribed burning is its production of smoke.

## *5.3 FIRE ECOLOGY*

Like the study of fire weather, the science of fire ecology isolates one set of processes from a complex system. The results are useful for understanding and perhaps essential for management, but they remain artificial. Fire is only one process of decomposition among many; adaptations to fire by individual organisms are rarely limited to the effects of fire alone, or to isolated traits; the impact of a fire upon ecosystems as a whole will vary widely among systems, between fires, and according to biological timing. Commonly these responses are analyzed in terms of *autecology,* the study of individual organisms; *synecology,* the study of communities; and the *fire regime* concept, the integration of ecosystems with fire history. The investigation of fire effects, broadly conceived, is complex because fire effects are not linear, but geometric; not singular, but plural; not simply completed, but extended over long periods of time.

Yet the isolation of fire as a process and of fire ecology as a distinct subdiscipline is not wholly arbitrary. For some ecosystems, as for some organisms, fire is so dominant a process that it does shape to an inordinate extent the structure, composition, and dynamics of the whole system. A number of such fire regimes have been relevant to the human history of the United States, and many remain of significance to the management of contemporary wildlands.

In general, fire works within ecosystems to decompose, recycle, and select. As an agent of decomposition, fire releases, in the form of a heat wave, the chemical energy stored in the available fuel, and it liberates, in slightly altered forms, many of the constituent biochemicals residing in the litter. Somehow an ecosystem must cope with this discharge of energy and chemicals, with the eradication of some organisms and the introduction of others, with the simultaneous processes of selective destruction and selective enhancement. So important is fire to many systems that if it did not exist, nature would have had to invent it. But fire does more than degrade: it recycles. On a microscale, it recycles nutrients, inorganic and biochemical both, liberating them from various biological reservoirs and rendering them accessible to different sorts of organisms. On a macroscale, fire recyles the biota itself, sustaining a system of different species, age classes, and physical arrangements. Over long periods of time fire can influence evolutionary developments through its selective actions.

The processes of fire ecology are complex, the outcomes often not repeatable, and the conclusions frequently contradictory. Yet some general principles do apply. And whether the object of study is considered as a biota or a fire ecosystem, whether fire is considered by itself or in conjunction with other processes, whether the administrative goal is the preservation of wilderness areas or the manipulation of productive wildlands, those principles underwrite contemporary planning in fire management and inform the operational guidelines that direct fire programs.

### *Autecology*

Few organisms are specifically adapted to fire as such. Most fire adaptations are general traits that represent responses to multiple selection pressures. For example, resprouting from roots is a means by which an organism can recover from any process or collection of processes that tend to strip off the abovesurface vegetation. The source of this denudation may be an herbivore, drought, or fire, or several agents in combination. The response, resprouting, is identical in all these cases, and it achieves the same effect. Nor are some traits typically isolated from other traits that react to similar processes. Commonly, organisms show a suite of traits that can respond to an array of selection pressures. In only a few environments is fire so domineering an event that it generates fire-specific adaptations, or so unique a process that it alone leads to particular ensembles of effects. Even in the case of fire-specific adaptations, moreover, organisms must respond to a range of fires—fires of varying type, timing, frequency, and intensity. Adaptation is a somewhat amorphous, often ambiguous concept.

*Flora.* Among the many adaptations shown to fire are those that protect an organism during a fire, those that stimulate reproduction as a result of a fire or the postfire environment, and those that seemingly use fire as a means of driving off competition. Some effects are almost wholly defensive, offering protection against the wave of killing heat released by the flaming front; others are more opportunistic, seizing on the nutrients released by the fire and the new habitats fashioned in its aftermath.

Fire can kill by consuming the organism as fuel, of course. But its heat wave alone can be fatal by means of crown scorch and cambium scorch (Figure 5–11). Crown scorch occurs when hot, rising gases dessicate foliage (Van Wagner, 1973). Scorch correlates with fireline intensity, hence with flame length. The effects of windspeed are mixed. Higher windspeeds increase fireline intensity, but they also cool the rising plume of hot gases even faster; the net outcome is a decrease in scorch height for a fixed value of fireline intensity. How much crown scorch can be tolerated varies with species. Common adaptations include the ability to recover from defoliation and the segregation of foliar fuels from surface fuels through rapid growth and selfpruning.

For protection against cambium scorch, bark is the most common adaptation. The penetration of heat into the cambium tends to be inversely proportional to the thickness of the bark. The killing effect of heat depends on its temperature and duration. Larch, ponderosa pine, sequoia—all shield the living cambium through the presence of heavy bark. Some species such as white fir acquire thicker bark with aging, making them susceptible to light fires during their younger years and impervious to fires of the same intensity during maturity. Others such as palms and palmettos, which lack a cambium layer, can resist heat well. For some organisms, it is enough to protect buds, through mechanisms like tufted crowns common in tree ferns and certain grasses; seeds, through dense, impermeable coatings typical of the legumes; or rhizomes, through shielding by soil. Typically, it is only necessary to offer protection during the passage of the flaming front. The presence of thick duff or heavy fuels with long residence times, however, can injure plants by long-term heat, overcoming the resistance to thermal conductivity offered by bark.

Among those traits that encourage rapid regen-

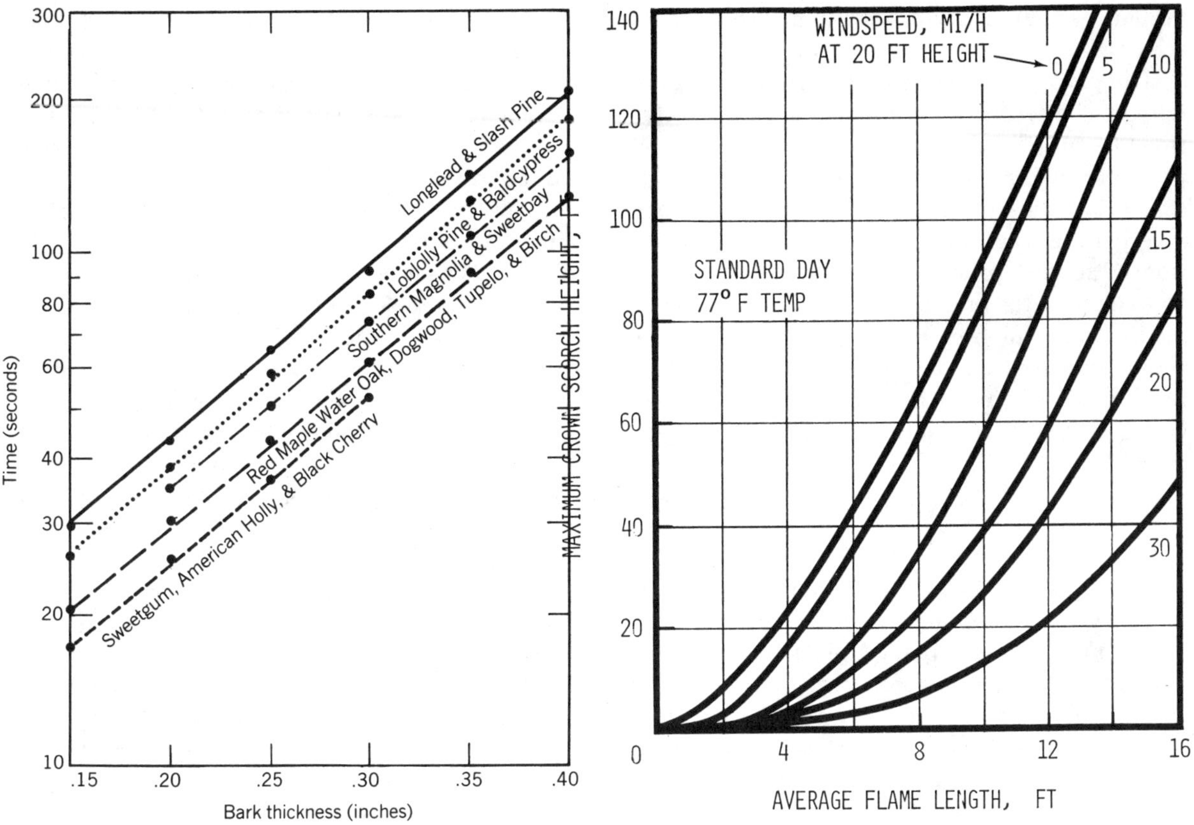


*Figure 5-11. Differential resistance to heating in trees. Graph at left shows the average time for the cambium to reach lethal temperatures (140°F) when heated by a propane torch. Lower graph includes tree diameter as well as bark thickness. Exposure of the cambium to high temperature is longer than exposure of the bark. Its insulating properties mean that bark first resists heat, then retains it. Lower heats of long duration can be as lethal as high heat sustained during passage of the flaming front. From Hare (1965). Graph at right correlates height of crown scorch to flame length, both of which quantities are predicted by fireline intensity and are therefore related. Values are given for a standard day, 77°F. With higher temperatures, higher crown scorch can be expected. Note the mixed effects of wind on flame length and scorch height. From Albini (1976).*

eration after a fire, sprouting is common. In virtually all fire regimes that are subject to frequent fire or to high-intensity fire or to both, there exist dominant species that rely on sprouting from the roots, like aspen; from the collar, like gambel oak; or from the bole, like pond pine. Such a mechanism obviously tends to perpetuate the existing biota, though there may be a period of 1–5 years during which other adventitious organisms move in until the process of reestablishment is complete. Only when the tops are killed or injured and when the aboveground site is open does sprouting from the roots occur.

A spectacular example among the conifers is the longleaf pine, which, early in its life cycle, experiences a grassy stage, with most of its nutrients stored in a large taproot. For a bole it exhibits a grass-like tuft of long needles, easily destroyed by the fires that sweep through the grasses with which it appears. The trunk then sprouts vigorously, putting the needles well above the range of normal grass fires. An example among deciduous trees is the aspen. Without new sprouting and in the absence of disturbances that keep out its taller competition (both conditions that can result from fire) aspen groves become senescent and eventually succeed to other trees.

Many fire regimes seem to favor rapid reproduction, intermittent reproduction, and rapid dispersal of seed. For some plants fire can stimulate flowering, either through site improvement, as with tallgrass prairies, or through the action of smoke, such as with certain species of laurel and orchid. For other fire regimes, annuals—with their capacity for intermittent reproduction and rapid dispersal—may be temporarily favored over perennials as new niches appear on the postburn site for a few years.

Plants that can produce seeds in a few years have an advantage in many fire regimes over those plants which require a longer time to reach reproductive maturity, and so do those plants capable of holding seeds for considerable periods of time until favorable conditions, as induced by a fire, result. Hard seeds lodged in litter are one example. Such seeds need heating, scarification, and a suitable open site before they germinate—a litter fire satisfies all these conditions. A more spectacular case involves trees that retain seeds until a fire releases them. In some instances, as with the Douglas-fir, fire can help cure the cones. In other instances, it promotes the liberation of the seed from its encasing shell. Such is the case with the serotinous cones of lodgepole, pond, and jack pines. Sealed with resin, the cone releases its seeds only when the seal is melted. Typically, fires in such species are crown fires that apply heat directly for short periods of time to the cones themselves, incinerate potential competitors, and prepare a warm, nutrient-rich site for germination. Fire regimes of this sort are characterized by relatively frequent, high-intensity fires and tend to be self-perpetuating.

In still other cases the presence of cleared sites presents opportunities for those species capable of rapidly dispersing seeds, often over some distance; this happens in the case of aspen and of Douglas-fir with its fine, winged seeds. Whether such species colonize a new site depends on how proximate seed trees or unburned refugia are to the site.

*Fire and Natural Selection.* It may be that fire behavior, through the medium of its fuels, itself represents an expression of past fire history, of the adaptation of the biota to the pattern of fire experienced on the site. Plants differ in their flammability. Some have a chemistry that favors hot fires. Some have traits that either accent or inhibit important properties of the fuel complex—fuel porosity, fuel load, the ratio of fine fuels to coarse, of live fuels to dead, of surface fuel to aerial. Because these fuel properties are the product of organisms, they are, in theory, amenable to natural selection. Since the outcome of excess fuel is often fire, fire may become an instrument of selection.

Several explanations for this association have been put forth, of which the Mutch hypothesis is perhaps the best known (Mutch, 1970). Most fire adaptations have traditionally been interpreted as defensive measures, as means to assure survival during and after a fire. The Mutch hypothesis goes somewhat further to argue that some plants encourage fire, that such fires work to the selective advantage of these plants, and that certain combinations of high flammability and quick reoccupation of a site constitute a form of adaptation to the reality of fire. Fire becomes a means of regeneration, a form of competitive struggle, and a process that favors some species over others. Fire not only sets in motion certain responses latent in the organism's genetic composition, but helps to shape that genetic base.

Hard data from population biology is difficult to come by, but the proposal is intuitively attractive in that it helps describe the extraordinary reciprocity that exists in many fire regimes between the character of their fires and the nature of their biota. In certain regimes some species are clearly favored by fire over other species: the continued presence of fire within a certain range of intensities and frequencies promotes and perpetuates these organisms. In some regimes the species favored by fire have acquired flammability properties that promote the desired pattern of fire. By undergoing a number of physiological changes, chamise chaparral increases the likelihood and intensity of a fire on about a 30-year cycle. Jack pine and lodgepole pine generate fuel complexes that favor high-intensity fires of exactly the sort which release the reproductive potential sealed in their serotinous cones. A roster of such organisms can be expanded easily. Those organisms that seem to show such special, fire-specific adaptations are referred to as *pyrophytes.* The concept does not mean that these organisms are adapted only to a fire frequency of short periods, but that they show special adaptations to the fires typical of their fire regime, a regime that they are responsible in some way for shaping.

The relationship between biotic cycles and fire frequency is fundamental to an appreciation of how fire can effect ecosystems. It is not necessary that fires come with a certain regular frequency but only that the important fires—the ones with significant biological effects—come at appropriate times within the life cycle of the organisms involved. For a plant dependent on fire in its competitive struggle, this means only that fire ought to appear between maturity and senescence, between the time the organism is capable of reproducing itself and its natural death. In the case of trees that may survive for hundreds of years, the fire frequency may be quite rare and still remain influential. In the case of tallgrass prairies, by contrast, the fire frequency must be short, or else trees will invade and displace the grasses. If succession occurs, one of the system characteristics that it changes is the set of parameters that determine its fire frequency.

Although the case of general fire adaptability is easily argued, the identification of specific fire-selected traits is more difficult. Consider the case of cone serotiny in lodgepole pine (Lotan, 1976). In natural populations there exist both serotinous and nonserotinous cones. Serotinous cones dominate in areas where the fire regime favors high-intensity, stand-replacing crown fires; but where fires tend to be surface fires of indeterminant frequency, a larger proportion of the lodgepole population shows nonserotinous cones. Moreover, high-intensity fires tend to occur where lodgepole pine faces stiff competition from other conifers—where ultimately, in the absence of disturbances, it will be replaced by other species. But the picture is complicated. The variability of serotiny in the population may be due to other environmental factors; the proportion of serotinous to nonserotinous cones in a single tree varies with its age, and the proportion in a community by fire frequency; and there exists a correlation between bark thickness and degree of serotiny, such that the prominence of the two are inverse proportion. Where bark is thick, serotiny is less.

Here there seems to exist a suite of traits which, among themselves, accomodate a range of fires that occur over a span of intensities and at different stages in the life cycle of the organism and the successional patterns of the community. Serotiny may be considered fire-selected in that it appears most prominently where high-intensity fires are most common, but serotiny contributes nothing more to the flammability of the system beyond what the organism already possesses and it is far from being the sole mechanism of adaptation to fire. Simply put, because of the character of fire in this regime, serotiny is here given extra prominence. By way of analogy, the fire regime itself is "adapted" to its fire history in the same way that a river is "adapted" to its flood history.

*Fauna.* Unlike flora, fauna do not contribute to a fire in the form of fuel. Instead animals respond to a fire, and, as one would expect, that response varies considerably. Part of the response is to the fire itself, and part, perhaps the largest part, to the changes in habitat brought about by the fire. Few animals are killed outright by fire, and most instances of animal fatalities are the result of unusual, high-intensity fires. For fires of lesser intensity, many organisms are attracted to the site of the burn. A few insects seek out smoke and heat, and others are wafted upward in the smoke plume, drawing bird insectivores who in turn attract raptors. Some small animals are driven by the flames, and predators search them out. Most small fauna simply retreat underground where they weather the fire in short order. Soil microorganisms suffer losses, depending on how deeply the litter is burned; so do a multitude of surface insects: mites, chiggers, and ticks, though not ants. Most fire effects on faunal populations, however, come indirectly through modification of the habitat, through the multiplication and loss of specific niches. Some fauna gain in population as a result, some decline, and some remain more or less constant (Lyon et al., 1978).

How drastically some populations change depends on how completely fire reshapes the total biota and on what possibilities there are for an exchange of species from burned and unburned habitats—a function of fire size and the presence of refugia. In general, there tends to be a slight increase in avifauna, and a relatively constant number of mammalian species; and in both these faunal types there tends to be an increase in the size of individual organisms. But to a remarkable extent, the overall number of species remains constant. Obviously the fauna of a fire regime show adaptations to it in a broad way, and in a few scrutinized instances (e.g., the Kirtland warbler) species seem to have specialized to the point of being dependent on a particular fire regime. Curiously, in many such examples the fire regime at issue is one sustained by anthropogenic fire practices.

For most regimes, however, the variable intensities of a fire ensure that a variety of biotic ensembles, a mosaic, persists. In ways not well understood, the constancy of faunal populations probably relies on this variegated habitat. As a generalization, fauna that seem to be especially adapted to a fire regime of frequent fire tend to possess generalized survival traits, high and variable birth rates, and a high dispersal rate. These are traits similar to those characteristic of the flora of such environments, traits that represent the ability to survive in ecosystems that are frequently disturbed by agencies that include, but are not limited to, fire.

### *Synecology*

Although fire tends to act on individual organisms, and natural selection only on individuals, wildland fires burn as synergistic events and they result in community effects. On a macroscale, fire determines species composition. That composition may simply reflect the proportions of species within a fixed total number inhabiting the site. This seems to be the case with fauna, where the total number of species on a site remains constant before and after a fire but the relative proportions of particular species change. Alternatively, that species composition may reflect small-scale geographic differences, expressed as a mosaic. Such a mosaic may represent spatial distributions of species types, such as sagebrush and grass in the Great Basin, differential densities of a single species such as the glade and grove complex typical of ponderosa pine in the Southwest, or of suites of species, as with aspen and pine in the Lake States. In all these cases fire simply determines the relative proportion of groups of a fixed number of species. Yet again, however, species change may reflect successional patterns that express a cycle of wholesale community modification. The geographic effect of succession may resemble that of a mosaic, but the changes in species composition signify changes over time rather than changes in space. They represent historical variances (developmental stages) rather than geographic variances (site differences).

*Mosaic Concept.* The expression *vegetative mosaic* is in one sense merely descriptive. It expresses the fact that nature tends not to exist in uniform monocultures but in blocks or patches of biota. Yet the phrase also conveys a set of concepts: that the tendency towards geographic ensembles represents an intrinsic part of natural systems, a part intimately connected with the character of its fire regime. Free-burning fire, so it is argued, is a primary mechanism for ensuring complexity, variety, and ultimately stability in natural systems.

Clearly, however, there are limits to this concept and its implications. What kind of mosaic exists depends in good measure on what sort of scale one employs. There are natural systems such as the grasslands of the High Plains or forests of jack pine, where fires tend to be large and stand-replacing. Massive blocks of even-aged vegetation do result. Nor are systems which show a variegated mosaic intrinsically more stable than those that do not, or necessarily more resistant to the development of very large fires. The mosaic that a fire regime exhibits is simply an attribute of that regime, no more applicable to other regimes than the particular parameters of its fuel complex. The chief value of the mosaic concept, as a concept, seems to be the contrast of certain natural fire regimes, which show a mosaic pattern on relatively small scales, to the modification of such regimes by anthropogenic activities—including plantation forestry and fire control—that have replaced interesting mosaics of cover with a numbing uniformity or even a forest monoculture.

In many instances the expression "mosaic" describes a biota in which different communities are intermixed and yet related. What unifies the ensemble is that one community will, over time, change into another community, and that this new community perhaps will change into another until all the particular communities in the regime are represented. All these biotas are members of a common, quasi-systematic series of changes over time. What creates a mosaic effect is that different communities exist simultaneously at different positions along these series. Given ample time ideally free from disturbances, all the different stages would eventually reach a common, uniform biota and the mosaic appearance would by and large disappear. This pattern of change over time is expressed in the concept of *biological succession.*

*Classical Concept of Succession.* In its classical formulation, the concept of succession extended the concept of a life cycle from the scale of an organism to that of a community. Biota grew, in effect, from simple forms (*pioneers* and *colonizers*) through a predictable sequence of stages *(sera)* until a complex association resulted (a *climax*) that resisted further change. The climax stage would persist in self-perpetuation until some external disturbance appeared that would reinitiate the cycle. The idea of a life cycle, moreover, saturated all forms of intellectual inquiry at the onset of the twentieth century and seemed plausible when applied to biotas. Frederic Clements, the chief architect of the succession concept, modeled his theory on the concept of a Geographical Cycle promoted by the geomorphologist, William Morris Davis.

In general, what drove the system from colony to climax was differential shade tolerance among those organisms that could potentially occupy the site. Following a massive disturbance such as a crown fire, shade-intolerant species such as grasses or aspen would invade the site. Over a period of years, by introducing shade, these organisms created an environment favorable for other, more shade-tolerant species; these, in turn, would give way or succeed to still other species, and so on, until a climax community was reached. All biotas shared this tendency to evolve in apparently collective forms, but the actual sequence varied from place to place. Tallgrass prairie showed one sequence, and lodgepole pine another.

The role of fire was to interrupt and retard the successional sequence. It reset the cycle back to a primitive stage. As a catastrophic event, it could set back even a climax to a pioneer stage. As a persistent, low-intensity episode, it could maintain a biota indefinitely at some intermediate *(subclimax)* stage. In effect the climax association represented a state of biological equilibrium. If this equilibrium was disturbed, the system would take predictable measures to restore it.

*Contemporary Concepts of Succession.* With the observation that, over time, a given biota experiences changes in species density and composition, there is little dispute. But with the explanation of those changes by appeal to the succession concept, as classically formulated, there are grave reservations. These doubts are both philosophical and factual. The simple, deterministic patterns of succession advocated by the concept are rarely found in nature. In part, the recognized complexities represent an astonishing increase in the quantity of fundamental data available. With scientific information doubling every 15–20 years or so, any conceptual system would become more complicated and show signs of breakdown. But equally the problems are conceptual and philosophical.

The specific case with fire ecology illustrates this difficulty very well. The history of a biota is stochastic, not deterministic. The problem is not of a fixed disturbance upsetting the equilibrium of a given system, but of a potential range of phenomena, fire, interacting with another potential range of phenomena, the biota. The response to fire obviously varies with the properties of the fire. The biota can respond not merely with one pathway but with many, depending on season, successional stage, fire history, and other biological parameters. Each of the potential pathways available may, in turn, feed back into the system in assorted ways. In such a view disturbances such as fire are inevitable, not exceptional events. Climax states are rare, not normal conditions. Disturbances occur episodically rather than predictably in nature.

And where humans interact with the biota, changes over time may be unique, idiographic. New, alien components may be introduced into the biota, like exotic organisms; and anthropogenic fire practices can replace natural fire episodes, altering fire frequency and intensity. No organism—not even *Sequoia gigantea*—nor any biota has a life cycle longer than the history of human presence in the lands of the United States. In few systems do the life cycles of climax species exceed the fire cycle of the regime, and in none do they exceed the events of human history. The recycling of nutrients and species must include the pathways of cultural systems; and a regime's life, fire, and successional "cycles," however conceived, must incorporate the acyclic processes of human endeavor.

None of these processes are self-contained within the ecosystem. They frequently originate from outside the system, and they often leave an inexpungable record of their presence. The succession concept requires that disturbances obliterate past history, or reset the successional cycle at particular points along its stages. But past events are not wiped clean, and they continue to distort the future direction of subsequent cycles. Whatever its theoretical correctness might be, the classical formulation of the successional concept simply loses its practical value amid this complexity. It describes an ideal type which nowhere exists in nature.

To cope with such new layers of complexity, the old successional model, based on an analogy to the life cycle, is being replaced by new models, broadly founded on the concepts of general systems theory. Instead of equilibrium as a state arrived at towards the conclusion of a life cycle, there exists a condition of dynamic equilibrium, for which fire may be important as an energy pathway and for which fire may be imagined less as an interruption of evolutionary progress than as a fundamental mechanism for recycling. Moreover, disturbances of any origin need not be interventions or undesirable intrusions. Human activity can be incorporated into the dynamics of the system. These assumptions are not merely academic. There are practical consequences to whether one chooses to say that, given time and freedom from disturbances, the ecosystem as a whole will reach a similar end stage, or whether one considers the array of successional states as representing a range of variance about a statistical mean. Such perceptions influence the range of practical management alternatives available for consideration.

This transformation in thinking is not simply the product of incomplete data. Rather it seems to represent a profound indeterminancy characteristic of the system. What happens after a fire is predictable only in probabilistic terms. To improve the resolution of forecasts of greater and greater detail, the quantity of information required is prodigious, and perhaps prohibitive. Yet the limitation is not merely one of data or the identification of the myriad responses that a system may make. It derives from the fact that such projections can only describe a range of responses, each of which has a certain probability of occurrence; that, as each response is made, other unpredictable events will occur to affect the subsequent history of the system; and that, even at its inception, no initiating cycle can exist as a *tabula rasa,* a blank page. Each ecosystem is burdened with and informed by its past history. Few fires burn with identical characteristics, few ecosystems experience fires with anything like identical biological traits, and few fires, as biological phenomena, are comparable.

To correct these deficiencies most contemporary models of successional tendencies show a different conceptual and philosophical foundation. To incorporate chance, models offer multiple pathways as alternative outcomes, each with a probability assigned to it. To overcome the problem of indeterminate future events, they appeal to simulation rather than prediction. To isolate the biological significance of a disturbance, they measure its effects on reproduction. To better define the physical attributes of the fire in question, they use fire behavior models. And to introduce better flexibility and to make use of the vast quantity of empirical data available, they appeal to computer modeling rather than to a fixed schema of successional sequences. Perhaps the most prominent of these models, because it can be adapted for computer use, is a version of the multiple pathways concept. In its more sophisticated forms the model resorts, in part, to techniques of gradient modeling to record environmental input and project its output (Kessell, 1979).

### *Fire Effects Models*

Models of fire effects may take several forms. They may be simply conceptual, or they may advance into more detailed diagrammatic representations. If sufficient quantitative information is available and the proper relationships are apparent, mathematical models can be developed. From mathematical models may come computer models. What makes modeling of any sort possible is that, for all the reservations about changes in an ecosystem over time, biota are not infinitely plastic, and the range of responses possible after a fire can, within acceptable limits, be designated. More or less regular patterns of succession do exist, mosaics do develop that reflect different stages of development within a system, and there does exist a working correspondence between a regime's fire history and its various biological rhythms.

Fires tend to be conservative rather than catastrophic in their effects. They tend to sustain an existing community rather than replace it, though a certain lag time may be needed to bring about this restoration. Succession tends to be more predictable in fire regimes that have a history of persistent, low-intensity surface fires (such as ponderosa pine forests) or of persistent, high-intensity crown fires (such as jack pine forests). In the first case, a mature forest persists, with little replacement by other species. In the second case, succession is consistently interrupted, and indefinitely suspended by the elimination of alternate pathways. Such facts simplify the problem of postfire effects, and make possible the creation of reasonable models.

*Contemporary Models.* All models begin with an inventory of the biota, especially those traits that relate to survivability after a fire. Against this biological potential is put a pattern of fire. Together they determine the number and nature of the successional pathways available after a fire. A diagrammatic way of expressing this information is through the use of *habitat classification* or the *fire group* scheme. The fire group concept applies to biota that show a similar response to fire and follow similar pathways in postfire succession. A pattern of *primary succession* is assumed, much as in the classical model of succession, but the trend it represents is thoroughly modified according to the frequency and intensity of its fire history. Numerous pathways for *secondary succession,* or variations of primary succession, are included. The model is generally qualitative, but has considerable use for land management on an extensive rather than intensive basis (Davis et al., 1980).

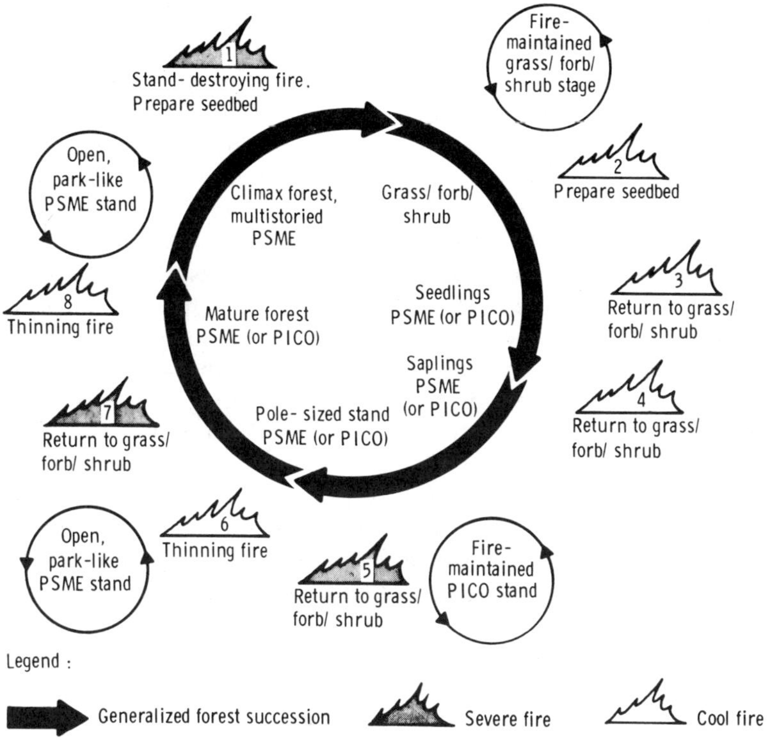


*Figure 5-12. Fire effects model: qualitative, low-resolution succession pattern. A diagrammatic summary of how fire events interact with gross biological properties. PSME indicates Douglas-fir; PICO, lodgepole pine. From Fischer and Clayton (1983).*

For an illustration, see Figure 5–12, which describes the moist Douglas-fir habitat typical of the Northern Rocky Mountains. The model recognizes two types of fire by intensity: cool and severe. It then traces theoretical pathways for succession relative to the frequency and timing of these two categories of fire. In the absence of fire, primary succession follows the dark circle of arrows, advancing from a biota of shrubs and herbs to conifer forests of mixed composition to, ultimately, a climax forest of pure but multistoried Douglas-fir. Yet even this broad trend may be fundamentally altered. Because the stand is mixed, containing both Douglas-fir and lodgepole pine, a severe fire that gives lodgepole a preferential advantage in reseeding can lead to a fire-maintained cycle from which Douglas-fir is largely excluded. Lodgepole that bear serotinous cones may provide such an advantage.

The severity and timing of fires is thus fundamental in shaping the actual path of succession. Intense fires at any stage will reset the cycle to the grass, forb, and shrub stage. But the timing of such fires (e.g., stage 5) may put subsequent successional development towards the Douglas-fir or lodgepole pine pattern. Light fires will only reset the entire cycle if they occur at a time when seedlings and saplings are just being established. Otherwise their effect is to prepare seedbeds, if they come prior to conifer establishment; to thin seedling patches, if they come after effective establishment; to sustain, in some mosaic pattern, stands of mature mixed conifers. A successional model of this type is not a forecast of the future, but a table of options that may occur should certain events happen. It does not forecast the likelihood of those events, and will not be able to do so unless long-term weather forecasts, among other things, become available.

Broad, diagrammatic models such as this can achieve significantly higher resolution by breaking down each component into identifiable subunits and by introducing at least first-order quantification. For purposes of fire effects modeling, the best developed program specifies four sets of species attributes that are fundamental to the survival of the organisms (Cattelino et al., 1979). One set relates to the method of arrival or persistence of propagules at the site following a disturbance. Rapid dispersal of seeds, storage of seeds in litter or protective organs, regeneration by sprouting—all are mechanisms of persistence.

A second set of attributes relates to the conditions of establishment, the site factors that promote survival to maturity. Essentially three varieties of traits are considered: those tolerant species able to establish themselves any time, regardless of competitive stress; those species intolerant of competition in mature communities, but able to colonize sites recently disturbed; and those species that require site factors, such as shade, which can only come in the presence of an established community.

A third set of vital attributes pertains to the life history of a species, those critical periods which define its reproductive capabilities. In order of appearance, four such events characterize the life cycle: the regeneration of adequate propagules to survive another disturbance, the establishment of maturity with normal powers to contribute propagules, senescence and removal from reproductive status, and extinction. Maturity and extinction relate to the survivability of the species, not merely the persistence of an individual organism. To the critical periods of a life cycle, numbers (ages) may be assigned for each species. When disturbances are measured against these periods, the relative survivability of species can be ascertained, the selection of a probable pathway for postdisturbance succession made, and a postfire vegetative mosaic projected.

Within the United States calculations of this sort have been made for selected fire groups in the Northern Rocky Mountains and Southern California, and they will almost certainly be attempted for other fire groups as well. For an illustration, see Figure 5-13. The model shows the successional pathways open to the same Douglas-fir and lodgepole pine group diagrammed in Figure 5-12. In this instance three states of fire are recognized: low-moderate, moderate-severe, and severe. The ideal primary succession path advances from a stage of grasses, forbs, and shrubs to old-growth, multistory Douglas-fir (A, B2, C3, D3, F2, G). A severe, stand-replacing fire at any time reverts the biota back to the initial condition (A). Note, however, how the model incorporates a successional pathway for lodgepole pine (C4, D4, E4) that culminates in a fire-maintained stand of lodgepole, and how there are two pathways indicated along which Douglas-fir biotas can transform into lodgepole (27, 31) and two paths for which lodgepole may be replaced by Douglas-fir. It only remains for actual ages to be placed at the various states for the model to acquire some quantitative rigor. The model could then be programmed for computer simulation.

Other refinements would include a fire system model, giving better quantitative data on the fire events, and a more refined inventory of species, including the ages of critical events in their life cycles. To improve the resolution of input data, important site and species characteristics can be entered into the model by means of environmental gradients rather than as discrete entities. For an illustration, consult Figure 5–14.

The foregoing technique of *gradient modeling* outlined above has led to complex computer models for fire planning within the context of land and resource management. One of the earliest examples was the BURN program developed for Glacier National Park. The U.S. Forest Service has developed similar, site-specific systems for Southern California (FBIIS) and the Northern Rockies, and has promoted a general postdisturbance model for forest planning, FORPLAN (Potter et al., 1979). FORPLAN considers any disturbance, fire included. Interestingly, to the degree that disturbances are predictable, simulation can phase into forecast. In an ideally managed land unit, disturbances will be deliberate rather than

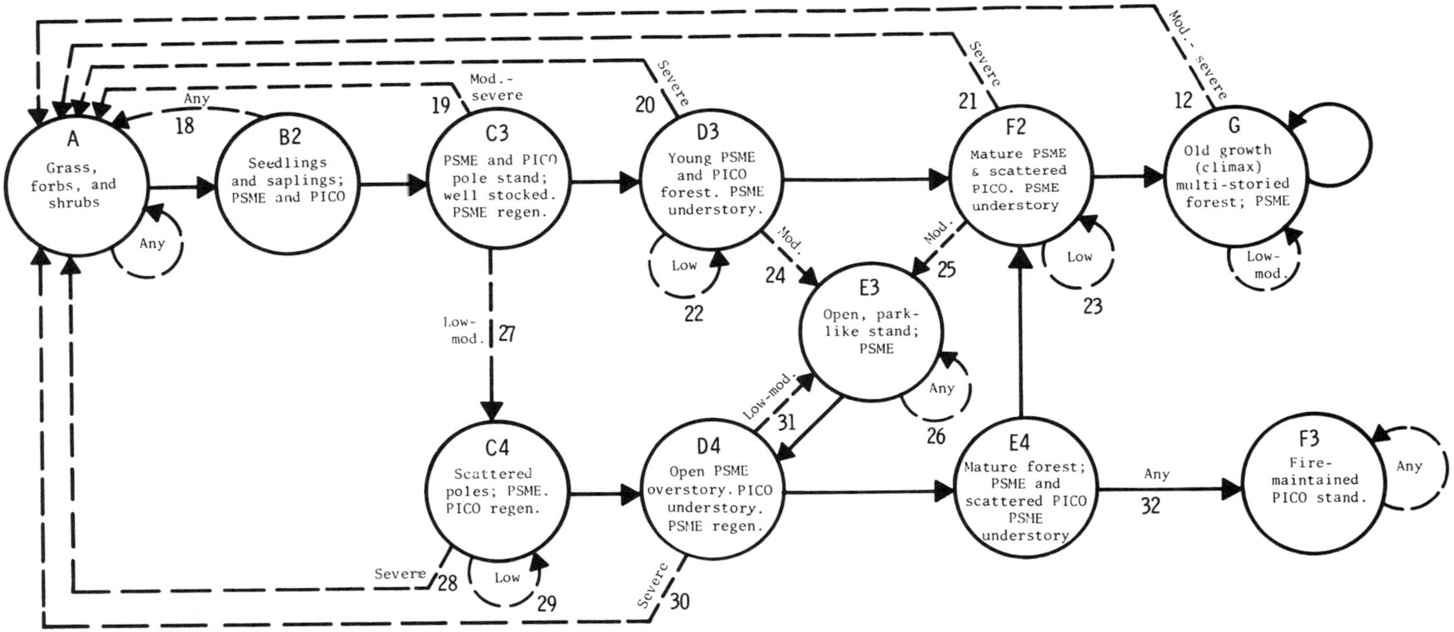


*Figure 5-13. Fire effects model: High-resolution successional pathway diagram, ready for quantification. Again, PSME indicates Douglas-fir; PICO, lodgepole pine. (→) Succession in absence of fire, and (⇢) response to fire. "Low" means cool or light surface fire. "Moderate" indicates a fire of intermediate severity. "Severe" refers to a hot, stand-destroying fire. For reference numbers, see text. From Fischer and Clayton (1983).*

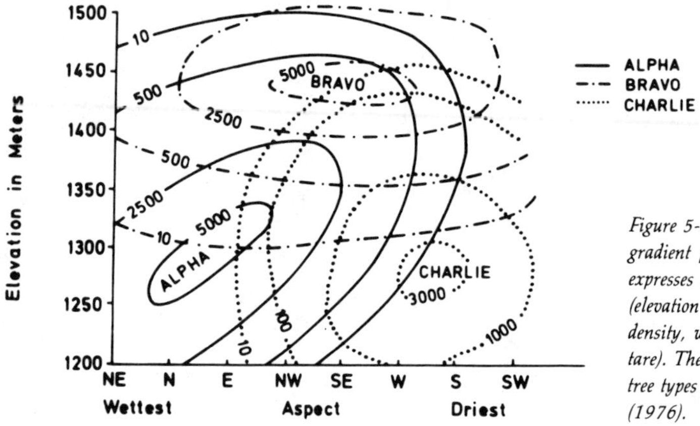

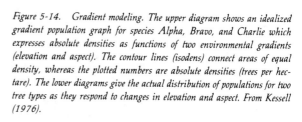

*Figure 5-14. Gradient modeling. The upper diagram shows an idealized gradient population graph for species Alpha, Bravo, and Charlie which expresses absolute densities as functions of two environmental gradients (elevation and aspect). The contour lines (isodens) connect areas of equal density, whereas the plotted numbers are absolute densities (trees per hectare). The lower diagrams give the actual distribution of populations for two tree types as they respond to changes in elevation and aspect. From Kessell (1976).*

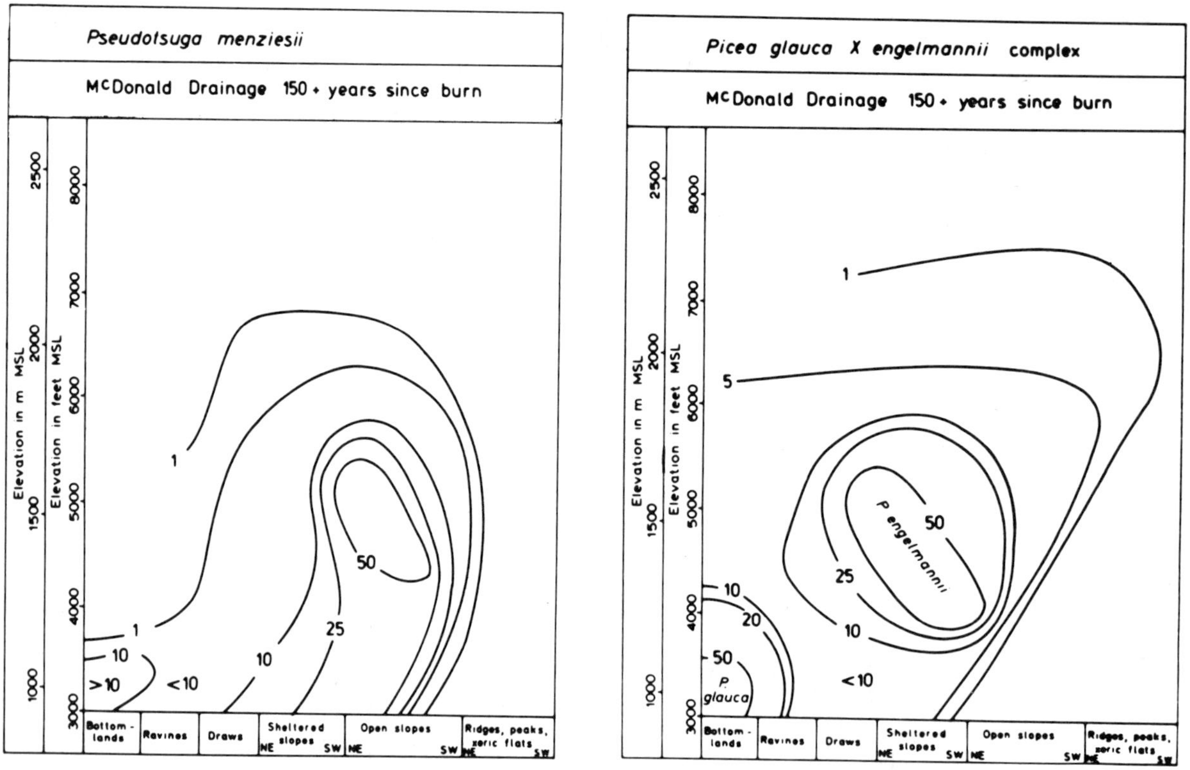


accidental, the result of logging and prescribed fire rather than windfall and wildfire. Simulation is also useful for scientific research, as a partial alternative to laboratories. The only other option, field experiments, demands time on the order of decades and control over variables that would prove nearly impossible under natural conditions. Simulation offers an alternative.

The FORPLAN model has, in turn, become a subunit of a large-scale model developed by the U.S. Forest Service to simulate fire effects throughout an ecosystem. The FIRELAMP (Fire and Land Management Planning) model is intended for use as a planning aid to simulate fire effects and fire management practices as these relate to the multiple-use doctrine of the national

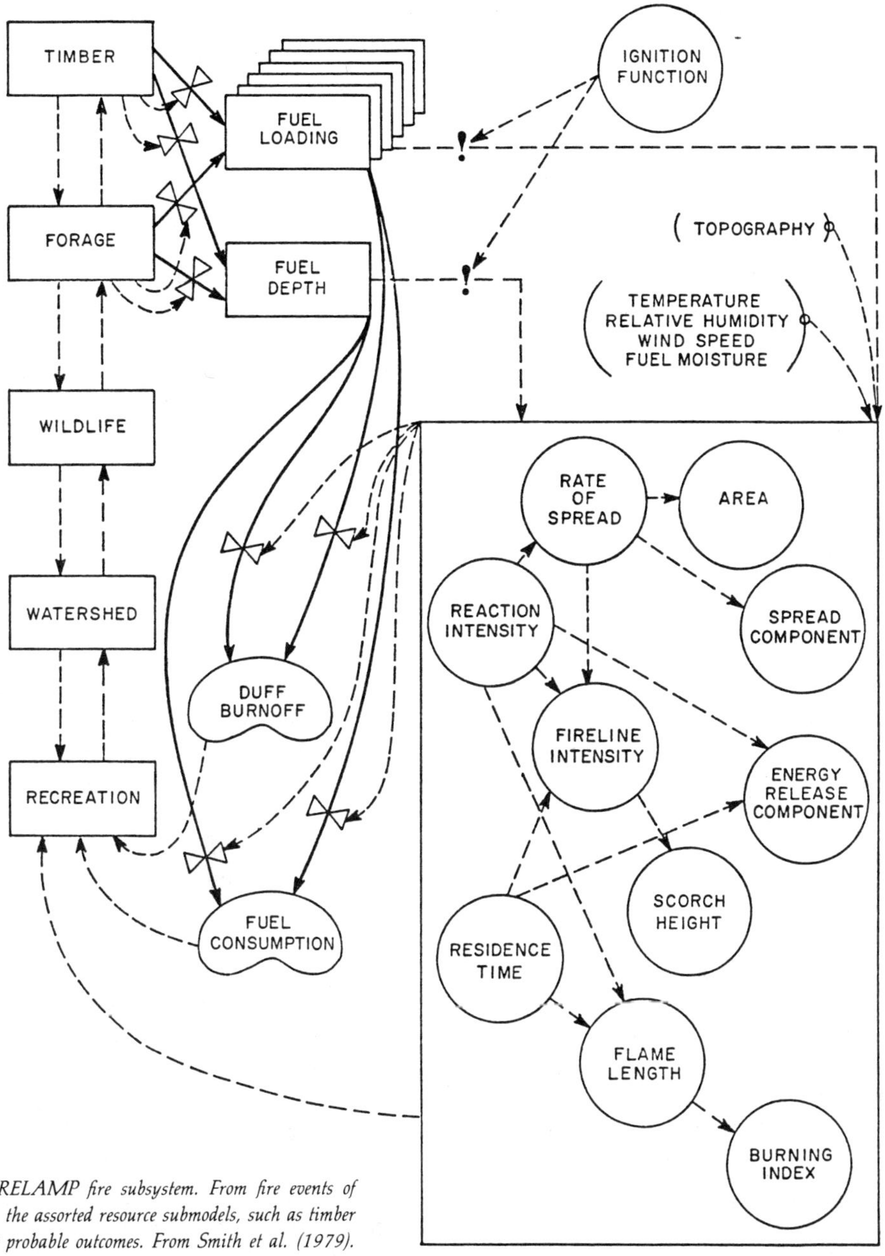


*Figure 5-15. FIRELAMP fire subsystem. From fire events of varying properties, the assorted resource submodels, such as timber and forage, project probable outcomes. From Smith et al. (1979).*

forest system (Figure 5–15). Its geographic scale is on the order of an average ranger district (200,000–400,000 acres; 81,000–162,000 hectares), and its historical scale on the order of a year. At its limits FIRELAMP can simulate about 5000 acres (2023 hectares) (the legal minimum for a wilderness area) or 200 years (about twice the life-span of government forestry in the United States). Among its input subsystems are weather, fuel, and fire behavior, with biotic information included in the form of fuels. By way of output there are subsystems for timber production, forage production, wildlife, water and sediment production, and recreational opportunity. A prototype was developed in the late 1970s for the Roosevelt–Arapaho National Forest in Colorado. Similar extensions will surely follow (Smith et al., 1979).

*Model Limitations.* For all its sophistication, problems nonetheless persist with modeling of this sort. These represent, in part, problems in the quantity and caliber of information demanded. The data requirements for undertakings of this order are, of course, enormous, and most models necessarily rely on various simplifying assumptions. Here considerations of cost, management objectives, and theoretical and practical limits on resolution all become constraints—just as they are for determining the appropriate level of fuel information. These limitations loom even larger since, as such models achieve higher resolution, they become more site-specific, less amenable to extrapolation to new areas. Not all administrative needs in fire management require high-resolution models.

In part, too, problems reflect difficulties with the models themselves, some internal to their programs and some simply the product of ignorance about ecological mechanisms. A single species may show multiple attributes, not merely one, with changes in site or biotic association. After a fire, for example, lodgepole pine can establish itself in dense stands, but once established its subsequent density based on reproduction will diminish. Subalpine fire is intolerant at low elevations, but tolerant at high elevations. And in part problems reflect uncertainties inherent in the submodels that make up the larger model. Some submodels, such as the one for dispersion, are deficient, and others, such as that for fire behavior, have an intrinsic element of uncertainty. When these models are compounded one after the other into hierarchies, the errors are not merely added but multiplied.

For most of these difficulties further research can supply adequate answers. The complexity of the models must be measured relative to the cor plexity of perceived needs, and for the near future the demands for information by managers of wildland systems will surely exceed the demand for data by low-resolution models. For problematic fire regimes, high-resolution modeling and computer simulation will become progressively more integral to both planning and operations. Much as fire behavior models, for all their limitations, now underwrite virtually all aspects of fire management, so will their companion models for fire ecology.

## *Fire Regime Concept*

Like the culture concept or the ecosystem concept, the concept of a fire regime is amorphous, intuitive rather than rigorous. What the fire climate concept proposed for the integration of fire traits with climate, the fire regime concept intends for fire and ecosystem. It seeks to reconcile the physical nature of fire with the biological context within which it burns. In physical terms, a fire regime consists of a particular complex of fuel, manifested as a biota, and a particular pattern of fire occurrence and behavior, described generally as its fire history. Within such a system fires exhibit certain more or less consistent properties—their rate of spread, intensity, frequency, and shape tend to be repeated. In biological terms, the ecosystem tends to show biotic assemblages that derive from and help to shape the pattern of fire. Change either the pattern of fire or the biota and a new fire regime will result.

The literature on fire ecology is confused, and often contradictory. Data are voluminous, but tend to be highly specific, descriptive of a particular fire at a particular time at a particular site. The reasons are simple. As physical events, fires are rarely comparable. Fire intensity can vary within the perimeter of a single flaming front, between different instants of a burning period, and by slight variations in fuels and microclimates. As biological disturbances, fires are far from uniform in cause, and no less variable in effect. The biological response will vary with the character of the biochemical and physical environment that results from the fire, with the genetic potential and ecological mechanisms latent within the system for coping with fire, and with the past and future history of disturbances. A fire that burns in the spring will elicit different responses than one with roughly the same physical traits that burns in the fall.

Yet general consistencies do exist, and the fire regime concept expresses this fact. As a biological entity, the fire regime does not consist of a state of nature so much as a set of natural processes whose response to fire operates within a range of probabilities. The looseness of the concept embraces this fact and, paradoxically, enhances the concept's usefulness.

Some attempts have, nonetheless, been made to develop the concept of a fire regime into a taxonomical tool. Perhaps the most elaborate example is the Heinselman classification scheme which identifies seven fire regimes on the basis of their fire history (see Figure 5–16). Regimes that have no natural fire are differentiated from those that have a pattern of frequent light surface fires (1–25-year return intervals), and regimes characterized by short return interval crown fires and severe surface fires in combination (25–100-year return intervals) from those that show very long return interval crown fires and severe surface fires in combination (return intervals over 300-years). An underlying assumption is that fire frequency and intensity automatically integrate the relevant biological information. These types have been applied to the classification of wildlands in the Lake States (Heinselman, 1981).

0 = No natural fire (or very little).
1 = Infrequent light surface fires (more than 25-year return intervals).
2 = Frequent light surface fires (1–25-year return intervals).
3 = Infrequent, severe surface fires (more than 25-year return intervals).
4 = Short return interval crown fires and severe surface fires in combination (25–100-year return intervals).
5 = Long return interval crown fires and severe surface fires in combination (100–300-year return intervals).
6 = Very long return interval crown fires and severe surface fires in combination (over 300-year return intervals).

*Figure 5-16. Classification of fire regimes by fire frequency and intensity. From Heinselmann (1981).*

The Heinselman taxonomy was really designed with natural systems in mind. The presumed source of ignition is lightning, and hence fire climate, fire ecosystem, and fire behavior all share a common origin in natural processes. Yet such systems are rare. Mankind is the most prolific source of ignition, human activities the most significant process in most fuel complexes, and anthropogenic fire practices the primary source for fire history. The assumption that fire history can be organized into "cycles" or "periods" is doubtful. Paradoxically, fire does not appear periodically in natural systems, for which the taxonomy was conceived, but it can approach periodicity in those regimes dominated by anthropogenic fire practices. A fire rotation, like the logging rotation of which it is an analogue, only occurs in systems under human management.

Still, the motivation behind the Heinselman schema—to provide guidelines for prescribed burning—and its recognition of fire frequency as a fundamental, perhaps defining component of fire regimes is sound. The fire regime concept must address fire not simply as a process but as history. What flaws most conceptualizations of a fire regime are not so much their submodels for fuel, fire behavior, and ecology, but their conception of history. Properly applied, the fire regime concept, through the medium of fire history, bridges fire behavior with fire ecology, and properly understood, it can reconcile fire as a natural event with fire as a cultural phenomenon.

## 5.4 SELECTED FIRE REGIMES: BIOLOGICAL CHARACTERISTICS

Most information published on fire ecology is specific rather than general. The data from one site are not easily extrapolated to other sites, nor are conclusions from one landscape readily generalized to other regions. The concepts of multiple succession and the computer models developed for simulation offer a degree of abstraction. But programs like these have been developed only for a handful of sites, and their outcomes more often reflect local conditions and fire history rather than universal patterns. By way of illustrating the dynamics of fire ecology, especially successional patterns, and of showing how fire management integrates with these processes, consider the following samples of fire regimes.

### *Tallgrass Prairie: The Midwest*

As a relatively pure form the tallgrass prairie historically extended over eastern Nebraska, Illinois, Iowa, and eastern Kansas. From here it graded into a forest–grass complex to the east and a shortgrass prairie to the west. Its biotic composition includes mixtures of grasses, forbs, shrubs, and trees. Soil tends to remain moist. In their origin, tallgrass prairies are principally shaped by fire and drought, either singly or in combination. Almost certainly humans were responsible for most of the fire. The maintenance of tallgrass prairies, once formed, is simpler and results from a somewhat larger array of processes, including drought, fire, herbivores, and anthropogenic practices such as mowing, plowing, and the introduction of exotic flora and fauna. Where fire has been eliminated through one means or another, tallgrass prairies succeed to shrubs and trees. In most managed sites fire is introduced on a 2–3 year cycle, though a 5–10 year cycle is probably adequate to maintain biotic integrity.

Fire affects prairie lands in two ways: through site modification, especially by altering pH and temperature, and through the elimination of potential site invaders—among them selected grasses and forbs, and most shrubs and trees. Of critical significance to its effects is the biological timing of the fire. If burned during a drought, when soil moisture is low, or if burned during midsummer, when plants are beginning seed production, postfire recovery is delayed. For this reason, plus the cycle of drying and curing in grass, anthropogenic fires are set in the spring and fall.

In general, big bluestem, little bluestem, and Indiangrass increase significantly in number and yield following a fire; prairie junegrass, sand dropseed, blue grama, and hairy grama show lesser increases; and buffalograss is unaffected. Annuals tend to proliferate briefly after a burn. But even on routinely fired sites perennials remain fundamental constituents of the biota, and they find in fire's consumption of the accumulated litter a means to avoid decadence. Introduced species, however, which initiate growth earlier in the spring than native species and continue active growth longer in the fall (cool-weather grasses), find themselves at a competitive disadvantage by properly timed spring and fall fires. For native perennials fire typically favors seed production, germination, and seedling establishment.

Forbs and shrubs tend to decrease following fires, though the composition of the forbs remains constant and shrubs are rarely purged from the biota. A few forbs, however, do increase through the action of fire in breaking seed dormancy: prairie sunflower, dotted gayfeather, Missouri goldenrod, false boneset, and silky prairieclover, for example; so do a few shrubs such as smoothleaf sumac, lead plant, and western snowberry. In the absence of fire, virtually all shrubs can show dramatic growth, and eventually trees may be established. Partly because of repeated anthropogenic disturbances, even the data needed to correlate fire history and biological rhythms is lacking. In broad terms, fire favors grasses and forbs over shrubs and trees, and frequent fire strongly favors native perennials. Where fire has been removed, the tallgrass prairie vanishes into one of several successional pathways.

No attempt at prairie restoration has succeeded without recourse to fire applied on a frequency of 1 to 3 years and timed so as to favor native perennials over cool-season exotics. With tallgrass prairies converted to farms, woods, and towns, wildfire is a concern only in those tiny remnants preserved as historical relicts. Almost exclusively fire management means the use of prescribed fire for the restoration and maintenance of such preserves (Wright and Bailey, 1982).

### *Chaparral: Southern California*

Chaparral is a generic expression for a group of brushy sclerophyllous species that thrive in California, Arizona, and parts of the central Rocky Mountains. There exists considerable biotic variation among these sites, with the Rocky Mountain chaparral experiencing an active growing season during the summer and the California chaparral (amidst a Mediterranean climate) during the winter. Even in California, chaparral can refer to three distinct brushlands: to those that coexist with coniferous forests, to those that form woodland–grass ecotones, and to those pervasive brushfields that give rise to periodic conflagrations and that replace themselves autogenically. The first type is found in northern California and in the middle slopes of the Sierra Nevada; the second, in the lower slopes of the Sierras and in much of the Coast Ranges; the last, in the mountains of Southern California.

The three types respond differently to fire. In the case of the first type, frequent burning can often replace forest with grass; less frequent burning can expand the range of brush at the expense of forest; fire exclusion can result in the establishment of reasonably well-stocked forests. For the second type, the elimination of fire tends to allow the woodland and brush to expand at the expense of grasslands. For the third type, the brush endures, with or without fire.

The ecology of this last type—the pure chaparral biota—has received considerable scrutiny. Dominating the composition are chamise, manzanita, ceanothus, and scrub (gambel) oak. All are vigorous sprouters, and some, like chamise, increase their fire adaptability by their seed strategy, producing seeds early in the season, storing accumulated seeds in litter over very long periods of time, and sealing the seeds in resistant shells that fire can liberate into germination. Other components of the biota, the grasses and forbs, similarly produce seeds that can lie dormant until a fire releases them and prepares a suitable site. Immediately after a fire endemic forbs grow profusely; within a year, annual grasses add to this flowering; but within five years, the grass and forb population rapidly slouches into decline, occupying only about 1% of the total cover and sustained only as seeds.

These adaptations pretty much account for the steady-state existence of this chaparral community. The life cycle of the dominant plants, the successional cycle of the community, the fire and fuel histories—all more or less coincide. Thus during the five or six years that chamise requires after a fire to bring itself back to maturity either through resprouting or germination from liberated seeds, annual grasses and forbs flourish. With biennial or annual firing this state can be maintained, though chamise (and oak) sprouting continues and is only controlled by heavy browsing and herbicides. During its mature phase, roughly between the ages of 6 and 20, the chamise biota has little ground cover either from grass or litter and hence small opportunity for surface fire. After about 20 years, however, chamise undergoes a series of physiological and structural changes that greatly increase its flammability. After 25 years the intensity of a fire will increase in predictable amounts. To a striking degree the record of large fires corresponds to this pattern of aging (Figure 3–23). The biota declines in productivity and variability, but it does not succeed to other species (Philpot, 1971; Minnick, 1983).

The pathway of succession, it has been observed, more resembles a process by which individuals of a particular species are eliminated than one in which the existing species are replaced by other species. Whenever a fire might occur within the period of chamise maturity, the effect is the same: fire resets the life and successional cycle of chamise, replacing old chamise with young chamise. The episodic nature of fire history does not affect this outcome.

Wildfire control under conditions involving mature chaparral fuels is notoriously difficult. Though foehn winds, droughts, long dry summers, and steep topography make for frequent periods of high fire danger, and although the mixture of people and chaparral often brings ignition at the worst possible times, the pattern of large wildfires tends to conform to the life cycle of the chaparral, a cycle largely set by preceding fire history. This coincidence of fuel and fire, however, suggests the means by which prescribed fire might be introduced for fuel reduction or habitat design. Ideally, prescribed fires should be set between the ages of 25 and 30 years. Thereafter, fire intensity makes control difficult, and commonly compels fuel preparations such as dessication and crushing. After treatment, fire can burn within the dead fuels without spreading beyond the zone of treatment.

The problems with prescribed fire are likewise formidable. Prescribed burning in mountainous terrain is everywhere difficult, but the water-repellent soils, loose surface debris, steep slopes, and watershed values of the mountains magnify the problems in Southern California. Air pollution brings other constraints, and the proximity to housing developments introduces still more hazards. Yet slowly, almost improbably, prescribed broadcast fire is increasing in use, not merely for isolated hazard reduction or the maintenance of fuelbreaks but for large-scale fuel reductions.

### *Ponderosa Pine: Southwest*

The ponderosa pine of the Cordilleran regions, like chaparral, can exist in both climax and seral states. Wherever it exists as one component of a mixed coniferous forest or as a species that predominates for a period of time along a successional pathway, its prevalence depends on the fire history of the site. Where it exists as a self-perpetuating community, fire tends to rejuvenate the biota by mechanisms similar to those which operate in chaparral—the replacement of old ponderosa pines by young ponderosa pines. In either instance, wherever ponderosa pine establishes itself over large areas, it creates a fire regime favorable to relatively frequent, low-intensity surface fires, a pattern to which it is well adapted. In the Southwest ponderosa pine forms an expansive cover type which, at lower elevations, grades into chaparral and, at higher elevations, merges into a mixed conifer stand for which ponderosa is seral. For the central zone, however, the life cycle of ponderosa pine, the successional patterns of the biota, and fire history all tend to converge. On the mechanics of this convergence there exists a considerable literature (Wright and Bailey, 1982).

The typical surface cover consists of grass with a smaller forb–shrub complex on a ratio of 2:1. If burned with great frequency, the entire biota may consist of only these components. But such regularity is difficult to maintain, and the same fires that eliminate very young seedlings are the very fires that prepare excellent seedbeds for new seedlings. For perhaps five years the ponderosa seedlings must compete vigorously with the grasses and are quite vulnerable to fire (Figure 5-17). Thereafter, the tree experiences a series of changes that increases its imperviousness to surface fires: it puts on a thick bark, it begins a long process of shedding its lowest branches, and it deposits a layer of needles, which gradually suppresses the grasses in its immediate vicinity. In this way the tree becomes less susceptible to fire while it alters the fuel complex and thereby the type of fire it must endure. Eventually the mature tree becomes senescent; its large shade canopy prevents rejuvenation by seedlings, even as its age makes it vulnerable to disease, beetles, and mistletoe, and its height to lightning. In the dry climate favored by the ponderosa pine decomposition, except by fire, is slow.

Yet the process is rarely limited to a single tree: it involves whole groves as well. For this the seeding habits of the ponderosa and the pattern of openings between trees are responsible. A congestion of seedlings typically results from a good seed year, and if undisturbed a throng of pole trees follows. In their appearance, and as a fuel complex, such thickets resemble brushfields. Light fires may thin the stand, thanks to varia-

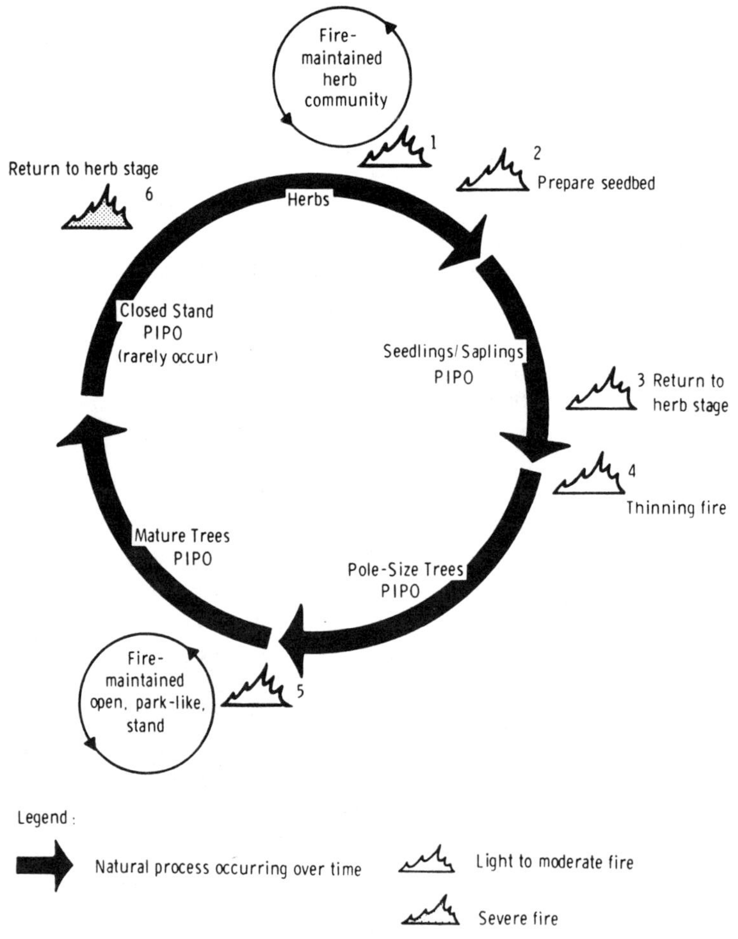


*Figure 5-17. Ponderosa pine fire type: successional tendencies. A simple system, somewhat resembling the chamise chaparral fire type in that succession is largely controlled by the life cycle of a single species. In the diagram, PIPO is an abbreviation for ponderosa pine. From Davis et al. (1980).*

tions of intensity by the fire and of adaptability by the trees. But in such a fuel complex even light fires may quickly escalate into crown fires that return the site to a grass and forb state. If the fire remains on the surface, the surviving trees endure to produce a spacious, relatively open stand, generally impervious to surface fires that flush the system of its accumulating litter. The end product is a mosaic of groves and glades.

As with chamise the successional pattern tends to correspond to the life cycle of the ponderosa pine. The various parcels of the regime's mosaic do not represent pockets of different species or different successional pathways coexisting at one time, but stages in the life cycle of a single species. The variability inherent in a biota with multiple pathways for succession is reduced to the variability within a single species and its manifestation at different stages of its life cycle. The absence of fire for this biota does not result in its replacement by a biota of different species, and a fire merely substitutes younger members of the existing species for older ones.

This scenario describes an ideal, steady-state system dominated by a single woody species and a single disturbance, fire. Real biotas are more complex; ponderosa pine is more often one member of a community than the sole constituent, more types of disturbance than fire occur, and humans have intervened in a host of ways. Yet the adaptations to fire are real, and the thrust of fire management is to substitute prescribed fire for wildfire wherever possible. Most wildfires, especially those ignited by summer thunderstorms, can be controlled while they are small. But some, in the spring and fall when the grass is dry, become large; where unthinned thickets of pine reproduction crowd sites with a dense netting of fine aerial fuels, fires can burn with great intensity and damage.

The tolerance by mature ponderosa for surface fires, however, means that the opportunities for broadcast underburning are excellent, and they are widely practiced. Presettlement fire history featured events on the order of 5–25 years. Prescribed burning comes with greater regularity, carefully timed with the important life and fuel histories of the biota. Too early in the fuel cycle of mature pine forests and there will be insufficient fuel to carry an effective fire; too late and ignition may result in very intense, lethal burns. In an arid region where regeneration is difficult and good seed years unpredictable, stand-replacing fires are hardly desirable. Yet in an area subjected to heavy lightning bombardment and frequent periods of high fire danger, some fuel management is essential, and prescribed fire on a cycle of 5–10 years is a useful solution.

### *Douglas-fir: Northern Rockies*

Like the ponderosa pine, Douglas-fir is a widely distributed conifer throughout the Cordillera and Northwest. Unlike the ponderosa, however, it favors moist rather than semiarid sites and it rarely inhabits biotas for which it is the sole conifer. Multiple pathways are thus possible for succession. In some scenarios Douglas-fir is one among many seral conifers that will, under the proper conditions, succeed to other species; in other projections, it is itself the climax species. Among the traits that make it compatible with certain forms of fire are a resistance to burning comparable to that enjoyed by ponderosa pine, fine winged seeds that may be cured by underburning and can carry long distances to saturate clearings, and a moderate shade tolerance that allows seedlings to thrive under conditions that are neither fully closed nor fully open. Douglas-fir can thus hold its own within fire regimes that show a mixture of conifers of varying fire resistance and adaptability, but it can, in more moist habitats, sustain itself without fire. Its response to fire intensities and frequencies will vary widely as a function of its context, the relative responsiveness of its competitors (Wright and Bailey, 1982).

Consider the situation in the Northern Rockies for which, under ideal circumstances, Douglas-fir stands form a climax (Davis et al., 1980). At the initiating level there is a biota of pinegrass and assorted common shrubs such as balkhip rose, spiraea, russet buffaloberry, serviceberry, and huckleberry. Frequent firing will maintain this condition indefinitely. Over a period of 5–20 years, seedlings will become established for ponderosa and lodgepole pine, western larch, and Douglas-fir. A fire in the early stages of establishment will return the biota back to a grass and shrub state; at a somewhat later time, it will thin out the stands of reproduction according to the particular concentrations of fuel available and the variable resistance to fire offered by the different species. In this regard larch, renowned for its thick bark and rapid self-pruning of lower branches, is supreme; ponderosa and Douglas-fir show somewhat lesser resistance, and the thin-barked lodgepole pine only slight resistance. As stands grow, fuel increases in load and porosity, often developing a ladder of fuels from the surface to the low crowns. A severe fire is possible; if it comes, it will return the site to the grass and shrub state again. Light fires will work to maintain a mosaic among the remaining species.

This fire-maintained mixture can persist indefinitely if fires occur with a frequency shorter than the shortest life cycle of the competing conifers. In this case the fires must occur with a frequency less than 220 years, the average life span of lodgepole pine. Larch and ponderosa have life spans around 400 years. But Douglas-fir can live to 1000 years, and in the continued absence of effective fires, it will gradually assume dominance over larch and ponderosa within a span of 500 years or so. Such long periods without disturbances of some sort—whether fire initiates the disturbance or follows upon it—are rare, and the climax condition is a theoretical one. Should fire enter into such a forest, say, under drought conditions and high winds, fires can be severe enough to reset the successional clock back to the grass and shrub state.

Which pathway or pathways are actually taken in the course of history is easy to explain in retrospect, and difficult to forecast in advance. The disturbing fires, for one thing, are not merely light or severe, but form a continuum of intensities. The possibility of returning a site to the same initial state it enjoyed previously is extremely low: there will always be slightly different initial conditions that will modify the new pathways. The schematic diagram (Figure 5-12) is thereby limited not only for its lack of quantitative resolution, but because it presents history as a two-dimensional cycle when in fact it is a four-dimensional cone of potential pathways. But as a first order approximation the schema is valuable and, for many of the management needs for which it was conceived, its level of resolution is more than ample.

Fire management in such a system is complex, and prescribed fire is, as yet, limited in scope. The variety of species and successional pathways means that, unlike chamise chaparral, large wildfires can occur at any time in the history of the biota and, unlike the case with pure ponderosa pine forests, the timing of prescribed fire must relate to the life cycle of more than one species. Large wildfires correspond to episodes of drought or large-scale ignitions (as with dry lightning storms) rather than to simple patterns of fuel history. Prescribed burning is widely practiced for hazard reduction and site preparation following logging, and to some degree within with programs of prescribed natural fire in wilderness areas. Broadcast underburning is not yet common.

### *Loblolly and Shortleaf Pine: The South*

The loblolly–shortleaf pine association is the largest timber group in the eastern United States, extending from Maryland to Texas. The biota is a rich one. For surface, understory, and overstory all, the variety of species is large, and the successional pathways many. In broad terms the successional trend is apparent. It progresses in the absence of fire from cleared site to a complex of grass, shrubs, and conifer seedlings; to dominance by one of several pines, with an understory of shrubs and hardwood seedlings (the rough); to gradual replacement of the pines with assorted hardwoods. For the earlier stages—through pine dominance—many of the particulars are known. But most of these biotas have developed from farm fields abandoned over the course of the past hundred years; there have been frequent, other anthropogenic interruptions and time has been inadequate to follow the process to its theoretical completion. In general, fire works in this regime as in others. A frequent fire favors shrubs over pine; less frequent fire, pine over hardwoods; infrequent fire, hardwoods over pine. In general, too, broadcast fire favors those natural resources most useful to humans, and prescribed fire has a long, productive history in the region (Wright and Bailey, 1982).

On cleared sites, grasses and forbs predominate with broomsedge bluestem and panicum as the important grasses; and composites, lespedeze, beggarwood, and partridge pea as the basic forbs. Conifer seedlings quickly invade the openings, though annual burning can suppress them. Almost always, however, some seedlings become established, pass through a temporary stage of vulnerability, and thrive on subsequent patterns of frequent surface fire. Under a maturing canopy of pine, the percentage of herbaceous cover may decrease dramatically, depending on fire frequency, to as low as 3–8%. Woody plants will form a heavy understory, perhaps 35–40% of total cover.

The principal pines are loblolly and shortleaf. Neither has the fire adaptability of the fabled longleaf, and in areas where longleaf competes with them, a pattern of frequent fire will favor the establishment of longleaf, and a pattern of less frequent fire, loblolly and shortleaf. All these species, however, find the postfire environment to be favorable for seedlings. Whether it is establishing itself in competition with shrubs or with longleaf, loblolly requires about 10 years of freedom from fire. By this time it will develop adequate bark thickness and crown height to survive surface fires; once a sufficient number of trees have acquired diameters in the range of 2–4 in. (5–10 cm), fires can successfully thin a stand without destroying the whole. Shortleaf is somewhat more fire-adaptive. It can often resprout when young and it occupies slightly more arid sites. With fire frequencies on the order of 5 years, shortleaf will triumph. With somewhat longer frequencies (10 years) or on slightly moister sites, loblolly—with its greater shade tolerance—will overtake shortleaf. Among other species sometimes found at this stage are Virginia pine and eastern red cedar. As the pines mature, a dense understory of hardwoods forms beneath them.

Given sufficient time and freedom from disturbance, the hardwoods will eventually eliminate the pines. Hickory and oak (a variety of species are present) are considered to be climax species for the regime. But this final stage is rare. Frequent fires while the rough is low in density and short in canopy will maintain the pine complex indefinitely; a pattern of fires every 10 years or so is ideal. But as the rough proliferates, the quantity of available fuel increases dramatically and it forms a complete matrix from ground to canopy. Fires at this stage can be intense, sweeping away pine as well as hardwoods. The steady accumulation of fuels and the probability of ignition are such that a hardwood (oak–hickory) stage is somewhat hypothetical, usually found only in wet areas or other sites protected from surface burning. How, among themselves, the various components of the hardwood complex compete is not well known: the important element for the fire regime is their collective competition with the pines.

Fire intensity is closely tied to the development of an understory, the rough. Since control of the rough is also essential for most of the objectives for which these wildlands are maintained, fire management through fuel reduction and wildland management through biotic manipulation can reinforce each other. The replacement of wildfire with prescribed fire is the point of intersection. Where the rough is low, wildfires remain surface fires, easily controlled. Where the rough blossoms into dense jungles of vegetation, crown fires are possible. With proper timing the pines can survive the surface fires that periodically flush out the rough. Most burns are ignited during the winter, when their intensity will be low. Their frequency will vary according to the particular species in question or the particular habitat desired. Broadcast fire is used for site preparation, cleaning up after logging, and for broadcast fuel reduction once loblolly and shortleaf have reached early maturity, with a burning rotation that will retard hardwood invasion without leading to high-intensity holocausts. Typically, maintenance burning means firing on a 3–6 year cycle.

### *Red, White, and Jack Pine: The Lake States*

Like the pine complex in the South, that in the Lake States offers a rich ensemble of species and, as a function of past history and site particulars, a labyrinth of successional pathways. In broad terms, the picture shows general correspondence to that in the South. There is an initial matrix of grass, forbs, moss, and shrubs; gradual dominance by pines, the relative proportions of which depend on local fire history; and ultimate climax with white spruce and balsam fir or assorted hardwoods such as basswood and maple. Into this arrangement must also be put such woody colonizers as aspen and birch. But apart from complexities intrinsic to the ecosystem, there is an important record of anthropogenic influence that makes extrapolation from recent successional patterns to a hypothetical "natural" one exceedingly difficult. For this reason, as well as a dearth of solid quantitative evidence, fire effects are known only in qualitative form.

For schematic purposes it is again convenient to assume the existence of a barren site. On it appear an abundance of grasses, forbs, and shrubs that includes, as dominants, big bluestem and little bluestem, hazel and edible berries such as blueberry, raspberry, huckleberry, and chokeberry. Light fires stimulate and maintain this complex; but intense fires, penetrating into and consuming the duff, can inhibit it, reducing the biotic composition to a minimum. A unique component is moss—a source of flammable cover that requires little nitrogen and that furnishes a fine seedbed for conifers. Barring a prohibitive history of fire, some tree species invade the site: aspen and paper birch with their light seeds are common. Once established, these colonizers can survive surface fires by basal sprouting, and in the case of aspen, by root suckering. Other conifers (especially pine seedlings) must compete with the tenacious hazel. All the trees favor sites that have been burned; hence, the significance of moss as a temporary ground cover. But if the system avoids fire for a 5–10 year period, the pines will establish themselves in a complex that includes red pine, white pine, and jack pine.

These three species show different adaptive traits relative to fire, and which species will predominate depends in good measure on the fire history at this early stage. In the absence of fire the moss recedes after 5–10 years. The aspen are removed only by a rare high-intensity fire or by the gradual process of succession to the conifers in the absence of fire; frequent surface fires tend to only perpetuate it. Of the three pines, jack pine shows the greatest all-around adaptability to fire; it may, in fact, be considered as a genuine pyrophyte. It possesses serotinous cones, produces seed at 10–15 years of age, and organizes a fuel complex of great flammability. In the event of crown fires or intense surface fires after its first seed crop, jack pine can successfully reestablish itself. Because it tends to promote hot fires, the jack pine biota may become self-perpetuating and gradually extend itself into adjacent areas; the pattern of fire thus established may be difficult to break. The ecological niche occupied by jack pine is thus analogous to that of lodgepole pine in the Rockies. Following massive conflagrations between 1870 and 1936, large portions of the Lake States reverted more or less permanently to aspen and jack pine.

If fire and jack pine and aspen can be kept out of the site for 50 years or so, then red pine and white pine come in for temporary dominance. Of the two, white pine enjoys many reproductive advantages, but red pine grows more rapidly and has a wider degree of resistivity—not only to fire but to diseases and root injury. A fire earlier than 50 years after their establishment, however, can be lethal to both species; thereafter, slight surface fires do little harm to either and intense surface fires on the order of 150–350 years work to the mutual advantage of both. Without large-scale disturbances, notably fire, the pines will falter before invading conifers and hardwoods. For boreal forest sites, the successors include white spruce and balsam fir on drained lands and white cedar and black spruce on bogs. For warmer sites, succession to climax would mean sugar maple, basswood, and hemlock. The time parameters for this process should exceed 350 years or so, but are not well known.

In the event of an intense, stand-replacing fire, the successional options, of course, vanish: the site is returned to grasses, forbs, and shrubs, or in the case of mature jack pine back to pine. Nearly all large stands of jack, red, and white pine have arisen on sites whose initiating state followed intense fires which, in turn, followed upon heavy logging. This initial state is not the blank slate hypothecated by theory, but a site contaminated by holdover species, seed trees, exotics, or one, if burning striped off all cover down to mineral soil, that lacked the biological potential to put successional tendencies into motion. Most contemporary Lake States forests have been shaped, directly or indirectly, by human activities.

During the era of major landclearing, fire was used widely to remove slash, and the peculiar character of many forests reflects that practice, especially when broadcast fires escaped to become conflagrations. Most wildfires began as controlled burns. With the general conversion of cover to slash, farms, and various postfire states dominated by aspen and shrubs, and with the segregation of farming and logging, an association that had brought ignition and fuels together, large fires vanished from the scene. Most wildfires now occur in the spring, prior to greenup of grass and conifer foliage, or in the late fall, after curing and the drying of large diameter fuels.

The commercialization of species such as jack pine and aspen, formerly without value as pulp or timber, and the emergence of recreational usage and wildlife management as primary uses of wildlands all have encouraged the use of prescribed fire. Site preparation and slash disposal remain principal uses, but broadcast fire is increasing as an instrument of silviculture, fuel management, and, in wilderness areas, of resource management. The cycle for hazard reduction fires depends on the logging rotation, but timing and proper intensities for other prescribed fire uses are still in experimental stages. Most burning is tied, like that in the South, to the life cycles of the pines. Seasonally, burning can be practiced in either the spring or fall, but more stable weather conditions in the fall make it the season of choice. The existence of parklands or refuges that are real or de facto islands makes prescribed natural fire an attractive option for wilderness management.

## *REFERENCES*

The literature on fire ecology is enormous and to tabulate even a portion would be both confusing and whimsical. Instead, apart from publications that deal with the conceptual foundations of fire ecology, the following sources represent bibliographies, symposia proceedings, and other consolidated sources of information.

An encyclopedic summary of fire ecology—full of management implications for prescribed burning—is Wright and Bailey (1982), *Fire Ecology.* Extremely informative are the series of "State-of-Knowledge Reviews" assembled by the U.S. Forest Service for its National Fire Effects Workshop (1978). These include: Lotan et al. (1981), "Effects of Fire on Flora"; Lyon et al. (1978), "Effects of Fire on Fauna"; Martin et al. (1979), "Effects of Fire on Fuels"; Sandberg et al. (1979), "Effects of Fire on Air"; Wells et al. (1979), "Effects on Fire on Soil"; and Tiedemann et al. (1979), "Effects of Fire on Water." To these may be added a related digest on water repellency: DeBano (1981), "Water Repellent Soils: A State-of-the-Art Review." A somewhat impressionistic survey, and one increasingly dated, is Kozlowski and Ahlgren (1974), *Fire and Ecosystems.* A summary of fire ecological principles, as applied to wilderness management, is contained in Heinselman (1978), "Fire in Wilderness Ecosystems." Spurr and Barnes (1980), *Forest Ecology* contains a discussion of fire within the context of forestry, as well as a useful summary of forest history since settlement.

Among the many symposia that have assembled information on fire ecology, the Tall Timbers Fire Ecology Conferences (1962–1976) were the most comprehensive and influential. For a guide to the 15 volumes of the proceedings, see Fischer (1980), "Index...." More recent conferences, also rich in fire ecology, include Mooney and Conrad (1977), "Proceedings of the Symposium on the Environmental Consequences of Fire and Fuel Management in Mediterranean Climate Ecosystems," and Mooney et al. (1981), "Proceedings of the Conference on Fire Regimes and Ecosystem Properties."

A computer-assisted information storage and retrieval service for all forms of fire publications, FIREBASE, is operated by the Cooperative Fire Program of the U.S. Forest Service. Abstracts of most accessioned publications is available. For a description of services, see Taylor (1977), "Transferring Fire-related Information to Resource Managers and the Public: FIREBASE."

For a popular overview of lightning, see Viemeister (1972), *The Lightning Book.* A summary of wildland lightning is available in Taylor (1973), "Ecological Aspects of Lightning in Forests," and much of Tall Timbers Fire Ecology Conference, Vol. 13 (1973). Schroeder and Buck (1970), *Fire Weather,* gives a well-illustrated, qualitative account of the electrical processes that accompany thunderstorm development. For the physics of lightning, consult Fuquay et al. (1967), "Characteristics of Seven Lightning Discharges That

Caused Forest Fires," and (1972), "Lightning Discharges That Caused Forest Fires"; and Uman (1973), "The Physical Parameters of Lightning and the Techniques by Which They Are Measured." The modeling of lightning fire probability is described in Fuquay et al. (1979), "A Model for Predicting Lightning-Fire Ignition in Wildland Fuels."

For the concepts behind many fire ecology models, there are several accounts. The multiple pathways concept is reviewed in Cattelino et al. (1979), "Predicting the Multiple Pathways of Plant Succession," and Vogl (1977), "Fire Frequency and Site Degradation." The fire (or habitat) type concept is well demonstrated in Davis et al. (1980), "Fire Ecology of Lolo National Forest Habitat Types." Many versions of the gradient modeling approach have been published, but Kessel (1979), *Gradient Modeling,* consolidates most of the information into a single volume. A careful, more popular account of how to construct multiple pathways models is given in Kessel and Fischer (1981), "Predicting Postfire Plant Succession for Fire Management Planning." An elaborate exercise in applying these concepts is available in Fisher and Clayton, 1983, "Fire Ecology of Montana Forest Habitat Types East of the Continental Divide." The FORPLAN program—which incorporates a fire effects model—is explained in Potter et al. (1979), "FORPLAN: A Forest Planning Language and Simulator." Also revealing is Smith et al., (1979), "Final Report. FIRELAMP."

For more specific references on fuels, consult the references in Chapter 3. For smoke composition, see Chapter 1, and for smoke management, Chapter 10. For the manipulation of fire ecology principles to management ends, see the prescribed burning examples in Chapter 10.

Ahlgren, I.F. and C.E. Ahlgren, 1960, "Ecological Effects of Forest Fires," *Botanical Review* **26:** 483–533.

———, 1965, "Effects of Prescribed Burning on Soil Microorganisms in a Minnesota jack pine forest," *Ecology* 46: 304–310.

Albini, Frank A., 1976, "Estimating Wildfire Behavior and Effects," U.S. Forest Service, General Technical Report INT-30.

Alexander, Martin E. and David V. Sandberg, 1976, "Fire Ecology and Historical Fire Occurrence in the Forest and Range Ecosystems of Colorado: A Bibliography," Department of Forest and Wood Science, Occasional Report (Fort Collins: Colorado State University).

Barrows, Jack S., 1978, "Lightning Fires in Southwestern Forests" Final report, Northern Forest Fire Lab. U.S. Forest Service.

Barrows, Jack S. et al., 1977, "Lightning Fires in Northern Rocky Mountain Forests," Final report, Northern Forest Fire Lab, U.S. Forest Service.

Botkin, Daniel B., 1977, "Life and Death in a Forest: The Computer as an Aid to Understanding," pp. 213–134, in C.A.S. Hall and J.W. Day (eds.), *Ecosystem Modeling in Theory and Practice: An Introduction With Case Histories* (New York: Wiley).

Bruce, David and R.M. Nelson, 1957, "Use and Effects of Fire on Southern Forests: Abstracts of Publications by the Southern and Southeastern Forest Experiment Stations, 1921–1955," *Fire Control Notes"* **18:** 67–96.

Cattelino, Peter J. et al., 1979, "Predicting the Multiple Pathways of Plant Succession," *Environmental Management* **3**(1): 41–50.

Cushwa, Charles, T., 1968, "Fire: A Summary of Literature in the United States from the mid-1920s to 1966," (Washington, D.C.: U.S. Forest Service)

Daubenmire, R.F., 1968, "Ecology of Fire in Grasslands," pp. 209–266, in J.B. Cragg (ed.), *Advances in Ecological Research*, Vol. 5 (New York: Academic).

Davis, Kathleen M. et al., 1980, "Fire Ecology of Lolo National Forest Habitat Types," U.S. Forest Service, General Technical Report INT-79.

DeBano, Leonard F., 1981, "Water Repellent Soils: A State-of-the-Art Review," U.S. Forest Service, General Technical Report PSW-46.

DeBano, Leonard F. et al., 1977, "Fire's Effect on Physical and Chemical Properties of Chaparral Soils," pp. 65–74, in H.A. Mooney and C. Eugene Conrad (coordinators), "Proceedings of the Symposium on the Environmental Consequences of Fire and Fuel Management in Mediterranean Climate Ecosystems," U.S. Forest Service, General Technical Report W0–3.

DuCharme, E.P., 1973, "Lightning—A Predator of Citrus Trees in Florida," pp. 483–496, in Tall Timbers Research Station, *Tall Timbers Fire Ecology Conference, Proceedings,* Vol. 13 (Tallahassee, Florida: Tall Timbers Research Station).

Fischer, William C., 1980, "Index to the Proceedings of the Tall Timbers Fire Ecology Conferences: Numbers 1–15, 1962–1976," U.S. Forest Service, General Technical Report INT-87.

Fischer, William C. and Bruce D. Clayton, 1983, "Fir Ecology of Montana Forest Types East of the Continental Divide," U.S. Forest Service, General Technical Report INT-141.

Fuquay, D.M. et al., 1967, "Characteristics of Seven Lightning Discharges That Caused Forest Fires," *Journal of Geophysical Research* **72**(24): 6371–6373.

———, 1972, "Lightning Discharges That Caused Forest Fires," *Journal of Geophysical Research* **77**(12): 2156–2158.

———, 1979, "A Model for Predicting Lightning-fire Ignition in Wildland Fuels," U.S. Forest Service, Research Paper INT-217.

Hare, Robert C., 1965, "Contribution of Bark to Fire Resistance of Southern Trees," *Journal of Forestry* **63**: 248–251.

Heinselman, Miron, 1978, "Fire in Wilderness Ecosystems," in John C. Hendee et al. (eds.), *Wilderness Management,* U.S. Forest Service, Miscellaneous Publication No. 1365.

———, 1981, "Fire Intensity and Frequency as Factors in the Distribution and Structure of Northern Ecosystems," pp. 7–57, in H.A. Mooney et al. (coordinators), "Proceedings of the Conference on Fire Regimes and Ecosystem Properties," U.S. Forest Service, General Technical Report WO-21.

Kessell, Stephen R., 1977, "Gradient Modeling: A New Approach to Fire Modeling and Resource Management," pp. 575–606, in C.A.S. Hall and J.W. Day (eds.), *Ecosystem Modeling in Theory and Practice: An Introduction With Case Histories* (New York: Wiley).

———, 1976,'Wildland Inventories and Fire Modeling by Gradient Analysis in Glacier National Park," pp. 115–162, in Tall Timbers Research Station, *Tall Timbers Fire Ecology Conference, Proceedings*, Vol. 14 (Tallahassee, Florida: Tall Timbers Research Station).

———, 1979, *Gradient Modeling. Resource and Fire Management* (New York: Springer-Verlag).

Kessell, Stephen R. and William C. Fischer, 1981, "Predicting Postfire Plant Succession for Fire Management Planning," U.S. Forest Service, General Technical Report INT-94.

Kessell, S.R. et al., 1978, "Analysis and Application of Forest Fuels Data," *Environmental Management* **2**(4): 347–363.

Kozlowski, T.T. and C.E. Ahlgren (eds.), 1974, *Fire and Ecosystems* (New York: Academic).

Larson, Signe M., 1969, "Fire in Far Northern Regions. A Bibliography," Department of the Interior, Departmental Library Bibliography Series No. 14 (Washington, D.C.: Department of the Interior).

Lotan, James E. 1976, "Cone Serotiny–Fire Relationships in Lodgepole Pine," in *Proceedings, Tall Timbers Fire Ecology Conference* **13**: 267–278 (Tallahassee, Florida: Tall Timbers Research Station).

Lotan, James E. et al., 1981, "Effects of Fire on Flora. A State-of-Knowledge Review," National Fire Effects Workshop, U.S. Forest Service, General Technical Report WO-16.

Lyon, L. Jack et al., 1978, "Effects of Fire on Fauna. A State-of-Knowledge Review," National Fire Effects Workshop. U.S. Forest Service, General Technical Report WO-6.

Martin, Robert et al., 1979, "Effects of Fire on Fuels. A State-of-Knowledge Review," National Fire Effects Workshop. U.S. Forest Service, General Technical Report WO-13.

Mooney, Harold A. and C. Eugene Conrad (coordinators), 1977, "Proceedings of the Symposium on the Environmental Consequences of Fire and Fuel Management in Mediterranean Climate Ecosystems," U.S. Forest Service, General Technical Report WO-3.

Mooney, H.A. et al. (coordinators), 1981, "Proceedings of the Conference on Fire Regimes and Ecosystem Properties," U.S. Forest Service, General Technical Report WO-21.

Munns, E.M., 1940, "Forest Protection. Forest Fires," in *A Selected Bibliography of North American Forestry,* 2 vols. (Washington, D.C.: Government Printing Office).

Mutch, R.W., 1970, "Wildland Fires and Ecosystems—a Hypothesis," *Ecology* **51**: 1046–1052.

Noble, I.R. and R.O. Slayter, 1977, "Post-Fire Succession in Plants in Mediterranean Ecosystems," pp.27–36, in H.A. Mooney and C. Eugene Conrad (coordinators), "Proceedings of the Symposium on the Environmental Consequences of Fire and Fuel Management in Mediterranean Climate Ecosystems," U.S. Forest Service, General Technical Report WO-3.

Potter, Meredith W. et al., 1979, "FORPLAN: A Forest Planning Language and Simulator," *Environmental Management* **3**(1): 59–72.

Sandberg, D.V. et al., 1979, "Effects of Fire on Air. A State-of-Knowledge Review," National Fire Effects Workshop, U.S. Forest Service, General Technical Report WO-9.

Schroeder, Mark J. and Charles C. Buck, 1970, *Fire Weather,* U.S. Forest Service, Agriculture Handbook No. 360.

Slaughter, Charles W. et al., 1971, "Fire in the Northern Environment—A Symposium," U.S. Forest Service, Pacific Northwest Forest and Range Experiment Station.

Southern Forest Fire Laboratory, 1976, "Southern Forestry Smoke Management Guidebook," U.S. Forest Service, General Technical Report SE-10.

Spurr, Stephen H. and Burton V. Barnes, 1980, *Forest Ecology,* 3rd ed. (New York: Wiley).

Tall Timbers Research Station, 1962–1976, *Tall Timbers Fire Ecology Conference, Proceedings,* Vol. 1–15, (Tallahassee, Florida: Tall Timbers Research Station).

Taylor, Alan R., 1973, "Ecological Aspects of Lightning in Forests," pp. 455–482, in Tall Timbers Research Station, *Tall Timbers Fire Ecology Conference, Proceedings,* Vol. 13 (Tallahassee, Florida: Tall Timbers Research Station).

———, 1977, "Transferring Fire-Related Information to Resource Managers and the Public: FIREBASE," pp. 215–219, in H.A. Mooney and C. Eugene Conrad (coordinators), "Proceedings of the Symposium on the Environmental Consequences of Fire and Fuel Management in Mediterranean Climate Ecosystems," U.S. Forest Service, General Technical Report WO-3.

Tiedemann, Arthur R. et al., 1979, "Effects of Fire on Water. A State-of-Knowledge Review," National Fire Effects Workshop, U.S. Forest Service, General Technical Report WO-10.

Uman, Martin A., 1973, "The Physical Parameters of Lightning and the Techniques by Which They Are Measured," pp. 429–454, in Tall Timbers Research Station, *Tall Timbers Fire Ecology Conference, Proceedings,"* Vol. 13 (Tallahassee, Florida: Tall Timbers Research Station).

U.S. Department of Agriculture, 1941, *Climate and Man.* Yearbook of Agriculture (Washington, D.C.: Government Printing Office).

Van Wagner, C. E., 1973, "Height of Crown Scorch in Forest Fires," *Canadian Journal of Forest Research* **3**(3): 373–378.

Viemeister, Peter E., 1972, *The Lightning Book* (Cambridge, MIT Press).

Vogl, R.L., 1977, "Fire Frequency and Site Degradation," pp. 193–201, in H.A. Mooney and C. Eugene Conrad (coordinators), "Proceedings of the Symposium on the Environmental Consequences of Fire and Fuel Management in Mediterranean Climate Ecosystms," U.S. Forest Service, General Technical Report WO-3.

Wade, Dale et al., 1980, "Fire in South Florida Ecosystems," U.S. Forest Service, General Technical Report SE-17.

Wells, Carol et al., 1979, "Effects of Fire on Soil. A State-of-Knowledge Review," U.S. Forest Service, General Technical Report WO-7.

Williams, Mildred, 1938, *Effects of Fire on Forests. A Bibliography. Annotated,* (Washington, D.C.: U.S. Forest Service).

Woodmansee, R.G. and L.S. Wallach, 1981, "Effects of Fire Regimes on Biogeochemical Cycles," pp. 379–400, in H.A. Mooney et al., Compilers, *Fire Regimes and Ecosystem Properties,* U.S. Forest Service, General Technical Report, WO-26.

Wright, Henry A. and Arthur W. Bailey, 1982, *Fire Ecology. United States and Southern Canada* (New York: Wiley-Interscience).

*Chapter Six*

# Fire and Culture

Everywhere, and from the earliest times, humans have altered the natural fire regimes they have entered. Mankind remains today the greatest source of ignition, the primary vector for the dissemination of fire into new landscapes, and the principal modifier of the fire environment, notably its fuels. As an organism, mankind may be considered as a species of pyrophyte, with the special property that he can initiate fire. Fire and mankind continue in reciprocal fashion to expand their shared realm. Fire is among the oldest of words, the most ancient of tools, and the most prominent of the means by which humanity has projected itself onto the landscape. No fire regime has entirely escaped the direct or indirect influence of anthropogenic fire practices, and some regimes, such as those in the tropics, owe their existence and perpetuation wholly to anthropogenic fire. Moreover, by using fire in conjunction with other activities, humans multiply the effects of fire. Fire used in shifting agriculture, for example, has different effects than simply underburning in the same forest. The larger effects of fire on Earth are really the effects of anthropogenic fire practices; most fire regimes are human artifacts, and most fire impacts derive from the use of fire in coordination with other human activities such as logging, farming, grazing.

A fire regime is a cultural as well as a biological system. Neither fire nor fire regimes are wholly natural nor wholly manmade. Fire as a process and event, and the fire regime as a concept and artifact—both have attributes that are equally natural and anthropogenic. The problem is not so much to sort out one from the other, but to see how nature and culture interact. The point of intersection is fire history. To study a fire regime without its history, especially its human history, is like studying a wildland fuel complex without reference to the fact that its physical chemistry is shaped by biological processes.

Fire is a multiple phenonemon, and mankind soon adapted it for specialized uses. The history of wildland fire describes only a portion of general fire history. Domestic fire, the torch and the hearth, changed eating habits, rendered edible many plants and meats otherwise too tough or poisonous for consumption, defined the necessity for and the design of shelters, and revised social structures in accordance with the need to supply fuelwood and to preserve fire. A family is defined as those who share a fireside; the familial gods are the gods of the hearth. As a hunter and gatherer, mankind used fire to hunt and harvest natural products, to ward off predators, and to maintain the habitat against the successional pressures that would convert it to other forms. With fire, early man had a means to driving off game, of baiting traps, and of creating a habitat favorable to those species he found most useful; fire hunting is among the almost universal purposes of broadcast burning. As a herder, mankind used fire to create, sustain, and improve pasture. Broadcast fire may well have helped the process by which wildlife was domesticated; wild herds could be moved from site to site as areas were burned and as palatable new growth appeared. The seasonal herding of livestock between two pastures (transhumance) is almost everywhere accompanied by firing for the improvement of range. As a farmer, mankind used fire in swidden agriculture in environments as diverse as the jungles of the Amazon Basin and the boreal forest of Finland. Fire assisted in site preparation, in the cultivation of cereal grains, berries, and nuts, in fertilization, and in the fumigation of fields, driving off many varieties of pest and vermin. As metallurgist, mankind resorted to fire mining in hardrock veins, to smelter fires for ore refinement, and to furnace fires for metal working.

As *Homo sapiens,* mankind found in fire a provocative source of myth, philosophy, religion, and science. Pyromancy was a form of divination. Ordeal by fire is a primitive form of law. The fire sermon has persisted from the days of Zoroaster. The office of "fire keeper" was perhaps among the earliest in human society, and by ancient times it had evolved into a position synonymous with the state. Fire myths are among the oldest of stories, fire ceremonies among the most ancient of rituals, and national fires, maintained in appropriate temples, among the most common expression of ancient civilizations.

Nearly all of the landscapes that primitive societies found most productive of food, fuel, and wildlife existed because of frequent fire. Only with the adaptations of some peoples to maritime or rivertine environments and to irrigation agriculture did the value of broadcast, wildland fire diminish, and then only to metamorphose into different forms. Without anthropogenic fire many environments were uninhabitable. Industrial man still remains dependent on fire, though in a different cycle and in more complex technological forms. Industrial fire—the fires of the furnace—founded ceramics, metallurgy, and chemistry; the industrial revolution had for its prime movers machines powered by fire.

The migration of people across the globe and their occupation of landscapes for tens of thousands, if not hundreds of thousands of years means that anthropogenic fire practices are a fundamental component of virtually all fire regimes. The scale of influence is vast. Even recent surveys of deforestation and desertification throughout the developing world cite the gathering of fuelwood, shifting agriculture, and heavy grazing on wooded lands as major causes. All of these impacts are forms of fire practices. The concept of natural fire cycles is a somewhat dubious, parochial one. Natural fires come episodically, not periodically. But independently of whether natural history is cyclic or progressive, it is a fact that human history, with which it is inextricably bound up, is neither cyclic nor progressive, but idiographic and that human history, although tied to the landscape in various ways, responds to impulses outside of natural history. The fire regime at a particular site may thus represent fire practices that originated at some distant time and place and that were brought to their present location by means such as migration, conquest, or technological diffusion which have little to do with native ecology.

The fire history of the United States illustrates this fact well. Fire regimes changed as new cultures replaced old ones, and as different cultures experienced internal evolution in their institutions and thought. The result was more often than not a succession of new fire regimes, not merely new cycles of fire with different periodicity. The concept of a fire regime implies a particular biota and a pattern of ignition: it does not demand that both biota and ignition be of only natural origin. Typically, they are not. For almost all fire regimes in the United States—whether contemporary or historical—mankind shaped both that biota and that pattern of ignition. The fire regime concept cannot be segregated from concepts of fire history, and fire history is inseparable from human history.

## 6.1 FIRE HISTORY

The reasons for administrative interest in fire history are many and simple. Planning for fire control requires an understanding of how fires have historically behaved and how fire seasons are distributed; by imitating natural fire patterns, prescribed fires may better achieve stated management goals; and the fuels of regimes under present management reflect their fire history. But how one acquires good data, how that data is processed and interpreted, and what management implications derive from those methods are complex questions. The idea of a fire history seems intuitively obvious: it is the record of fires in a particular place. But in fact the record of individual fires, or of fire seasons, is less valuable than the chronicle of fire regimes, or the study of how fire regimes, and the fire practices that have shaped them, have evolved. For such questions it is often necessary to turn to the methodology and data of history rather than to that of science.

### *Historical Information*

The data for a fire history can derive from all components of a fire regime: from its physical systems, in the form of charcoal sediments and assorted geologic measures of climate; from its biological systems, with techniques of palynology, paleoecology, dendrochronology, and forest inventories; from its cultural systems, with written documents and photographic records. Each source of information by itself has serious limitations, and there are limitations, too, inherent in the methods available for processing each type of information; the most reliable histories will include data of many sources. Of special need is some means of identifying significant, or "effective" fires from the general population of fires, and some means of establishing the historical boundaries for fire regimes shaped by climate change and anthropogenic fire practices.

*Natural Fire History.* A useful method that relates vegetation and fire history to the geologic record is stratigraphy. In some lakes and channels, fine-layered varves provide an annual record of deposition, one often laced with charcoal and pollen. Similar deposits occur in bogs and peat beds. Stratigraphic correlation can establish the position of these varves in geologic time; palynology, the composition of the surrounding vegetation; paleoecology, the dynamics of such biota. Charcoal varves provide a measure of fire frequency. The advantages of such analysis is that fire history can be correlated with vegetation history and, in broad terms, to climate history. Its disadvantage is that deposition is restricted to certain environments and that the processing of data is complicated (Figure 6-1).

More common—since most fire history research is done by foresters or by land management agencies having responsibility for the management of forests—are techniques that relate to trees. An inventory of the forest by age classes may record important episodes in the history of the biota. These events may mark the emergence of forests from other cover types such as the reclamation of abandoned farm fields or reforestation of grasslands, the reestablishment of forests after some stand-replacing event such as a fire, or the shape of a forest mosaic, reflecting successional stages and pathways, again often the product of fires. Such an inventory can be made by cruising, aerial photography, or even remote sensing with satellites, provided that some ground truth is available.

Where fires are surface burns rather than stand-replacing crown fires, the silvicultural record can be found in *fire scars,* burn injuries seared into the pattern of annual growth rings of appropriate trees. Fire scars can be located easily, processed by techniques common to forestry, and related, somewhat like sedimentary varves, to larger climate history through the techniques of dendrochronology. The tendency for a tree, once

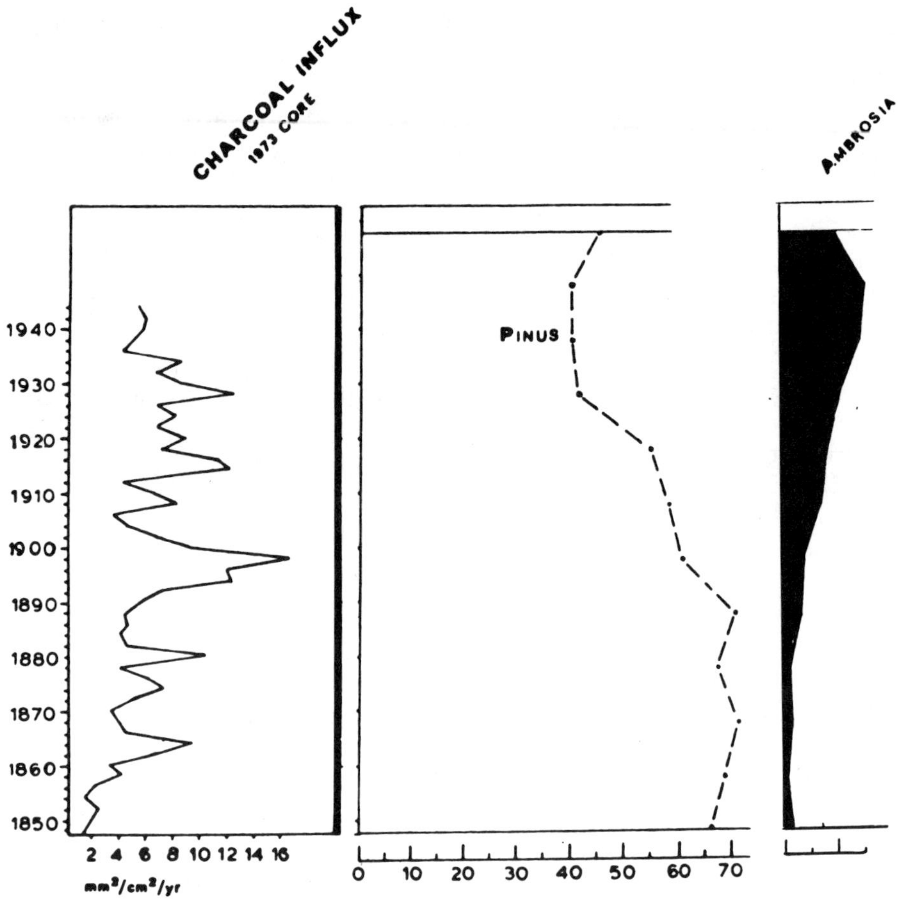

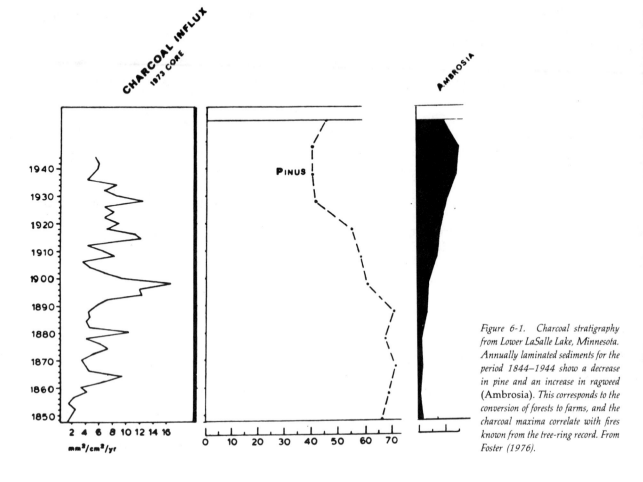

*Figure 6-1. Charcoal stratigraphy from Lower LaSalle Lake, Minnesota. Annually laminated sediments for the period 1844–1944 show a decrease in pine and an increase in ragweed* (Ambrosia). *This corresponds to the conversion of forests to farms, and the charcoal maxima correlate with fires known from the tree-ring record. From Foster (1976).*

scarred, to scar more easily from subsequent fires can make data collection rapid: a few trees will record many fires (Figure 6-2). The simplicity and directness of the method accounts for its popularity. It offers a tangible means of documenting fire events (Arno and Sneck, 1977).

But there are problems associated with the construction of fire histories from fire scars. Tree rings do not invariably relate to annual growth, and chronologies constructed solely of tree rings are not always reliable. In some years, a tree may put on double rings, and in other years, no ring at all. The randomness increases with trees such as those in the Southwest, that experience frequent stress; these environments, often prone to drought, are also prime sites for fire. It is imperative that tree-ring data be cross-dated before accepting them into a master chronology.

Perhaps more fundamentally, there is the question of just what sort of events are recorded. In part, this reflects the nature of scarring: scars record certain kinds of events in certain kinds of trees. The mechanism of fire scarring is part of a continuum of responses that a tree may make to fires of varying intensities. A fire of greater intensity would kill or consume the tree; a less intense fire would replace scarring with discoloration or charring (see Figure 6-3; Gill, 1974). Only a certain range of fire will be recorded in scars, and only a certain variety of trees can be routinely scarred. Some environments are thus more likely to have scarrable materials, and some trees such

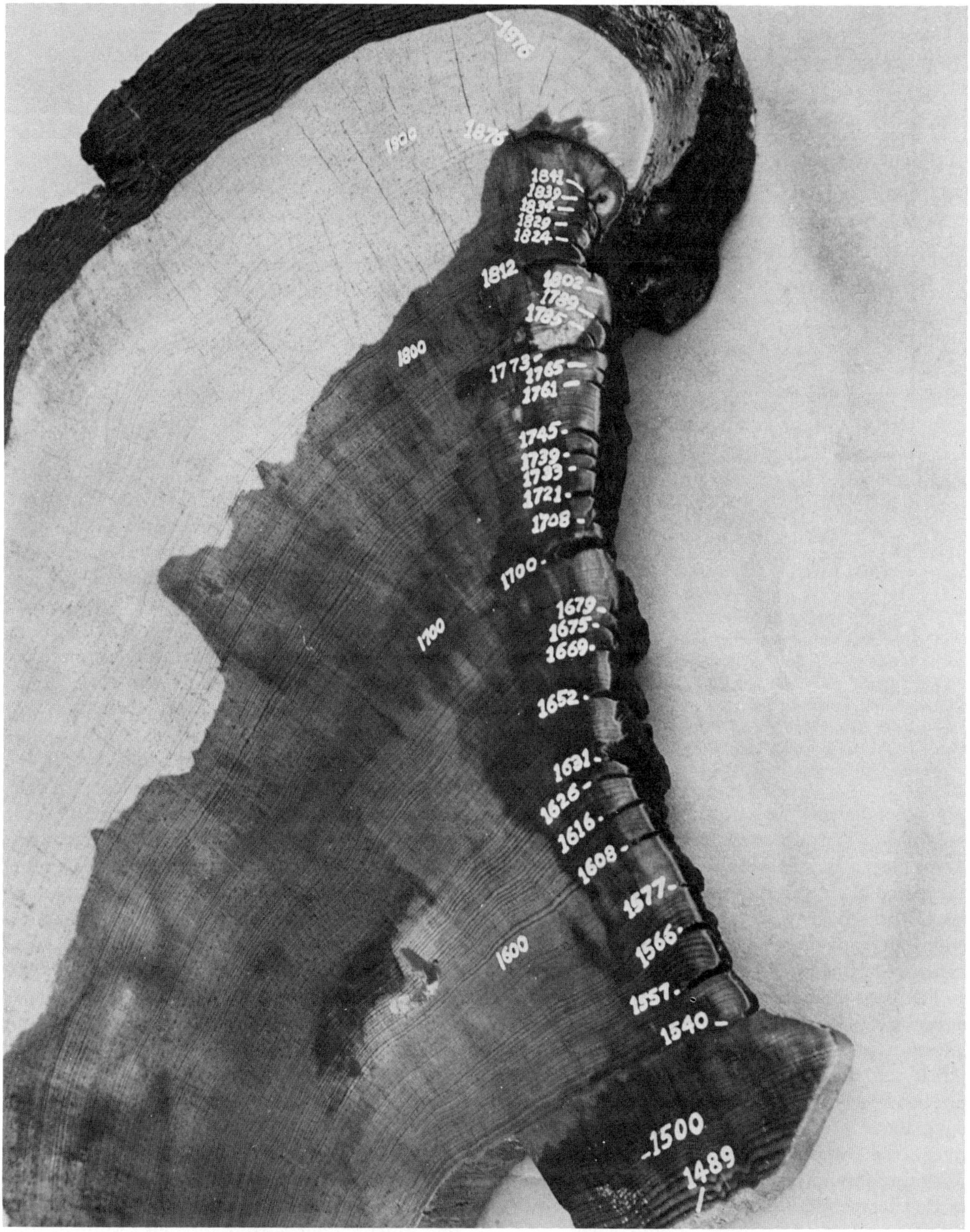


*Figure 6-2. Fire-scarred ponderosa pine, Arizona. Note the cessation of fires with white settlement in the late nineteenth century. Courtesy U.S. Forest Service, Rocky Mountain Experiment Station.*

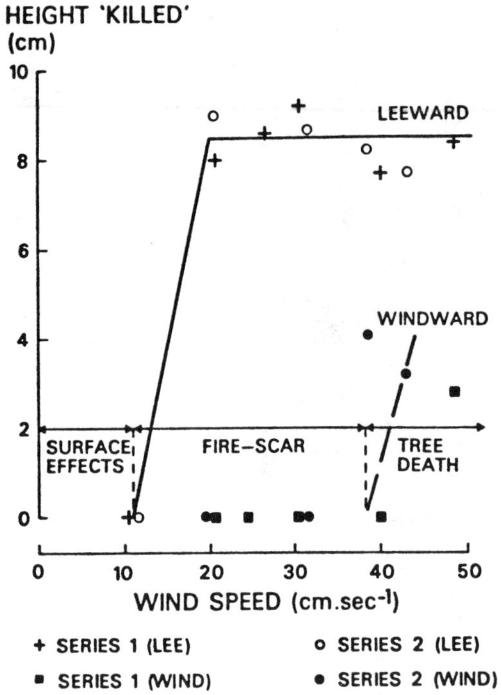


*Figure 6-3. Simulation of fire-scarring mechanism. Three categories of fire-tree interaction are recognized: surface effects (no scar), fire-scar (tree killed on one side), and tree death (killed on all sides). The experiments differentiate between windward and lee sides, since the interaction of wind and bole shapes flame differently around the trunk. Note that leeward flames are much taller than windward flames. From Gill (1974).*

as the ponderosa pine are more prone to take and hold scars. The tendency for trees to rescar is, ironically, a drawback because it further biases the record in favor of certain, fire-adapted species. So is the tendency to record surface fires: the larger, perhaps more significant fires destroy the evidence of their passage by consuming the biota.

There are difficulties, too, with sampling techniques, since the trees are not randomly selected; questions about the appropriate size of a sampled forest and the comparability of the constructed fire histories; dilemmas, so common to fire ecology, about possible feedback mechanisms between the record (the tree) and the agency of recording (fire). Mean fire intervals are inversely related to sample unit size. Large units (such as stands) lead to short intervals and small units (such as trees) lead to long intervals. Unit size is thus fundamental to the design of a sampling project and to the interpretation of its results (Arno and Peterson, 1983). Moreover, there can exist a reciprocity between fire and tree, not apparent in a geologic record, that can bias fire history. To generalize from fire scars alone is a complicated business, analogous to the reconstruction of a Triassic landscape from those few artifacts such as bones that happened to be left by a minute portion of the ecosystem—a portion that survived in a special environment suitable for the preservation of such remains. The point is not to dismiss fire scars as data but to reinforce the need for corroborative evidence and to qualify their interpretation.

*Human Fire History.* For the history of human fire practices, there are numerous cultural documents. Land survey notes, explorer journals, ethnographic reports, newspaper accounts, maps, pioneer diaries, and, for many twentieth century lands, formal fire and agency reports all provide evidence for past fire practices and historic fires (Figure 6-4). A simple approach, growing in popularity, is rephotography. Most post-Civil War expeditions to the American West included landscape photographers. By rephotographing the identical scenes a century later, a synopsis of vegetation history is possible. Since most of these projects reveal an increase in woody vegetation, they record a change in fire regimes—a transformation that largely results from a reduction in broadcast fire, though the sources of fire control are varied. Figures 6-5 and 6-6 show changes from the Black Hills, South Dakota and the Sierra Nevada, California. For a correlation of evidence with fire scar data, study Figure 6-7.

General histories, not specifically concerned with fire, are useful to unravel patterns of settlement, land use, the introduction of exotic flora and fauna, and the policies of land management agencies having direct control over fire use and suppression. A change in human use from hunting buffalo to raising cattle, for example, may not find much evidence from biological sources (un-

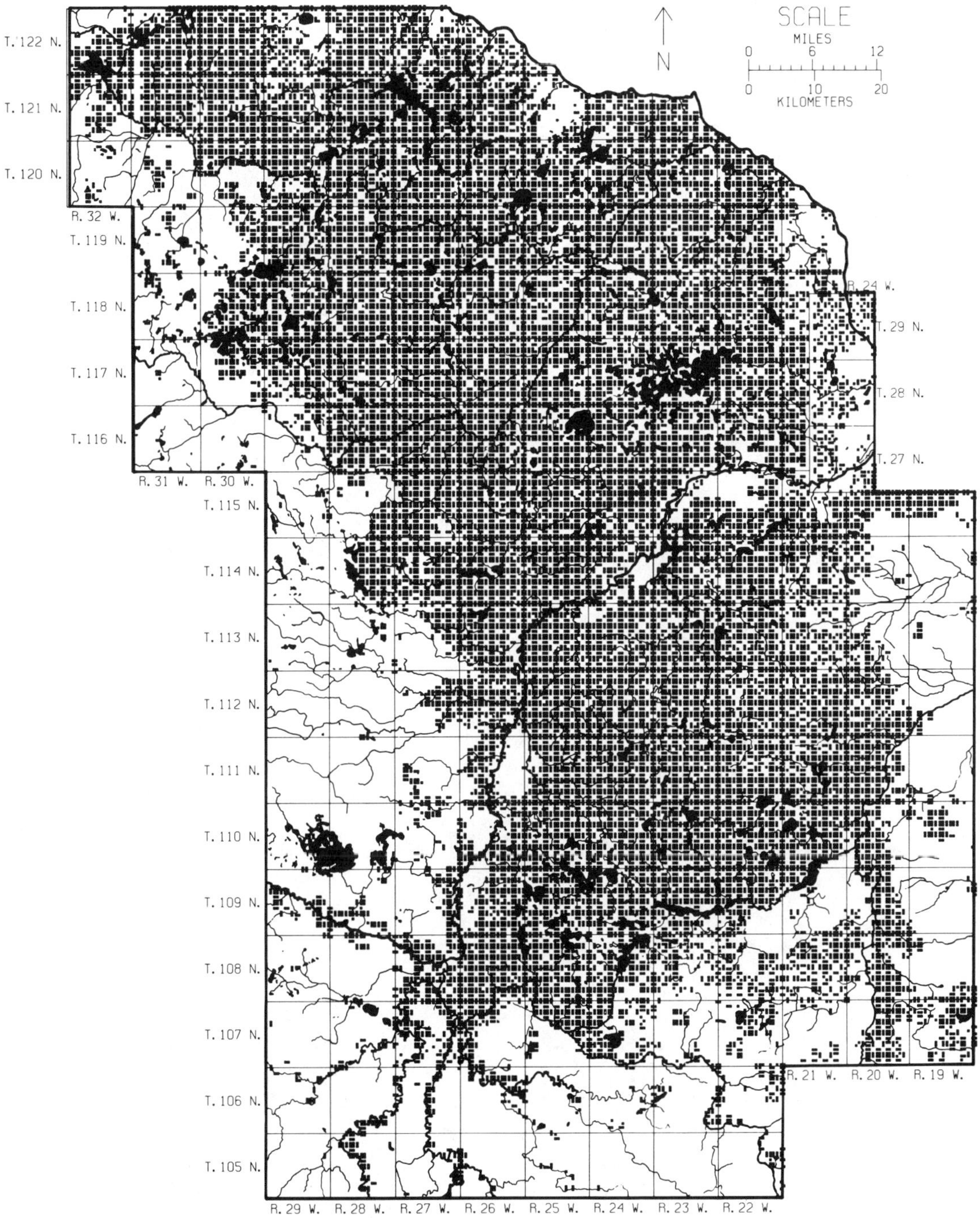


*Figure 6-4. Forest reconstruction from land survey records. Map shows trees in the Big Woods area of Minnesota. Note the effect of streams in protecting trees from prairie fires sweeping out of the south and west. From Grimm (1979).*

*Figure 6-5.* *Black Hills, South Dakota. The upper photo was taken during the Custer Expedition (1874), and the lower, from the identical spot 100 years later. From Progulske (1974).*

*Figure 6-6.* *Sequoia forest, Confederate Group, Yosemite National Park. The upper photo (1890) shows an open, grassy understory—the product of frequent surface fires by lightning and Indians. The lower photo (1960) documents the buildup of an understory and the development of ladder fuels as a result of successful fire exclusion. Photo by George Reichel (historical documentation by Mary and Bill Hood); NPS photo by Dan Taylor.*

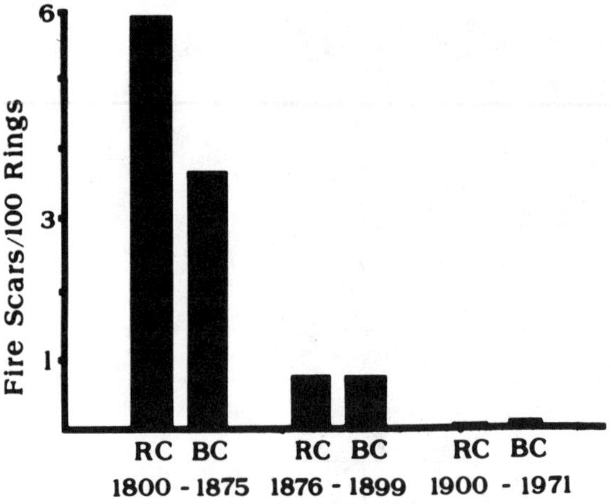


*Figure 6-7. Fire scar analysis for two sites in Southern Sierra Nevada. Three historical periods are recognized: (1) 1800–1875 includes fires from lightning, Indians, and miners; (2) 1876–1899 includes lightning and settler-set fires, but no Indian burning and limited fire protection; and (3) 1900–1971 defines the period of organized fire protection. RC-Redwood Creek, number of fires-37; BC-Bearskin Creek, n-183. From Kilgore and Taylor (1979).*

less the site was overgrazed). But there may be a dramatic change in fire practices, and the documentation may be found in material normally identified as history rather than as forestry or fire. Similarly, political reform movements, otherwise indifferent to fire practices, may create bureaus to manage land; scientific discoveries in fields other than forestry may revise policies about proper fire management programs; growth or decay in the national economy may result in changes in fire practices; warfare may introduce otherwise unexpected fire sources and practices.

Moreover, general histories provide a means of discriminating the cultural impact of fires or fire regimes. A large fire by physical standards may not be a major fire by biological standards. In the same way, an important fire as measured by ecological effects may not be influential in its cultural effects. Burned area is not the same as significance. There is no biological record for the range fires that burned in Texas during the 1880s, but these fires led to legislation which had the effect of reforming range fire practices and which were responsible for exporting a large herding economy to the Southwest. Many fire scar analyses in the Northern Rockies fail to include the 1910 fires, an episode which, as much as anything, led to a national program of fire control on wildlands in the public domain. By physical and biological standards, the 1937 Blackwater fire in Wyoming was undistinguished, but that it claimed the lives of 15 firefighters gave it importance in determining how fire crews could be used. Most fire regimes are the product, directly or indirectly, of human activities; their fire history cannot be separated from human history. To consider only certain silvicultural evidence such as fire scars is like trying to construct a history of a nation from the biographies of only those people who published books.

## *Concepts of Fire History*

Getting evidence of fire, of fire regimes, or of fire practices is one problem. Knowing how best to interpret that information is something else. The problem, in part, reflects a lack of consensus in terminology; in part, an uncertainty over what statistical techniques and organizing concepts best apply to the data at hand; and in part, confusion in historiography, over how to characterize natural and human history. The question of terminology will be solved by continued use; of data processing, by the management needs for which the history is constructed; of historiography, by a philosophy of history. The problems affect all aspects of fire history, as a record of physical, biological, and cultural events.

All history consists of events, and for fire history each such event is often referred to as a *fire occurrence.* How to describe the relation between occurrences, however, immediately plunges one into a wilderness of terminology and a thicket of concepts more or less related to one another. Perhaps the most elaborate lexicons have been developed for fire scar analysis, here briefly summarized. For a specified area, the number of years between two successive fire occurrences is considered a *fire interval.* An average of all fire intervals within a specified area and designated time period is a *mean fire interval.* The number of fire occurrences per unit time in some specified area is the *fire frequency.* Thus two descriptions of fire frequency are in common use: simple fire frequency (number of fires/total age) and the fire return interval (the average value for all the intervals between scars). When all fire occurrences are assembled into a general time frame, the product is a *master fire chronology.* From the master chronology derive such measurements as mean fire interval and fire frequency (Stokes and Dieterich, 1980). Recall, however, that the size of the sampled area —from which the master chronology is constructed—is inversely related to the mean fire interval.

*Fire Cycle Concept.* Underlying this whole enterprise, however, is a concept variously titled the *fire cycle, natural fire rotation,* or *fire period.* Fundamental to this notion is the measurement of burned areas as distinct from the recording of fire occurrences. Thus a fire cycle for a designated area expresses the number of years required for that area to burn over once. Some portions of the area may burn more than once during this period, and some may not burn at all; but the total burned acres will, during the time of the cycle, equal the total area under consideration. To relate an area's fire occurrence to its fire cycle requires more evidence than that provided by fire scars. The idea for a fire occurrence chronology derived from the study of fire scars, typical of areas that experienced underburning; the concept of a fire cycle derived from the study of age–class distributions, typical of areas that experienced stand-replacing fires. Often, however, the two expressions are used interchangeably.

What both concepts share is a common management need to determine an appropriate level of prescribed burning, and it is natural that, for foresters, such a program should be expressed by analogy to the rotation cycle applied to sustained-yield logging. Hence burning should proceed according to some chronological pattern, a cycle or rotation or period. The concept of a fire cycle has obvious management appeal. For wilderness areas, it suggests that nature obeys a certain rhythm of fire, a perception that justifies the reintroduction of fire through natural or anthropogenic means. For other wildlands, the presence of a fire cycle suggests a pattern by which programs of prescribed burning can imitate natural processes. That such activities will occur cyclically means that they can be programmed for budgets, staffed at planned levels of crews and equipment, and integrated into larger land management programs in ways that wildfire, episodic and unpredictable, cannot.

The difficulties in terminology are not merely failures in linguistic protocol, like a grammatical mistake or a malapropism, but signify underlying uncertainties about research objectives in fire history, about appropriate data, and about explanatory concepts. A fire occurrence as recorded in a fire scar, for example, does not distinguish a typical fire from an atypical fire, a biologically significant fire from an ineffective fire, a natural from an anthropogenic fire. Without outside corroboration, such as records of lightning and human activity, fire intervals and their averaging may mask more information than they reveal. As yet the statistical processing of fire occurrence data remains primitive, and fire history an almost ingenuous form of historical writing.

*Fire as a Natural Event.* As a record of physical events, the fire history of a regime more closely resembles the flood history of a watershed than it does the life cycle of a tree. As physical events wild fires are distributed along a frequency–intensity curve, not about a mean frequency; they show a logarithmic distribution, not a Gaussian (normal) distribution. Like floods, fires may be characterized as 10, 20, 50, or 100-year events, not as a mean fire frequency. Rather than comparing fire frequency to the growth rings of a tree, which show a normal distribution, or to a logging rotation, a planned anthropogenic activity, they should be considered as analogues of natural energy eruptions—stochastic processes such as floods, windstorms, and earthquakes. The analysis of age–class distributions that resulted from stand-replacing fires and the distribution of natural fires by frequency and size show this characteristic very strongly (Figure 6-8). Because of its close relationship to meteorological phenomena,

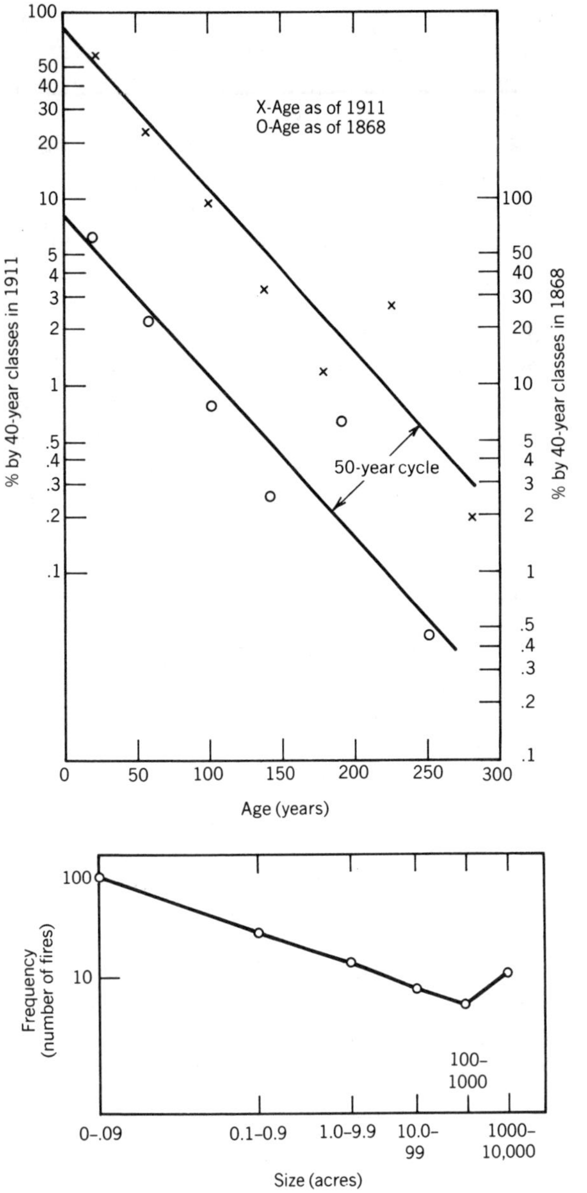


*Figure 6-8. (Top) Age–class structure of Boundary Waters Canoe Area as of two dates, 1911 and 1868. Two widely differing estimates for a 50-year fire cycle are given. "Fire cycle," as used here, refers to the time required to burn an amount equal to the entire area under consideration. The fires of record are high-intensity stand-replacing fires; no evidence of less severe fires is included. From Van Wagner (1978). (Bottom) Frequency-size distribution of natural fires in Yellowstone National Park, 1972–1982. The scope of the line will vary by fire regime. The slight increase in the number of large fires is the result of compounding effects—multiple runs made by a single fire or the merger of two fires into one. Data courtesy Yellowstone National Park.*

it makes more sense to consider fire as an episodic process for which, under special circumstances, fire history approximates a cycle. Curiously, those instances for which fire history is most cyclic are those for which fire history is most manipulated by humans.

As a biological event, the frequency of fire or the size of burned area is less important than the relationship between fire frequency and intensity and the biological timing of fires. Not all fires are significant in that not all exert much change on the ecosystem. The important fires show a certain intensity as well as a certain frequency, and they may be effective within a wide time frame that is set by the life cycle of the species in question and the successional options open to the system. Frequency alone does not indicate importance. Much as with floods, big events cause proportionately more change than do many small events. Age-class distributions, as determined by fires, do not oscillate about a mean age, but show a logarithmic distribution which corresponds to patterns of fire behavior. Moreover, because succession does not mimic a life cycle but leads to a braided tangle of successional pathways, the meaning of a simple mean fire frequency as a fundamental measure of fire history is still further complicated.

*Fire as a Human Event.* When fire history is compounded with human history, as it must be, the shape of fire history is changed. Anthropogenic fire practices transform fire history just as they reshape fire seasons. In one respect, fire history becomes even more idiographic, mirroring the larger movements of human history. But, in another respect, fire history becomes more cyclic and predictable, as humans deliberately make it so. Again, the analogy to floods is appropriate. Both fires and floods result from the interaction of weather, topography, and cover type, with weather predominating. Both are stochastic processes whose history shows a large number of small-intensity events, with an exponentially decreasing number of high-intensity events. Yet for rivers that are heavily managed—confined by dams and levees—the fluctuations are dampened, the discharge levels out, and variations in water releases ("floods") follow a daily and annual cycle. During times of extreme rainfall, this pattern may be temporarily swamped, but it is quickly restored.

It is likely that fire history behaves in much the same way. Humans have modified both the frequency and intensity of fires by modifying fuels and by intervening in the ignition and spread of fire (Figure 6-9). The general effect is to make fires more predictable. Within the twentieth century this tendency has been masked by an administrative emphasis, through fire suppression, on breaking inherited patterns of fire use and occurrence. But as a new pattern of fire use and withdrawal is defined, fire will take on more definite, even cyclic traits. These characteristics will better approximate certain annual rhythms of nature such as the growing season, and annual cycles of management such as the fiscal year which sets budget priorities. As wildland is converted from extensive to intensive management, moreover, this tendency will become increasingly pronounced. Having a fire cycle that resonates with a logging rotation plan makes sense in those areas that are heavily logged.

*Fire History and Fire Management.* The implications of fire history studies for management are thus varied. Where the intention is to preserve natural systems, the episodic nature of fires can, in general, be tolerated. Some intervention is likely, but there is no reason to twist stochastic processes into deterministic processes. It is only necessary to preserve the natural processes themselves: an appropriate fire occurrence will follow. Where the intention is to manage wildlands more intensively, then fire suppression and prescribed fire will tend to make fire history more uniform. Whether or not wildlands conform to a cycle, management agencies do obey periods—an annual cycle of reviews and budgets, for starters. Management needs are even carrying the process one step further by supplementing (and perhaps replacing) the study of fire history with the simulation of fire futures through computer gaming. To the degree that plans will be made on the basis of these scenarios, natural history comes to more closely approximate the characterizations humans make of it. It has often been assumed that human history is, or ought to be, modeled on natural history. Ironically, it seems that human history is becoming the model for fire history.

## 6.2 U.S. FIRE HISTORY: A SYNOPSIS

Wildland fire came to the landscape of the United States from three sources, and it has assumed four general patterns. It came from nature, in the form of lightning; from Asia, at the hands of the American Indian; and from Europe, through the practices of a host of immigrants. By itself light-

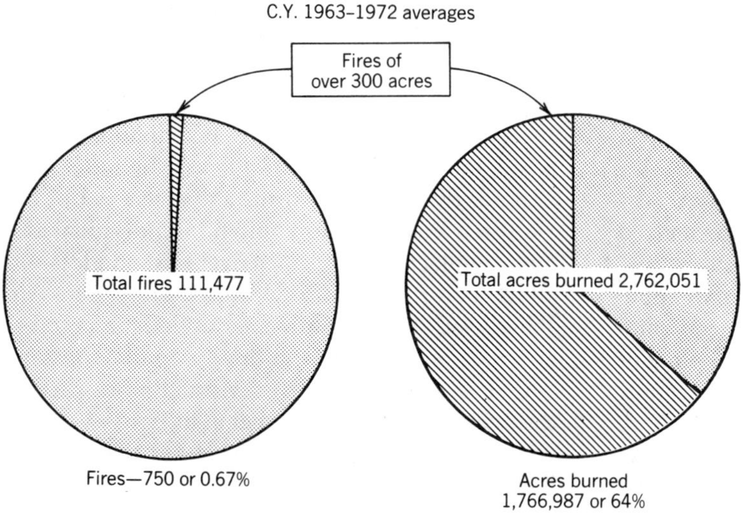


*Figure 6-9. Fires and acres burned on protected lands. Less than 1% of all fires have accounted for 64% of all burned acreage. The effect of suppression is to eliminate the average size of steady-state fires and to reduce the number of fires that achieve large fire status. But the fundamental distribution is set by environmental variables. From U.S. Forest Service, Cooperative Fire Program.*

ning can shape a natural fire regime. But to this ideal state must be added three other patterns that have shaped the fire regimes known to American history: a pattern of fire in the service of hunting and gathering, a pattern for shifting cultivation and sedentary agriculture, and a pattern for an industrialized landscape. Sometimes these patterns existed separately in space and time, but often they were simultaneously compounded one with another at a single site or coexisted in a geographic ensemble.

### *Indian Fire Practices*

It is difficult now to recapture the degree to which the economy of the American Indian was dependent on fire. In its domesticated forms, fire was used for cooking, light, and heat. It made possible ceramics and a primitive metallurgy. Its smoke was used for communications and, in smudge pots, to drive off insects. It felled trees and shaped canoes. It formed part of ritual and served as spectacle for public entertainment. In broadcast forms fire was applied to the cultivation and harvest of natural grasses such as wild rice, of forbs such as the sunflower, of berries such as the blueberry, and of nuts such as the acorn and mesquite bean. Oregon tribes used smoke to harvest pandora moths from infested trees; California Indians, to drive away mistletoe infecting those oak and mesquite groves that supplied mast. Broadcast burning along the coastal gulf plains and in the Alaskan interior drove off mosquitoes and flies; the mosquito, it is often said, was responsible for the destruction of more Alaskan timber than any other cause. Fire could be used indiscriminately to kill off broad expanses of forest, which might then be harvested for firewood, or, more circumspectly, to produce delicacies like the carmelized confection that resulted from burning sugar pine cones on the ground. It was used as a weapon: tactically, to drive off or confound an enemy, and strategically, as part of a scorched earth policy or to create an open field of fire around encampments and reduce the possibility of ambush. And, of course, fire could be distributed carelessly. Parties of warriors and hunters rarely extinguished campfires and signal fires when they were away from base camp. It was common practice, for example, to set fire to a downed tree for a campfire and then to leave it, where it might continue to burn for days. Some tribes such as the Apache set large fires in the hopes that the conflagration would induce rain.

For many areas the combination of broadcast fire and native crops proved highly productive, and may have discouraged the adoption of sedentary agriculture. Where yield was slight, the natural fertility could be maintained almost indefinitely by setting fires to recycle unused residue. But where yield was high, as with maize culture, additional fertilizer had to be added. Typically this came about through a pattern of slash-and-burn agriculture, with new sites periodically cleared and fired as the old sites declined in fertility. The cycle of fire varied according to the crop and the intensity of yield: cereal grasses were fired annually; basket grasses and nuts, every 3 years; brush, perhaps every 7 to 10 years; large timber, for swidden cultivation, on a cycle of 30 years or more. After harvest broadcast fire served to rid the fields of vermin and disease. Once cleared for swidden cultivation, moreover, the site could be maintained in grass and browse—and thereby as a hunting site—by continued firing.

Of all Indian uses for fire, the most widespread was probably the most ancient: fire for hunting. At night torches spotlighted deer and drew fish close enough to canoes to be speared. Smoke flushed bees from their hives, raccoons out of their dens, and bears out of their caves. Recognizing that new grass sprouting on a freshly burned site would attract grazers by its superior palatability, Indians could control largescale movement of herds by alternately firing and greening up sites and, on a smaller scale, could hunt individual animals by placing snare traps over burned plots—in effect, baiting the trap with fired grass. The Apaches used smoke to lure deer driven mad by flies and mosquitoes. Fire hunting, properly speaking —that is, the strategy of surrounding or driving the principal grazers of a region by fire—was universal. In the East it was used for deer; in the Everglades, for alligators; on the prairies, for buffalo; along the tules of the Colorado River, for rabbits and wood rats; in Utah and the Cordillera, for deer and antelope; in the Great Basin, for grasshoppers; in California and the Southwest, for rabbits; in Alaska, for muskrats and moose. It is, in fact, a remarkable coincidence that virtually all animals and birds prized as game come from environments that are sustained by periodic fire. Special hunting sites, often extensive, were maintained and visited during an annual cycle which typically included fires in autumn.

Nor was the Indian restricted to natural sites. It was possible, through broadcast fire, to create new habitats for such game. None of these creatures were true forest fauna; some could live in extensive grasslands, but all required a vegetative mosaic that included at least some open prairies and browse. In the absence of fire that mosaic tended to disappear. Fire hunting thus not only served as a tactic of harvesting but as a means of sustaining (and even creating) the habitat upon which it depended. Tribes burned nearly all areas that were naturally drained, leaving many "barrens" or savannahs or, where forests persisted, relatively open woods free of underbrush. Early explorers and settlers widely reported such grassy "deserts," culminating in the "Great American Desert" of the central plains. The pattern of Indian burning, moreover, changed the biological timing of fire. Natural fires came with the summer storm season, whereas Indian fires followed the life cycle of grasses, with burning in the early spring and late fall.

At first blush it seems impossible that relatively small numbers of seminomadic hunters could have exerted much influence over their environment, but fire, properly used, has a multiplicative effect. It propagates, and intelligent beings armed with fire can apply it at critical times for maximum spread and effect. It is likely, too, that the larger changes from forest to grassland were a cumulative product of fire and warm climatic episodes. Once established it was simple, through broadcast fire, to maintain such sites in grass even against climatic pressures that could support forests.

Such fire practices were not unique to the American Indian. The Seminole practice of hunting alligators with fire is identical to the Indonesian method of hunting crocodiles; the Iroquois practiced swidden agriculture in much the same way as do modern peasants in northern Thailand. Pasturage fires for wildlife in America are indistinguishable from those set on the steppes of central Asia, on the veldt of East Africa, or on kangaroo ranges in Australia. Ancient Scythians on the Asian steppes and aborigines along the Australian coast used broadcast fire as a weapon in much the same way that the Sioux used it on the Great Plains. Of course, even within North America, there were variations in fire practices among tribes and variations within the history of a single tribe, as it acquired maize or the horse, as it drew near European settlements or was driven away from them, as it moved from forest to grassland or grassland to desert. Yet this general panorama of Indian fire practices is almost certainly correct, and it was from the American Indian that European frontiersmen acquired many of their own fire practices and learned how to live amid the fire regimes of their newly discovered lands.

### *European Fire Practices: The Reclamation*

The discovery and settlement of America belongs, as political history, with the expansion of Europe and, as fire history, with the agricultural reclamation of Europe. What fire practices people brought with them to the New World depended largely on when and from where in the Old World they emigrated. Folk from the Mediterranean basin such as Basques were more prone to use fire in the service of herding, and to use it broadly; folk from Scandinavia, to use fire for slash-and-burn agriculture; from western Europe, for landclearing and annual stubble removal. By the sixteenth century most of the reclamation of Europe was accomplished; by the eighteenth century, it had reached geographic limits, extending up mountain slopes in the Alps and sandy barrens along the Baltic. For much of Western Europe—France, Germany, and England—widespread firing had accompanied landclearing and large fires had vanished when that landclearing ceased.

Much as the medieval church assimilated pagan fire festivals by Christianizing them into a liturgical calendar, so European agriculture had broken down most relic fire practices and reincorporated them into a cycle suitable for its particular agricultural pursuits. By the eighteenth century, with the cessation of landclearing, widescale burning persisted only along the fringes of Europe, among peoples for whom sedentary agriculture was an exception—the swaling fires of Finland, heath burning in northern Scotland, pastoral brush burning along the Mediterranean, grassland fires set for range improvement on the Spanish plateau and the Russian steppes. In recapitulating the reclamation of Europe, Americans rediscovered many of the same needs for fire as their ancestors, but they often had to look elsewhere, notably to the American Indian, to recapture a suitable technology of fire.

### *American Fire Practices: Reclamation and Counterreclamation*

Many Indian fire practices and the landscapes they shaped were well-suited to the settlement frontier that crowded upon them. From the forests most affected by fire, pine and oak, settlers found their timber. With grassland corridors maintained by fire, like the Shenandoah Valley, they had a thoroughfare for rapid dispersion. With the wildlife that flourished in the mosaic of woods and grasses, frontiersmen found sustenance to carry them through their first years. On the grassy hunting grounds of the Indian, settlers raised cattle; on abandoned Indian fields, they planted new farms. The transfer of the torch greatly assisted in the exchange of landholders from American Indian to American European.

But such conditions were only temporary. They demanded a state of mobility, almost a seminomadism, that would vanish as these lands passed into sedentary agriculture. Where reclamation became more settled, fire codes and custom regulated fire uses; volunteer crews suppressed unwanted fire; a new cycle of field and woods fires appeared. Domesticated grasses replaced wild grasses; orchards, wild forests; pastures, wild prairies; cattle and hogs, wild buffalo and bear; agricultural fire, its wilder predecessors. Where landclearing occurred on a prodigious scale with massive disturbances of fuels, as in the Lake States in the late nineteenth century, conflagrations could result. But consensus held that such holocausts were only a temporary phenomenon of transition. With time wildlands would be reclaimed as arable land, and wild fires would disappear.

Control of wildland fire per se was considered impossible in a technical sense, indefensible in economic terms, and undesirable on environmental grounds. Local juries prosecuted only the most flagrant abuses, reluctant to reduce accessibility to fire, an often essential tool for the maintenance of frontier economies. But where the frontier persisted either in whole, as in portions of the West, or in part, as with the open range of the South, there frontier fire practices continued. (Even in the twentieth century 105% of Florida was reported burned in one year, the result of firing many areas both in the spring and fall.) When the time came to relearn the techniques of controlled burning in the mid-twentieth century, it became necessary to recover them out of Indian reservations in the West and the piney woods of the South.

The American reclamation differed significantly from those that had preceded it in both the New World and the Old. It differed from the European reclamation in both geographic scale and historical tempo, compressing into less than 200 years a millenium of European activity. It differed from that much smaller reclamation by the Indian in that it opened up many areas, through systematic drainage, which the Indian had not been able to penetrate; that it promoted a sedentary agriculture, complete with land titles; that it introduced a near plague of new flora and fauna, which changed the susceptibility of the land to fire and the value of burning. Some floral introductions such as Kentucky bluegrass showed less fire-adaptability; some such as cheatgrass were extremely fire-prone, driving off less fire-hardy natives; others such as domestic cereal grasses had a fire cycle entirely circumscribed by close cultivation. Faunal exotics induced even greater changes. Cattle, sheep, burros, horses, goats, hogs—all directly modified the grasses and shrubs upon which Indian fire practices largely depended. The American reclamation differed from both Indian and European reclamation, too, in that its greatest period of expansion coincided with the industrial revolution. The agricultural reclamation barely had time to consolidate itself before industrialization encouraged a counterreclamation.

The concept of a *counterreclamation* encompasses the variety of processes by which, with the spread of industrialism, land was reserved from agricultural settlement and by which once-arable land reverted to wild or forested land. In some cases, the process was deliberate—the national parks and forest reserves being an outstanding example. In other cases, the process came inadvertently. For example, steamboats and railroads, fossil fuels, and industrial fertilizers made possible the transfer of farms from areas that had once been forested to areas that were originally grasslands; the abandoned lands reforested, and many were put into industrial production or were maintained as recreational wildlands. The process of abandonment and relocation continues. Some of the abandoned lands go into urban development, but large portions continue to enter the general category of wildlands. In terms of vegetative cover, the European occupation of America has resulted in an expansion of woody vegetation at the expense of grass. For the lands of the counterreclamation some form of administration was necessary, and a primary charge for such administration would be fire protection, especially control over laissez-faire fire practices characteristic of the frontier.

But the exclusion of traditional fire practices also meant the exclusion of traditional means of fire control. The new administrators would have to invent new policies to replace rural fire codes, develop new sources of manpower to replace rural fire brigades, and encourage a new science of wildland fire to supercede the accumulated folk wisdom about fire use and control. In brief, such an administration would have to occupy often vast areas of public domain in the name of fire protection, and in time fire control became a primary institution in the settlement of the American West. The first manifestation of this problem came in 1885 and 1886 when New York created its Adirondacks Forest Preserve and the U.S. Army assumed control of Yellowstone National Park, in good part to bring fire control. Private timber protective associations developed, especially in the northwest. But the key lands in question were the enormous forest reserves carved by presidential proclamation out of the public domain in the West after 1891.

### *U.S. Forest Service*

Impressed by the Army example, a National Academy of Sciences committee in 1896 recommended that, at least for the present, the Army assume control over all these lands and that forestry be taught at West Point. Instead the Transfer Act of 1905 gave these lands to an obscure government agency then headed by Gifford Pinchot, the Bureau of Forestry, soon renamed the U.S. Forest Service. Modern fire protection as a national enterprise dates from this event. "Like the question of slavery," warned Pinchot, "the question of forest fires may be shelved for some time, at enormous cost in the end, but sooner or later it must be met." The Forest Service assumed a central institutional and intellectual role in fire programs at all levels of national life—institutionally, by its control of the national forest system and by its promotion of cooperative fire protection programs with other federal agencies, the states, and industry; intellectually, by introducing the standards of professional European forestry into the debate about fire policy. From its origin in technology transfer from Europe, fire expertise in the United States would evolve from within the intellectual confines of forestry and the institutional framework of its principal arm, the U.S. Forest Service. That evolution was not always simple. European forestry was better suited to the management of woodlots and estates than of whole mountain ranges and vast, wilderness watersheds; especially with respect to fire, European forestry often offered precious little precedent.

An administrative history can be described in terms of four problem fire types which at different times have challenged the fire protection mandate of the Service and have evoked different responses from it (Figure 6-10). All aspects of fire protection tended to cluster around the particular concerns of the reigning problem fire, and each fire type brought with it both a strategic concept for its control and a certain tactical emphasis, a means by which the larger strategy might best be achieved. These problem fires might be charcterized as the frontier fire (1910–1929), the backcountry fire (1930–1949), the mass fire (1950–1970), and the wilderness fire (1970–present). Moreover, the Forest Service sought to extend its policies and solutions to its cooperators, primarily, the state foresters. In fact, it may be said that fire was usually the first, and often the only, basis for the vast cooperative programs in forestry that have evolved nationwide.

Each of these periods may also be characterized in another way: as a response to certain types of abundances that became suddenly available to fire protection. For the era of the frontier fire, the surplus was one of land (the result of the Transfer Act and Weeks Act) and of money (a consequence of the Act of 1908). For the era of the backcountry fire, there was an overabundance of money and manpower as a result of New Deal conservation programs, especially the Civilian Conservation Corps. This confluence was replaced during the era of mass fire by a military and civil defense liaison that brought large quantities of war surplus to fire control. By the 1970s an almost abnormal, or at least unassimilable, amount of information had been produced on the subject of fire, particularly on fire ecology, and policy reforms can be imagined as a means of coping with this "surplus." The exploitation of these abundances did not involve simply the opportunistic acquisition of new means: the means at hand were often so powerful as to dictate to some degree the ends to which they might be applied. They upset an evolving internal equilibrium between policy and programs.

Each of these fire types suggested, too, a potential range for controlled burning and proscribed, as well as prescribed, certain fire uses. Originally, the Forest Service believed that fire protection only required breaking down the pattern of fire use characteristic of frontier agriculture. It did not appreciate that breaking down that pattern of fire use also destroyed the basis for fire control in the same environment, and that the larger problem

| | | | Fire Control | | |
|---|---|---|---|---|---|
| Date | Problem Fire | Policy | Strategic Concept | Tactical Emphasis | Research |
| 1910–1929 | Frontier fire | Economic theory | Systematic fire protection | Administration | Fire as forestry<br>Economics, planning, statistics of fire |
| 1930–1949 | Backcountry fire | 10 AM Policy | Hour control | Manpower | |
| 1950–1970 | Mass fire | 10 AM Policy | Conflagration control | Mechanization | Fire as physics<br>Laboratory<br>Field experimentation |
| 1971–present | Wilderness fire | Fire by prescription | Fuel modification | Prescribed (broadcast) fire | Fire as biology<br>Natural laboratories<br>Simulation experiments |

*Figure 6-10. Wildland Fire Protection: The U.S. Forest Service Experience. From Pyne (1982).*

required not merely that the old cycle be broken but that a new one be constituted which would include fire use as well as fire suppression. During the frontier fire period, slash burning in piles was acceptable, but broadcast burning was not—it too closely resembled the light burning of the frontier and the careless disposal of logging slash by land-clearing fires. As systematic fire protection extended its limits, so too were limits set on traditional fire uses; as systematic protection was successfully achieved, as in the South, so it accomodated certain uses of free-burning fire. During the era of mass fire, a prescription for mass fire was a prominent objective of fire research, and the application of mass fires for military purposes was tested. The era of wilderness fire, however, divided all the fires into either wild or prescribed fires, and it sought to expand the range of prescribed burning.

The issue, that is, was never one of simple fire exclusion but of fire use only within the confines and context of the dominant problem fire type. For the periods of frontier, backcountry, and mass fire the ruling concern was control over these particular types of wildfire, and it had the effect of limiting many potential uses of controlled fire. With wilderness fire the overriding concern became the dissemination of a particular set of fire applications, and it had the effect of limiting certain forms of fire suppression.

*Frontier Fire.* The magnitude of its fire problem did not become apparent to the Forest Service until the summer of 1910, a scant seven months after its charismatic prophet, Pinchot, had been fired for insubordination. That August the Forest Service confronted two literal trials by fire. In California the long-smoldering "light burning" controversy—an attempt by landowners to adapt frontier fire practices for timber management—flared into public view, while throughout the West a complex of wildfires burned some 5 million acres (2.03 million hectares), 3 million (1.21 million hectares) in the Northern Rockies alone. The fires traumatized the young Forest Service. Forestry had made fire control a foundation of professional management and promised that the national forest system could provide it; the presence or absence of conflagrations became public tests on these premises. The fires, too, were the first crisis faced by Henry Graves as chief forester, and they were an indelible reference point in the memory of two future chiefs, William Greeley and Ferdinand Silcox, regional and assistant regional foresters, respectively, at the time. Within a year yet another event added urgency to the fire question. With the Weeks Act of 1911 the Forest Service was authorized to acquire land for national forests in the watersheds of the eastern United States and to enter into cooperative fire protection programs with the states. Almost immediately on its new forests in Florida another version of light burning threatened to overwhelm Forest Service administration. The 1910 fires challenged Forest Service competence to control fire, the light burning debate its ability to formulate a useful policy on fire, and the Weeks Act its capacity to establish national standards for fire protection.

Yet the rancor of the light-burning controversy, the high drama of the 1910 fires, the call to arms for a war on nature, the reform-minded idealism of forester-progressives stamped in the mold of Gifford Pinchot—all gave the debate about fire policy an intensity seemingly out of proportion to the problem at hand. There could be little compromise with fire. In the end it was perhaps the suddenness of what had happened that was most significant: the need to manage millions of acres overnight, the shock of the 1910 holocaust, and the heady liberation of firefighting monies made possible by the Forest Fire Emergency Fund Act of 1908. The act allowed for deficit spending in the event of fire emergencies; its first test came, naturally, with the 1910 fires when the Forest Service acquired a debt of over 1 million (1910) dollars. But the act was sustained, the Service was spared bankruptcy, and the need grew to provide some kind of guidelines for the expenditure of money on this scale. The most important response to these almost simultaneous pressures was the development in California of a singular idea—at once a concept and a program—referred to as "systematic fire protection." Its architect was regional forester Coert duBois, its basic document his *Systematic Fire Protection in the California Forests* (1914), and its program a blend of comprehensive planning, special research, and a gutsy enthusiasm that could only have come from an organization composed, as the Forest Service then was, almost wholly of young men.

From 1911 to 1921 considerable debate occurred nationwide about a suitable solution to the fire question. In one form or another all the variants appeared in California—light burning, let burning, and systematic fire protection among them. Systematic fire protection was justified, initially, only for frontcountry lands of demonstrable value such as those containing mature timber. For other lands and other values, there existed no method for setting appropriate levels of protection. Without forest fire insurance, moreover, land agencies lacked an economic standard on the model of the grading schedule known to urban fire services. Especially in the context of the Act of 1908, which made deficit spending possible during fire emergencies, some program of fiscal control was necessary. The accepted compromise was the "economic" or "least cost plus loss" theory, first proposed in 1916, which stated simply that total costs of suppression and fire damages should be held to a minimum and that the investment in fire control should be commensurate with the value of the resources under protection.

All of these concerns led in 1921 to the first servicewide conference on any topic sponsored by the Forest Service. Its subject was fire; its location, California; and its effect, to consolidate all aspects of fire protection, from planning objectives and financing to the design of shovels and fire report forms. The Mather Field Conference led, in general, to approval for the duBois program and the economic theory. Within a decade after its publication, systematic fire protection had triumphed in California, and within another decade it would become the basis for fire planning nationwide. Within the next few years, too, the other political challenges to Service fire policy were also turned aside. The California Forestry Commission, after several years of tests, condemned light burning; the Clarke–McNary Act of 1924 greatly extended the cooperative fire program initiated under the Weeks Act; an interagency Forest Protection Board consolidated Forest Service hegemony among federal land holders, and the McSweeney–McNary Act of 1928 confirmed its supremacy in fire and forest research. By the end of the 1920s the Service established a demonstration program at the Shasta National Forest, modeled on the administrative principles of systematic fire protection. How widely that program would be imitated depended on the resources at hand. As long as the Service confronted frontier fire practices on frontcountry timber lands it found systematic fire protection an adequate statement of techniques and the economic theory an apporopriate summary of goals.

*Backcountry Fire.* The term *backcountry fire* includes fires occurring on forested lands remote in space and time—that is, lands in the undeveloped backcountry of the public domain and lands, such as in the South and Lake States, that had been cut over and for which fire protection was an investment in the future. Thanks to duBois' protege, S.B. Show, the precepts of systematic fire protection were refined into a program of *hour control,* and Show's collaborator, Edward Kotok, made these ideas fundamental to the Copeland Report (1933), the Service's ambitious scheme to reforest America. With the coming of the New Deal, the careful equilibrium of means and ends that had informed the expansion of fire protection underwent an upheaval.

The conservation programs of the Roosevelt Administration proved both a boon and a burden to Forest Service fire control. They brought an expansion in the amount and type of lands under protection, an enlargement of incentives, and a magnification of means. They pushed fire protection physically into the backcountry—not only of the national forest system, but of the nation—by resettling marginal farmland and by acquiring cutover forests, tax delinquent lands, and deteriorated grasslands. The lands under Forest Service jurisdiction in the eastern United States quadrupled in size, and many states acquired extensive state forests. Despite their lack of immediate values, such lands were to be protected as fully as frontcountry lands. For political reasons, that is, the economic theory was rendered worthless as a guide; it could not apply well to lands that did not have an immediate value, nor could it incorporate the "social" goals that were important to the new legislation.

The new programs pushed protection into the backcountry backhandedly, too, by requiring that the abundance of money and manpower made available under the CCC, ECW, and WPA be put to use. By dumping money and manpower in larger quantities than could often be used, the New Deal swept away economic objections to an expansive program of fire control. By packaging that support in the form of labor gangs, it put a special emphasis on labor-intensive programs and the development of organized fire crews. It was amid this unprecedented outpouring of men and materiel that the Forest Service launched a complete modernization of its fire program. The amplification of means demanded a similar amplification of ends. In the three years between 1935 and 1937 the Service sought to complete in full the program with which it had struggled piecemeal for three decades.

In 1935, as an "experiment on a continental scale," the Forest Service adopted the 10 AM Policy, which stipulated control by 10 AM the morning following the report of a fire, or, failing that, control by 10 AM the day following, ad infinitum. The policy applied to all lands under Service jurisdiction. Thanks to the invention of a fire-danger rating meter, the emergency fire fund was expanded to include presuppression expenditures. After a round of debates not unlike that which had accompanied the adoption of systematic fire protection earlier, the Service brushed aside proposals for let burning in remote lowvalue lands, for allowing fires to burn in wilderness areas, or for compromising with lingering frontier fire practices. Savage droughts had led to widespread fires in the early 1930s, and the 10 AM Policy may be compared to other panic legislation enacted at the time to cope with soil erosion and floods.

The one exception to this hard line was the South. Here some form of controlled burning was necessary, part of the price of admitting the cutover pineries of the region into industrial forestry. The Forest Service paid that price, but as stubbornly and quietly as possible, even suppressing the publication of results from its own research stations that contradicted official policy. The Service worried that revelations about the value of controlled burning might compromise its fight against laissez-faire woodsburning, embarrass its allies in state forestry departments, and call into question the assumptions of organized conservation. Even after prescribed burning became accepted silvicultural practice in the South, it was considered as an oddity, not generally suitable for export.

When the Forest Service established a separate Division of Forest Fire Research in late 1948, its objective was fundamentally restorative. The division would eliminate duplication among regional research stations, and it would return research to its traditional goals, conducted by forester-engineers and forester-economists. But already a new problem fire was emerging that promised to redirect the goals of fire protection, to promote new methodologies for fire research, and to realign the institutional basis for national fire management.

*Mass Fire.* The firebombings of World War II convinced observers that the next war would be a fire war. When in 1949 the Soviet Union exploded an atomic bomb, an incendiary weapon, it became necessary to better understand the physics of firestorms. On the behavior and control of *mass fire,* there developed an unprecedented fusion of interests among the Forest Service, the Office of Civil Defense, the military, and urban fire services. Fire control became a part of national defense, and as its policy and research were again reorganized, the Forest Service seemingly entered into a cold war on fire. Calculations on fire effects tended to be made on the assumption that the fire under consideration was "hostile" fire. The hypothetical scenarios of thermonuclear war and the practical problems associated with the fire regime developing in Southern California—where nearly annual conflagrations were thought to imitate the conditions of fire warfare—moved the Forest Service out of the backcountry and into the urban fringe.

The foundation for this activity grew out of World War II, nearly all of whose images are of fire. Even the two national fire prevention programs, Keep America Green and Smokey Bear, emerged out of wartime propaganda. The search for fire weaponry stimulated considerable interest in fire behavior. Inspired by a sense of emergency and the shortage of firefighters, the Forest Service strengthened its cooperative program with the signing in 1943 of a mutual defense pact between the departments of Agriculture and Interior. Forest Service liaisons with the military and civil defense were even more fundamental. Civil defense organized fire crews, and there were mutual aid agreements between national forests and military bases. But it was the campaign against Japanese fire balloons by the Forest Service and the Army that became a precedent for later research and cooperation on the subject of mass fire control, under the assumption that such fires would result from atomic weapons. Executive wartime orders directed the Forest Service to assume control over a program of fire defense on both rural and wild lands, another precedent. In postwar years rural fire defense would be expanded through the Clarke–McNary program to become the basis for the Rural Community Fire Protection programs (authorized in 1972); and with the help of new funding arising out of the interest in mass fire, the Forest Service established a Division of Fire Research, which by 1963 included three fire laboratories.

No less fundamental was a directive that gave the Forest Service priority access under the federal excess equipment program to acquire and distribute materiel to its cooperators. The excess equipment program led to the rapid mechanization of fire control by the mid-1950s when large quantities of surplus military equipment became available. The support given fire protection by the emergency conservation program of the New Deal was, in effect, replaced by a wartime alliance with the military and with civil defense, a connection perpetuated by the Cold War with its spectre of atomic weaponry. Fire protection soon expanded to include territories and protectorates; interstate compacts appeared, with the Forest Service as a cosigner, to integrate state protection; like other manifestations of American influence, the principles of fire protection were disseminated globally to political and military allies. By the mid-1960s the Forest Service position in fire protection within the United States was supreme by virtue of its political alliances, its control over suppression resources, and its near monopoly of fire research.

The controlling doctrine of this era was a concept of *conflagration control.* The idea was to isolate hot areas and to concentrate resources in such a way that small fires could not escalate into holocausts. One means was to contain problem areas through a system of conflagration barriers, principally fuelbreaks. This was done most extensively in California. Another means was to create a kind of rapid deployment force, which the Forest Service did in 1961 with the development of the interregional fire suppression crew. The IR, or hot shot crew, could move within hours to any trouble spot in the national forest system. And perhaps most importantly, the Forest Service followed the example of a mechanized military, supplementing its old emphasis on organized crews with the abundance of war surplus equipment. This transfer was often deliberate. Immediately after the war, the Forest Service and military participated in mutual tests on the conversion of bombers and helicopters for fire suppression; the experiments were inconclusive, though army engineers and foresters joined together again in the mid1950s for successful experiments on mass-fire control that involved aircraft. In 1956 both air tankers and helitack became fireline realities. The

Forest Service established two equipment development centers to explore other means of conversion and to work out special pieces of equipment such as those required by smokejumping that were unique to firefighting.

The search for an understanding of mass fire utterly reoriented wildland fire science. With urban conflagrations largely a thing of the past, the Forest Service was practically the sole authority on free-burning fires, experiencing dozens annually and supplying much of the expertise for handling the complex urban–wildland fire regime that had emerged in Southern California. Forest Service fire research came into a commanding role for the study of mass fire in all environments. In 1954 a one-year crash program called Operation Firestop, situated in Southern California, explored the behavior and control of mass fire. Perpetuating an alliance between civil defense and the Forest Service, Firestop led four years later to the chartering of a Committee on Fire Research within the National Academy of Sciences-National Research Council. The committee would advise Civil Defense and DOD on their investments in fire research. After international tensions rose in the early 1960s, particularly following the Cuban missile crisis, considerable funding became available for fire behavior studies. By 1962 OCD was able to hold annual conferences among its fire contractors, and the Forest Service had established three fire research labs. Within a few years virtually the entire fire research program of the Service was concentrated inside these facilities.

There were some unexpected consequences. Fire research examined fire as a physical phenomenon rather than fire as a problem in silviculture or forest administration. In so doing fire at last left the confines of forestry, but it also left the traditional means of evaluation by foresters. It became a subject for physics, chemistry, meteorology, and new forms of operations research; it often abandoned the backwoods for the lab; it experimented with new research methodologies such as large-scale field tests for mass fire behavior. Prescribed fire, too, took on some new forms. It was used for conflagration control, especially in the construction of fuelbreaks following mechanical or chemical pretreatments, and it was used for conflagration initiation. Both techniques were appropriated by the Army for use in the Vietnam war. But the environmental movement in the United States looked at the possibilities of fire in quite different terms, and beginning with the passage of the Wilderness Act in 1964, new legislation brought another problem fire to national attention.

*Wilderness Fire.* In 1970, despite several decades of intensive fire research, equipment development, and healthy financing, fire complexes in Washington and California burned more acreage on the national forest system than in any year since 1910. Like other large fires before them, these catalyzed a broad reorganization. But this time the center of attention was the question of natural fire in wilderness areas. The recognition had grown that lightning fires in natural areas were part of the native ecology and that, in order to maintain the primitive state of these areas, it was necessary to tolerate such fires or to introduce surrogate, prescribed fires. There began wholesale experimentation to determine how far such beneficial fires could be extended and for how many objectives of land management. The mass fire experiments, moreover, had concluded that fuel complexes were the greatest determinant in fire intensity, and fuel modification replaced conflagration control as a ruling concept for wildfire control; the preferred tool was prescribed, broadcast fire.

The era witnessed the breakup of Forest Service hegemony over wildland fire. Reorganization occurred in all aspects of fire management, though the Service remained by far the most important single agency involved. The Service saw its connections with military fire ended in 1972 by a Senate resolution forbidding such exchanges; with urban fire, by the establishment of the U.S. Fire Administration in 1974; with the Committee on Fire Research, by the transfer of some fire research to the Bureau of Standards and by the dissolution of the Committee in 1977. Partly as a result of ample funding and partly because of different legislative mandates, other federal agencies such as the National Park Service and Bureau of Land Management developed more or less independent fire programs.

Accordingly there occurred a reorganization of interagency relationships. The Boise Interagency Fire Center (1969) furnished coordinated logistical support for fire suppression; the National Wildfire Coordinating Group (1976) provided a forum for the exchange of ideas and concerns; the National Advanced Resources Technology Center (1974) gave training in accordance with a national fire qualifications rating system; in Southern California, the Firescope program brought together all levels of government for coordinated attacks on fire and other emergencies. On the international scene, treaties with Canada and Mexico transferred some of the Forest Service's interest in global fire problems to the State Department; domestically, funding for the Clarke–McNary program began to wither away. The proliferation of fire institutions led to the doctrine of *total mobility*—the complete interchangeability of crews and equipment between coordinating agencies.

Even in fire research the Forest Service found competitors. The National Park Service launched a program in fire ecology; the National Science Foundation gave the modeling of fire ecology high priority in the Forest Biome Project, broadening fire research to include academics; and the Tall Timbers Research Station, a private lab in Florida, sponsored a series of annual fire ecology conferences (1962–1976) on the subject of fire biology and prescribed burning. Similar global conferences on fire brought together researchers from many countries; against the example of European forestry, to which Americans had traditionally looked, there now appeared dozens of counterexamples, most of which relied on some form of prescribed fire. Even within its own house the Forest Service faced major realignments as Congress poured out environmental legislation madating particular programs and as the Service vigorously sought to subordinate fire management to the larger objectives of land management.

In 1971 the Service formally amended the 10 AM Policy. The revisions were intended to liberalize fire use, especially on wilderness areas, yet simultaneously to stiffen fire control by setting 10 acres (4 hectares) as a presuppression goal for the containment of wildfire. The 10 Acre Policy resulted in a wild surge of emergency spending without any reduction in the cost of fire suppression. Even by the mid-1960s virtually the whole apparatus of federal fire management was financed by a system of deficit spending. Review by the Office of Management and Budget and the Forest Service resulted in the elimination of the emergency presuppression fund and its replacement by a more accountable Fire Management Fund. In 1978 the Service scrapped its amended 10 AM Policy in favor of an entirely new one, a de facto policy of fire by prescription. Fires were divided into either wildfires, which were to be suppressed, or prescribed fires, which were to be supported. To consolidate the new policy, fire research launched new programs for general fire management (1974) and assembled state-of-the-art knowledge with a National Fire Effects Workshop (1978).

In the early years of the century, foresters declared that industrial forestry and the protection of reserved watersheds would be impossible unless surface fires were eliminated. By 1980 they insisted that forestry was impossible without such fires. The effects of suppression were considered, in many areas, as undesirable as the effects of uncontrolled fire. A new set of fire practices had arrived, one that excluded the lingering fire habits of preindustrial economies and yet incorporated forms of controlled fire suitable to the landscape of the counterreclamation and to the cycle of industrial forestry. No less significant was the fact that fire protection had completed its historical pattern of growth. Geographically, there were no new lands to put under protection, and institutionally, the pattern was one of realignment and retrenchment rather than of expansion. For fire management in general and the Forest Service in particular the coming era would emphasize reform by consolidation.

The past era, however, had been one of great adventure and accomplishment and surprise. Experience finally showed the Service that, contrary to early beliefs, fire control was not a one-time affair, that wildland fire was ineradicable so long as wildlands existed. Ironically, precisely because its fire-protection mission did not wither away, the Service acquired a special source of strength. Fire management had given the Service a charge that, unlike those of many bureaus, would not fade away and that helped make the Service dynamic long after its formative zeal had passed by. The relationship between fire and the Forest Service has been curiously symbiotic: it was forestry, and especially the Forest Service, that brought systematic fire protection to America, but equally it was the need for fire protection on the reserved lands of the public domain that had created the need for foresters. The Forest Service's greatest nemesis had, in many respects, been its best friend.

## *6.3 SELECTED FIRE REGIMES: HISTORICAL CHARACTERISTICS*

A fire regime has historical boundaries as much as it has geographic boundaries, and a regime is an expression of its past history as fully as its present dynamics. No regime begins from nothing, and no regime has its entire past ever wiped clean; some form of that past remains to influence the future. This may mean nothing more than the persistence of seeds and rhizomes, or the preservation of refugia by which a former biota transforms itself into a future biota. Typically, there is more continuity than change, and the history of fire regimes tends to be conservative rather than revolutionary. Even fundamental modifications in climate occur over long periods, though they may fall within the life histories of many trees.

Fires, too, may be conservative in their effects. Much of the research on fire history seeks to demonstrate how fires, especially in wilderness sites, work to maintain a regime. Though there may be substantial time lags involved, the effect of most fires is to recycle an existing biota rather than to remove it. But whether change occurs suddenly or slowly, whether its effects are broadly or narrowly felt, new fire regimes retain some elements of their predecessors and they, in turn, pass on some of their new properties to the regimes that succeed them.

Fire history can describe not only the pattern of fires within an established regime, but the transformation in patterns that defines the evolution of regimes themselves. The point of interest here is not so much a record of fires but a record of fire regimes, and, because few such changes occur without human influence, a record of fire practices. Fire history becomes not merely the study of fire occurrences, but of largescale metamorphoses in the character of the regime itself. Fire history becomes inseparable from general human history. For the United States the record shows profound changes in regimes, often within well-circumscribed historical limits, as a product of changing human activities. Few of these activities are directly concerned with fire; most involve general land use or social, political, intellectual, and economic developments which, nonetheless, impinge on fuel and fire management.

### *Eastern Great Plains*

Whether the tallgrass prairies originated from fire or not is undetermined. The lands of the tallgrass prairie are well-dissected by streams, so lightning fires would have had to be more abundant than at present for fire to be effective. The tallgrass complex may well have developed in response to soil and climate conditions, and in the process acquired adaptations which were also suited to survive fire. But regardless of origin, the tallgrass complex was almost certainly sustained by fire. This fire was anthropogenic, and the strong possibility exists that the origin of the entire biota was similarly anthropogenic. Fire came from tribes of American Indians; when both Indians and fire were removed, the tallgrass fire regime vanished, replaced by farms, towns, and forests.

The spread of the tallgrass prairie may have come from fire and humans radiating out from the ancient migration corridor down the High Plains. The biota was probably fashioned during a warm period, then maintained by broadcast fire as the climate turned colder and wetter and eventually extended over the Ohio Valley. This much is speculation. But early explorer records indicate the existence of such a prairie and refer to a spring–fall burning cycle by the Indian tribes who inhabited the region. Here more uncertainties enter, for the character of the tribes was rapidly changing. Already there was a general migration west—in part, because of a desire to exploit the large population of ungulates on the grasslands, a harvesting made possible by the introduction of the horse, and in part, because of displacement by European settlers to the east. In short order several waves of tribes passed through the region, and they experienced—as a result of the horse and other European trade goods—an internal revolution in their social economy. Unfortunately, this period of contact and translocation was also the period of the first written reports, so the long-term pattern of ignition is difficult to determine. Yet the character of Indian burning for fire hunting, for example, is so uniform across the United States that, regardless of which tribe inhabited the tallgrass prairies at one time, the fire cycle included an almost annual burning or combined spring–fall burning. In the largest sense this fire was associated with hunting.

With the advent of European and American settlers in the early twentieth century, an economy based on hunting and subsistence agriculture gave way to one based on sedentary farming. In place of wild grasses, corn and wheat were planted; in place of wild animals, domesticated cattle and hogs became common. Instead of wild or at best semidomesticated prairies, a pattern of roads, small towns, and farms cut on a rectangular grid appeared. By 1846 this process had extended so far westward that Iowa had a sufficient population to warrant statehood. The breakup of fuels by roads and plowed fuels, the close cropping of grass by livestock, primarily cattle, the substitution of special, often exotic grasses for wild pasture, of domesticated tallgrasses for wild tallgrasses, of trees for grasses of all sorts, and of intensive, commodity agriculture for the harvesting of natural products all demanded that wildfire be suppressed in favor of a cycle of burning tied to agricultural pursuits.

This was done. Fire codes and custom dictated when burning could be conducted, and rural fire brigades protected farms and towns. A wildland fire regime was transformed into a rural fire regime. The transitional period, during which there existed a mixture of wild and domesticated lands and of the fire practices of both settler and Indian, was the most volatile period. What was a controlled fire for an Indian burning off a hunting ground was, for a farmer, a wildfire in his field. The seminomadism required by Indian fire practices was replaced by sedentary settlements, with much greater constraints on broadcast fire.

The fire management problems of the region now are predominantly those of rural fire protection—the protection of structures on farms and small towns. Some burning of fields and woodlands remains, but intensive, high-technology agriculture has found replacements for the effects of open burning. Instead the main arena for free-burning fire is prairie restoration and management. Where areas have been set aside for the reestablishment of tallgrass prairie, broadcast fire has, along with other techniques, become an essential tool of restoration and management. In recreating a past biota it has been necessary to recreate equally the former fire regime, and thereby a pattern of prescribed broadcast fire.

### *Southern California*

The fire regime surrounding the broader Los Angeles basin is notorious for its volatile fuels (chaparral), its rugged topography, recurrent droughts, and episodic eruptions of strong foehn winds (Santa Ana). Charcoal preserved among varves within the Santa Barbara channel records fires as far back as 2 million years. Yet the circumstances surrounding those fires have not remained constant. Although basics such as the Mediterranean climate and foehn winds and probably some chaparral-like fuel complex have persisted, the regime itself is different. For this difference anthropogenic fire practices are a likely reason. For the extraordinarily intransigent fire regime that has developed since the 1950s, human activity is almost wholly responsible.

The historical record suggests widespread Indian firing, primarily for hunting and harvesting. Spanish and Mexican settlements, clustered around missions and presidios, did not fundamentally alter this pattern, though officials did make an attempt to control promiscuous Indian burning. With the expansion of herding by American settlers, the fire pattern continued—maintaining grass where possible, and keeping chaparral at a low stage of development by burning at the successional stage during which brush was just achieving ascendancy over grasses and forbs. A mixture of Indian and herding fire practices continued well into the late nineteenth century.

Big changes came after the Forest Reserve Act (1891) allowed the President to create forest reserves out of the public domain. At the insistence of irrigation agriculturalists and urbanites—both desperate for secure watersheds in the mountains—Southern California acquired some of the earliest reserves. A program of fire control and grazing control promptly came into being. With each decade the effectiveness of this program, largely under the jurisdiction of the U.S. Forest Service, improved, with steady decreases in the total number of acres burned and equally steady increases in fuel loads and the uniformity of chaparral cover. Large wildfires continued, intimately tied to this new fuel cycle but having radically new sources of ignition, both from people and equipment. Into this regime in the 1950s came dramatic changes in both the fuel complex and the patterns of ignition. The present fire regime dates almost precisely from these events.

To the old chaparral fuels were added houses, often outfitted with highly flammable wooden roofs, and to the old cycle of ignitions were added other sources from powerlines, power equipment, children, and arsonists. The product is a regime that behaves much like a wildland regime characterized by large, episodic wildfires, but whose values and whose sources of ignition resemble urban fire regimes. There is a corresponding mixture of urban and wildland firefighting techniques and jurisdictions. This regime is only partially the product of natural processes, and largely an outcome of historical events that resulted in a demographic shift to California after World War II. The contemporary fire regime is the product of two general processes: the reservation of mountain wildlands and the subsequent urbanization of lands that either border or interpenetrate those reservations. If either were transformed, so would the fire regime.

But unlike the tallgrass prairies, neither the reserved wildlands nor the settlements will eventually eliminate the other. The transitional period of mixing the two persists, and as long as it continues so will the pattern of fire characteristic of this regime. Its largely anthropogenic nature is reinforced by a definite pattern of arson fire, applied to areas and at times when lightning fire would be extremely unlikely. Some prescribed fire is used as a method of constructing fuelbreaks and of broadcast fuel reduction, and experiments in broadcast burning for fuel reduction are underway. But such fires often require mechanical or chemical pretreatment of fuels, and they challenge restraints on the production of air pollutants. A more general solution will require zoning, building codes, and, in the broadest sense, a change in the demographics that have inflated Southern California real estate values.

### *Lake States*

Early records from the Lake States report two broad biotas. To the south, there flowed a tallgrass prairie, and to the north, oak and then pine and then boreal forest. This forested region has an ancient record of fire—of large, episodic fires tied to the appearance of droughts, and, in more recent centuries, of a mixture of crown and surface fires associated with anthropogenic activities. Fire scars, stand age analysis, and palynological residues in bogs furnish a basic chronology for large fires and fire seasons. Documents provide some insight into Indian burning practices. Until the conclusion of the Civil War, the fire regime of the north woods was the product of lightning and Indian burning.

The advent of the railroad, the gradual shift of industrial logging from the northeast to the pineries of the Lake States, and hunger for farmland all came together by the late 1860s to produce a new fire regime. Farmers rushed onto clearcut lands: the plow, it was proclaimed, would follow the axe. The railroads that brought out lumber would later bring out agricultural products. But the process of conversion often failed to come in logical sequence and never became anything like complete. Instead, a mixture of logging slash, rough wooden towns, and half-cleared farms created a fuel complex of terrific loads and hazardous proportions, and a medley of landclearing fires, field burnings, hunting fires, and sparks from locomotive smokestacks and brakes established a new pattern of ignition. When the new fuel complex and the new ignition sources were brought together under drought conditions, the result was a fire regime of often deadly holocausts. As an historic event, the regime took shape around 1870, when all the basic components first came together, and it expired during the 1930s, when its sustaining elements were disentangled.

The era witnessed the largest and certainly the most destructive fires of American history. Somewhat analogous to the incomplete transformation of brush to suburb in Southern California, the conversion of forest to farm in the Lake States was never completed. Ultimately, however, industrial logging migrated to other regions; through a combination of economic incentives and forcible enclosures farming left the forests for former grasslands. With the development of robust state forestry departments and of mechanized firefighting equipment during the 1930s, this historic fire regime metamorphosed into a recognizably different regime. There were new forests, but forests of different composition such as jack pine and aspen, growing on old sites. The land was put to new uses—some industrial logging for pulp, but mostly devoted to recreational development. There was a new administrative presence in the form of state and federal agencies to furnish fire protection for rural and wild lands. Prescribed fire reappeared as a means of habitat management and wilderness preservation.

Thanks to human activities, however, the new fire regime as a whole was dramatically different from those that had preceded it. Certain elements remained, of course, but the fire regime as a whole —as a regime not merely a mosaic of fire environments—had a new character. For nearly 50 years the only large fires to appear came in 1976, the product of lightning, drought, and administrative confusion over fire policy.

### *Northern Rockies*

Lightning has historically been a primary ignition source in the Northern Rocky Mountains. Its effects have been—and still are—profound. But for many areas this source of fire has been compounded, and its effects modified, by anthropogenic fire practices. Most of these fire practices in the hands of the American Indian were concentrated in valleys and along routes of travel, though such fires could spread outward and upward into the mountains for considerable distances under the proper circumstances. Tied to grasslands, Indian burning appeared during the spring and especially the fall. The spread of anthropogenic fire was apparently enhanced by the advent of European trade goods and the horse, and it was greatly accelerated by the advent of American miners, ranchers, and homesteaders during the latter half of the nineteenth century. But by the 1890s the Indian reservation system isolated Indian fire practices, and the forest reservation system eliminated many frontier fire practices and began to confine even lightning fire. The resulting fire regime was distinctive not so much by its rearrangement of fuels or its introduction of new sources of ignition, but by its suppression of prior sources of ignition. The effects of removing fire could be as dramatic as the effects of introducing it.

The size of the forest reservation system made the potential impact of such a program (largely under the auspices of the U.S. Forest Service) extensive. But that very scale made the execution of such a program stubbornly problematic. Much of the backcountry of the Rockies was, in effect, settled in the name of fire control. The effectiveness of fire suppression varied, of course, with the density and sophistication of that administrative presence. Gradually, even control over lightning fires, including lightning fire busts on the order of several hundred fires in the course of a few days and arrayed deep in the mountainous backcountry, became relatively effective. For this outcome the development of aerial fire control such as smokejumping was especially responsible. Finally, however, the very size and remoteness of much of the system encouraged its designation as wilderness. For wilderness sites lightning fire was often considered an asset.

Moreover, there were fuel loads to consider. For most of the Inland Empire the success of fire suppression and the acceleration of logging had generated an alarming buildup of fuels, activity fuels and broadcast natural fuels both. For all these sites prescribed fire is encouraged—as prescribed natural fire in wilderness areas, and as prescribed management fires on activity sites and for wildlife habitat. Thus the appearance of effective fire suppression, often in advance of real settlement by Americans, transformed the preexisting fire regime. That regime is, in turn, being changed into one that—at least in areas—promotes various kinds of broadcast fire for specified objectives in wildland management.

The history of the Northern Rockies has some other peculiarities that bear on the question of fire history. The importance of a fire or of a fire regime will depend on the cultural context in which it occurs as much as on its ecological context. The 1910 fires had enormous consequences for the evolution of a national fire policy and fire program, much of which subsequently determined the character of the Northern Rockies fire regime. Yet this was only one complex of fires, and a somewhat anomalous one, whose presence is very frequently missed in the study of fire scars from the region. The significance of the fires, however, was less in their position within a master fire chronology or their biotic impact than in their timing within the historical history of American politics and the administrative context of the U.S. Forest Service. Most of that context came from outside the Northern Rockies proper, and the conclusions and programs drawn from the 1910 fires were in turn applied to other regions that had been unaffected by the holocausts. The fire thus not only brought about a biological response, releasing the potential in the biota, but encouraged a cultural response, liberating ideas and programs existing or latent in American society. As the 1910 episode shows, the impact of a fire may come indirectly through its influences on human institutions rather than directly on the biota.

### *The South*

Though lightning fires are prevalent in portions of the South (notably Florida), the fire regime that existed during the period of initial colonization by Europe was shaped by Indian firing practices. The basic division in lands was between those that were naturally drained, and hence fired on a one to three year cycle, and those that were naturally wet, and hence susceptible to fire only during severe droughts. Pockets of prairie (called "deserts" by European explorers) were common, and the oak and pine forests were typically free of heavy rough. Obviously, mature forests did not exist everywhere, or to the exclusion of a vegetative mosaic showing a mixture of many successional stages, but the record of a savannah-like forest is too extensive to ignore. Into this fire regime came European settlement.

The consequences of settlement were mixed, for the land was occupied for many purposes. Early frontiersmen adopted many Indian fire practices: fire for hunting, fire for habitat and range improvement, fire for slash-and-burn subsistence farming. Herding and hunting were common economies. Plantation crops such as sugar cane, indigo, and rice were initially tied to wetlands. But with the advent of cotton culture (especially after the invention of the cotton gin and steamboat) large-scale reclamation took place.

Clearing and cultivation changed the fuel complex, but it also drove frontier fire practices into the less easily cultivated piney woods. After the Civil War southern agrarianism persisted in various subsistence forms, surrounding small islands of industrialization and abandoned wildland. By the 1890s, however, industrial logging arrived in the South: the southern pineries reached peak production around 1909 and collapsed during the 1920s. By then the boll weevil swept southern cotton fields. Vast amounts of land were abandoned to tax-delinquency even before the Great Depression; much of the cutover lands and deserted fields were maintained for pasture or hunting through the periodic application of broadcast fire. Much reforestation had occurred during the depopulation following the Civil War, but much was also prevented on the newly opened lands because of heavy firing.

At the same time, as cutover lands and abandoned farmland appeared in quantity, so did formal fire protection. The U.S. Forest Service began acquiring lands after the Weeks Act of 1911 and, to a lesser degree, so did state foresters, with the added charge that they furnish rural fire protection. Foresters argued for fire control as a means of reforestation—and reforestation, via sustained-yield logging, as a means of industrialization. During the Depression large acreages were absorbed into federal and state protection and into corporate forests; techniques were developed to convert the southern pine into pulp, instead of saw timber; fire protection increased in intensity. The forests came back, but so did the southern rough, with the prospect of replacing commercially valuable pine with less valuable hardwoods and with the reality that fuel loads were increasing to the point that fire control was becoming almost impossible.

One outcome was the rapid reintroduction of fire in the form of prescribed burning. Prescribed fire was advocated for wildlife habitat maintenance, for range improvement, for silvicultural management, for control over successional tendencies, and for fuel reduction. But the application of broadcast fire was not restoring a previous environment, except perhaps in isolated pockets. The longleaf pine, for example, was by and large replaced by loblolly and slash pine. Such fires, moreover, did not occur at the hands of local farmers, herders, or woods-dwellers, but within the context of substantial fire protection organizations. Its history did, however, leave the South with a tradition of woodsburning. The rapidity of settlement during the days of King Cotton and the legacy of the Civil War had left large portions of the South in a quasi-frontier state, still dependent on herding and hunting and on their associated fire practices. The establishment of prescribed burning under industrial conditions has come not merely as a substitute for wildfire but as a competitor for rural woodsburning.

These sudden changes in ignition source and pattern, a process far from complete, have created the region's modern fire regime. The fire problem results from the persistence of a certain fuel complex, the southern rough, and of a certain socio-economic pattern of life, southern agrarianism. The significant events of southern land management—from the demand for cotton culture to the introduction of industrial forestry—generally originated from outside the region, but often continued past their time of real applicability.

The management of fire regimes in the South has thus depended on both the transformation of socio-economic conditions and on the control over the very rapid tendency towards biological succession. The first has tended to eliminate a legacy of woodsburning, and the second to substitute a pattern of prescribed burning. The mosaic that makes up the fire regime of the modern South does not represent simply states of biological succession but of socio-economic development. That regime is as much a record of cultural as of natural history.

### *Southwest*

The fire regimes of the Southwest are, more than those of any other region, dominated by lightning fire. The extent of lightning fire is practically identical with the range of the ponderosa pine. In part, this fire pattern reflects a unique geography of deserts, mountains, and a summer rainy season, and in part, a sparsely settled landscape. Yet the region has not been without anthropogenic fire. In historic times this meant primarily the Apache Indian.

The Apache was both drawn and driven to the Southwest. The acquisition of the horse by the Comanche cut off the western Apaches from their Plains relatives. At the same time the presence of Spanish livestock in abundant herds provided an attractive alternative to hunting wildlife on the Plains. In the process of relocation the Apache never really lost their heritage as prehorse hunters on the grasslands. They were as liable to eat a horse or mule as to ride one. They merely exchanged the grasslands of the Great Plains for those in the mountains of the Southwest, and the wild grazers of the Plains for domesticated livestock that the Spanish herders encouraged along the northern frontier of New Spain. Their fire practices were essentially those of grassland tribes, and in the Southwest the Apache found ample grasslands: grasslands flooded the intermountain basins, carpeted the savannahlike floor of ponderosa pine forests, and flowed into deserts during wet years.

The combination of Apache and lightning fire appeared during the sixteenth century and became entrenched by the middle of the nineteenth. Despite a war of extermination waged, first, by other tribes, then by the Spanish, then by the Mexicans, and (following the Mexican War) finally by the Americans, the Apache survived. Livestock, which elsewhere in the United States had drastically altered biotas and fuel complexes, were kept to low population levels by intense predation on the part of the Apache. But by the 1870s the Apaches were broken; by the 1880s only a few fugitive bands remained to be hunted down. Herds of cattle moved in from Texas and herds of sheep and goats from Utah and California. Grazing was intense on the precise component most essential to the old fire regime: grass.

With heavy grazing, with the reservation of virtually all forest lands into protected areas during the late nineteenth century, with demands for watershed protection by irrigation agriculturalists, deliberate fire control came to the Southwest, and with it a new fire regime. As in the Northern Rockies, this regime was not shaped by adding new ignition sources so much as by removing old ones. In some instances, active fire control was the means; more often, simply the reduction in grass cover brought about by grazing. The outcome was a dramatic recession in grasslands. Succulent deserts encroached from lower elevations; brush crept over hillsides; ponderosa forests swelled into dense dog-hair thickets and pinyon–juniper groves splashed across the landscape. Range deteriorated, site productivity diminished, and fuels built up to alarming quantities.

This regime is now changing once again as prescribed fire—natural and scheduled both—is introduced. In wilderness areas the abundant lightning fire of the region again burns. In the ponderosa pine belt across the Mogollon Rim broadcast fire is applied during the late fall for fuel reduction. In brush-infested grasslands, broadcast fire is used, along with chemical supplements, to control the spread of woody plants, and on brushy watersheds similar techniques attempt a type conversion back to grass. On areas invaded by pinyon and juniper the trees are knocked down, left to dry, and then burned with the expectation that grasses will once again establish themselves.

All of this reintroduction of fire, however, comes with a significant difference from what went before. Here it is conducted within the confines of a program that can, within limits, choose how much fire and what kind of fire it wants, and it comes into a biota and fuel complex rather different from that which preceded it. As this program reaches a new equilibrium of fire use and fire suppression, it will define yet another historic fire regime. For the parameters of this new regime the history of land management agencies and of national land politics will be more significant than the natural history of climate, vegetation, and southwestern geology.

## *REFERENCES*

A survey of fire history techniques and results is given in Stokes and Dieterich (1980), "Proceedings of the Fire History Workshop." Appended to the reports is a lexicon of common terms, defined according to the consensus of participants. For the particular techniques of fire scar analysis, see Arno and Sneck (1977), "A Method for Determining Fire History in Coniferous Forests of the Mountain West." For palynology and sedimentation, consult Wright (1981), "Role of Fire in Land/Water Interactions." For land survey records, see Spurr and Barnes (1980), *Forest Ecology,* and Grimm (1979), "An Ecological and Paleoecological Study of Vegetation in the Big Woods Region of Minnesota." Many rephotographic surveys have been conducted. For the Great Plains, see Phillips (1963), "Photographic Documentation. Vegetation Changes in Northern Great Plains"; for the Black Hills, Progulske (1974), "Yellow Ore. Yellow Hair. Yellow Pine"; for the Northern Rockies, Gruell (1979), "Fire's Influence on Wildlife Habitat on the Bridger–Teton National Forest, Wyoming," 2 vols; for the Southwest grasslands, Humphrey (1963), "The Role of Fire in the Desert and Desert Grasslands Areas of Arizona."

Most national parks, and many national forests, have undertaken fire history surveys. Some of these results have been published, and others kept in-house. Most of the major symposia include papers that describe fire history either directly or indirectly. See Mooney and Conrad (1977), "Proceedings of the Symposium on the Environmental Consequences of Fire and Fuel Management in Mediterranean Climate Ecosystems"; Mooney et al. (1981), "Proceedings of the Symposium on Fire Regimes and Ecosystem Properties"; Wright and Heinselman (1973), "Ecological Role of Fire in Natural Conifer Forests of Western and Northern America."

The cultural history of fire has been surveyed in Pyne (1982), *Fire in America: A Cultural History of Wildland and Rural Fire.* The book's bibliographic essay gives a reasonably comprehensive survey of sources—for agency policies as well as regional histories. For the impact of wilderness on contemporary land management, see Nash (1982), *Wilderness and the American Mind,* 3rd ed; Allin (1981), *The Politics of Wilderness Preservation*; Dana and Fairfax (1979), *Forest and Range Policy,* 2nd ed.; Hendee et al. (1979), *Wilderness Management.* For an overview of the Forest Service, consult Steen (1976), *The U.S. Forest Service: A History.*

Allin, Craig, 1981, *The Politics of Wilderness Preservation* (Westport, Connecticut: Greenwood Press).

Arno, Stephen F., 1980, "Forest Fire History in the Northern Rockies," *Journal of Forestry* **78**(8): 460–465.

Arno, Stephen F. and Terry D. Peterson, 1983, "Variation in Estimates of Fire Intervals: A Closer Look at Fire History on the Bitterroot National Forest," U.S. Forest Service, Research Paper INT-301.

Arno, Stephen F. and K.M. Sneck, 1977, "A Method For Determining Fire History in Coniferous Forests of the Mountain West," U.S. Forest Service, General Technical Report INT-42.

Clar, C. Raymond, 1959–1969, *California Government and Forestry,* 2 vols. (Sacramento: California Department of Forestry).

Clepper, Henry, 1971, *Professional Forestry in the United States* (Baltimore: Johns Hopkins University Press).

Dana, Samuel T., 1956, *Forest and Range Policy. Its Development in the United States* (New York: McGraw-Hill).

Dana, Samuel T. and Sally Fairfax, 1979, *Forest and Range Policy,* rev. ed. (New York: McGraw-Hill).

DuBois, Coert, 1914, *Systematic Fire Protection in the California Forests* (Washington: U.S. Forest Service).

Foster, D.C., 1976, "Lower LaSalle Lake, Minnesota: Sedimentation and Recent Fire and Vegetation History" (M.S. thesis, University of Minnesota).

Gill, A. Malcolm, 1974, "Toward an Understanding of Fire-Scar Formation: Field Observation and Laboratory Simulation," *Forest Science* **20**: 198–205.

Grimm, E.C., 1979, "An Ecological and Paleoecological Study of the Vegetation in the Big Woods Region of Minnesota," (PhD thesis, University of Minnesota).

Gruell, George E., 1979, "Fire's Influence on Wildlife Habitat on the Bridger–Teton National Forest, Wyoming," Vol. I, "Photographic Record and Analysis," U.S. Forest Service, Research Paper INT-235.

———, 1979, "Fire's Influence on Wildlife Habitat on the Bridger–Teton National Forest, Wyoming," Vol. II, "Changes and Causes, Management Implications," U.S. Forest Service, Research Paper INT-252.

Heinselman, Miron, 1981, "Fire Intensity and Frequency as Factors in the Distribution and Structure of Northern Ecosystems," pp.7–57, in Mooney et al., "Proceedings of the Conference on Fire Regimes and Ecosystem Properties," U.S. Forest Service, General Technical Report WO-21.

Hendee, John C. et al., 1979, *Wilderness Management,* U.S. Forest Service, Miscellaneous Publication No. 1365 (Washington, D.C.: Government Printing Office).

Hernby, Lloyd A., 1936, "Fire Control Planning in the Northern Rocky Mountain Region" (Misseula, Montana: U.S. Forest Service).

Humphrey, Robert R., 1963, "The Role of Fire in the Desert and Desert Grassland Areas of Arizona," pp. 44–61, in Tall Timbers Research Station, *Tall Timbers Fire Ecology Conference, Proceedings,* Vol. 2 (Tallahassee, Florida: Tall Timbers Research Station).

Kilgore, Bruce, 1970, "Restoring Fire to the Sequoias," *National Parks and Conservation Magazine* **44**(277): 16–22.

Kilgore, Bruce and Dale Taylor, 1979, "Fire History of a Sequoia Mixed Conifer Forest," *Ecology* **60**(1): 129–142.

Mooney, H.A. and C. Eugene Conrad (coordinators), 1977, "Proceedings of the Symposium on the Environmental Consequences of Fire and Fuel Management in Mediterranean Climate Ecosystems," U.S. Forest Service, General Technical Report WO-3.

Mooney, H.A. et al. (coordinators), 1981, "Proceedings of the Symposium on Fire Regimes and Ecosystem Properties," U.S. Forest Service, General Technical Report WO-21.

Nash, Roderick, 1978, "Historical Roots of Wilderness Management," pp. 27–40 in John Hendee et al., *Wilderness Management,* U.S. Forest Service, Miscellaneous Publication No. 1365 (Washington, D.C.: Government Printing Office).

———, 1982, *Wilderness and the American Mind,* 3rd ed. (New Haven, Connecticut: Yale University Press).

Phillips, Walter S., 1963, "Photographic Documentation. Vegetation Changes in Northern Great Plains," Agricultural Experiment Station, Report 214 (Tucson: University of Arizona).

Progulske, Donald R., 1974, "Yellow Ore, Yellow Hair, Yellow Pine. A Photographic Study of a Century of Forest Ecology," Agricultural Experiment Station, Bulletin 616 (Brookings: South Dakota State University).

Pyne, Stephen J., 1982, *Fire in America: A Cultural History of Wildland and Rural Fire* (Princeton, New Jersey: Princeton University Press).

Schiff, Ashley, 1962, *Fire and Water: Scientific Heresy in the Forest Service* (Cambridge, Massachusetts: Harvard University Press).

Spurr, Stephen H. and Burton Barnes, 1980, *Forest Ecology,* 3rd ed. (New York: Wiley).

Steen, Harold K., 1976, *The U.S. Forest Service: A History* (Seattle: University of Washington Press).

Stokes, Marvin and John Dieterich (coordinators), 1980, "Proceedings of the Fire History Workshop," U.S. Forest Service, General Technical Report RM-81.

Tall Timbers Research Station, 1962–1976, *Tall Timbers Fire Ecology Conferences, Proceedings,* Vol. 1–15 (Tallahassee, Florida: Tall Timbers Research Station).

Taylor, Dale L., 1981, "Fire History and Fire Records for Everglades National Park, 1948–1979," South Florida Research Center, Report T-619 (Homestead: National Park Service).

Van Wagner, C.E., 1978, "Age Class Distribution and the Forest Fire Cycle," *Canadian Journal of Forest Research* **6**: 220–227.

Widner, Ralph (ed.), 1968, *Forests and Forestry in the American States: A Reference Anthology* (Washington, D.C.: National Association of State Foresters)

Winters, Robert K. (ed.), 1950, *Fifty Years of Forestry in the U.S.A.* (Washington, D.C.: Society of American Foresters).

Wright, H.E., 1981, "The Role of Fire in Land/Water Interactions," pp. 421–444, in Mooney et al., "Proceedings of the Symposium on Fire Regimes and Ecosystem Properties," U.S. Forest Service, General Technical Report WO-21.

Wright, H.E. and M.L. Heinselman (eds.), 1973, "Ecological Role of Fire in Natural Conifer Forests of Western and Northern America," *Quaternary Research* **3**(3).

*Chapter Seven*

# *The Administration of Fire Regimes*

A fire regime is typically heterogeneous in geography and history, and the concept of a fire regime is in reality a statistical composite. Just as a fuel complex is composed of smaller elements or ensembles of elements, so a fire regime consists of smaller components which, taken as a whole, show a recognizable unity. The concept of a fire regime may be expanded into rather large, amorphous landscapes or restricted to tiny units with more definite boundaries. The choice of scale is somewhat arbitrary. As a practical guide, a fire regime consists of a sufficiently large portion of the country such that the individually distinct elements of the vegetative mosaic become small relative to the whole ensemble under consideration and such that fires behave relative to the mosaic as a whole rather than to the separate biotic components within it. Fire climates and ecosystems describe certain natural boundaries, while fire history and fire practices identify cultural boundaries.

The study of those natural boundaries belongs in the field of fire ecology. A fire regime may even be defined in terms of its shared biological processes and its common history. But the study of cultural boundaries involves the study of those institutions through which fire practices are disseminated or regulated. These boundaries are both geographical and historical. For twentieth-century America they coincide with the administrative history of the major fire management agencies. Since fire regimes result from patterns of fire use and control and since patterns of use and withdrawal are set by institutions, these agencies have the power to shape fire regimes.

Between a human institution and a fire regime, then, there develops a certain reciprocity. On one hand, a fire regime may provide a convenient unit of management for administrative purposes; on the other, the imposition of an administrative unit can shape a regime. In practice fire regimes rarely coincide with administrative boundaries. In some cases an administrative unit may only contain a portion of a single regime. In most instances, an administrative unit will contain several regimes, or a regime will be managed according to several units of the same or different agencies. The management problem, in brief, is not simply a question of reconciling fire practices to the fire regime at hand but of reconciling—through appropriate fire practices—the fire regime to the administrative context in which it occurs.

That adjustment is not easy. It is through administrative institutions that objectives for fire management are set, that plans are developed to meet objectives, and that the future characteristics of the fire regime will be determined. Land which would in certain circumstances belong to a single fire regime may show dramatically different properties because of its management by different institutions, differences sufficiently great to break up one fire regime into several. Less common but possible is the situation whereby various fire regimes become more homogeneous because of the application of similar fire practices by an single institution, ultimately taking on the properties of a single regime.

The administration of fire regimes includes the study of how anthropogenic fire practices are organized, by what organs of society and government they are applied, by what concepts that application is directed. As more and more land is put under organized fire management and as the impress of management programs intensifies, the study of fire regimes becomes to a greater and greater extent the study of anthropogenic fire practices, and the study of fire practices the study of the institutions that direct them. Certainly the history of fire regimes in twentieth-century America is very largely the history of those institutions responsible for managing them. These institutions did more than maintain a status quo: through the use and withdrawal of fire they actively shaped the fire regimes under their jurisdiction. Paradoxically, however arbitrary the designation of administrative units might be relative to natural fire regimes, the natural and human systems tend to become more identical over time.

The administration of fire regimes has several dimensions. It includes stated or implied objectives. It advances plans, or methodologies of planning, to satisfy those objectives. It expresses itself as a structure of organizations, interrelated to a greater or lesser extent. And it has a history. Plans, objectives, organizational structure—all appeared simultaneously in the United States, all evolved synchronously, and all bear the imprint of that origin. Some of the characteristics of fire administration in the United States derive from the unique properties of fire as a natural process, and others from the unique circumstances by which fire protection came about.

## *7.1 OBJECTIVES OF FIRE MANAGEMENT*

At the foundation to all thinking about fire management objectives and plans is the fact that mankind has the ability to stop and start fires. Any statement of goals or theory about the adequacy of fire management must originate with the techniques by which accidental ignition can be prevented, wildfire suppressed, and prescribed fire substituted for wildfire. Without these capabilities, fire protection would resemble flood control or tornado warning systems, a program of prediction and defense. Instead fire management can exercise some control over how fires start, how they spread, and how they can be used for human purposes.

From its beginning fire protection has pursued a four-part strategy: to prevent unwanted ignition, to modify fire behavior and effects by altering the environment in which a fire burns, to suppress wildfire, and to exploit controlled fire for use. All of these approaches are interrelated. In one respect this is simply a program of domestication, by which controlled fire is substituted for wildfire. Nothing quite so simple exists in actual practice because a controlled fire does not breed true; there is always the possibility that it will transform into a wildfire in ways that are not predictable. For this reason fundamental changes come through modifications of the fire regime, not simply by the actions taken with respect to individual fires. That regime, of course, includes people, and changing the cultural context of fire practices will bring about changes as surely as converting its fuels. For humans to intervene into the pattern of fire starts and fire spread that are characteristic of a fire regime is to change that regime. But it is equally the case that by changing the regime, it is possible to bring about changes in the patterns of fire starts and spread.

Common to all fire practices are certain physical properties of wildland fire. That fires begin as point sources means that both prescribed fire and the control of wildfire are possible. Only ignition is needed to initiate a fire, the fire will propagate on its own. Conversely, all wildfires can be extinguished simply, provided that they are attacked while they remain small. Wildfire spread and intensity—and hence fire effects, fire damages, and fire suppression costs—cannot be expressed as a linear function. The secret to controlling fire is to stop small fires before they can make the transition to large fires. It is true that routine surface fires can be devastating to reforestation programs, whether or not they show large fire characteristics. But even in this case, steady-state fires are simpler to control while their perimeter is small than when they grow in size. All this puts a premium on speed of attack.

Similarly fires, and fire seasons, are not distributed uniformly or in a simple binomial mode. Most fire effects, biological and economic both, occur from large fires and from large fire seasons. Through the use of historical records, it is possible to determine a level of manning suitable to the regime, but it is not yet possible to determine which year will present abnormal problems for fire control or exceptional opportunities for fire use. The result is a pattern for which averaged conditions are not a useful guide. The exceptional years account for most of the costs and damages.

These properties are inherent to wildland fire. The unpredictability of fire growth—largely a function of weather—means that the simple solution to fire management is to ignite fires under reasonably controlled conditions, often with fuel preparations; to attack all wildfires while they remain small; and to plan for adequate attack forces to meet an average worst year. The agency of attack has always included firefighters. They may be supplemented by other devices, but ultimately the management of fire has meant the management of firefighters, the creation of plans to better organize firefighters, and the development of transportation systems to more quickly move them to the scene of a fire. This emphasis makes the management of wildland fire different from the management of other natural processes such as floods, windstorms, or drought.

But fire management for its own sake has never been a sufficient justification for a program of fire management, and the techniques of fire management by themselves have never constituted a satisfactory theory of administration. Always there have been other considerations: historical, political, administrative, and economic. These considerations have shaped the institutions actually involved in fire management, determined their policies, and interceded in the accomodation of means and ends that is the essence of fire planning.

### *Historical Considerations*

Many features of fire management in the United States derive from its unique historical role. Organized fire protection arose from the transformation of an agricultural to an industrial society. Its historical mission was, first, to break down an agricultural cycle of fire and, second, to extend the principles of organized protection to as many lands as possible. Its historical experience, that is, has been one of growth. Naturally, fire protection emphasized prevention and suppression programs as a means of breaking the old pattern of ignition, and nearly all theories of fire protection have correspondingly argued for an increase in suppression resources. It is another historical fact that these resources have, in general, been adequate to support this strategy. It was never necessary, for example, to pursue alternate strategies of fire protection—fuel reduction by prescribed broadcast fire, zoning to segregate fire hazards from fire ignitions, volunteer fire brigades rather than paid forces, and so forth. This growth in fire protection involved processes of intensification and extension, a strengthening of control in those areas already under organized protection and an expansion into areas not yet protected. Only when organized fire protection completed this mission—when the lands under protection reached saturation—did the question of diminishing returns and limited resources really become a problem and did other means of manage-

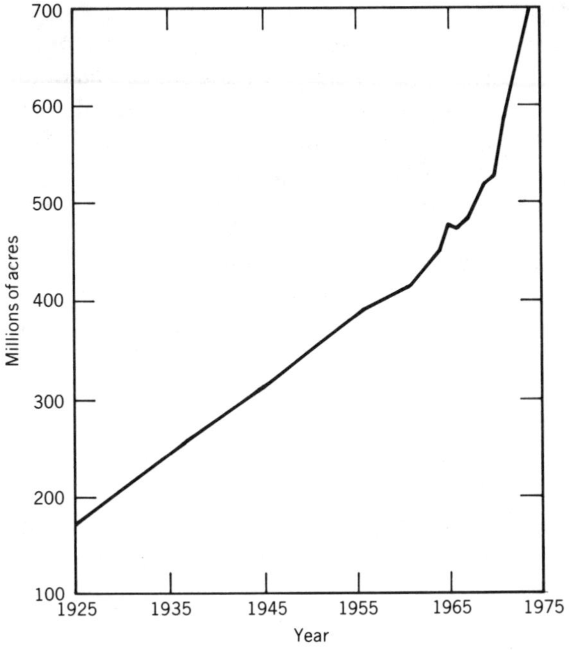


*Figure 7-1. Growth in protected lands. From U.S. Forest Service, Cooperative Fire Program.*

ment become serious alternatives (see Figures 7–1 to 7–4).

There were other features to the evolution of fire protection, also unique. That foresters controlled the machinery of wildland fire protection meant that theories of fire protection would develop out of the tradition of academic forestry and that institutions for fire management would develop out of industrial forests and government agencies charged with forest management. To realize that organized fire protection grew from these beginnings explains much about its character in the United States today: its goals, techniques of planning, style of research, and institutional arrangements. Equally, however, fire protection had to respond to events originating outside its particular sphere. Time and again episodes of major significance to American history intruded into the internal development of fire protection. Sometimes good, sometimes ambiguous in its consequences, always requiring assimilation into the existing theoretical and institutional structures—such episodes continually

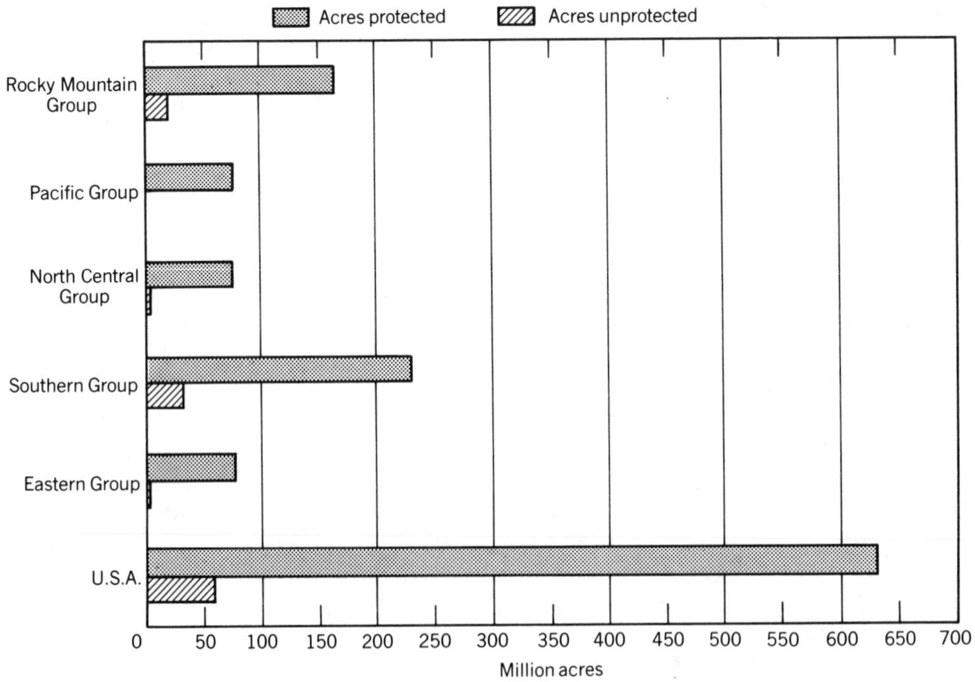


*Figure 7-2. Protected and unprotected lands, 1972. From U.S. Forest Service, Cooperative Fire Program.*

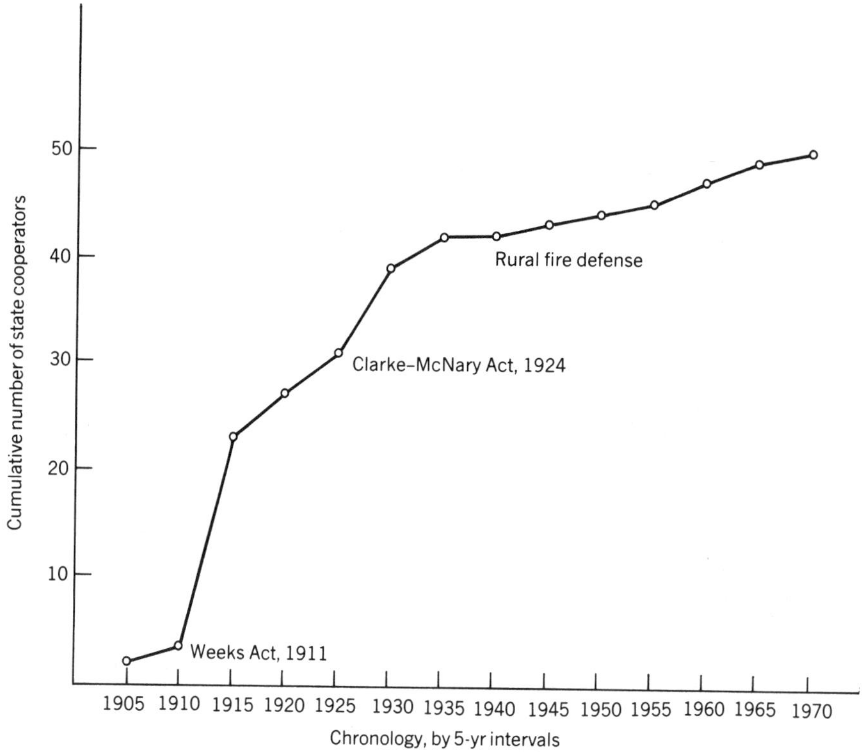


*Figure 7-3. Growth in cooperative fire protection. From 1911–1924 covers the era of the Weeks Act; from 1924 to the present, the era of the Clarke–McNary Act. The Cooperative Forestry Act of 1978 began the withdrawal of federal funding. Data from U.S. Forest Service, Cooperative Fire Program.*

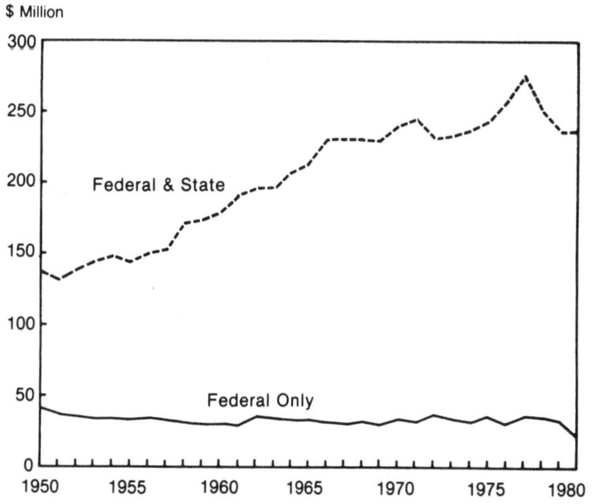


*Figure 7-4. Growth in rural fire protection. From Myles (1981).*

modified the purposes to which fire protection was put and altered its techniques. Perhaps the most fundamental manifestation is the paradox under which wildland fire protection originated. Wildlands were not reserved and then protected as their value dictated, but were set aside in order to protect them.

The contrast to urban fire protection is striking. Unlike urban fire services (which began in large measure as an economic institution, a response to fire insurance) wildland fire protection began as a political institution, necessitated by the reservation of public lands from settlement. Much of the argument for reserving these lands was the desirability of protecting them from fire. With the establishment of emergency funding for firefighting in 1908, the programs were even further removed from a strict economic rationale. These emergency monies could be applied to suppression (and later presuppression) activities, reinforcing the belief that the best way to fashion a new fire regime was to break the ignition pattern of the old one.

Its history then explains much about the confusion of goals and plans apparent in agency histories. The determination of objectives was always complicated by the larger imperative to expand the system into new, unprotected areas; the development of plans, by the need to break the old cycle of fire use; the creation of institutions, by their evolution from forestry and a national forest reservation system. That history was often spasmodic, like the history of fire itself. Agencies struggled to cope with problems (or possibilities) imposed from outside as a result of such things as war, depression, affluence, social philosophies, popular movements, and political reforms. The creation of the CCC, for example, had profound effects on Forest Service fire programs and policies. The story was never simply an internal evolution, the progressive reconciliation of a fire regime with the tools available to manage it or the policies posited for it. Instead a coevolution developed, a rough correspondence of means and ends. As the institutional and environmental situations changed, so did theories of fire management and concepts of fire economics.

The case of fire economic theory illustrates, again, that this history was emphatically not a case of applying, with progressively better fit, an ideal theory to a fixed situation. Rather it involved the correspondence of one variable, fire economics, to another variable, fire management programs. Often fire protection resulted from noneconomic events, such as legislation to which economic criteria had to be applied after the fact. Often, too, areas mandated for protection had little if any economic value, as measured by a market economy; rather scenic, recreational, cultural, historic, and aesthetic criteria dominated. Occasionally, too, fire agencies were presented with special "surpluses"—resources available for suppression—which had to be used, and for which economic criteria were of secondary concern in determining how they were used.

Overriding all these considerations, moreover, was the need to actually manage fire, to act. Especially as new lands were brought under protection, the need for action was immediate. The realities of field administration prevailed over theories, and programs dominated planning. Programs tended to be empirical rather than abstract, local rather than typical. Not until there existed a more stable pattern of administration, largely brought about by an end to the expansion of the system, could there develop close economic analysis, and then for the purposes of allocating fixed resources within an agency rather than between agencies.

With the completion of that geographic and institutional expansion, there has occurred a reexamination of the goals and theories behind fire management systems. The sources of this reevaluation are varied—political pressures, in the form of legislation; economics, as protection programs encountered the point of diminishing returns; strategic, the shift from fire suppression after the fact to fuel modification before the fact. There were administrative pressures to consolidate programs between agencies, and a powerful campaign to limit the autonomy of fire programs within agencies. The intention was to bring fire management into alignment with larger political and administrative goals. A degree of consolidation has resulted: the integration of fire suppression with fire management, of fire management with land management, of fire management institutions with interagency bodies having mutual programs and shared resources.

For convenience the important goals (or determinants of goals) may be grouped somewhat arbitrarily into three categories: political, administrative, and economic. Only rarely have such considerations been logically compatible among themselves, and frequently they conflict with the peculiar properties of fire, fire regimes, and fire management. In practice a rough accomodation takes place, whereby the various intentions behind policies are brought into a pragmatic working balance rather than a strict logical consistency.

### *Political Considerations*

Many of the goals of fire management have been, and will continue to be, mandated by political acts. Essentially this means legislation and executive orders. Executive orders were important in establishing reservations out of the public domain, principally through the Forest Reserve Act (1891) which led to the national forest and grassland system and the Antiquities Act (1906) which resulted in many national monuments. In this case fire protection came to selected areas by a process of land reclassification, of moving land from unprotected to reserved status. Another approach was to expand the effective jurisdiction of a fire agency or statute. An example is the executive order that gave the Forest Service responsibility for rural fire defense during World War II.

But the most pervasive form of political intervention has been legislation. Statutes may specify a particular goal or they may direct an agency to satisfy general intentions such as providing adequate protection to the resources under the agency's jurisdiction. The latter process is the most common, and it is usually embodied in the organic act that established the agency. It leaves to the particular agency the determination of policy by which to meet the stated objectives. This introduces an important flexibility into the program, but it also requires considerable innovation on the part of the agency. Few legislative acts translate directly into a program of fire management. In still other cases, special directions for overall land management and procedures for planning may be legislated that demand the revision of existing programs. Examples include the Forest and Rangelands Renewal Resources Act (1974) and National Forest Management Act (1976) that required the Forest Service to produce a Forest Plan for each national forest. The fire plan would follow from the Forest Plan.

As more statutes give legislative directions, the effects may thus be mixed. In some areas, the agency may acquire statutory authority to expand its program; while in other areas, it may be restrained. An example of the first is the Weeks Act (1911), later succeeded by the Clarke-McNary Act (1924) which allowed for the expansion of the national forest system to lands other than those currently within the public domain and provided for a program of cooperative fire protection with the states. A case of the latter is the Wilderness Act (1964) which had the effect of prohibiting certain kinds of fire management techniques within designated wilderness areas.

As legislation proliferates, moreover, the possibility of contradictory goals may also increase. Thus the need for prescribed fire in a wilderness area, as interpreted under the provisions of the Wilderness Act, may conflict with the goal of air quality, as set by the Clean Air Act; the Forest Service, or National Park Service, which administers the first, may find itself in dispute with the Environmental Protection Agency, which oversees the latter through state air quality boards. Very little of such legislation is made from the standpoint of fire management considered in itself, or strictly from the needs of particular agencies. The acts represent national or state perspectives, representative of a political constituency which is generally not involved in fire management. It is the business of the fire agencies to work out some accomodation.

This can be a complex process. Different land agencies exist because they have different missions. Even for the same land, or the same kind of land, radically different programs and policies may be set in motion because of provisions in the organic acts of the agencies involved. The U.S. Forest Service must administer land for multiple-use purposes; the National Park Service, for the preservation and enjoyment of the natural scene; the Bureau of Indian Affairs, as an instrument of tribes who are themselves considered wards of the state. That all three agencies are charged with fire protection may mean that they develop remarkably different fire programs which seek to harmonize the agency charter, its fire mission, and the characteristics of the fire regime under management. For otherwise homogeneous land, the presence of these agencies will result in different fire practices and, over time, in different fire regimes. This consideration will be reinforced by the different land use programs each agency supports, projects that will affect fuels.

Apart from direct forms of political intervention, there are other forms of political influence, no less compelling in the actual development of programs. The budget of fire agencies is set by legislative act. So are civil service regulations, which guide the recruitment and hiring of personnel. Other programs—originating outside of fire considerations per se—can also enter into fire management through the political process. The presence of the Civilian Conservation Corps, for example, made possible a tremendous expansion in permanent fire protection facilities and in temporary capabilities for staffing them. World War II set up an alliance between fire agencies and the military to defend the United States against anticipated enemy incendiary attacks, a liaison that persisted with widespread consequences after the war. Legislation allowing priority access to surplus military equipment, and statutes encouraging interstate fire compacts altered the institutional landscape of fire management. Popular movements for conservation, reforestation, and wilderness have all projected themselves, through political constituencies, onto fire management programs. It is not an exaggeration to say that fire protection by federal and state agencies is a political institution. The influence of legislation on fire programs is typically indirect, but often powerful. The great reforms in fire policy and practice characteristic of the 1970s came hand in glove with a slough of environmental legislation (Barney and Aldrich, 1980).

How the parameters of the political process are reconciled with the constraints imposed by the properties of fire and fire regimes belongs in the domain of fire management theory, planning, and agency policy. The physical properties of fire cannot be ignored. Because fire propagates, the process of fire protection must also spread. It must continually expand to new frontiers to protect its old ones. Lacking a structure of forest insurance but confronted with the reality of public reservations that required some administration, political institutions, not economic ones, became the medium for that expansion. Many public lands, moreover, were protected for reasons that could not be easily translated into economic terms. Then, too, the process began at the federal level. Prior to the reservation system there was little interest in, or mechanisms for, the protection of wildlands *qua* wildlands. The basic infrastructure was developed by federal agencies. All American fire agencies are in some respect dependent on that federal presence.

### *Administrative Considerations*

All fire management institutions show similarities that are attributable to fundamental properties of wildland fire behavior and the dynamics of fire regimes. But they also show profound differences as a result of their charges, as embodied in their organic acts; of their history, both their evolution as a fire agency and their experience as a bureaucracy; of their structure, traditions, self-perception, and governmental context. Differences in their bureaucratic environment mean that, even for a common fire regime, there may exist startlingly different programs. Although legislation establishes the general goals, and in some cases the methods, appropriate to the agency, the particulars—strategies, policies, plans—are set by the responsible agency. These goals are in turn filtered down layer by layer through the institution. No agency exists solely for the purpose of fire protection, so each must develop its own institutional organ for implementing the fire programs it elects. At some level the concepts and policies are applied. It is at this level that the tension between thought and action, policy and program, intention and deed will be most keenly felt.

The exact location of this tension will vary by agency, but it will exist in all of them. In one part of the agency, the need for action will be overwhelming: fire is at hand, and some decision or some activity is mandatory. Not to act is itself an act. In another part of the agency, the fire program exists only as a conceptual problem. Between its mandated purposes and its imperative to act, the agency must function. Accordingly, it establishes its own subgoals, expressed as agency objectives, its special methodologies, manifest as programs or planning guidelines, and its economic decisions, reflected in its budget. These procedures are commonly assembled into a manual. The intention behind this standardization is, or ought to be, to free the lower field units for action. Among federal bureaus, the fire programs of the Forest Service, Bureau of Land Management, National Park Service, Fish and Wildlife Service, and Bureau of Indian Affairs are distinct, despite shared fire regimes and mutual suppression resources. Similar differences characterize state programs, which show among themselves a great variety of forms and purposes.

General intentions are set by the Congress, but actual programs are worked out at the agency level or within the fire management division of the agency. To a strong degree the development of fire programs for a particular area reflects agency history and character, not simply local history. Agency experiences in one part of the country may be transferred, for better or worse, to other parts. Despite these distinctions, a national infrastructure for fire management does exist as a result of legislation, interagency and interstate agreements, and the need to meet common fire emergencies with shared resources.

All agencies must act within constraints set by outside interests, but all must act. Unlike many programs, the response to a fire cannot be deferred. Thus all fire agencies share common strategies, exploit the same suppression resources, exchange information, and tend to judge success according to similar critieria. Regardless of theory or mandated objectives, the reality of fire behavior and fire management technique has led all fire agencies time and again to measure their success by *performance criteria.* Fire effects have been too nebulous to use as a standard, and economic criteria, too imprecise. It is almost impossible to determine ecological benefits and losses or to measure liabilities and damages. But it is possible to measure certain properties of the fire such as burned area that bear some relationship to fire loss, to observe certain properties of the suppression organization such as response time that bear some relationship to its efficiency, and to record certain characteristics of prescribed burning such as acres burned that relate to the success of the program. The great constant in all programs is that something must be done, some action taken. Actions speak louder than theories. Fire prevention programs are thus measured by the number of fire starts, fuel modification programs by the number of acres treated, fire control programs by amount of burned acreage or the time of control.

There are historical reasons for this outcome, too. Theories about fire damages and fire protection followed from the facts of fire control. Protection did not arise out of theory, but out of the need to control and use fire. Theories about fire damages and fire protection objectives followed after the fact. Protection was mandated first, then goals posited after the programs were in operation. The number of fires and the number of acres burned were obvious, easily measured criteria. Somewhat arbitrarily the U.S. Forest Service early established an allowable burn of 0.1% of the total acreage under protection. This was for sites with high-value timber; for other lands, the allowable burn was increased by some ratio to this standard. For nontimber resources no calculus of values existed. Similarly, it was known that control of any fire was easier the sooner suppression crews arrived at the scene. Taking typical rates of spread for fires that occurred on high-value timberlands, it was possible to calculate an ideal time of control. The elapsed time between start and initial attack provided a measure of organizational efficiency. Obviously, more efficient initial attack implied fewer burned acres. But the exact relationship is difficult to establish—even more so the relationship between such measurements and the damage caused by the fire or averted by fire control. Similar questions apply to fuel modification programs, to prescribed burning projects, and to prevention programs. The general solution was, for many years, to devise programs that would reduce the number of fire starts, the number of burned acres, and the amount of money spent in fire protection, especially in fire suppression.

Again and again this meant control of wildfire within a certain period of time. The hour control program, the 10 AM Policy, the concept of conflagration control—all focused on time criteria that would, on the average, translate into smaller burned acreages, damages, and costs. In this way one readily measurable criteria, time of control, could substitute for many related, but amorphous and unmeasurable criteria such as ecological damage and economic effectiveness. The simplicity of the system made it bureaucratically useful. But as costs escalated, as the potential for fire to bring benefits as well as damages became apparent, as many agencies with different mandates tried to work out the terms for mutual cooperation, as new legislation established new goals and procedures, and as fire protection accomplished its historic expansion into previously unprotected areas, more complex criteria were demanded.

The particulars of this evolution are less important than the fact that such criteria were set within the agency or between agencies. Legislation only established broad goals; the agency determined its particular missions, allocated resources within its many divisions, set priorities and established guidelines for their execution, and made the criteria by which to evaluate success. Much of American fire policy, in brief, resides within agencies, a product of bureaucratic structure, personnel, and history.

### *Economic Considerations*

The sense that fire management is, or ought to be, guided by sound economic considerations is strong. It arises, in part, from the intellectual origins of fire protection in forestry, with its emphasis on forest economics; in part, from the assumptions of the young U.S. Forest Service that its programs, fire protection among them, would be managed as a business; and in part, from the need to control the large, unpredictable expenditures made possible by the pattern of fire occurrence and the peculiar structure of funding available to fire control. Some wildlands are economic institutions by their very nature—corporate forests and private tree farms, for example. But most wildlands are managed by political institutions. Nonetheless, virtually all fire agencies at least imply economic criteria as a basis of policy. Where legislation has demanded compulsory protection for private lands, where funds must be allocated among competing programs or lands within a given agency, and where appropriate levels of distribution among members of cooperative programs must be determined, some concept of economic behavior is required. By the 1970s Congress was specifically mandating at least some economic considerations for fire protection into law.

But against these ambitions, which define the desirability of fire economics, there are certain realities of fire behavior and fire administration which define its usability. Some of these objections relate to historical considerations, the way in which fire protection was first established in the United States and the means by which it has grown. Many protected lands were set aside originally for intangible values such as spiritual uplift or recreational use; others, for undetermined usage by future generations; some, for reforestation at a time when plenty of uncut timber made reforestation far from economically reasonable. Initially reserved for noneconomic reasons, many protected lands remain insulated from a free market economy. Most federal lands, in general, were reserved in order that they might be protected. The historic mission of fire protection has been one of growth—growth directed by needs and perceptions often quite outside economic analysis. In particular, the funding of fire emergencies through a special, deficit account has had dramatic consequences for any attempt at economic control over fire programs. Other objections relate to the difficulty of obtaining respectable economic data: the values of the resources at risk, the damages or benefits that result from fire, the true nature of liability. No private system of wildland insurance has succeeded in the United States, originally because of the magnitude of the fire problem, and then because of the success of fire protection by other means.

Within such a context, economic theories were understandably weak. Apart from their internal problems (the difficulty of determining the important production functions), they could not expect to translate into practical guidelines for the conduct of programs. Wildland fire does not behave like economic man in its competition for resources or money, and fire administration shows many fundamental characteristics that make it distinct from a true economic institution. For most of its evolution, fire management has been dominated by historical circumstances, political mandates, and noneconomic criteria. Economic analysis could rationalize this process, but it could neither direct it nor substitute for it. Economic theory could occasionally advise, but it could not command. No federal fire administration has gone out of business because of poor economic judgment; no federally managed fire has been abandoned for lack of funds; no American fire regime has been withdrawn from protection because of strict economic analysis.

Perhaps nothing goes to the heart of the matter better than the existence of the emergency firefighting fund (FFF) among federal agencies. Beginning in 1908 the Forest Service was allowed to engage in deficit spending during active fire suppression, then apply for a supplemental appropriation after the fire season to cover its expenditures. In time this account was expanded to include a wide range of costs—presuppression activities, replacement of damaged equipment or expended supplies, rehabilitation of burned areas. In part, the fund was a sensible solution to the problem of predicting suppression needs in advance of the season. But in part, too, it was justified because the lands under protection demanded protection for reasons that were not strictly, or even largely, economic.

Partly because the subject is intrinsically complex and partly because the question was often academic, hard data for, and suitable theories about, fire economics were not particularly well-developed. Most concepts have served only as general guides to fiscal restraints for fire officers, and as expressions of good faith on the part of fire agencies. Values at risk, resource losses and gains attributable to fire, the nature of liability due to wildland fire—not all are relevant to the lands under protection, but they have all stubbornly resisted an easy calculus. Perhaps more serious is the ignorance of the production functions that economic theories need, the relationship between investment in fire programs and their net cost and returns.

Intuitively, it was understood that an ounce of prevention money was worth a pound of suppression funds, that increased presuppression activity should reduce suppression activities, that stronger initial attack would diminish the need for large fire organizations. As long as fire protection continued to expand into unprotected, or obviously underprotected lands, this relationship was generally true. But when expansion into new lands ceased, when consolidation between agencies and the sharing of resources became dominant trends, when the FFF account was divorced from presuppression expenditures, the question of diminishing returns became apparent and the prospect improved for marginal analysis among the various components of a fire program.

The degree of ignorance about the relationship between investment and return was painfully brought home to the U.S. Forest Service during the 1970s when planning guidelines encouraged great drafts from the FFF account in the name of presuppression. Expenditures rose dramatically, but showed no evidence of affecting suppression expenditures, which also rose. In fairness to fire management it should be noted that nearly *all* federal programs showed a similar, uncontrolled growth during this period, but its mechanism of growth was peculiar to fire management and it highlighted the uncertainties of fire economic analysis.

With the political and administrative changes that occurred in fire management during the late 1970s, the opportunity brightened for economic analysis of at least some portions of fire management. The reasons were many. Legislation such as the Forest and Rangelands Renewable Resources Planning Act (RPA) and the National Forest Management Act specified some economic criteria, thus mandating economic analysis through the mechanisms of politics. The abolition of presuppression FFF money for the Forest Service shrank the range of FFF influence, with a corresponding expansion in the realm of programmed accounts. The change of agency policies and the transformation of fire management philosophy which underwrote that change encouraged the use of strategies other than simple suppression. This reorientation required economic theories that could go beyond the damages of wildfires and the costs of large fire suppression. A mix of fire programs was needed, presuppression funds were budgeted, and economic information was desired that could evaluate the relative payoffs of different program mixes.

Perhaps most influential of all was the fact that organized fire protection had, by and large, completed its historic mission. Nearly all lands in need of protection were protected, and at least a rudimentary physical plant for fire management was nearly everywhere in place. In some areas that plant, and the crews that staffed it, have reached great intensity and sophistication. Since the historic strategy towards fire protection had been to break the previous fire cycle by suppression, and since suppression had received generous funding, the historic emphasis of presuppression programs had been to prepare for suppression; this was made clear even in its name. By the 1970s that process of preparation was finished. A certain amount of capital investment in renovation and modernization would be needed, along with a level of maintenance funding. But the first-time establishment of the plant was complete.

An analogy perhaps is to flood protection. As long as unmanaged rivers and watersheds remained, the principal thrust of a national program was to put in primary facilities—dams, levees, canals. Eventually, all the important flood plains acquired some degree of protection. Among the remaining candidates for dams some would not be worth the investment, and some would have been rezoned for wilderness or scenic uses, thereby eliminating the need for facilities. Because fire protection, unlike flood control, is dependent on firefighters, it has to staff its facilities, not merely build a physical plant. Its operating costs remain high. But something of this quandary afflicted fire protection during the 1970s, manifested in a classic case of diminishing returns. As a result the prospects for real economic analysis of at least some portions of fire management have perhaps improved.

## *7.2 THEORIES OF FIRE ECONOMICS*

Of all federal agencies, none has been so assiduous in its search for an economic theory of fire management than has the U.S. Forest Service. In part, this reflects its academic origins in forestry, and, in part, the insistence of its founders that the Service would be run like an efficient business, a public utility. None of these efforts has really succeeded in developing a comprehensive theory of fire economics, and because of the historical circumstances under which they were constructed perhaps no such theory was really demanded. Yet the theories represented systems of economic accountability, a mechanism of cost accounting by the agency in the absence of outside political checks, expressions of good faith on the part of field officers, and the intellectual stuff out of which contemporary economic thinking has evolved.

### *Concepts and Funds*

*Concepts of fire economics.* In general, economic theories fall into three basic groups: theories based on the concept of adequate protection, theories proposing a standard of minimum damage, and theories seeking to minimize the sum of costs and damages. All of these types are integral theories, in that they propose to describe in economic concepts the whole range of fire protection programs. In more recent years, as diminishing returns has become apparent, several other models have been proposed: cost-benefit analysis, which has the advantage of comparing fire control to other forms of fire management; decision analysis, which relies on simulation of policy alternatives; and marginal analysis, which can isolate particular components of the total fire program for comparative study. All rely, too, on simulation experiments that can control variables, rather than on historical statistics that cannot, and all presuppose one of the integral theories as a theoretical context.

In the early days—when the need for any kind of organized fire protection was so great—it was common to accept several theories of economic behavior or program standards without any sense of contradiction. Fundamental statements of all three integral theories appeared between 1916 and 1928, and all were generally well received and, in some respect, reconciled. *Adequate protection* argued that fire protection should, in effect, act as a surrogate for forest insurance (Flint, 1928). The level of investment in fire protection should equal that which a prudent person would be willing to invest for fire insurance. Developed for high-value timberlands, the concept suffered when applied to lands without market value. But it recognized that some losses would occur and that presuppression dollars were, or ought to be, limited by the value of the lands under protection.

*Minimum damage* proposed that the damages due to fire be held to a minimum, usually expressed in terms of burned acres (Show and Kotok, 1923). The minimum standard might vary according to land values, but as a rule of thumb losses should not exceed 0.1% of the total acres under protection, and no more than 15% of all fires should reach sizes greater than 10 acres (4 hectares).

The *least-cost-plus-loss* theory (LCPL), meanwhile, tried to incorporate the costs due to suppression into the total economic context (Headley, 1916). It recognized that, thanks to the FFF account, it was possible for suppression costs to greatly exceed the resource losses that might result from allowing the fire to burn or the sum of costs and losses that might result from other strategies of control. Instead the theory sought to identify the point at which the cumulative investment in presuppression and suppression and the losses due to fire would reach a minimum.

With all these theories there were serious practical problems, and most early policy analysts felt intuitively that all were really restatements of the same problem. Given the nature of fire control, they believed that a minimum loss standard was a simple means to express adequate control, and that strong initial attack was, on the average, the surest way to ensure a least-cost-plus-loss result. The minimum damage theory put a strong emphasis on initial attack, recognizing that the fundamental problem was to stop big fires and that the FFF account offered the means by which to do it. Outfitted with the concepts of systematic fire protection developed by duBois, Show, and Kotok, the minimum damage theory was an ideal program for putting in a protection system where none existed before and for assessing the effectiveness of that system with a single, simple measurement—burned acreage.

By contrast the LCPL theory did not recognize the quasi-discontinuities between small and large fires, implictly assumed the existence of a working protection system, and sought to regulate the FFF account. Since it required complex measurements of damages, the tendency was to emphasize the cost of suppression, which could be determined easily. Because of its concern with suppression costs as an integral component of the total economic package, the LCPL theory became known as the "economic theory." The LCPL was first proposed in 1916, then restated in graphic form in 1925 (Sparhawk, 1925).

Useful as a general philosophy, the concept proved less helpful as a practical guide and became the subject of endless permutations and recalculations. There was, first of all, no calculus by which to estimate fire damages. Nor did the concept appreciate that the national forest system was a political institution first, and only secondarily an economic one. Even more damaging, wildfires simply did not behave in a way required by the theory. Individual fires did not wax and wane in proportion to the amount of resistance offered by suppression forces; they tended to simmer or to blow up in explosive runs. Blowup fires were either controlled while small or they were not controlled at all. Wildfire, in short, did not behave as though it were in economic competition with the Forest Service. Nonetheless, nearly all subsequent theories of fire economics are in some way restatements of the LCPL concept. And as circumstances surrounding fire management evolved, especially during the late 1970s, the utility of the theory improved.

*Fire Financing.* The costs of fire management are paid in two ways. For normal operating expenses there is a programmed account: a line item in the annual budget of the agency. For actual fire suppression, however, costs are paid out of a special emergency account. For federal agencies, this "account" is nothing more than an authorization for deficit spending. After the end of fire season, Congress will approve supplemental appropriations to cover the expenses incurred by fire suppression. The Forest Fire Fighting (FFF) fund was first established for the Forest Service by the Act of 1908. The practice was eventually extended to the Department of the Interior through annual appropriation bills. In time the FFF account was expanded to cover presuppression costs and rehabilitation activities as well. In the late 1970s, this embellishment of FFF uses was ended for the Forest Service, though not for the Department of the Interior. For the Forest Service, FFF funds could no longer be used for presuppression activities; instead an expanded Fire Management Fund (FMF) was established to pay for all expenses other than those related to suppression and rehabilitation.

For the States, which by law cannot engage in deficit spending, an emergency fund is generally set up out of which fire agencies draft money as needed during fire season. After each fire season, the State legislature will appropriate monies to bring the emergency fund back up to its authorized amount. Unlike the federal situation, there is an upper limit to seasonal drafts from this fund.

Especially within the federal fire establishment, the consequences of the FFF account are pervasive. One effect is to emphasize suppression, where funding is ample, over those other fire programs whose funding must compete among themselves and with other appropriated line items in a budget. Another outcome is to grant to fire protection a special degree of autonomy within a land agency because it enjoys an important source of funding from outside normal bureaucratic channels. By being outside the budget, the suppression function cannot be influenced to the degree that other programs can. Yet another effect is to segregate the costs of suppression from the costs of other fire programs, including presuppression. Spending more money in presuppression activities, for example, does not reduce the amount of money available for suppression. There exists no meaningful economic mechanism by which to link the two funds. The soul of economics—the allocation of limited resources—does not really apply. For certain fire programs (e.g., prevention, detection, fuels modification, and prescribed fire) economic considerations can work because the quantity of funds is fixed. The mixture of programs and the level of investment in each program can be analyzed by economic techniques and expressed in a budget.

But suppression funds are not competitive with other program funds, are not coupled to them in meaningful ways, and are, for practical purposes, unlimited. The argument that increased expenditures for prevention and presuppression can reduce suppression costs is largely academic. The two funds are not joined in such a way that a fire officer, or a fire agency, has to decide between one or the other. From a fiscal standpoint the best strategy is to have as many fire costs as possible paid out of the suppression (FFF) account. In a sense a program with a large number of fires pays for itself in ways that a successful program of fire prevention or prescribed fire does not. Wildfire is not competitive with prescribed fire for funding. In fact, much of the infrastructure that sustains prescribed burning is the product of the suppression program, and no areas or agencies that are poor at suppression are good at prescribed burning. The FFF account, moreover, does provide a means of adjusting costs to fires without having to forecast long-term weather patterns. It is a fiscal reality that theories of fire economics must accomodate, one made stubborn by the fact that agencies have grown up with, and developed around, this pattern of funding for many decades.

### *The LCPL Theory*

Since it was first proposed, the LCPL theory has been endlessly modified according to the prevailing strategies of fire protection. In this regard it resembles the history of all theories of fire economics. The issue was never that of a fixed problem (fire protection) interacting with progressively improved theories (such as the LCPL concept), but of a changing problem evaluated by changing theories against a background of constantly changing fire costs and resource values. In essence, the LCPL theory assumes relationships between the investment in fire protection (variously measured) and returns (resource damages and suppression costs). As the investment in presuppression and prevention increase, damages and control costs generally decrease in some way. The optimum protection level occurs when the combined costs of fire damages, presuppression activities, and suppression are at a minimum. Too little fire protection results in unacceptable damages (losses); too much only reaches a point of diminishing returns (costs).

Unfortunately, the production functions that relate investment to return are difficult to identify, and they often make unrealistic assumptions about fire behavior or the distribution of fire seasons. The size of a fire is not proportional to the size of the suppression force sent to contain it; a small force can control a small fire, but no force can halt a blowup fire once the transition to large fire is made. The theory describes average conditions, or average worst conditions, when the problem is to account for the exceptional event, the episodic large fire or fire season. An "average fire" was an acceptable concept when no fire protection existed, but as large numbers of small fires were easily brought under control, it was necessary to cope with the question of the big fire. To pursue the flood analogy, the problem is to design a levee or bridge to withstand a 50-year flood, not an average flood.

*Historical Survey.* Perhaps the fundamental problem with the theory is that it has never expressed a fixed condition, but an evolution of conditions. As circumstances have changed, so have strategies of protection and, correspondingly, so has the choice of an independent variable in the production function. A quick historical review will illustrate this point (Figure 7-5).

The LCPL concept was first proposed during the time when systematic fire protection confronted light burning as an alternative management system for high-value timberlands. Damages could be expressed in terms of timber losses. Investments in prevention and presuppression coincided with the construction of a physical plant for fire control—roads, trails, communication networks, and so forth. If difficult to specify in dollars, the return was intuitively obvious: fire protection now existed where, for the most part, it had not existed before. As systematic fire protection expanded geographically, the emphasis continued to be on presupppresssion investment of this sort, the need to create for the first time a physical plant for fire control.

With the advent of the 10 AM Policy, however, the relationships that the economic theory needed to express were changed. The new policy made all lands equivalent in value for purposes of fire protection, so the emphasis could rightly shift from the determination of losses (damages) to burned acres or time of control—two complementary relationships which described the efficiency of the initial attack program. Presuppression monies became available through various emergency funding mechanisms, the FFF account and the CCC program. The levels of funding were set by outside political considerations, so there was little point in considering presuppression expenditures as a controlled investment. Paradoxically, the only economic quantity that could be adequately measured in the backcountry and cutover lands, because it was the only one expressed in dollars, was the cost of presuppression and suppression. Rapid initial attack and quick suppression would, in the long run, prevent drawn-out campaign fires with their enormous costs. The precise relation-

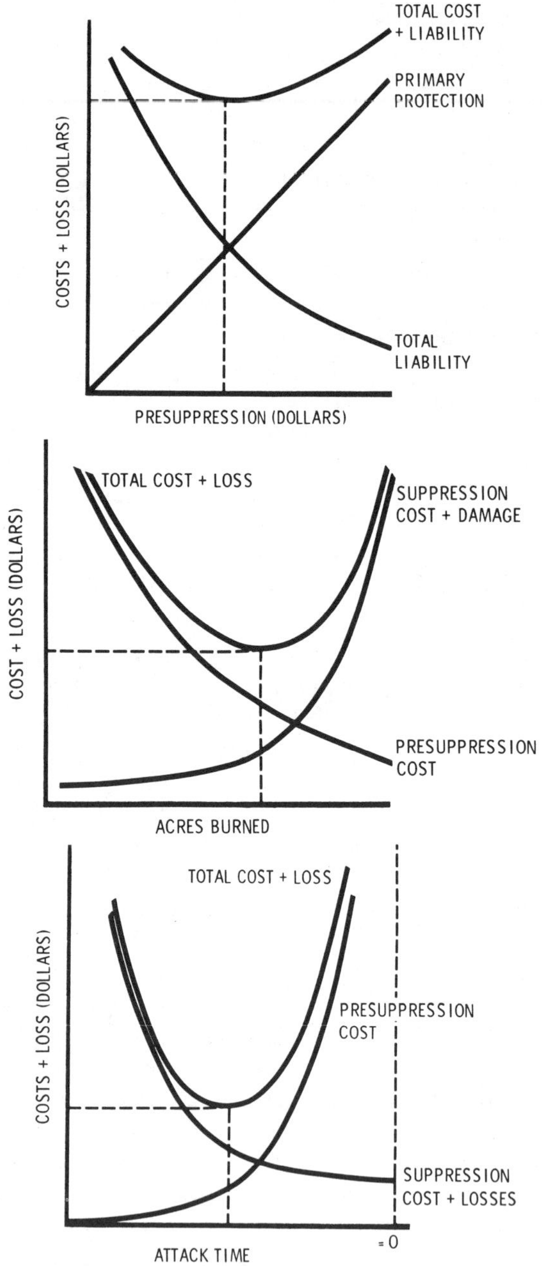


*Figure 7-5.* *Evolution of LCPL theory of fire economics. Upper graph gives the curves derived from Sparhawk (1925); middle graph, those from Hornby (1936); lower graph, those from Arnold (1949). The principal difference is in how presuppression is defined, and in how differences in presuppression investment translate into reduced damages and suppression costs. The steepening of the presuppression function curve reflects the growth of a first-order system for fire control. From Gorte and Gorte (1979).*

ships between such costs and the fire damages averted were unknown, but neither were they really relevant in the face of suppression resources like the CCC and the FFF account.

When conflagration control became a ruling strategy, the LCPL theory used attack time or suppression-force size as its independent variable. Presuppression investment was less in the physical plant for fire control than in equipment and crews, with their success measured by response time. A program (Increased Manning Experiment) which increased manpower as a presuppression experiment seemed to show a decrease in suppression costs. In the initial years of fire protection, the production function for suppression success required that firefighters simply get to the fire. This could be expressed by the investment in presuppression structures such as roads. Later the question of control took on the character of a two-dimensional competition in perimeter control, between the growth of the fire and the construction of fireline. Burned acreage described this condition well. For conflagration control the problem was to apply sufficient suppression resources to the fire before it could make the transition to mass fire. Size of force and response time ("force enough fast enough") were thus appropriate measurements of presuppression investment.

As fire by prescription became a dominant strategy of fire management, the LCPL theory again metamorphosed. Clearly a point of diminishing returns had been reached in terms of fire protection's physical plant and its suppression resources; most fire damages, burned acres, and suppression costs came from a small, seemingly irreducible percentage of total fires. An all-out program of presuppression in the 1970s (in accordance with the 10 Acre Policy) failed to eliminate this quantum of damaging fires. The realization grew, too, that the very success of fire control was, in some areas, leading to an implacable buildup of fuels which made intensive fire suppression somewhat self-defeating. The historical mission of fire protection—its expansion into new areas—was complete.

*Contemporary Developments.* As other strategies of fire management were proposed, so were revisions of the LCPL concept. For its presuppression variable, the LCPL theory followed the popular strategy of fire protection that sought to modify the fire environment, especially in the form of fuel management programs. Suppression costs (and fire damages) could be contained, in theory, by reducing the intensity of wildfire rather than by a continued increase in the magnitude of suppression resources or the means to move them quickly to a fire. In other cases, as with wilderness areas, fires were simply transferred from the status of wildfire to that of prescribed fire. The motive behind this transfer, it should be reiterated, was not economic, but a matter of policy as a consequence of the Wilderness Act. The abolition of the emergency presuppression account in favor of a Fire Management Fund (FMF) reinforced the tendency to look for alternative methods of presuppression.

Against the traditional emphasis on fire damages, theories incorporated fire benefits; against the control of wildfire, they encouraged prescribed fire; against increased initial attack resources, they promoted a mixture of presuppression activities which collectively could reduce the problem of wild fires. Decision theory sought to quantify the alternatives of presuppression programs, like fuels management. Marginal analysis investigated the mixture and level of fire programs. Simulation promised to hold variables constant, in ways not possible with historical data, to determine economic outcomes of different investment strategies. Equity analysis held the prospect for a more sophisticated determination of who should pay costs, and reap benefits, from fire management programs. The issue was particularly relevant for the Clarke–McNary program, which allocates federal funds to the states.

The trend towards a pluralism of management responses will probably result in a proliferation of economic theories of fire administration. Even where one theory maintains its dominance, such as the LCPL concept, its integral approach will be broken down into smaller components for better comparative analysis. But it must be remembered that large changes in the strategy of fire management have repeatedly come from processes outside a market economy, and for reasons beyond economic analysis. Economic theories are one means among many for harmonizing mandated objectives to the resources at hand. They can no more determine fire policy by themselves than fire behavior, considered alone, can dictate an appropriate strategy of fire suppression, or fire ecology, a program of prescribed fire. As a method evaluation, however, they offer a unique means to shape, though not drive, fire management systems.

### *Economic Models: Contemporary Examples*

Powered by agency determination to make the economics of fire management conform to its new policies and by congressional mandates to evaluate fire programs at least partially according to criteria of economic efficiency, the U.S. Forest Service has developed sophisticated computer models for simulating fires and fire management systems as economic events. In general, these models resemble the growing category of fire effects models, though for economic not ecological analysis. Rather ambitiously, they seek to determine appropriate fire management budgets, the proper allocation of expenditures between various fire management functions, and the costs, damages, and benefits of wildland fire and of fire management programs (Baumgartner and Simard, 1982). To update the LCPL theory, whose theoretical validity they accept, so that it reflects contemporary appreciation for potential fire benefits as well as damages, they redefine "loss" to read "net value change" (NVC).

The models were originally developed for the national forest system (Schweitzer et al., 1980). But they have been extended to certain State programs with considerable success (U.S. Forest Service, 1982). Out of these experiences a general model, the Fire Economics and Evaluation System (FEES), is being created (Mills, 1979). It is hoped that FEES will generate, in some cases for the first time, the kind of economic information that agency policy and congressional mandate require for fire planning. For each fire management program option, FEES will produce three kinds of information: economic efficiency, through a restated version of the LCPL theory; effects of the program on resource outputs, measured as both benefits and losses; and risk, or the probability that certain costs or resource changes will occur as a result of fire events (see Figure 7-6). The model is designed, moreover, to be situation-specific, not agency-specific or fire-specific. It can evaluate total levels of program investment, or the relative allocation of a total investment among the components of a given program.

The new economic models do successfully address some new issues in economic terms. They make fire programs responsive to the values at risk in the lands under protection, thus reflecting the 1978 fire policy adopted by the Service. By having resource values drive the investment in fire management, the models help evaluate fire programs in the light of Forest Plans, which are also driven by resources. By simulation the models can generate curves for the LCPL theory that are more realistic and less hypothetical than those used in the past. The models, moreover, tend to agree with commonly held perceptions. Test runs show that the national forest system is, by and large, well-protected, and that the State programs could accommodate increased investments in fire protection before reaching a point of diminishing returns.

But the models also fail to resolve some old questions, and they solve some old problems that no longer exist. Numbers meaningful to the revised LCPL theory are often as hard to come by for these programs as for those that had gone before. It was as difficult to determine fire benefits

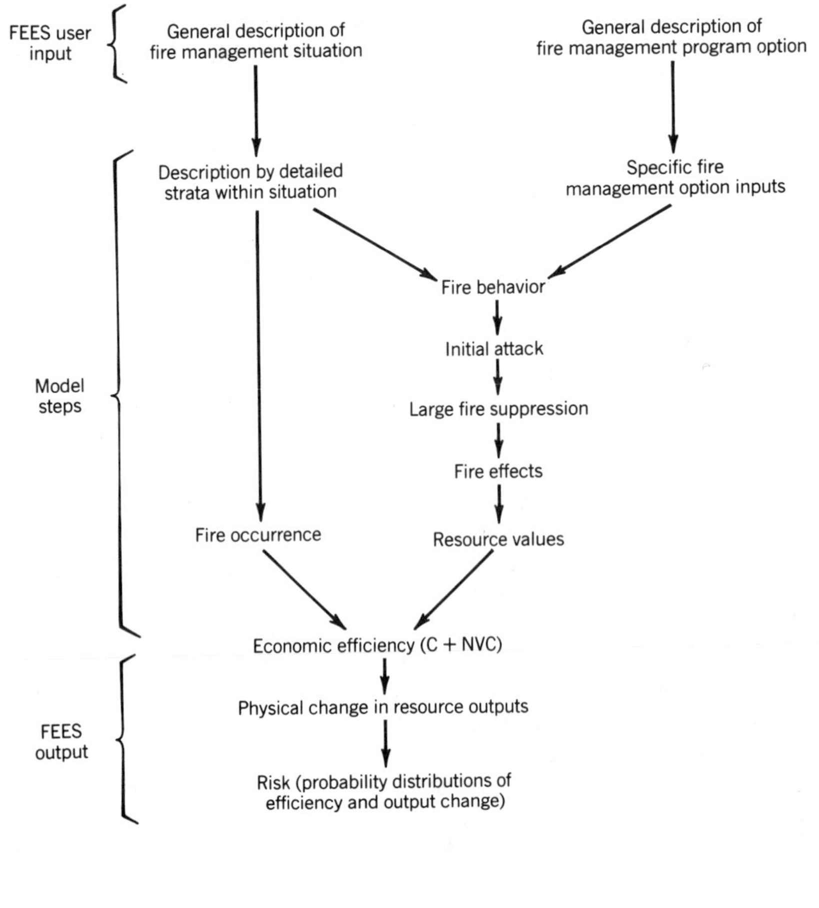


*Figure 7-6. A contemporary economic model: FEES. From Mills and Bratten (in progress).*

as fire damages. Nor could the models resolve the nagging question of equity, of the relative contributions federal and state agencies should make to a national fire program. Analysis, indeed, confirmed that the first order construction of initial attack facilities had been accomplished by the 1970s, and that further heavy investment in initial attack capabilities could be questioned *vis-a-vis* other fire management programs. But the explosion of emergency presuppression expenditures during the 1970s had ended prior to the new round of analysis.

At the same time economic theory still failed to integrate the suppression function within the total fire management program and it failed to account for the full costs of prescribed fire. Those presuppression programs, such as fuels management or initial attack buildups, which competed for the same monies could be compared as relative investments. But suppression was funded from an account that was not competitive with it. Moreover, the scale of economic analysis had to be considered. Most suppression resources are shared within and among agencies. Each administrative unit does not have to gear up for large fires within its own capabilities; rather it draws from larger regional or even national pools. This system of shared resouces is a means of sharing risk, of distributing the cost of large, perhaps infrequent episodes among a bigger group. Fire caches, mobile crews, heavy equipment, air tankers—all are shared. Air tanker use is scheduled by calendar year to coincide with the staggered appearance of fire seasons among different regions. No one region has to absorb the entire cost of needed suppression resources, yet those resources are available to all.

Somewhat paradoxical has been the attempt to incorporate prescribed fire into economic modeling and to compare it as a strategy of fire management to suppression. Part of the problem is a false dichotomy between fire control and fire use. Even considered in operational terms, each supports the other. Prescribed fire programs, for example, rely on the historic infrastructure of fire management created by decades of fire suppression, though these overhead costs are rarely considered in calculating the benefits of prescribed fire. Nonetheless, prescribed fire programs have implicitly assumed that fuel treatment is an alternative to large fire suppression, that prescribed fire can enhance resource values in many places, and that prescribed fire can substitute for wildfire. In some areas, it can. But in other areas, for many reasons, the replacement will be difficult and, if possible, costly.

One complication is the fact that the conversion of wild to prescribed fire can only come about through a change in land use, especially from extensive to intensive management. Yet the creating legislation for many wildlands stipulates that these lands be kept as wildlands and some of them as wilderness. Wildfire will be an inevitable component of any such landscape.

Another dilemma is that the very nature of prescribed fire makes it more amenable than wildfire to economic analysis. The situation is analogous to that which plagues fire ecology research. Most studies require controlled experiments, which means controlled fire, and the outcome of fire effects studies really applies to these controlled fires, not to wildfires. Similarly, economic models assume a reasonably controlled fire. With prescribed fire it is possible to set fire intensity in order to maximize benefits and to predetermine fire size to allow for an economy of scale. Prescribed fires, in short, behave "economically" in ways that wildfires do not. There is a demonstrable production function between the investment and return: between site preparation and burning operations, on one hand, and fire behavior, size, and effects, on the other. In the case of wildfire, fire size, costs, and damages are all uncontrollable. Inevitably, economic theories favor controlled fires over uncontrolled fires, and prescribed fire programs over suppression programs. The theories simply work best in those situations where prescribed fire is possible.

By way of example, consider the economic analysis of fire management programs in two very different types of national forests. In areas where domesticated fire had more or less substituted for wild fire, as on the Francis Marion National Forest in South Carolina, the LCPL theory, and marginal analysis based upon it, can work well. In areas dominated by intangible values and wildfires, mostly lightning-caused, such as characterize the Tonto National Forest in Arizona, the theory's limitations become apparent (see Figure 7-7).

Clearly, economic analysis has its value, and thanks to legislation and policy it has a mandate for use. Its applicability may be restricted to a narrow range of subjects and geographic areas, but the prospects for including more topics and more regions within that range are improving. The subject shows great intellectual ferment. Fire economic analysis must be considered as one means among many for determining policy, one technique among several for evaluating program success, and one process among many that contributes towards that reconciliation of means and ends which constitutes fire planning.

## *7.3 STRUCTURE OF FIRE MANAGEMENT IN THE UNITED STATES*

Fire management in the United States is both fragmented and integrated. The need to place small units in the field to meet local needs and improve effectiveness against fires, which are local events, often conflicts with the economy of scale offered by large agencies organized for the attainment of national goals. In practice, the organization of fire management shows the same hierarchy evident in the American political system: federal agencies furnish a basic infrastructure which, through the medium of state agencies, reaches to local counties, fire protection districts, and even cities. In part this reflects the tendency of fire protection to expand in order to protect areas already under protection, and in part, historical events which established fire control for the protection of lands reserved from the public domain and which charged the U.S. Forest Service with the administration of those lands, including their protection from fire. From that point of origin the politics of American federalism has evolved the present system.

In any particular region different levels of government may have the primary responsibility for fire protection, but through interlocking agreements and contracts virtually every level of government shares fire protection responsibilities with every other level. Nor is the enterprise

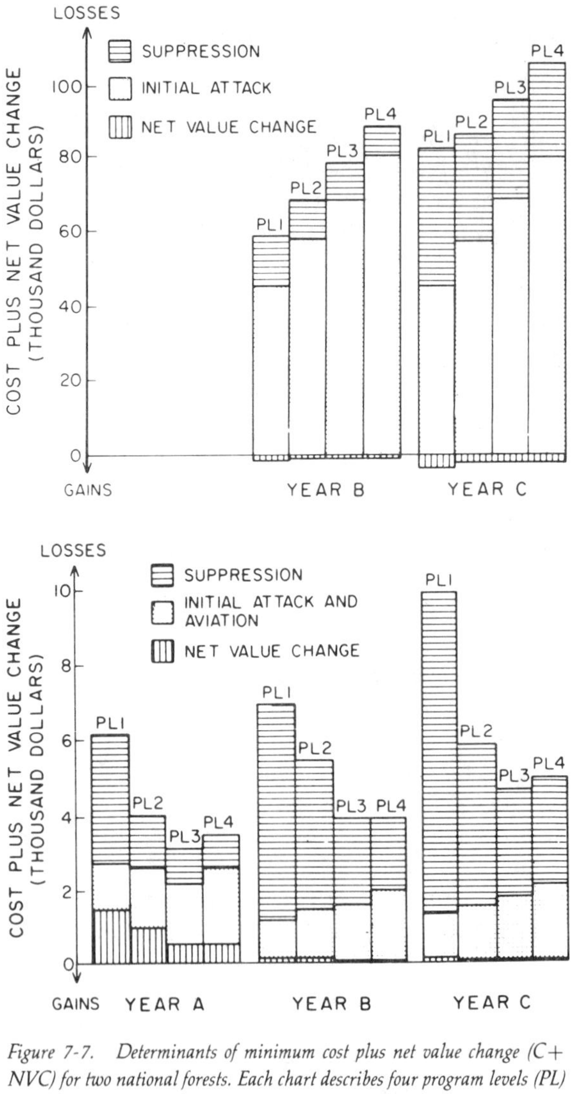


*Figure 7-7. Determinants of minimum cost plus net value change (C+NVC) for two national forests. Each chart describes four program levels (PL) for three years of different fire loads. Upper chart applies to the Francis Marion National Forest, South Carolina; lower chart, to the Tonto National Forest, Arizona. From Schweitzer et al. (1982).*

strictly governmental. Private associations exist for the protection of their members' wildlands, commercial companies contract with governments for fire services, private laboratories and research organizations produce information under contract, industry supports fire brigades on company lands and a national fire prevention campaign, and fire councils provide a forum for the discussion of mutual problems in an environment outside the restraints of formal government. Nor, again, is the national fire protection system even limited to the United States. International treaties stipulate conditions for mutual aid on fires, and various United Nations agencies sponsor research programs, study tours, and symposia to help integrate fire management in the United States with similar programs throughout the world.

The impossibility of confining fire to the lands of one particular administrative unit, the value of shared suppression resources, the similarity of fire management problems due to the commonality of fire behavior—all have argued for a degree of administrative synthesis. For the United States the agent of synthesis has consistently been the U.S. Forest Service. The present structure of fire management in the United States is largely the story of Forest Service involvement, directly or indirectly. No program of note has succeeded without its help, and none has failed for lack of that support.

### *Federal Fire Management*

All of the federal landholding agencies exercise at least some fire protection responsibility over the lands under their jurisdiction. This includes the U.S. Forest Service, located within the Department of Agriculture; the Bureau of Land Management (BLM), the National Park Service (NPS), the Bureau of Indian Affairs (BIA), and the Fish and Wildlife Service (FWS), all within the Department of the Interior; and the military services, within the Department of Defense (DOD). For the Forest Service, protected federal lands include the national forests and grasslands. For the BLM, the lands include otherwise unclaimed public domain in the Far West and Alaska and the O&C lands in Oregon; for the NPS, national parks and recreation areas; for the BIA, Indian reservations; for the FWS, wildlife refuges. For the Defense Department, responsibilities extend to military reservations. In most instances, the responsible agency provides its own fire protection; in other cases, it contracts with additional agencies, or cooperates under various mutual assistance agreements. All have supported at least some research on fire management questions, though here—as in other matters—the Forest Service dominates the field.

To these agencies, whose fire mission derives from their land management charge, must be added others, who offer support services of various sorts. Though the Forest Service has commanded the overwhelming bulk of wildland fire research, some funding has come through the National Bureau of Standards (NBS), the Office of Civil Defense (OCD), Federal Emergency Management Agency (FEMA), the Department of Defense, and the National Science Foundation (NSF). From 1958 to 1977 a special Committee on Fire Research within the National Academy of Sciences-National Research Council (NAS-NRS) oversaw much of the federal investment in fire research. Here again a large amount went to the Forest Service either in the form of direct awards or indirectly through contracts with the other agencies. After the Fire Prevention and Control Act of 1974, the National Bureau of Standards established a Center for Fire Research, though little effort has gone into the subject of wildland fire. Beyond research, other bureaus showing an active concern in fire management include Civil Defense, or its progeny such as FEMA, with an interest in rural fire protection; the U.S. Fire Administration (USFA), also involved in overlapping jurisdictions in the rural fire area; and the National Weather Service (NWS), with its long-maintained Fire Weather Service.

Historically, a degree of integration between all these agencies came by association with the Forest Service. The primary instrument for these liaisons was the mutual aid agreement, which specified the conditions under which one agency would extend suppression resources to another. Such agreements existed at all levels of organization, from departments to agencies to districts. Where lands of mixed jurisdiction were involved, one agency frequently contracted with another to provide protection. For a brief time (1928–1933) a Forest Protection Board was established by executive order to coordinate programs among federal agencies.

During the 1970s, however, interagency cooperation acquired a certain autonomy; Forest Service participation remained essential, but the instrument of coordination existed outside the Service. The drive for cooperative fire programs was in many ways innate to fire protection. As costs escalated, resources multiplied, and cooperative opportunities intensified among field operatives, the need grew for better mechanisms and criteria for cooperation—a standardization of equipment, of operating procedures, of personnel certification. In some areas a *closest forces* doctrine came into play, under which the suppression unit nearest to a fire, regardless of agency affiliation, responded. In other areas, dual dispatching systems integrated responses. In Southern California Firescope formalized terminology and organizational structure. On a national level there developed the concept of *total mobility,* the complete interchangeability of all suppression resources.

As a result some instruments of coordination outside the control of any one agency had to be created. Perhaps the two most important institutions were the National Wildlife Coordinating Group (NWCG), a consortium of federal agencies which exercised advisory capabilities on all aspects of fire management, and the Boise Interagency Fire Center (BIFC), which consolidated logistical support for going wildfires. At the request of the Forest Service, NWCG included a representative from the National Association of State Foresters. Under the inspiration of these bodies, and often on their recommendation, a host of interagency programs developed to make total mobility a reality. There emerged a National Interagency Fire Qualifications System (NIFQS), which standardized the certification of personnel for fireline positions; a National Advanced Resources Technology Center (NARTC), which offered interagency training according to a large curriculum of fire management courses; a National Fire Equipment System, which standardized suppression hardware; cooperative programs in prevention, fuel management, and detection. The concept of mutual aid expanded from its origin in fire suppression to include cooperative management of prescribed natural fires and fuels. Virtually all of these programs developed out of Forest Service precedents, though they acquired a new, supraorganizational identity on their own.

The role of the Forest Service is not only fundamental but unique. From its inception the Service has sought a kind of vertical monopoly on the management of forest lands, and the structure of fire management in the United States testifies to its relative success in at least one area of that self-appointed mission. The Service is organized into three operational areas, each of which has fire responsibilities. For the national forest system, responsible for the administration of the national forests and grasslands, this includes the Aviation and Fire Management staff (AFM). For the state and private forestry branch, an extension service that distributes information and aid to the states, it includes the Cooperative Fire Program (CFP) staff. For research, it means the Forest Fire and Atmospheric Sciences Research (FFASR) staff.

All three branches, moreover, serve to extend Forest Service influence and examples. AFM does it through mutual aid agreements, contracts, the transfer of personnel to other agencies, the maintenance of equipment development centers, and training programs. CFP acts through the Clarke–McNary program, the Cooperative Forest Fire Prevention program (Smokey Bear), participation in interstate compacts, rural fire protection, the federal excess equipment programs, and Firetip (Firefighting Technologies Implementation Project). FFARS supports research programs that have state as well as federal significance. Everywhere the tendency towards homogeneity is reinforced by the selection of professional foresters to administer fire programs. No other agency, federal or cooperative, can match the intensity or breadth of the Forest Service commitment to fire management. If, during the 1970s, its hegemony was challenged, the Service nonetheless retained a lion's share of suppression resources, control over most of the apparatus of fire research, and the richest liaisons with state and private fire agencies.

### *State Fire Management*

Most wildland and rural fires in the United States are fought by state agencies. Fire protection came to each state somewhat differently. Some states had long antecedents, with statutes for rural protection even in the eighteenth century. Other states took on fire protection responsibilities when they acquired lands for state forests. Others —perhaps the majority—responded to stimulants emanating from federal programs such as the Weeks Act and Clarke–McNary Act (Pierce and Stahl, 1964). Everywhere the instrument of fire protction was the office of state forester, and everywhere fire protection provided a nexus with federal programs and a prod for other forest management activities. Every state has a state forester, charged with at least some responsibility for both rural and wildland fire protection, the last state joining the network in 1965. With the 1978 Cooperative Forestry Act Congress began withdrawing the federal contribution on the theory that substantial subsidies were no longer necessary.

The extent of jurisdiction varies by state. For some states responsibility resides in state lands such as forests; for most states, legislation mandates that the state forester, in effect, protect all unincorporated lands; in the case of Alaska, the establishment of fire protection districts is stipulated by the state constitution. Typically counties contract with the state forester for fire protection; state resources are comingled with county resources, such as rural or volunteer fire brigades; funding comes from a mixture of state revenue, county taxes, and federal grants-in-aid under cooperative programs. Where rural fire is a primary concern, the office of a state fire marshal may cooperate with the state forester. The fire marshal bridges rural and urban fire services in the way that the state forester does rural and wildland, and the office provides a source of federal assistance through the U.S. Fire Administration much as the office of state forester does through the U.S. Forest Service. This liaison, however, is advisory: legal responsibility for protection resides with the counties and state foresters, and it is through the state forester that counties and unincorporated towns participate with federal agencies in fire control.

What the rural fire problem accents, however, is the relationship between state and federal agencies. The wildland and rural fire protection schemes of the various states assume a federal superstructure in much the same way that state and country road systems assume the existence of a national highway system. For certain areas of rural fire protection, the U.S. Fire Administration assists with planning grants, model fire codes, and training. But overwhelmingly the federal connection is through the U.S. Forest Service.

There is, first, the Clarke–McNary program, operating in every state. The program provides for a system of matching grants-in-aid from the federal government, allocated by a complex formula which balances state need for assistance with the degree of state commitment, as manifest by the state budget. Those states that need the funds most and those that have invested most heavily in fire protection receive the greatest support. Upon this program rests the Rural Fire Protection Program, which extends training, planning, and equipment to rural fire units; the Federal Excess Equipment Program, by which the Forest Service can pass on surplus equipment (usually military) to its cooperators; cooperative research and planning endeavors such as Firescope; the Cooperative Forest Fire Prevention program (Smokey Bear, Mr. Burnit); and participation by the states in national fire planning through membership in the NWCG. On more local levels there exist mutual aid agreements and contracts, with the state sometimes providing protection on federal lands and a federal agency sometimes giving coverage to state lands. States share in training programs, and insofar as state laws permit they may accept the standards of the NIFQS. The states maintain their own training programs, of course, and some show considerable sophistication, but the CFP makes possible access to federal programs as well. In certain instances federal disaster funds may be available for catastrophic fires.

Apart from shared federal programs and resources, however, the states have organized among themselves. The National Association of State Foresters (1920) provided one model, and one of the provisions of the Weeks Act gave another, by allowing for the states to form fire protection compacts among themselves. Beginning in 1949 several regions have adopted such interstate fire compacts for mutual training and support on fires. The compacts are most powerful in those regions where the federal presence is less dominant. The first was the Northeastern States Fire Compact (1949); the southeastern and south central states adopted similar compacts in 1954, the mid-Atlantic states in 1956, the Pacific Northwest states and California in 1978. To bring even more states at least informally into cooperation, two still larger groups were created: the Northeast Forest Fire Supervisors (1973) and the Southern Forest Fire Chiefs Association (1974). The former included the Lake States and north central states, as well as the states under the northeastern and mid-Atlantic compacts; the latter, both the southeastern and the south central compact states. In keeping with its cooperative mission, the Forest Service signed as well. In 1972 the first intercompact agreement was signed between the mid-Atlantic and southeastern states (Society of American Foresters, n.d.).

Access to federal (especially military) surplus has stimulated another form of interstate association. In 1972 the northeastern states agreed to underwrite the Roscommon Equipment Center (REC) on the site of the Michigan Forest Fire Experiment Station for a trial period, and the arrangement was extended in 1974 for another five years. The center emphasized plans for the conversion of excess military hardware such as 6 x 6 trucks, into equipment suitable for state and rural fire protection.

*Private Agencies*

In some areas of the United States fire protection developed first through the private sector. Historically, the techniques of wildland fire protection evolved out of rural fire protection, as developed along the agricultural frontier. In a few instances, however, private agencies developed in response to statutes requiring compulsory fire protection; the individual landowners could either provide a certain level of protection for themselves or submit to taxes with which the state could furnish protection. In most instances, however, private agencies emerged out of a shared perception that fire control was necessary, that fire protection had to be cooperative in scope and broad in scale in order to be effective, and that governmental remedies were either not forthcoming or would be incompatible with company needs. In time many of these bodies were replaced by state or federal institutions, but not all. Some continue to flourish, especially in the Pacific Northwest, and some additional fire management problems have been added to their original mission in fire control.

Perhaps the most celebrated of the fire protection associations is the Western Forestry and Conservation Association (WFCA), chartered in the Pacific Northwest in 1909. The WFCA was an umbrella organization, bestowing a collective voice to its members and setting a national example for the conduct of such a program. Eventually, it agitated for better fire research as well as better fire control, and it sponsored an array of fire councils through which fire managers of a region could coordinate and disseminate information about programs. Under its Western Forest Fire Committee, participants include the California–Nevada, Intermountain, Northwest, Rocky Mountain, Southwest, and Alaska fire councils.

These early fire (or timber) protective associations were cooperatives, providing a service for members. A more commercial approach—a fire control service available to anyone willing to pay for it—is epitomized by an Arizona company, Rural/Metro (Jekel, 1979). Originally conceived to furnish low-cost rural fire protection, Rural/Metro has expanded to include primary fire protection for one city (Scottsdale) and supplementary protection for a nearby national forest (Tonto). There is some precedent for expanding these services further into wildland fire. Most of the costly fire suppression resources (air tankers and helicopters, to name two) are rented from commercial vendors, and many resource activities such as timber thinning, are contracted out. How much incentive will exist to acquire other fire suppression services is hard to determine, further complicated by the economics of seasonal use. The most likely area for expansion for such companies is in the field of rural fire protection, an alternative to full-time and volunteer companies.

There are other areas of private involvement in addition to suppression. The Keep America Green fire prevention program is managed through the American Forest Institute, an organization representing the major forest products companies. And in the field of fire research, private organizations have often exerted an influence beyond their apparent resources. The WFCA promoted equipment development, and agitated for certain forms of practical research. In the 1960s and 1970s, the Tall Timbers Research Station—a privately endowed lab in Florida—became a major focus for research on the biology of fire (Komarek, 1977). In recent years symposia on fire have been held under the aegis of the East–West Environment and Policy Institute in Honolulu, and the Science Committee for Problems of the Environment (SCOPE). And then there are the universities, a relatively minor but sporadically significant source of research, though one both strengthened and compromised by connections to the research program of the Forest Service through the practice of cooperative agreements.

The importance of private research bodies, even more than private fire protection agencies, is that they can contribute an alternative voice and may investigate unpopular topics. With so much research conducted by the federal government (and that almost wholly by the Forest Service), with so much of that research conducted by foresters, and with such close institutional liaisons between federal and state forestry—the value of even minor programs outside of this framework is real, and their voice useful.

*International Agencies*

Both fire research and fire management in the United States have acquired international dimensions. The United States has treaties with Canada and Mexico for assistance on wildfires. BIFC has helped supply logistical support for large fires in Manitoba and Ontario. Quebec and New Brunswick belong to the Northeastern States Fire Compact; British Columbia has joined the WFCA. Under the auspices of the Food and Agriculture Organization (FAO) of the U.N., the United States has been the object of study tours by fire management specialists from other countries, and United States observers (mostly Forest Service) have toured Australia, Canada, and Mexico. The State Department has sponsored a fire tour to the Soviet Union. The Department of Defense helped sponsor a joint research program among the United States, Great Britain, Australia, and Canada on mass fires. The Peace Corps and the Agency for International Development (AID) have sent American advisers to numerous countries. The North American Forestry Commission (NAFC), an FAO subsidiary, includes a Fire Management Study Group (Sorenson, 1979). The NAFC sponsors an informative periodical on global fire problems, *Forest Fire News,* published by the Forest Service. Through the FAO, the Forest Service, private organizations such as SCOPE, and UNESCO—especially its Man and the Biosphere (MAB) program—important symposia have been conducted on fire problems in several parts of the world. The Mediterranean Basin, and by analogy lands subject to a Mediterranean-like climate, have been a special focus.

As more developing nations make the transition to industrialization, the prospects for technology transfer increase and fire management will be prominent among the desired skills. This need will be felt with special keenness in those semi-arid regions, long deforested, that are attempting to reestablish plantation forests and in the tropics, so long and extensively subjected to non-industrial practices. For the transfer to be effective, fire planning must pay special attention to local political systems, cultural history, and past fire practices, or else the transfer of bureaucratic structures and fire management practices will have little more to recommend it than did that earlier transfer from Europe to the United States. Moreover, for many countries wildland fire protection will develop from scratch, *ab initio,* not like current planning in the Unites States from within a well-established system. In this transition the role of prescribed fire will probably be crucial.

## 7.4 SELECTED FIRE REGIMES: ADMINISTRATIVE CHARACTERISTICS

The elementary dynamics of fire behavior, the fundamental principles of fire ecology, the shared experience in the techniques of fire management —all give fire agencies a common appearance as well as a common focus. But fire agencies belong in a bureaucratic environment whose mission extends beyond fire mangement, and it is within that environment that fire management must be conducted. As agencies acquire new and unique responsibilities, as they develop interlocking programs among themselves, and as they address fire within the constraints of their legislative mandate and their particular history, the character of their administration over fire regimes will differ. Consider, by way of example, the following sampling.

### *Lolo National Forest: The U.S. Forest Service*

Situated in the Northern Rockies, nearly surrounded by other similar units, the Lolo National Forest is in many ways typical of the multiple-use public lands administered by the U.S. Forest Service in the West. For Region One (Northern Rockies) the administration of fire reaches back to the Transfer Act; its lands were the scene for some of the great 1910 fires; large fires in 1934 influenced the adoption of the 10 AM Policy; the lightning bust of 1967 steered the Forest Service into interagency cooperation at BIFC; the early drive within the Forest Service to reintroduce natural fire into wilderness areas found here some of its first expressions. The region acquired an impressive assortment of fire resources along the way, mostly concentrated in Missoula, Montana: a large fire cache and smokejumper loft at the Aerial Fire Depot; the Missoula Equipment Development Center (MEDC); and the Northern Forest Fire Laboratory (NFFL). The Lolo National Forest was involved, in some way, with all these episodes of the region's fire history, and with its administrative headquarters located in Missoula it shares in the region's present-day wealth of resources.

As an administrative unit the Lolo National Forest is one national forest among 13 that make up Forest Service Region One (Figure 7-8). For the region as a whole there is a fire management officer (FMO) and a staff responsible for such tasks as fuels management, planning, aviation, fire prevention, and logistics. The Lolo is in turn subdivided into districts. The forest as a whole has a fire management officer with a small staff; each district has an FMO with assistants (AFMOs). Fire management proceeds in accordance with a fire management plan, which accounts for fire use as well as fire suppression. The fire plan, in turn, follows from the forest plan with its grounding in resource values and land management objectives.

In addition to administrative units such as the district, the plan divides the forest into fire management areas (FMA), each of which may be guided by different sets of responses. For some units natural fires will be tolerated, as long as they stay within specified geographic boundaries and fire behavior parameters. For other units a mixture of suppression and prescribed fire is in order, depending on season and management objectives. For still other units, suppression is the standing order—though even here, if initial attack should fail, fire officers must make an escaped fire analysis to determine what response is best suited to the particular circumstances of this fire.

For all fires, agency response conforms to predetermined procedures approved by the Service. On those FMAs designated for prescribed natural fire, the choice of action follows from decision charts that relate fire behavior, effects, and location to land management objectives. In areas marked for suppression, response follows predetermined manning and action guides as applied by experienced dispatchers. On those sites intended for prescribed fire, implementation follows from predetermined burning plans. For a wild (or escaped) fire three responses are allowed. The fire may be confined within predetermined boundaries, contained along its active perimeter, or controlled through extinguishment.

Wildfires are reported by lookouts at fixed towers and by aerial reconnaissance, the latter used mostly during times of high fire danger. The forest relies on a mixture of suppression resources: ground tankers, smokejumpers, smokechasers, helitack crews, and air tankers. The key positions in the fire organization—down to the crew boss level—enjoy permanent civil service employment status. Most of the personnel and equipment, however, are hired for the season, though many return for more than one season. For the advanced positions, especially beyond the district level, a degree in forestry is useful, often obligatory. Where these units are located and how they are deployed is specified by the fire management plan, which matches suppression strength against fire danger rating. Actual response to a wildfire falls under the direction of a fire dispatcher, who maintains a master file on available resources and consults manning guides and decision charts.

If initial attack fails, the dispatcher can request more aid in a steady escalation—from adjacent forests, from regional fire cache resources, from cooperators, from BIFC, and through BIFC and the interlocking agreements that form the fabric of fire management in the United States from virtually any suppression unit in the country. The degree to which the Forest Service controls the land base in the Northern Rockies, however, keeps the operation largely within its own purview. Conversely, should one of its cooperators in the area (the State of Montana or the National Park Service, for example) need assistance on a fire, the Forest Service would respond to the extent possible.

### *North Carolina: A State Operation*

Under the state forester, the North Carolina Forest Service has jurisdiction over all forest fires within the state. Exceptions to this mandate include fires within incorporated cities, for which urban fire services exist, and within federal lands, for which federal agencies are responsible. The extent of legal authority is equally great. Statutes provide for the prevention of fires through enforceable codes and even, with authorization from the governor, through the temporary closure of wildlands to recreational and industrial use; for the suppression of fires by active control measures; and for the control over prescribed burning through the issuance of permits. The charge to suppress wildfire grants to the state forester and his agents permission to commandeer equipment (e.g., crawler tractors) and able-bodied citizens between the ages of 18 and 45 for fire control activities, though such provisions are rarely invoked; state resources and their integration with rural fire departments provide ample manpower and equipment. State officials are absolved from charges of trespass when engaged in fire control. Even when functioning outside of state-owned lands, fire officers have the authority to engage in backfiring, line construction, and other measures required for suppression. It is a broad legislative mandate. North Carolina is a heavily wooded state, and its administration is typical of state responsibilities and operations in general, especially in the South where the federal government is not a major landholder.

Administratively, the North Carolina Forest Service belongs under the Department of Natural Resources and Economic Development. Although fire protection is a dominant assignment, the Ser-

KEY MAP

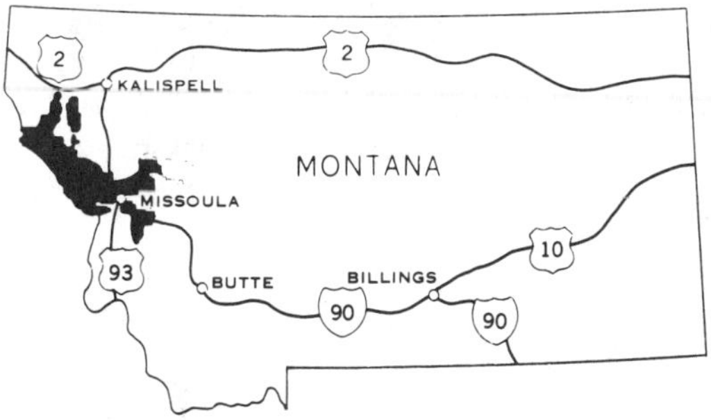


*Figure 7-8. National Forest System. The Lolo National Forest surrounds Missoula, Montana. From U.S. Forest Service.*

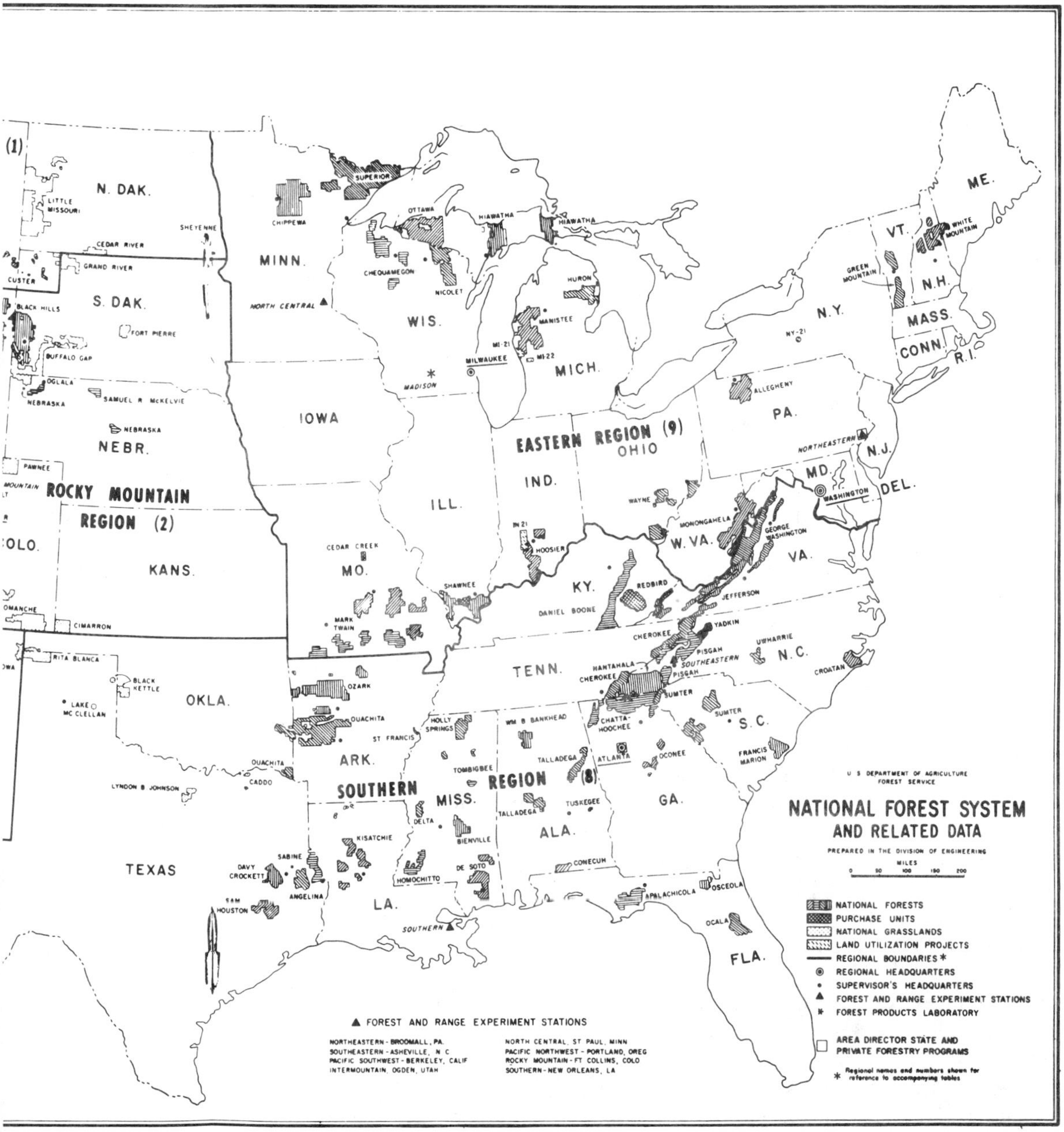
U S DEPARTMENT OF AGRICULTURE
FOREST SERVICE
NATIONAL FOREST SYSTEM
AND RELATED DATA
PREPARED IN THE DIVISION OF ENGINEERING
MILES
0 50 100 150 200
NATIONAL FORESTS
PURCHASE UNITS
NATIONAL GRASSLANDS
LAND UTILIZATION PROJECTS
REGIONAL BOUNDARIES *
REGIONAL HEADQUARTERS
SUPERVISOR'S HEADQUARTERS
FOREST AND RANGE EXPERIMENT STATIONS
FOREST PRODUCTS LABORATORY
AREA DIRECTOR STATE AND PRIVATE FORESTRY PROGRAMS
* Regional names and numbers shown for reference to accompanying tables
FOREST AND RANGE EXPERIMENT STATIONS
NORTHEASTERN - BROOMALL, PA.
SOUTHEASTERN - ASHEVILLE, N C
PACIFIC SOUTHWEST - BERKELEY, CALIF
INTERMOUNTAIN, OGDEN, UTAH
NORTH CENTRAL, ST PAUL, MINN
PACIFIC NORTHWEST - PORTLAND, OREG
ROCKY MOUNTAIN - FT COLLINS, COLO
SOUTHERN - NEW ORLEANS, LA
ROCKY MOUNTAIN REGION (2)
EASTERN REGION (9)
SOUTHERN REGION (8)
(1)
N. DAK.
S. DAK.
NEBR.
KANS.
COLO.
OKLA.
TEXAS
MINN.
IOWA
MO.
ARK.
LA.
WIS.
ILL.
MISS.
MICH.
IND.
OHIO
KY.
TENN.
ALA.
GA.
FLA.
W. VA.
VA.
N. C.
S. C.
PA.
N.Y.
MD.
DEL.
N.J.
VT.
N.H.
ME.
MASS.
CONN.
R.I.
LITTLE MISSOURI
CEDAR RIVER
SHEYENNE
GRAND RIVER
CUSTER
BLACK HILLS
FORT PIERRE
BUFFALO GAP
OGLALA
NEBRASKA
SAMUEL R McKELVIE
PAWNEE
CIMARRON
COMANCHE
RITA BLANCA
BLACK KETTLE
LAKE MC CLELLAN
LYNDON B JOHNSON
CADDO
SUPERIOR
CHIPPEWA
NORTH CENTRAL
CHEQUAMEGON
NICOLET
OTTAWA
HIAWATHA
HURON
MANISTEE
MI-21
MI-22
MILWAUKEE
MADISON
CEDAR CREEK
MARK TWAIN
SHAWNEE
HOOSIER
IN 21
WAYNE
MONONGAHELA
GEORGE WASHINGTON
JEFFERSON
REDBIRD
DANIEL BOONE
ALLEGHENY
NORTHEASTERN
WASHINGTON
NY-21
GREEN MOUNTAIN
WHITE MOUNTAIN
OZARK
OUACHITA
ST FRANCIS
HOLLY SPRINGS
TOMBIGBEE
DELTA
BIENVILLE
DE SOTO
HOMOCHITTO
KISATCHIE
SABINE
DAVY CROCKETT
ANGELINA
SAM HOUSTON
SOUTHERN
WM B BANKHEAD
TALLADEGA
TUSKEGEE
CONECUH
CHEROKEE
NANTAHALA
PISGAH
SOUTHEASTERN
YADKIN
UWHARRIE
CROATAN
CHATTAHOOCHEE
ATLANTA
OCONEE
SUMTER
FRANCIS MARION
APALACHICOLA
OSCEOLA
OCALA

vice does have other responsibilities—among them the management of state forests and a tree nursery. For its fire mission, the Service has divided North Carolina into three regions, mirroring the physiographic provinces of the state: the Coastal Plains, the Piedmont, and the Appalachian Mountains (Figure 7-9). Each region has different fuels, different problem fires, and different suppression resources. The regions are, in turn, subdivided into a total of 12 districts, each district composed of several counties. For each county there is a county ranger who has the direct responsibility for issuing burning permits and controlling wildfires. In some counties, where fires are large and the population small, the Forest Service controls fire with its own units. In others, the county ranger (and subordinates) coordinate and supplement local resources, usually in the form of rural fire departments.

If a fire grows, larger units of the chain of command are brought to bear. For conflagrations, a Big Fire Organization—predesignated by the state office at Raleigh—is activated to assume control over suppression. Each unit maintains a Readiness Plan, also predetermined, to guide its responses; such plans are tied to fire danger ratings supplied by the Raleigh office. All members of the Service undergo fire training. The state has developed training aids (including films) to supplement national courses, and three days of annual maneuvers each spring tests equipment and the preparedness of the Big Fire Organization.

The North Carolina Forest Services enjoys considerable suppression resources. The tractor-plow continues to be the mainstay of Southern fire control. In the Appalachians, where the plow cannot function well, ground tankers and helicopters are used; in the Piedmont, plows are supplemented by rural fire engines; along the Coastal Plain, high-flotation vehicles and airtankers form an effective initial attack unit for broad expanses of organic soils, industrial plantation forests, and pocosins. To these units the Service can add many others as a consequence of cooperative agreements with the U.S. Forest Service, forest industries, rural fire departments, other state agencies (e.g., Highway Commission), and the Southeastern States Forest Fire Compact. The agreements, of course, cut two ways, and the North Carolina Forest Service has dispatched heavy equipment and crews to support federal agencies in places such as Michigan and California. For fire research the state relies, in general, on the programs of the U.S. Forest Service, though it has occasionally supported research on topics of special interest to it such as organic soils or equipment development using the state forests as testing grounds.

### *Southern California: An Interagency Ensemble*

Its jumble of jurisdictions, its hybridization of urban and wildland firefighting techniques, and its history of almost annual conflagrations have brought a special form of fire management to the Los Angeles Basin and its environs. Firescope was conceived after the devastating fires of 1970. Primary support came from special congressional appropriations, with research conducted at the U.S. Forest Service Riverside Fire Lab, and portions became operational during the large fires of 1977. Firescope sought to integrate, on an operational level, the response to fire by city, county, state, and federal suppression agencies. Each unit had an autonomous history, and each had more or less complete capabilities for normal fire loads within its own jurisdiction. But the fire regime did not coincide with legal boundaries, and the pattern of mutual aid agreements that had grown up during the previous decades simply proved inadequate during exceptionally heavy fire loads. What was needed to oversee large fire complexes was a superorganization which could be activated during fire emergencies and which had the capacity to orchestrate fully the pool of available resources (Figure 7-10). This was the province of Firescope.

Among the participating agencies, one is an urban fire service, the Los Angeles Fire Department. Though it has long had to grapple with wildland fire situations (brushy hills incorporated into the city as parks or residential suburbs), the Department approached the subject from the perspective of municipal fire services, grafting on

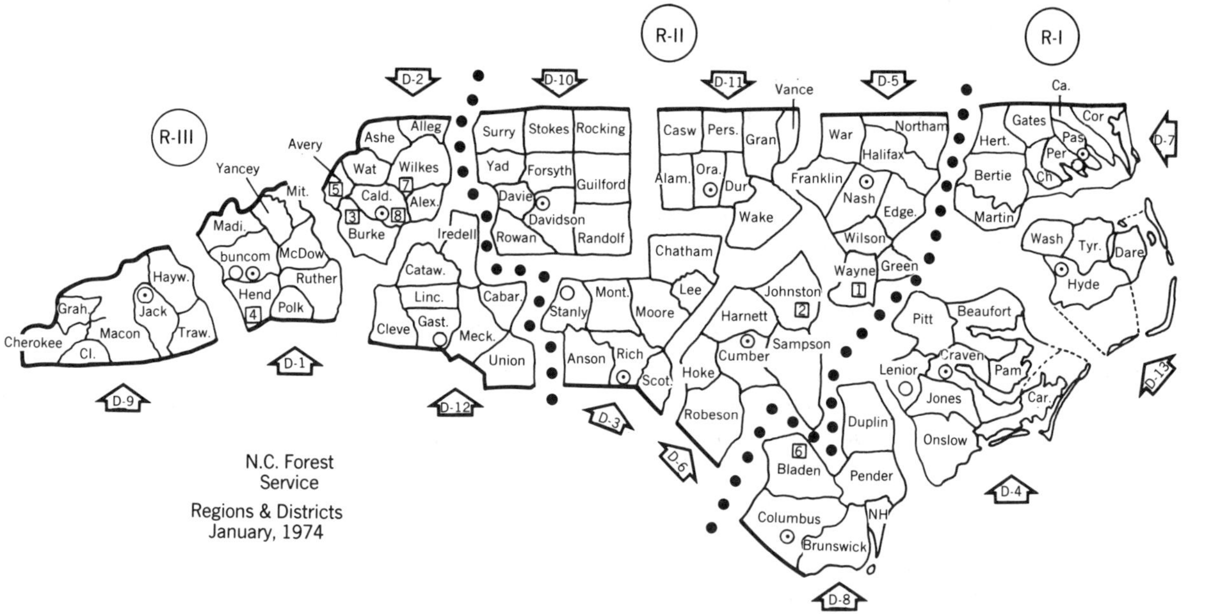


| | | | |
|---|---|---|---|
| ○ Region I | — Kinston | ⊙ District 6 | — Fayetteville |
| Region II | — Albemarle | District 7 | — Elizabeth City |
| Region III | — Asheville | District 8 | — Whiteville |
| ⊙ District 1 | — Asheville | District 9 | — Sylva |
| District 2 | — Lenoir | District 10 | — Lexington |
| District 3 | — Rockingham | District 11 | — Hillsborough |
| District 4 | — New Bern | District 12 | — Mt. Holly |
| District 5 | — Rocky Mount | District 13 | — Fairfield |

1. Goldsboro Forestry Center
2. Clayton Forestry Center
3. Morganton Forestry Center
4. Holmes State Forest—Hendersonville
5. Gill State Forest—Crossnore
6. Bladen Lakes State Forest—Elizabethtown
7. Rendezvous Mountain State Forest—N. Wilkesboro
8. Tuttle State Forest—Hartland

*Figure 7-9. North Carolina fire districts. From North Carolina Forest Service.*

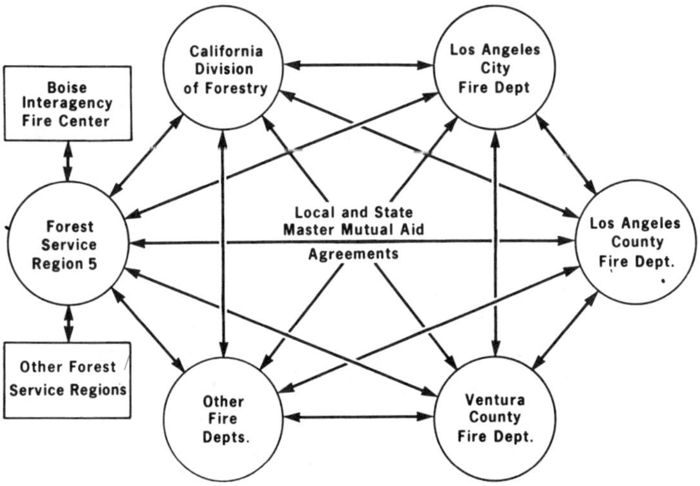


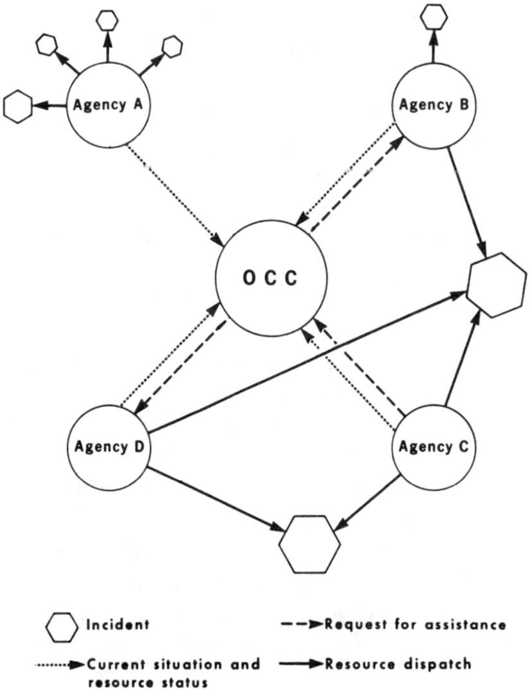

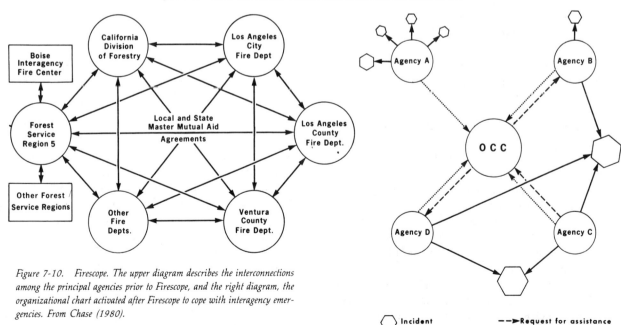

*Figure 7-10. Firescope. The upper diagram describes the interconnections among the principal agencies prior to Firescope, and the right diagram, the organizational chart activated after Firescope to cope with interagency emergencies. From Chase (1980).*

some special apparatus as the situation dictated, but otherwise maintaining its normal training, personnel, and procedures.

The three county fire departments—Los Angeles, Ventura, and Santa Barbara—that participated showed greater hybridization. Early in the century California had allowed counties to contract with the state for fire protection or to provide it themselves; five counties opted for separate protection programs, four of them in Southern California. Originally, this meant rural fire protection, but with explosive urbanization, the programs evolved into a combination of wildland and municipal fire services. Los Angeles County Fire Department, for example, furnishes primary urban fire protection under contract for numerous cities as well as for outlying wildlands and unincorporated developments. For the most part, the counties follow the model of an urban fire department, thus suiting their techniques to most of their values at risk.

The U.S. Forest Service, which administers the mountain watersheds surrounding the basin, follows the general pattern described for the Lolo National Forest. There are no natural fire zones, and for the most part prescribed fire has been limited to the maintenance of fuelbreaks, although a prototype program of broadcast burning is being conducted with the State. Otherwise the design and administration of the national forests resembles that apparent elsewhere in Region Five (California) and the West, in general. The program differs in intensity, not in kind.

The remaining participant is the State of California, both its Office of Emergency Services and its Department of Forestry (CDF). State responsibility under CDF resembles that under the North Carolina Forest Service. Even with large urban centers and the magnitude of federal land in California, state jurisdiction extends to over 24 million acres. The comingling of state and federal lands in portions of the state has led to contracts whereby the state furnishes primary fire protection to 4 million acres of federal lands and federal agencies protect nearly 5 milion acres of state lands. A State Fire Disaster Plan provides for mutual assistance among municipal, county, and state resources during emergencies, commonly invoked when large numbers of structures are threatened. For training CDF operates a Fire Academy; for information, it often assists U.S. Forest Service research programs and occasionally sponsors research of its own; and for additional manpower it operates a network of Conservation Camps in cooperation with the California Department of Corrections.

The state, moreover, has assumed greater and greater responsibility for rural fire protection as more development moves into former "wildlands." This it does by contract, providing either full-time protection or seasonal protection; the latter possibility occurs because the wildland fire problem is largely confined to certain months, freeing equipment for other duties in the offseason. One consequence is the development of dual-purpose equipment, notably ground tankers (engines). When not used for wildland fire, this equipment can be leased by rural departments; when most rural departments acquire equipment, they do so under the guidance of the state, ensuring a degree of standardization and cost efficiency. Under law the state has entered into mutual aid agreements with all the legally constituted firefighting agencies in California (Alaska Department of Education, 1977).

Firescope sought to integrate this array of agencies for response to specific, shared emergencies (Chase, 1980). The program did not propose a new model of general fire management, but a new methodology for coordinated fire suppression. At its core are two independent but ultimately interrelated systems: an Incident Command System (ICS), for handling individual incidents (such as a fire), and a Multiagency Coordination System (MACS), for processing various incidents that affect more than one agency. The result, known as an Operations Coordination Center (OCC), more resembles a clearing house than a command center. The Firescope concept requires uniform terminology, compatible communications systems, shared procedures (where possible), and mutually agreed upon training and certification. The organization proposed by ICS thus applies to any incident, not merely to an urban or wildland fire. Conversely, the consolidation proposed by MACS does not supercede the jurisdiction of any agency: each agency responds, wherever possible, to incidents within its own bailiwick. What it does do is process mutual assistance requests in a rapid, logical way according to criteria and procedures agreed to by all the participating agencies. The concept, in brief, does both more and less than what it seems to do.

The program demands considerable information and it looks to computer assistance for processing. The Firescope Information Management System (FIMS) organizes a comprehensive database linked to automatic weather stations, infrared mapping vehicles (where applicable), and 28 agencies. The system can furnish virtually instantaneous status reports on suppression resources, incident status, fire behavior predictions, and initial attack assessments. The same information, after the fact, gives a report and the same programs, before an incident, can be used for preplanning. The R&D program for Firescope centered in the Riverside Fire Lab, but there were numerous subcontracts to private research outfits that specialized in systems analysis.

At least portions of the Firescope concept are destined for widespread use elsewhere. The ICS component will likely underwrite future forms of the Large Fire Organization, notably the National Interagency Incident Management System (NIIMS); the certification criteria, an updated NIFQS; the Firecast module and Initial Attack Assessment Model, the automation of dispatching in areas of heavy fire loads; FIMS, mapping projects; MACS, interagency coordination at BIFC and in various regions. To assist the process of technology transfer, the U.S. Forest Service has inaugurated a special program, Firetip. But whether it disseminates nationally or remains in Southern California, whether its presence is felt as a whole or in parts, whether the national need is for better suppression capabilities or for better integrated programs of fire management, Firescope will continue to oversee a unique fire problem in a unique fire regime in a unique administrative ensemble.

### *Everglades National Park: Wilderness Management*

The administration of its fire regimes qualifies Everglades National Park as an exemplar of wilderness fire management. Its techniques of planning and field manipulation are typical of those used in primitive and wilderness lands. For the Everglades fire protection began in 1947, over a decade after establishment of the reserve. From the start there were serious problems—a source of ignition from lightning and incendiarism that made fires common, both in winter and summer; drainage and diversion of surface waters that magnified fuel conditions; a forbidding landscape for which suppression equipment developed elsewhere was not particularly well-suited and was often damaging; exotic flora that aggravated fire control and the agency mandate to maintain the land in a state of "naturalness." The difficulty of fire control, the prospects for controlled burning to reduce hardwood invasion of pinelands and grassy glades, early research into the character of fire ecology in the area and enhanced environmental research that appeared after Miami proposed to construct a giant airport in the vicinity, and rethinking of Park Service programs after the Leopold Report (1963) all conspired to create a new fire management plan for Everglades in 1972. By then prescribed natural fire had been proposed for wilderness areas in other regions and Descon showed how fires of any origin might be incorporated into a fire management plan.

The park fire plan recognizes three fire management units: an upland region of pine, a broad expanse of grasses interspersed with hardwood hammocks (islands), and a coastal belt of mangrove swamps (Figure 7-11). For a fire in any of these units, three responses are available: suppression, containment, or observation. What particular action results depends on the existing prescription, a formula especially sensitive to the soil moisture that determines so much of fire's effects in this biota, and to smoke, with its consequences for visitation and the pollution of nearby metropolitan centers. For every reported fire, a fire boss is assigned. The fire boss assesses the fire, declares it to be either a wildfire or a "management" (pre-

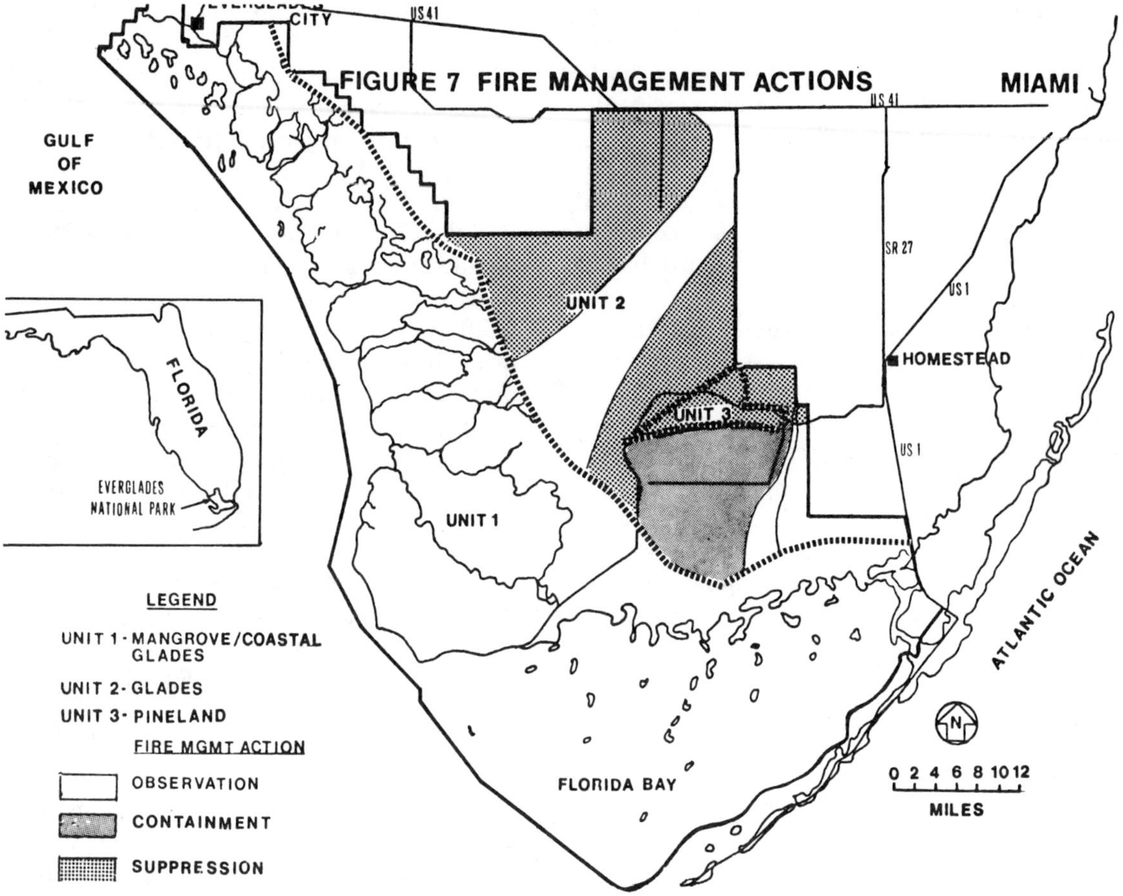


*Figure 7-11. Everglades National Park. Three fire management areas are recognized: (1), mangrove/coastal glades; (2), glades; (3), pine uplands. From Bancroft (1976).*

scribed) fire, and assigns to the fire a fire behavior officer and a fire researcher. Prescribed burning is widely practiced—in the pine uplands to retard hardwood incursion, in the sawgrass prairies to prevent decadence, and in pockets of land invaded by exotics. Most natural fires appear in the summer, and most prescribed burning is conducted during the winter. One result is a year-round fire program.

Fire management in the National Park Service is far less structured than in the Forest Service. Each park enjoys greater autonomy but has less backup from its parent organization. For Everglades a fire management officer runs the fire program under the aegis of resource management. Staff advisors work closely with fire management, and under the FMO there are technicians to measure prescriptions, monitor fires, begin prescribed burns, and suppress wildfires (Bancroft, 1976). Most of the crews are hired seasonally. For support on larger fires, Everglades must look to the U.S. Forest Service, the Florida Forest Service, and the interagency programs at BIFC. Most of the Park Service commitment to fire resides in its large western parks, from which Everglades is geographically remote. But the Park Service is a member of the NWCG and a participant in most national fire programs, and Everglades is far from isolated with respect to the national fire community.

Nor does it lack for information. Within the park itself there is the South Florida Research Center, part of whose mission is fire ecology. The park maintains a resident fire ecologist on its resource management staff. There are numerous research contracts and cooperative agreements made with universities. And because of its interest in smoke management in the South, the U.S. Forest Service has sponsored a special work unit for fire in South Florida within its Southern Fire Lab (Wade et al., 1980).

Far from being an anomaly, Everglades is in many ways typical of a whole class of lands—wilderness areas, biological preserves, wildlife refuges. In all these cases the perceived need is more for the planned introduction of fire than for its control. The strategy of multiple responses is a means of incorporating as many favorable ignitions as possible. Obviously, suppression capabilities are needed; an unwanted fire can often do more damage than a desired fire can do good, and prescribed burning assumes the capacity to control a fire should weather conditions change. But to satisfy the larger objectives of the park—the maintenance of a particular biota, or the processes that shape that biota—fire is essential. Wherever possible, natural fire is used. Where fuels have built up to unacceptable limits or exotics have infested an area, however, prescribed fire may be needed to prepare the site for natural fire. Where natural fire without controls cannot be tolerated, as around developments, prescribed burning will substitute for natural fire indefinitely.

The Everglades fire program must confront some difficult problems. The invasion of park lands by exotic pyrophytes such as maleluca, the continued diversion of waters from the glades, and the potential for air pollution over Miami as a consequence of prescribed fire all complicate the fire scene. Its management goals, too, are made needlessly complex by a desire to "restore" fire to an imagined pre-European state. Yet Everglades enjoys advantages as well—an environment uncomplicated by economic considerations, multiple-use objectives, routine interagency activities, or rural fire needs. Like its ecosystem, the Everglades fire program is unique. But its administrative principles are widely practiced, and it illuminates the paradox that, with regard to fire, even wilderness requires active management.

## *REFERENCES*

Surprisingly, there is no systematic survey of wildland fire administration in the United States. Perhaps the most useful introduction is Pyne (1982), *Fire in America,* though agency detail is not great and the references are scattered through a general historical narrative. For a sense of federal agency structure and mission, see the Praeger Library series, especially Frome (1971), *The Forest Service,* Everhart (1983), *The National Park Service,* and Clawson (1974), *The Bureau of Land Management.* An introduction to other federal agencies can begin with the Office of the Federal Register (1982), *U.S. Government Manual.* To bring these works up to date, consult Dana and Fairfax (1979), *Forest and Range Policy,* 2nd ed. The first edition (Dana, 1956) includes many more references to fire policy than does the second. A specific look at how the new legislation affects fire programs is given in Barney and Aldrich (1980), "Land Management—Fire Management Policies, Directives, and Guides in the National Forest System: A Review and Commentary."

The story of individual states is even more scattered. A general, but somewhat anecdotal digest is given in Widner (1968), *Forests and Forestry in the American States.* Most of the State bureaus have good descriptions of their programs in their files, but have not published them. For specific information, specific inquiries are needed. An overview of state organizations is listed in Society of American Foresters (n.d.), "Interstate Wildfire Organization Directory." The Clarke–McNary program through the 1950s is ably documented in

Pierce and Stahl (1964), "Cooperative Forest Fire Control: A History of Its Origin and Development Under the Weeks and Clarke–McNary Acts." Also see Dana (1956).

The international scene is fragmented. Winters (1980), "International Forestry in the U.S. Department of Agriculture," summarizes the involvement of the U.S. Forest Service. Sorensen (1979), "Seventeen Years of Progress Through International Cooperation, 1962–1978," describes the North American Forestry Commission. Wilson (1975), "Developing a Global Programme in Integrated Fire Management," surveys some international trends. Most of the major symposia of recent years have been global in scope and include some descriptions of fire relationships. *Forest Fire News,* published by the U.S. Forest Service for the NAFC, describes international developments.

The literature on fire economics is large, but a recent synopsis of economic theory is given in Gorte and Gorte (1979), "Application of Economic Techniques to Fire Management—A Status Review and Evaluation," and an excellent reference work has been published in Baumgartner and Simard (1982), "Wildland Fire Management Economics: A State of the Art Review and Bibliography." For more recent economic modeling, see Schweitzer et al. (1982), "Economic Efficiency of Fire Management Programs at Six National Forests", which tested forms of economic inquiry; U.S. Forest Service (1982), "Fire Protection on Non-Federal Wildlands," which extended the techniques to representative state operations; and Mills (1979), "Economic Evaluation of Alternative Fire Management Programs," which outlines the direction FEES is taking.

Alaska Department of Education, 1977, "Applicability of Wildlands Firefighting Techniques for Structural Fires," (Anchorage: Alaska Department of Education).

Allin, Craig, 1981, *The Politics of Wilderness Preservation* (Westport, Connecticut: Greenwood Press).

Arnold, Keith, 1949, "Economic and Social Determinants of an Adequate Level of Forest Fire Control," (PhD dissertation, University of Michigan).

Bancroft, Larry, 1976, "Natural Fire in the Everglades," pp. 47–60, in U.S. Forest Service (R-8), "Proceedings. Fire by Prescription Symposium," U.S. Forest Service, Southern Region.

Barney, Richard and David F. Aldrich, 1980, "Land Management—Fire Management Policies, Directives, and Guides in the National Forest System: A Review and Commentary," U.S. Forest Service, General Technical Report INT-76.

Baumgartner, David C. and Albert J. Simard, 1982, "Wildland Fire Management Economics: A State of the Art Review and Bibliography," U.S. Forest Service, General Technical Report NC-72.

Biddison, Lynn R., 1980, "Impacts on Federal Agencies," pp. 34–47, in Richard J. Barney (ed.), *Fire Control for the 80s* (Missoula, Montana: Intermountain Fire Council).

Brown, Arthur A. and Kenneth P. Davis, 1973, *Forest Fire: Control and Use,* 2nd ed. (New York: McGraw-Hill).

Chase, Richard A., 1980, "Firescope: A New Concept in Multiagency Fire Suppression Coordination," U.S. Forest Service, General Technical Report PSW-40.

Clar, C. Raymond, 1959–1969, *California Government and Forestry,* 2 vols. (Sacramento: California Department of Forestry)

Clawson, Marion, 1974, *The Bureau of Land Management* (New York: Praeger).

Clawson, Marion and Burnell Held, 1957, *The Federal Lands: Their Use and Management* (Lincoln: University of Nebraska Press).

Crosby, John S., 1977, "A Guide to the Appraisal of: Wildfire Damages, Benefits, and Resource Values Protected," U.S. Forest Service, Research Paper NC-142.

Dana, Samuel T, 1956, *Forest and Range Policy. Its Development in the United States* (New York: McGraw-Hill).

Dana, S.T. and Sally Fairfax, 1979, *Forest and Range Policy,* rev. ed. (New York: McGraw-Hill).

Davis, Lawrence S., 1971, "The Economics of Fire Management," pp. 60–69, in Southwest Interagency Fire Council, *Planning in Fire Management Proceedings* (Phoenix: SWIFCO).

Everhart, William C., 1983, *The National Park Service* (Boulder, Colorado: Westview Press)

Flint, Howard R., 1928, "Adequate Fire Control," *Journal of Forestry* **26**(5): 624–638.

Frome, Michael, 1971, *The Forest Service* (New York: Praeger).

Gale, Robert D., 1977, "Evaluation of Fire Management Activities on the National Forests," Policy Analysis Staff Report, U.S. Forest Service.

———, 1980, "The What, Why and Where," pp. 28–33, in Richard J. Barney (ed.), *Fire Control for the 80s* (Missoula, Montana: Intermountain Fire Council).

Gorte, Julie K. and Ross W. Gorte, 1979, "Application of Economic Techniques to Fire Management—A Status Review and Evaluation," U.S. Forest Service, General Technical Report INT-53.

Headley, Roy, 1916, "Fire Suppression. District 5," U.S. Forest Service.

———, 1943, "Rethinking Forest Fire Control," U.S. Forest Service, Northern Rocky Mountain Forest and Range Experiment Station, Research Paper M-5123.

Hendee, John et al., 1979, *Wilderness Management,* U.S. Forest Service, Miscellaneous Publication No. 1365.

Hornby, L.G., 1936, "Fire Control Planning in the Northern Rocky Mountain Region," U.S. Forest Service, Region One.

Jekel, Louis G., 1979, "Rural/Metro—A Commercial Approach," pp. 227–234, in Richard J. Barney (ed.), *Fire Control in the 80s. Proceedings of a Symposium,* (Missoula, Montana: Intermountain Fire Council).

Kelly, Asher W., 1980, "Impacts on State Agencies," pp. 48–62, in Richard J. Barney (ed.), *Fire Control for the 80s* (Missoula, Montana: Intermountain Fire Council).

Komarek, E.V., 1977, "A Quest for Ecological Understanding. The Secretary's Report," Tallahassee, Florida: Tall Timbers Research Station.

Marty, Robert J. and Richard J. Barney, 1981, "Fire Costs, Losses, and Benefits: An Economic Valuation Procedure," U.S. Forest Service, General Technical Report INT-108.

Mills, Thomas J., 1979, "Economic Evaluation of Alternative Fire Management Programs," pp. 75–89, in Richard J. Barney (ed.), *Fire Control in the 80s. Proceedings of a Symposium* (Missoula, Montana: Intermountain Fire Council).

Mills, Thomas J. and Frederick W. Bratten, in progress, "Design of a Fire Economics Evaluation System: FEES," U.S. Forest Service, General Technical Report.

Myles, George A., 1981, "Trends in Rural Fire Protection and Control—Expenditures, Acres Protected, and Number of Fires," *Fire Management Notes* **42**(3): 10–12.

North Carolina Forest Service, n.d., "State Fire Readiness Plan."

Noste, Nonan V. and James B. Davis, 1975, "A Critical Look at Fire Damage Appraisal," *Journal of Forestry* **73**(11): 715–719.

Office of the Federal Register et al., 1982, *1981/82 United States Government Manual* (Washington, D.C.: Government Printing Office).

Pierce, Earl and William Stahl, 1964, "Cooperative Forest Fire Control: A History of Its Origin and Development Under the Weeks and Clarke–McNary Acts," U.S. Forest Service, Cooperative Fire Program.

Pyne, Stephen J., 1982, *Fire in America: A Cultural History of Wildland and Rural Fire* (Princeton, New Jersey: Princeton University Press).

Schweitzer, Dennis J. et al., 1982, "Economic Efficiency of Fire Management Programs at Six National Forests," U.S. Forest Service, Research Paper PSW-157.

Show, S.B. and E.I. Kotok, 1923, "Forest Fires in California, 1911–1920," U.S. Forest Service, Agriculture Department Circular 243.

Simard, Albert J., 1976, "Wildland fire Management: the Economics of Policy Alternatives," Canadian Forest Service, Technical Report 15.

Society of American Foresters, n.d., "Interstate Wildfire Organization Directory."

Sorensen, James, 1979, "Seventeen Years of Progress Through International Cooperation, 1962–1978," U.S. Forest Service.

Sparhawk, W.R., 1925, "The Use of Liability Ratings in Planning Forest Fire Protection," *Journal of Agricultural Research* **30**(8): 693–762.

Steen, Harold K., 1976, *The U.S. Forest Service: A History* (Seattle: University of Washington Press).

Tall Timber Research Station, 1976, *Tall Timbers Fire Ecology Conference Proceedings,* Vol. 14 (Tallahassee, Florida: Tall Timbers Research Station).

U.S. Forest Service, 1980, "National Fire Planning Handbook," U.S. Forest Service.

———, 1976, "Evaluating National Fire Planning Methods and Measuring Effectiveness of Presuppression Expenditures," U.S. Forest Service, Report to the Chief.

———, 1978a, "Forest Service Manual, Title 5100—Fire Control"

———, 1978b, "The Principal Laws Relating to Forest Service Activities," U.S. Forest Service, Agriculture Handbook No. 453.

———, 1980, "National Forest System Fire Management Fire Budget Analysis," U.S. Forest Service.

———, 1982, "Fire Protection on Non-Federal Wildlands. A Report on the Efficiency, National Interest, and Forest Service Role in Fire Protection on Non-Federal Wildlands," U.S. Forest Service, Cooperative Fire Protection.

Wade, Dale et al., 1980, "Fire in South Florida Ecosystems," U.S. Forest Service, General Technical Report SE-17.

*Western Wildlands* (Summer 1974) 1. The entire issue is devoted to fire management issues.

Widner, Ralph, 1968, *Forests and Forestry in the American States. A Reference Anthology* (Washington, D.C.: National Association of State Foresters).

Winters, Robert K., 1980, "International Forestry in the U.S. Department of Agriculture" (Washington, D.C.: U.S. Forest Service).

Wilson, Carl, 1975, "Developing a Global Programme in Integrated Fire Management," in "Global Forestry and the Western Role," *Proceedings of the Permanent Association Committees* (Portland, Oregon: Western Forestry and Conservation Association).

Zivnuska, John A., 1972, "Economic Tradeoffs in Fire Management," pp. 69–74, in U.S. Forest Service, "Proceedings, Fire in the Environment Symposium." U.S. Forest Service, FS-276.

*Part Three*

# *Fire Management*

*Chapter Eight*

# *Programs for Fire Management*

The strategy of fire control, it will be recalled, has three components: to prevent ignition, to modify the environment in which a fire burns, and to suppress small fires before they can make the transition to large fire status. For fire agencies of any size these components result in separate but interrelated programs. There will be a prevention program, a fuel modification program, and a complex of activities—referred to as a presuppression program—which culminate in and support suppression. The *presuppression* concept is an elastic one. It can refer to such capital improvements as roads, lookout towers, and fire caches, to fuel management projects, or to such short-term activities as adding aerial reconnaissance patrols and beefing up suppression crews during periods of high fire danger. Suppression itself divides into two more or less separate organizations—one for initial attack and one for large fires.

The strategy behind fire management, as distinct from fire control, adds to these components a fourth: the substitution of prescribed fire for wildfire. Control and use are reciprocal activities, and programs for prescribed fire tend to build upon, and mimic, programs for fire control. In part, this reflects the history of fire protection in the United States, with its original imperative to break down old fire practices; in part, it resides in the logic imposed by fire behavior. No fire can be considered prescribed unless it can be considered controlled. Presuppression projects have a counterpart in preprescription projects. Both rely on similar preplanning considerations, both share similar objectives in land management, and both appeal to fire research for vital information.

That both fire use and fire control look to organized research is not accidental. Virtually every aspect of fire management is amenable to some form of scientific inquiry, and many dimensions (an ever-increasing number) depend on fundamental knowledge about fire behavior and fire ecology. How a program acquires information and how it applies that knowledge is basic to its mission. It may sponsor original research through its own facilities or by contract to others, or it may transfer information from another source, with appropriate adaptations, to its needs. Fortunately, the very nature of cooperative fire protection brings to its participants a flow of research information as well as ideas, equipment, and grants-in-aid. Not every source of data or every form of expertise derives from scientific investigation, but any program without a firm linkage to scientific research will soon lapse into ritual and administrative decadence, progressively isolated from the larger society which it serves. In the United States fire research and fire administration are inextricably intertwined in nearly all activities and at nearly all levels of management.

All of these activities and all these levels of management require planning. Especially as fire management enters a period of consolidation, plans by which to integrate program with program, agency with agency, region with region will acquire even greater importance. The complexity of the whole increases geometrically, and planning must keep pace. Much of the environmental legislation enacted during the 1970s, moreover, has demanded a substantial increase in the number and quality of plans. For many agencies the planning process assumes a separate function, commands research to meet its needs, and even develops plans by which planning may proceed.

## *8.1 FIRE PREVENTION*

The logic behind prevention programs is obvious: an ounce of fire prevention can be worth many pounds of fire damages and fire suppression expenses. But the establishment of suitable objectives and the evaluation of success, to say nothing of actual execution, can be troublesome. The problem, first of all, is not to eliminate all fires, but only those that are considered undesirable. This requires policy decisions at a level far above a functional program such as prevention. Many problematic fires, moreover, originate from causes seemingly outside the control of a fire agency. Such is the case with lightning, for example, and those instances of woodsburning where, as in Appalachian feuds, the motive lies buried in history. Solutions to ignition sources such as these demand techniques that may well be inaccessible to fire managers.

Even where the problem is more traditional and techniques are available, their success is difficult to describe. Measures of statistics on fire starts are reasonably good, but their interpretation is complex. How many fires were really prevented? How do these nonevents relate to damages averted or suppression costs avoided? Do fires, as a population of events, exhibit some predictable behavior such as epidemics against which the success of prevention programs can be measured? What ought to be measured, moreover, is not simply the number of starts but a ratio of actual starts to prevented starts. The number of starts has increased over the years, but it is likely that the ratio of starts to nonstarts has diminished. Although the actual ratio remains unknown, it best describes the success of fire prevention as a concept and a program.

Part of this conundrum derives from the way in which the questions are asked. Fire prevention can be identified as a distinct activity of fire management, but its accomplishments should perhaps best be evaluated within the total context of the success or failure of fire management as a whole. Its relationship to suppression, for example, is not as an inverse, but an obverse. Prevention and suppression programs are not substitutes, one for the other in mutual exclusion, but complements. Not all fires are preventable even in theory, and a single large fire can wipe out the gains of many years of successful prevention. Suppression capability is necessary to protect the gains made by a prevention program and, in many areas, to give that program a credible deterrent. Equally, no suppression program can hope to be cost-effective or ultimately successful without some means of regulating the number and timing of starts. Historically, prevention and suppression programs shared responsibility for dissembling the matrix of frontier fire habits. The prevention of wildfire, furthermore, made possible the introduction of prescribed fire, just as the ability to control fire has made possible the use of fire. It was not the exclusion of fire that was at issue, but the transfer of control over fire practices from one group to another. Both suppression and prevention were means of shaping this transfer.

Paradoxically, the success of any suppression and prevention program can contribute to the problems of the other. Suppression added new incentives for deliberate fire—job-hunting, harassment of government agencies, and burning for hazard reduction among them. Similarly, prevention programs can relieve some of the burdens on suppression programs by reducing the total fire load, but may enhance others by allowing fuels to accumulate and by complicating the process of legitimizing other forms of controlled burning. Suppression and prevention remain complementary: neither is complete in itself, and both are mediated by related programs of fuels management and presuppression.

### *Strategy of Fire Prevention*

Fire prevention seeks to eliminate the unplanned ignition, the accidental fire. By its very nature such events cannot be removed totally. Some accidental fires will always exist. Control over fire sources requires, in effect, control over people, machinery, and lightning, none of which is wholly predictable or governable. Chance will always intervene; the probabilities may be low, but they exist. The general strategy of prevention is to reduce this probability by separating ignition (risk) from fuels (hazards). This requires that the nature of the ignition be understood, that techniques be developed to reduce the probability of ignition, and that programs be established that can apply those techniques.

As an event a fire can be imagined as the conjunction of a source, a place, and a time. To reduce the likelihood of any component is to reduce the likelihood of the whole. The modification of source is the special charge of fire prevention programs; the modification of place, the work of fuels management; the modification of time, the duty of presuppression planning as manifested by fire danger ratings. The modification of place accepts the likelihood of ignition, but seeks to segregate it from fuels, to disentangle risk from hazard. The modification of time accepts the long-term existence of both ignition and fuels, but attempts to prevent their conjunction during periods of high fire danger. The modification of source acknowledges the reality of fuels and high fire periods, but tries to abolish, as far as possible, chronic ignition sources. Thus fuels management along a railroad right-of-way reduces the problem of place, whereas increasing suppression resources during peak burning seasons reduces the problem of time. Most fire programs will include a mixture of responses, but the prevention of ignition source is especially attractive because its effect is multiplied by reducing to a vanishing point the likelihood of all three elements coming together at once.

Only in rare events are all three elements responsive to the same process. When this occurs, however, the prospects of fire prevention and control are notoriously difficult. An example is a high wind like the Santa Ana. Here the winds simultaneously enhance the hazard of each component—place, by drying fuels; timing, by driving fires to large sizes quickly; source, by causing powerlines to sway and eventually perhaps to arc, break, and ignite ground fuels. Similar dangers attend arson fires under Santa Ana conditions, where the magnified hazard to place and time is apparently an incentive to introduce a source. Superior design can mitigate against powerline failure under such conditions, but only rapid suppression can, as yet, effectively protect against this form of incendiarism.

This is not to say that source is readily identifiable or easily controlled. On the contrary, few ignition sources, once created, disappear. Rather their number is ever compounded, like the acquisition of books in a library. Most ancient uses of fire persist, though often in diminutive forms—a pile of burning leaves instead of a broadcast burn through the woods, for example. Most fires result from ignition sources of great antiquity. To these sources industrial machinery continually adds others, and the mobility of modern life adds opportunities for the conjunction of source, place, and time. Fire prevention, in short, does not deal with a relatively constant ignition source interacting with a reasonably constant fuel, but with changing risks, hazards, and opportunities for their association. Historically, fire prevention was one means among many of breaking down fire practices of the American reclamation and replacing them with a more appropriate set. Many fire programs were unique to that historical situation.

Fire management is not simply a case of fire agencies interacting with fire regimes, but of societies interacting with land through the medium of fire agencies. Many fire prevention problems cannot be solved by instruments such as fire agencies but only by societal reform. An obvious case is the proliferating mixture of suburbs and wildlands, often with inadequate building and fire codes. A simple solution is zoning, the planned segregation of urban and wild lands. Yet zoning, which may restrict property values, is unacceptable to many communities. Another example is the persistence of frontier fire practices within portions of the South. Endemic woodsburning reflects a socioeconomic history of intermittent industrialism which intermingled agrarian and industrial landscapes. Like the mixture of wild and urban lands in Southern California, the compound has been volatile. No simple solution is possible, and no program of fire prevention alone can solve it.

Although fire agencies may be vigorous advocates of such programs, they can neither create nor enforce remedies by themselves. They can only express society's will in the forms allowed them. The refusal to prohibit fireworks, to eliminate wooden roofs, to regulate debris-burning, or to condemn vestigial woodsburning—all are statements that society is willing to tolerate some fires for the sake of other values. Even apart from accidental fires, that is, all fires are not preventable because fire management conflicts with other social goals. Solutions lie outside the capabilities of the fire agency.

The encouragement of prescribed fire, like the promotion of suppression in an earlier age, has had ambivalent consequences for a prevention program. It eliminates certain categories of fire by reclassifying them as prescribed (or management) fires rather than as wildfires. But just as suppression added new incentives for ignition such as job-hunting, so prescribed burning programs have increased another category of wildfire, escapes from prescribed burns. Prescribed fire programs have not eliminated a source of ignition. Rather they have transferred the torch from one group of people to another and altered the timing of ignition. In a sense, prescribed burning modifies source by modifying timing, and in some instances, by modifying place in the form of fuel preparation. But it also compounds the problem of source when controlled burns escape. By 1980 some of the largest wildfires were escape fires from prescribed burns; although prescribed fire planning is not generally regarded as the purview of the prevention program, it might be so considered.

The point is that fire prevention is not an isolated function. It participates in a general program of fire management, whose separate components are divided for administrative convenience, but each function of which is interrelated with all the other functions. Should any one component change, so will all the rest. The success of any part is measured by the success of the whole.

### *Techniques of Fire Prevention*

Effective prevention begins with the identification of problem fires. This is not so simple as it seems. Statistics help, but they are limited, on one hand, by the accuracy of the fire investigation and, on the other, by the processing given them. Classification into groups—the U.S. Forest Service, which oversees national statistics, recognizes eight groupings—gives a broad brush analysis

1. *Equipment.* A fire resulting from use of equipment.
2. *Forest Utilization.* A fire resulting directly from timber harvesting, harvesting other forest products, and forest and range management except use of equipment, smoking, and recreation as related to the above activities.
3. *Incendiary.* A fire willfully set by anyone to burn vegetation or property not owned or controlled by him and without consent of the owner or his agent.
4. *Land Occupancy.* A fire started as result of land occupancy for agricultural purposes, industrial establishment, construction, maintenance, and use of rights of way and residences except use of equipment and smoking.
5. *Lightning.* A fire caused directly or indirectly by lightning.
6. *Recreation.* A fire resulting from recreation use except smoking.
7. *Smoking.* A fire caused by smokers, matches, or by burning tobacco in any form.
8. *Miscellaneous.* A fire of known cause that cannot be properly classified under any of the other seven standard causes.

*Figure 8-1. Categories of fire cause. From U.S. Forest Service, Cooperative Fire Program.*

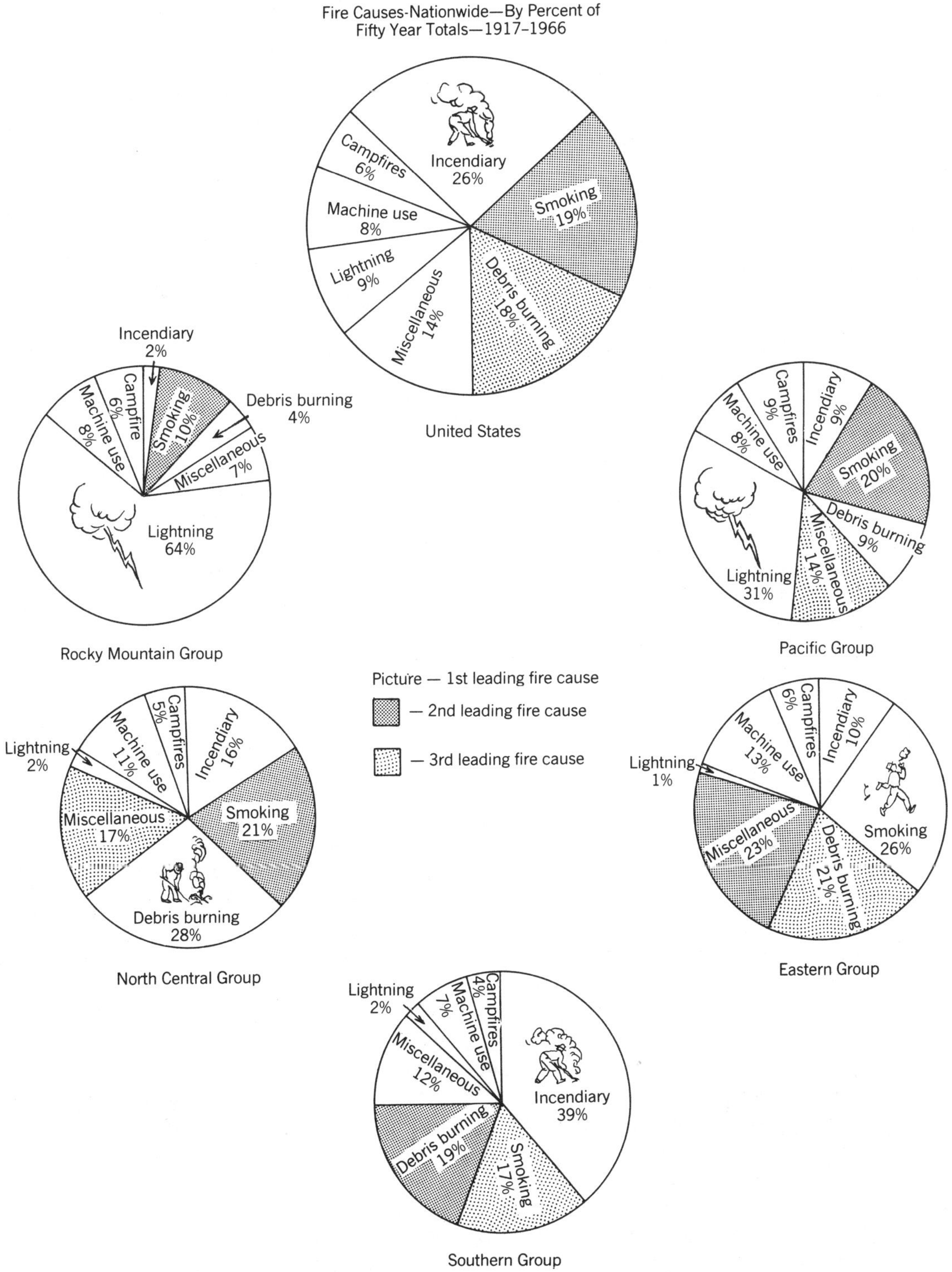


*Figure 8-2. Fire causes. Fire occurrence nationally (top), and by region (bottom). From U.S. Forest Service, Cooperative Fire Program.*

(Figures 8–1 and 8–2). Yet for solutions to be effective, they need to be specific; precise data as to source, place, and time are essential. For some areas fire reports may contain adequate information. Where the source of ignition is persistent, or locally intractable, formal research may be brought to bear.

Such research may take many forms: legal, to investigate suitable statutes; sociological, to discover the motives behind endemic carelessness or incendiarism; scientific, to engineer arrestors for spark-producing machinery; administrative, to design programs with better advertising appeal or public relations value. Research in prevention goes back to the origins of systematic fire protection, and all fire agencies have engaged in it. Repeatedly, states have sponsored inquiries for more precise identification of fire origins and better methodologies for their prevention. Private protection organizations, like the WFCA, have undertaken programs for public education about fire. And the U.S. Forest Service has exercised a degree of national leadership.

Through the Service are gathered national statistics on fire. Through its Cooperative Forest Fire Prevention (CFFP) program is coordinated the Smokey Bear campaign. From its equipment centers come engineering devices for the control of machinery-related ignitions. And from its research organization come special projects for particularly troublesome prevention problems. Prominent examples include the prevention analysis unit established at its Pacific Southwest Experiment Station (California), which addresses man-related fires peculiar to the West; a unit managed as a cooperative program with the Social Sciences Research Center at Mississippi State University, which inquires into persistent incendiarism in the South; and Project Skyfire, administered out of the Intermountain Experiment Station, which had for its ultimate goal the investigation and suppression of lightning fire.

Still other channels exist for the dissemination of prevention research and techniques. In keeping with the interagency reorganizations that characterized the 1970s, the NWCG sponsors a Prevention Committee to coordinate national goals and disseminate information. The NARTC offers courses in fire prevention. Occupational periodicals (such as *Fire Management Notes)* routinely describe state, private, and federal prevention programs, and assorted guidelines and handbooks are distributed through interlocking administrative channels. A summary digest is available in the NWCG *Fire Prevention Handbook.*

Once a specific cause is identified, specific remedies can be proposed. Traditionally, prevention specialists select from a range of options identified as the "three Es: education, engineering, enforcement. The best programs include mixtures of responses, perhaps escalating in intensity as the need arises. With such techniques fire administrators seek to eliminate the accidental fire and to replace the incendiary fire with prescribed fire. Most fires continue to originate from traditional sources, misplaced burning, or ignorance and incompetence in the handling of deliberate burns. What prevention often addresses, in macrocosm, is a change in fire practices that promises to result in a change of fire regimes. It may be as inappropriate to prevent lightning fires in selected wilderness sites as to introduce promiscuous woodsburning into a pine plantation.

*Education.* Education as a technique of fire prevention refers to that ensemble of processes by which people are informed and persuaded. Its domain is partly instructional, partly promotional, and wholly motivational; its devices may merge indistinguishably with those of commercial advertising. Under this rubric people are instructed in proper fire practices, informed about fire damages and costs, warned abaout legal responsibilities under fire codes, and made aware about fire danger in particular wildlands.

Instruction may consist of showing the proper methods of building and dousing a campfire, or describing wise practices in managing the vegetation about a rural dwelling. Information may come from media broadcasts, personal contact, posters, or fire danger warning signs along a roadway. Fire prevention messages have been successfully carried into classrooms, community organizations, and shopping centers, and they have been packaged into exhibits and traveling shows. The actual design of a project depends on the specific character of the problem: rural dwellers, consistently careless with debris burning, for example; children, playing with matches in flammable suburbs; transients, like hunters and campers, unfamiliar with proper fire practices or indifferent to land not their own. In all cases, however, the ultimate objective is to rally self-interest, which will bring with it self-restraint.

Suiting the message to the problem and delivering it to the right group is essential. In this regard fire prevention resembles the marketing of books. Only a few books—the blockbuster species—deserve mass market advertisement, or will benefit from it. Most books sell to particular, often well-defined groups. Promoting a scholarly monograph as though it were a best seller will waste money and probably miss the real target audience. Sending brochures to select mailing lists is far more productive. Moreover, most promotion comes by word-of-mouth, by the example of significant opinion-makers. An important strategy, then, is to get endorsements from appropriate sources such as reviewers. Much the same considerations guide successful prevention programs. Matching particular messages to particular audiences is fundamental, and where problem ignitions result from human activity, the identification and conversion of local opinion-makers is essential. A few well-placed allies among an indigenous population will buy more results than endless television advertisements.

Still, nationally inspired campaigns with widely broadcast messages are useful, and they help to keep the fire agency, if not the fire prevention message, in the public eye. In the long run such publicity can enhance the prevention program by keeping the fire program as a whole (or its host institution) in the public eye. Two national media campaigns, in particular, have proven successful. Keep America Green is supported by the private sector, the American Forest Institute, and it is administered through state branches, often in cooperation with the office of the state forester. The Cooperative Forest Fire Prevention program, with its centerpiece, Smokey Bear, is overseen by the Cooperative Fire Program of the U.S. Forest Service, again with distribution through the state foresters.

Both campaigns rely on basic advertising techniques: an identifiable logo or emblem, a preference for pictures over words, the repetition of slogans, and an appeal to emotion, not merely to reason. The Smokey Bear program aims its message primarily to preadolescents. So successful has the program been that Smokey (or in some cases Smokey-surrogates) have been exported to many countries outside the United States. Building on its example, the southern states and the Forest Service devised a complementary program, centering on Mr. Burnit, to address incendiary woodsburning by adults. Professional advertising agencies have contributed services *pro bono* through the auspices of the Advertising Council.

*Engineering.* By engineering as a technique of fire prevention is meant the separation of risk from hazard by deliberate design. Educational techniques involve people and traditional fire practices as a source of risk; engineering also considers machinery. Railroads start fires from sparks thrown out of smokestacks or emitted from faulty brakes, powerlines can break or arc in high winds, chain saws and bulldozers can discharge sparks from their exhaust ports the roster can be extended indefinitely, with new sources of industrial ignition inexorably added. The design of rural houses, of fire grates in campgrounds, and of right-of-ways for spark-prone machinery all can be engineered to reduce the problem of fire starts. In some cases the problem can be solved by new equipment such as a superior spark arrestor; in still others, by routine inspection to check for the faulty installation or poor maintenance of existing arrestors.

Considered historically from the perspective of fire in general, not merely wildland fire, the trend has been to replace free-burning fire with confined fire. With the industrial revolution fire was even housed by machinery (heat engines). The results have been ambiguous. In some instances the evolution of efficient machinery has reduced risk. When originally invented, the steam locomotive was a prolific source of ignition. The conversion of engines from wood to coal fuel, however, reduced fire starts considerably, and from coal to diesel still more. Railroads continue to start a disproportionate number of fires in certain regions such as the Northeast and Lake States, but the problem is less serious than it was a century ago. In other instances new technology introduces additional risk. The development of catalytic converters for automobiles, for example, may have introduced an unwanted fire source to grasslands.

As a partial solution permits are required in order to use machinery in state and federal forests, with the permit contingent upon an inspection. During periods of high fire danger, closure of wildlands to machinery may be invoked. The U.S. Forest Service has promoted research into engineering solutions to machine fires at its San Dimas and Missoula Equipment Development Centers. The NWCG distributes field guides with standards for powerlines, railroad locomotives and right-of-ways, industrial machinery, and residential dwellings.

*Enforcement.* Education and engineering are techniques for coping with accidental fire; enforcement, with arson or with fires occurring under such high-risk circumstances that they may be considered incendiary. The enforcement option attempts to regulate behavior under the threat of legal action. Ultimately, conformity rests on the consent of the governed, on getting people to want to do what they have to do. But a fire prevention program based solely on such expectations is hopelessly utopian. Chance will always enter into the equation, whether it be in the form of accident or carelessness. Some people, true pyromaniacs, will not respond to reason or deterrence; their pattern of starts will show a randomness tied to their inner compulsions. And there will be criminals who simply defy the law, arsonists intent on calculated, malicious violence. Engineering may produce a spark-arrestor, but good will does not guarantee that it will be used.

If only held in reserve, then, fire prevention programs must include the force of law for their credibility. During critical periods, risk and hazard must be forcibly segregated. Enforcement includes a range of responses, not a single act; it is a last resort, not a quick fix; but its power is fundamental. Taken in the broadest sense—in the form of fire codes and enabling legislation for fire agencies—legal acts underwrite the entire capacity of a public institution to prevent and suppress fires for the public good.

Statutory powers include the right of administrators of federal lands and of state governors for state lands to limit access to those lands during times of high fire danger. The degree of closure can vary. It may deny smoking privileges along trails or open fires at campsites, close roads to recreationists or loggers, shorten the hunting season or suspend the traditional season for debris burning, or close an entire forest to all use. Typically, the degree of closure will depend on the magnitude of the fire danger, with an escalation of closures accompanying the advancing fire season. Even where the lands themselves are not under agency control, activities on those lands such as hunting or debris-burning may be subject to agency regulation. Most states, for example, permit debris-burning only during specified times or under permit from proper authorities.

In these matters, as in others pertaining to fire codes, both criminal and civil penalties exist for violations. Negligent fire-setting (a form of trespass) is recognized as a misdemeanor; willful fire-setting, as a felony. Apart from criminal intent or culpability, people who start fires that burn beyond their private property incur civil liabilities. They may be sued for damages, a form of restitution that harks back to ancient times. For fires on public lands, the responsible party may be charged with fire trespass and sued for damages, including reasonable costs for suppression. The responsible party may be an individual or a corporation. A railroad company, for example, may be held liable for fires that escape their right-of-ways. The number of cases prosecuted under such legislation is not large, but its deterrent value does not require many cases.

Fire codes, moreover, work two ways. They proscribe and they prescribe. They may restrain the behavior of certain parties, by restricting their fire practices. But they also direct the behavior of others such as fire agencies by commanding them to engage in certain activities for the prevention and control of fires. The Supreme Court has ruled that government agencies can be held liable for negligence, and thereby sued for damages, even when they perform in a uniquely governmental capacity as a public firefighter. The particular case involved a large fire on a national forest that was controlled, then patrolled without being mopped up. The fire eventually blew up and overran surrounding private lands (Rayonier Inc., 1957).

Ironically, some problems of fire starts reside uniquely with fire agencies. Escape fires from prescribed burns, supposedly suppressed fires that reignite and escape old control lines, and fires set for employment or overtime pay all are internal to a fire protection system. Job-hunting fires—a kind of arson-for-profit—have plagued all fire agencies from their origin, though reliance on large fire caches and highly mobile fire crews has reduced considerably the incentive for local residents to set fires for economic gain. In contrast to early days, when local men were hired as firefighters and local merchants were paid to furnish supplies, suppression funds do not typically go to those areas where the fire occurs. By segregating opportunity from incentive, risk from hazard, the interregional fire suppression organization serves a fire prevention function.

### *Weather Modification*

In the Far West, Alaska, and Florida, lightning is a prominent source of fire. Nationally, about one fire in ten results from lightning. For the national forest system, lightning causes an average of 10,000 fires annually, and as many as 15,000 in troublesome years. For many primitive areas, lightning may be the source for virtually all fires. Lightning is not distributed with geographic randomness, nor are lightning fires. Semiarid regions, areas that experience a dry season, and areas that are prone to dry lightning storms are more likely to experience heavy lightning fire loads even though other areas may have lightning more frequently. Problem fires come in clusters (known as "busts"), overloading efforts to control dozens or hundreds of fires. They concentrate in remote and mountainous areas that lack easy accessibility to suppression units and do not have resident populations that can control other kinds of accidental starts such as those associated with machines and people. Thus summer lightning storms may compound source, place, and time with disastrous results. The costs of maintaining suppression capabilities for these overloads—resources that are heavily dependent on aircraft—are such that a small diminution in lightning fire starts in selected areas and at critical times could return big dividends.

Lightning fires are also responsible for the paradox of fire management in wilderness areas. They compel a decision. The source of fire does not reside solely in human hands, though the outcome of lightning fire can be modified by human intervention. That lightning kindles a fire which is simultaneously a wildfire and a natural process confronts wilderness managers with something that is, at one and the same time, a problem to control and a process to exploit. For all these reasons an understanding of lightning fire mechanics is desirable, and control over the process, a useful technique of fire prevention.

Research results are equivocal, however. An adequate theory for electrical charge generation and separation within a storm is still elusive. Consequently, research has concentrated on the modification of electrical development by the modification of cloud development. Inhibition of vertical development, enhancement of precipitation, or dissipation of electrical charge buildup within a cloud all can modify lightning by modifying its medium, the cloud. Preliminary data based on massive seeding in individual storms have suggested that cloud-formation can be inhibited, that the total frequency of lightning strikes can be diminished, and that the character of the lightning discharges that do result are less favorable to ignition. The U.S. Forest Service conducted most of these experiments during its Project Skyfire program in the 1950s and 1960s (Fuquay and Baughman, 1969). The Bureau of Land Management contracted with companies during the 1970s to perform field experiments in Alaska. The outcomes were sometimes promising, but not precise enough to warrant continuance of the program, and the trials, as such, were abandoned (Bureau of Land Management, 1969–1973).

The outcome of lightning suppression experiments based on cloud modification thus follows the trend with weather modification in general. Experiments in cloud-seeding to induce rain, to diminish hurricanes, and to dissipate severe thunderstorms have all ended in scientific uncertainty. Social and legal concerns about weather modification have further complicated trials. Instead the Forest Service has redirected its lightning research project from the mechanisms of lightning suppression to models for lightning fire forecasts suitable for use within the risk component of the NFDRS. Attention has shifted from the question of source to the problem of timing, from the early suppression of lightning flashes to the rapid suppression of lightning fire.

## 8.2 DETECTION AND COMMUNICATION

For suppression to be effective it must be rapid. Ideally a fire should be controllable by the smallest suppression unit, and control forces should arrive before a fire makes the transition from initiating to steady-state fire, certainly before it escalates into large fire status. The first step in this chain is to discover the smoke quickly and to report its location accurately. The appropriate elapsed time for the detection function will, of course, vary with timing and place, for little about an administrative unit is uniform. Hazards are distributed unequally, fuel and weather conditions fluctuate widely, and ignition is intermittent rather than continuous. What kind of detection system is most suitable and what sort of standards best apply will be set, ultimately, by the larger objectives of fire management for the administrative unit in question. As circumstances change, so will the character of the detection function, though each new system will inevitably build upon its predecessors.

The detection program, moreover, relates not only to suppression but to all aspects of fire mangement. A toughened presuppression goal (e.g., a 10 Acre Policy), the inclusion of structures within protection boundaries, the use of prescribed natural fire—all change the standards for discovery. Smoke from prescribed burning reduces visibility generally, making it more difficult to detect wildfire smoke or to discriminate approved from escaped fires within a prescribed burn. Spot programs in prevention, fuels management, and patrols by suppression units may all be necessary to cover lapses in the detection network—areas temporarily or permanently blind to an existing system of lookouts.

In the design of a detection network, the question of diminishing returns is ever present. To avoid it, detection programs must rely on a mixture of techniques that adjust their particular capabilities to the probable distribution of starts by source, place, and time. Fundamental to all detection programs are the requirements that it be speedy and accurate. Detection must be systematic and it must have the capability of reporting its discoveries quickly to dispatchers.

### Detection Methods

*Patrol.* The oldest form of detection relied on systematic patrols by firefighters and on reports by cooperators, including those people using the lands under permit or enjoying it for recreation. The defects of such a system are obvious. But at a time when suppression was less specialized than it has become, when the same persons who discovered a fire would also fight it, and when the mere presence of a ranger (with authority to make arrests for fire trespass) was a form of prevention, the systematic use of patrols offered many advantages. In particular circumstances it still does.

Beefing up patrols in areas of high hazard or high risk can supplement other detection methods, especially when the problem will exist for only a finite period of time; such would be the case with logging slash or holiday tourist traffic. Patrols can cover areas otherwise missed from regular lookout posts. This can be especially helpful during periods of high fire danger. By reducing travel time, patrols can compensate somewhat for their restricted range of coverage. Similar considerations can apply to the use of cooperators. Air traffic over the Grand Canyon, for example, is intensive, with much of the Canyon routinely flown several times daily by sightseeing services. Commercial flights report a majority of Canyon smokes in swift order; the remainder of the park detection system is built around this predictable source of observation.

*Fixed Ground Detection.* The hour control program demanded more systematic coverage, however. As a standard it required that a detection unit sight a 1/8 acre fire within the elapsed time prescribed for that fuel, that the administrative unit have a fixed network of observation posts which could cover 75–80% of the land under protection, and that 60% of that coverage come from two or more lookouts. To assist an individual lookout, devices (firefinders) were installed that gave azimuth readings (and with the Osborne model, a distance calculation). To weld individual posts into an interconnected system, with the possibility of cross-readings that could accurately fix the location of a smoke, special planning methods were devised. Calculations were simplest for level terrain, in which every lookout had an effective coverage described by a circle. For mountainous landscapes, planning became more laborious, although it often attained a high degree of sophistication.

Typically the design included a network of fixed observation posts. Three categories were recognized: primary posts, manned continuously during fire season; secondary posts, manned intermittently as fire danger or hazard demanded; and temporary posts, used during emergencies or by firefighters en route to a fire to better pinpoint the fire's location. Broadcast coverage could come from a small handful of lookouts, but covering blind spots could quickly inflate the required number of lookouts beyond the point of diminishing returns. A hierarchy of posts, some unemployed intermittently, offered a partial solution (Figure 8-3).

Most of the lookout towers installed throughout the United States, or the locations for such lookouts, emerged during the era of hour control planning (Show et al., 1937). They thus share in the problems and successes of that concept. The system assumed that detection, like suppression, would develop on the ground. A system of permanent, fixed lookouts worked best when it was matched with a landscape of fixed hazards and risks, where source tended to coincide with place, as modulated by timing. It worked splendidly for coverage of fuels and ignition that resulted from intensive human activity. Where fires came randomly, intermittent as to time and place, the system could be efficient but not so cost-effective.

To retain the values of a fixed lookout without its fixed high costs, several modifications have been advanced. One is the portable lookout, a trailer with observation post. The facility can be moved to areas of high hazard such as logged terrain, and then relocated when the worst fuel conditions have abated. Another experiment is to substitute television cameras for an observer, hoping to lower costs through automation. In theory the concept is identical to those security systems for buildings that cover sensitive areas with cameras. But TV does not distinguish smoke as readily as the human eye, costs remain high, and application appears limited. A combination of scanner with infrared sensor remains a possible compromise.

The relative success of hour control planning meant that the fire detection system of the United States was largely based on fixed lookouts and

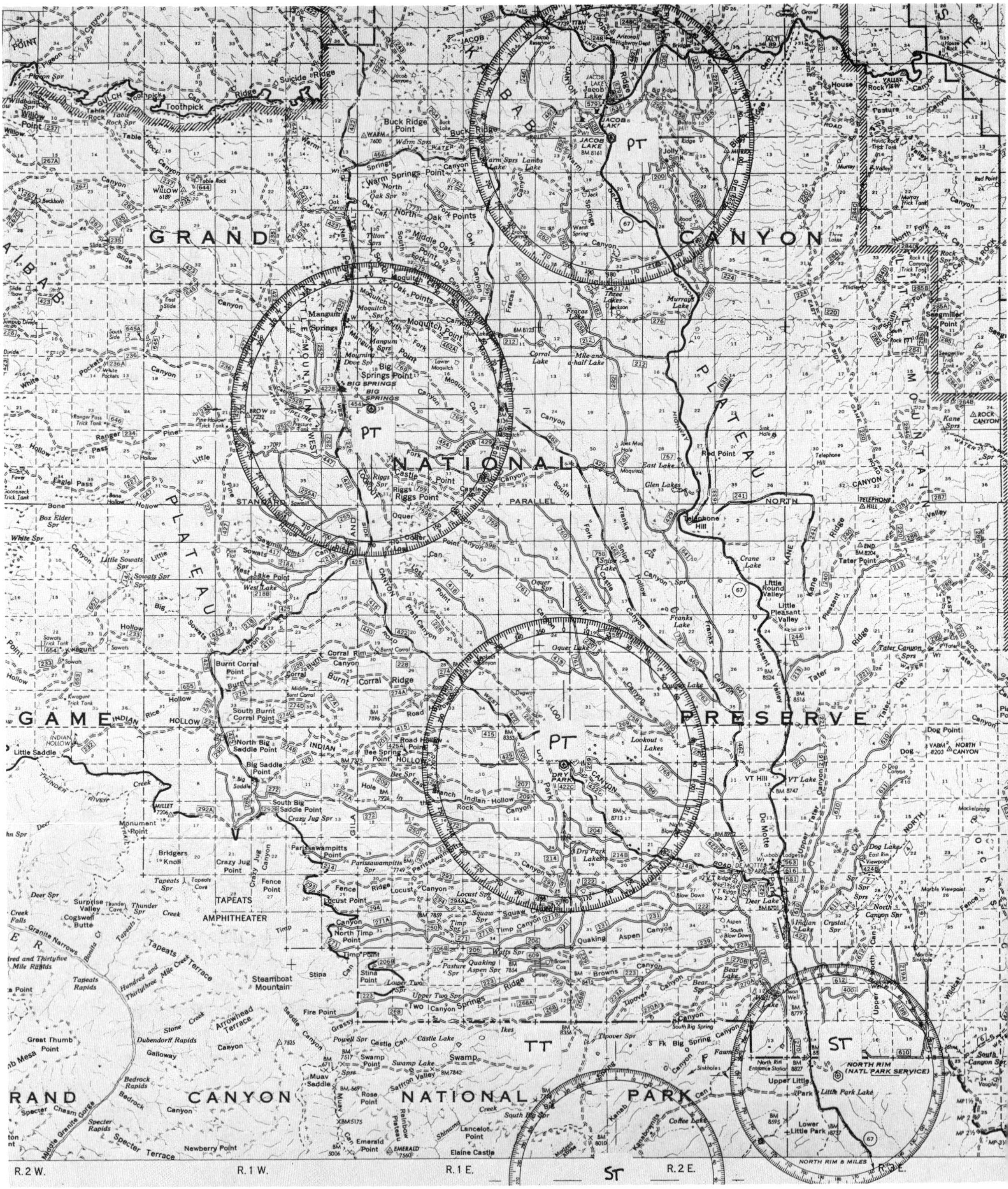
GRAND CANYON
NATIONAL
GAME PRESERVE
KAIBAB PLATEAU
PLATEAU
GRAND CANYON NATIONAL PARK
PT
PT
PT
ST
ST
TT
Jacob Lake
Buck Ridge Point
Warm Springs Point
North Oak Points
Moquitch Point
Big Springs
Castle Point
Riggs Point
Telephone Hill
Red Point
Tater Point
Crane Lake
Little Round Valley
Little Pleasant Valley
Franks Lake
Lookout Lakes
Dry Park Lakes
Deer Lake
Indian Lake
VT Hill
VT Lake
De Motte Park
Dog Lake
Dog Point
Burnt Corral Point
South Burnt Corral Point
North Big Saddle Point
Big Saddle Point
South Big Saddle Point
Monument Point
Bridgers Knoll
Crazy Jug Point
Fence Point
Locust Point
North Timp Point
Timp Point
Stina Point
Fire Point
Steamboat Mountain
Tapeats Amphitheater
Great Thumb Point
Dubendorff Rapids
Galloway Canyon
Bedrock Rapids
Specter Rapids
Specter Terrace
Newberry Point
Rose Point
Emerald Point
Lancelot Point
Elaine Castle
Muav Saddle
Swamp Point
Powell Spr
Castle Lake
Swamp Lake
Saffron Valley
Little Park Lake
Coffee Lake
NORTH RIM (NATL PARK SERVICE)
North Rim Entrance Station
Upper Little Park
Lower Little Park
NORTH RIM 6 MILES
Toothpick Ridge
Suicide Ridge
Table Rock Canyon
Little Saddle
Sowats Point
West Lake Point
Sawmill Point
Pine Hollow
Ranger Pass
Eagle Pass
Box Elder Spr
Kwagunt
Rice Hollow
R. 2 W.
R. 1 W.
R. 1 E.
R. 2 E.
R. 3 E.

*Figure 8-3. Fixed lookout. Metal tower, originally emplaced by CCC (above), and its location (ST) relative to the tower network that grids portions of Grand Canyon National Park and the Kaibab National Forest, Arizona (facing page). PT indicates a primary tower; ST, a secondary tower; TT, a tree tower. Map courtesy U.S. Forest Service; photo courtesy Leonard Dems.*

that the present system of fixed lookouts has been inherited from the past. Few areas offer either the opportunity or need for redesign. Most changes are by way of deletion, the replacement of fixed lookouts with alternative methods. But fixed lookouts will remain as a permanent feature of detection in some capacity, and as the replacement of old facilities is needed, the possibility for relocation arises. The development of a computer program, VIEWIT, for determining what portions of a landscape can be seen from set locations promises to assist this process. The program was originally developed by the U.S. Army to read topographic maps. It was adapted by the U.S. Forest Service for help in determining the visual impact of activities on national forest lands, and adapted once again to help select good observation posts. The program shows what landscape is visible from a particular vantage point (Travis et al., 1975).

*Aerial Detection.* The chief alternative to fixed detection is aerial detection. The history of aerial reconnaissance is as old as aircraft, and joint patrols by the Army Air Service and the U.S. Forest Service created a sensation in 1919. As advantages, aerial detection can boast of its mobility, of the superior information it can provide about a fire, and of the multiple purposes to which it can be put. Aerial observers can spot fires quickly, determine their location accurately, give information about the burning characteristics of the fire, and even guide suppression units to the fire's location. If the fire escapes initial attack, an aerial observer can give invaluable information about a fire's behavior and location while an adequate suppression force builds up.

Aerial detection can provide coverage that is relatively continuous or intermittent. In the case of fixed lookouts, economics favors use where continuous coverage is demanded; in the case of aerial detection, where intermittent coverage is adequate. For the most part this restricts the use of aerial observation to lightning fires, to times of heavy (e.g., holiday) visitation or extreme fire danger, and to places not covered by fixed lookouts. Few areas now lack aerial detection capabilities, and over the past few decades the national trend has been to replace fixed lookouts with aerial observers wherever possible.

The chief limitation on aerial detection is cost. In those areas that have lots of land and lots of lightning fires, aircraft can compete economically with fixed lookouts. Scheduling flight time according to fire danger rating further reduces costs. Integrating fire reconnaissance with other uses of the aircraft and coordinating flights used for other purposes with fire detection can also bring down fixed costs. The State of Texas, for example, has for years relied on a cooperative program with the

Civil Air Patrol to report rural fires. To date, drone aircraft, such as those used in military observation, have not been adapted to fire reconnaissance, but they might prove effective, particularly if outfitted with remote sensing equipment. In some circumstances reconnaissance is combined with suppression capability—the capacity to make water drops, for example, or to deliver smokejumpers.

But the trend in fire protection seems to be away from such multiple-purpose use. At one time a lookout tower also housed firefighters, and the patrolman functioned as smokechaser. Such combination roles are effective for small operations that must cope with only a limited number of events, but they are less effective for large organizations and heavy fire loads. To be efficient detection must be more or less continuous; it cannot be abandoned during high fire periods, even temporarily, to man individual smokes.

*Remote Sensing.* Whether by fixed lookout or aerial reconnaissance, detection relies on human observers. Recent technological developments in remote sensing, however, suggest ways in which special equipment might reduce reliance on human observers and the heavy cost of housing or transporting them. At present remote sensing takes two forms: infrared (IR) scanning for fire detection and mapping, and lightning detection for possible ignition souces. Infrared detectors record radiation in the lower, long-wavelength electromagnetic spectrum, distinguishing a heat source such as fire from its environs (Wilson et al., 1971). IR radiation cannot penetrate water (hence, clouds) but it can pass through smoke. IR scanners can be used to map fires whose perimeter is otherwise obscured. Handheld IR devices are used to locate hot spots during mopup operations; airborne IR scanners map fires, instantly sending the information via telemetry hookups to planners; other devices will soon supplement detection networks, mounted either to fixed lookouts or to aircraft. High-altitude aircraft (such as the U-2) have been used to measure fire effects. Costs and needs will decide how widespread such practices become. Geosynchronous satellites—already an ideal platform for many forms of remote sensing—may find applications, although low resolution and high costs make it a detection technology for the not-too-immediate future (Figure 8-4).

Lightning fire detection through remote sensing has developed principally through the BLM (Vance, 1978). The Forest Service traditionally used its network of fixed lookout to track lightning storms, but the BLM in the Great Basin and the interior of Alaska lacked any such network or the justification for installing one. Instead it looked for an automated surrogate. By 1978 it successfully emplaced a basic system for the detection of lightning strikes. The network relied on remote automatic weather stations (RAWS) outfitted with sferics-sensing instruments (Figure 8-5). Information about lightning quantity and direction, along with weather data, is relayed to some 52 BLM field locations. When this data is combined with fire behavior models for computer processing, the result was a dispatching program known as the Initial Attack Management System (IAMS; McBride, 1981).

IAMS furnishes to the fire dispatcher information about lightning strike location, storm movement patterns, weather, the probable number of strikes that could result in survivable fires, and preliminary fire behavior forecasts for those fires. In brief, the system provides short-term information about fire risk, based on actual lightning events, not (as with the NFDRS risk component) from past history. Ultimately the system forsees some 350 remote automatic weather stations for data collection. Its basic information is in such demand, however, that the BLM expects to distribute that data to perhaps a thousand users outside its own field network. The principles for designing a system of automatic detection are fundamentally identical to those laid down in an earlier era for detection based on fixed lookouts.

*Rural Fire Reporting.* Where structures are under protection, a detection program faces special difficulties. Structural fires escalate more rapidly and more predictably than do wildland fires; response

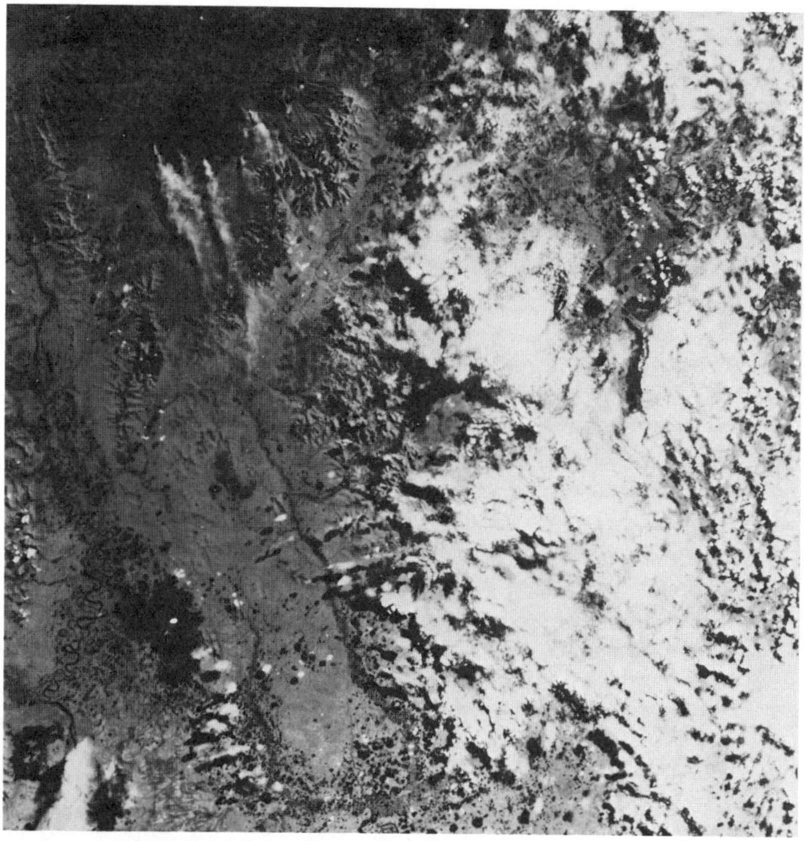

*Figure 8-4. Infrared sensing. From satellites, IR devices can record large fires such as these in Alaska (above); from planes, they can map firelines with a resolution adequate to assist suppression planning (below). The image recorded is from the Sundance fire. Compare to map (Figure 2-27). Courtesy NASA and U.S. Forest Service, respectively.*

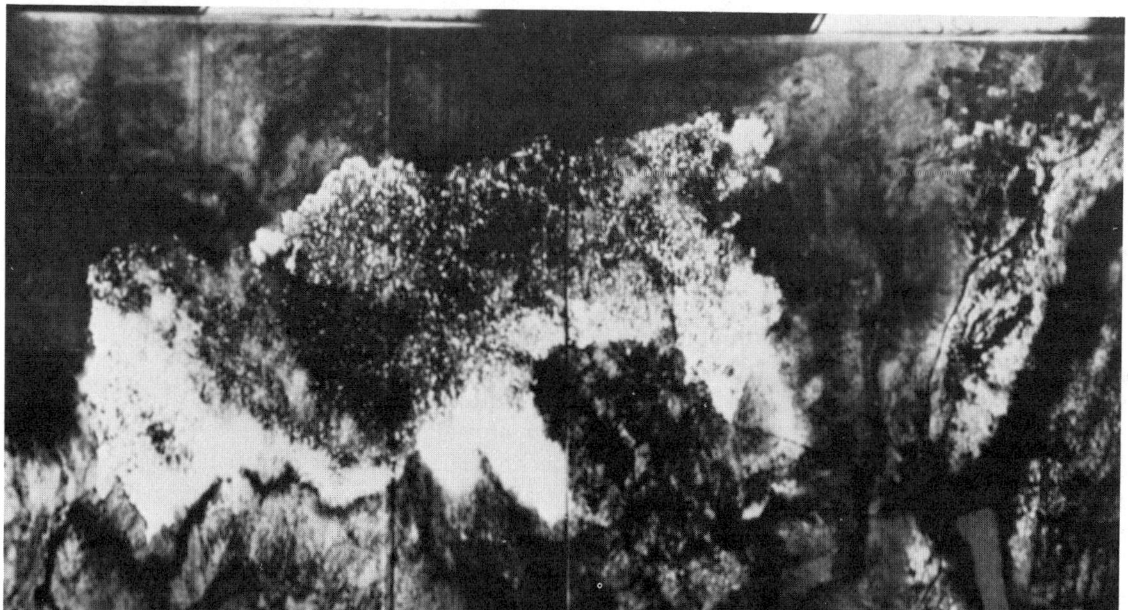

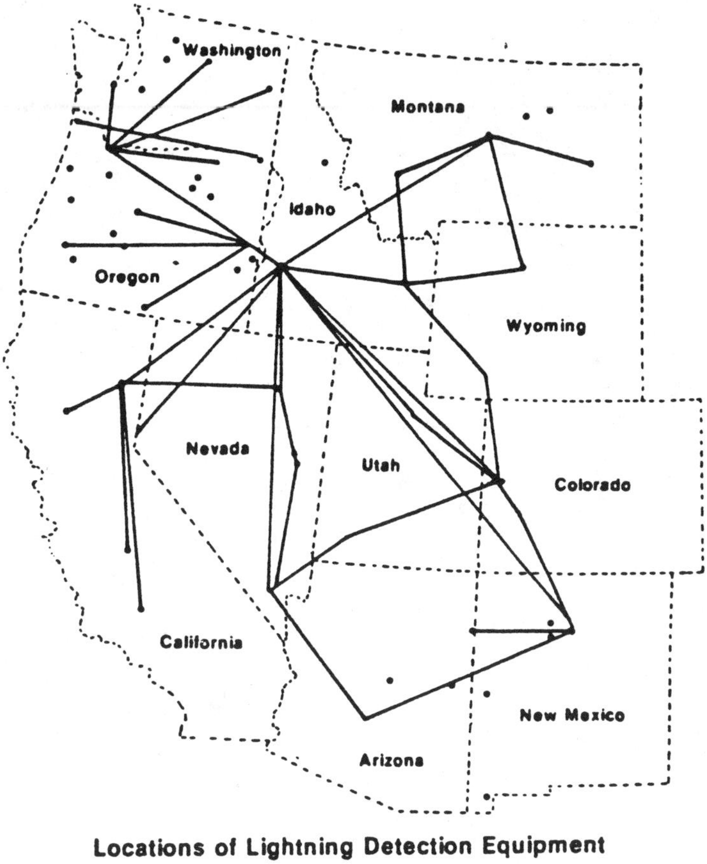

*Figure 8-5. Automated lightning detection. RAWS unit (right) in Nevada, and general map of lightning-detection grid in western U.S. Courtesy BLM.*

time must be universally quick, even though distances may be long. The rural fire must be attacked like an urban fire but without the apparatus for detection enjoyed by urban fire services. For wildland agencies, however, the problem with rural fire is less with detection than with reporting. By definition rural fires occur in populated areas; people will report the fire, just as in cities; the breakdown in response typically comes in the mechanism for communicating the fire to suppression units. The problem resides less in the establishment of a detection system than in the integration of fire reports with fire response. The rural fire scene shows, once again, how wildland fire protection developed from the paradox that by excluding settlement the reservation system also excluded rural methods of fire control. What would be done automatically in a populated area had to be created deliberately for unpopulated areas.

Nearly all systems of fire detection represent mixtures of techniques. The choice of the right proportions will be made pragmatically. That risk and hazard are constantly changing, that policy periodically sets new objectives, that the relative costs for different techniques ever vary, that new technologies add new capabilities—all argue against dogmatic solutions and in favor of particular responses to particularized needs. It is unlikely, however, that any future mechanism will dissolve the existing structure entirely. On the contrary, for better or worse, any new development will be grafted onto the existing arrangement of lookouts, patrols, aerial observers, and automatic sensors. The likely trend is for more specialization, not less; more integration with information technology, not less.

### *Communication Systems*

In the early days the limiting factor on most detection systems was their ability to report a smoke. Few areas had a communications system in place that could be adapted to fire control; in nearly every instance, some network had to be deliberately installed. One result was a weird concoction of techniques—heliographs, semaphores, carrier pigeons, messengers, telephones, to name a few. The ground-return phone system became the communications system of choice, however, and smokechasers carried portable phones with which they could tap into lines. This was haphazard and expensive and it lacked the flexibility promised by radio. During the late 1920s the U.S. Forest Service established a radio laboratory, conducted fundamental experiments in the transmission of radio waves over mountainous terrain, and developed quasi-portable radios (Gray, 1982).

Phones and radios remain the basis for fire communications. Their current technological sophistication is great and their presence ubiquitous. They influence every stage of presuppression and suppression from the broadcasting of fire danger ratings to fire reports, from dispatches to fireline command. BIFC maintains a National Radio Cache which contains radio systems for all levels of large fire suppression. For communication between dispatch centers, most organizations rely on teletype.

The trend in fire communciations follows the national example that links communications systems with electronic information-processing devices. Remote sensors and automatic weather stations collect information, satellite telemetry broadcasts it to central collection points, and computers process the data. Programs such as Affirms transmit their information over phone lines, yet allow for any portion of the network to acquire information from any other portion. In the past, communications was thought of as a service function, something designed to support detection or suppression. But so great is the capacity of modern communications systems to move information, and so powerful is the process of computer-assisted decisionmaking, that the communications system may drive the other functions rather than serve them. An example perhaps is Firescope, which is, in the final analysis, only a vast communications network.

As with detection systems, future communications systems will be grafted onto existing ones, the telephone (with its ability to transmit digital information from computer terminal to terminal) being a good example. The major reforms in fire planning and fire management are inextricably bound up with contemporary communications systems. One can more easily imagine a future fire management program without an advanced program of fuels management or forms of suppression or presuppression than without information technology.

## 8.3 FUELS MANAGEMENT

The concept behind fuels management is simple. To modify fuels is to modulate fire behavior, fire effects, and the costs of fire suppression. Fuel modification may come through processes of reduction, in which the load of available fuel is decreased; of conversion, by which certain fuels are replaced by others with different flammability; and of isolation, through which large expanses of fuels are broken up with fuelbreaks or greenbelts. The available techniques are many, and they naturally merge one with the other.

Fuels treatment has always been a part of fire protection, though few fire regimes have justified separate fuel management programs. Especially where grass and shrubs were involved, protective burning as a means of fuel control was a common reason given by Indians and European-American settlers for broadcast burning. The introduction of exotic flora and fauna in the form of crops and livestock drastically revamped fire regimes by reconstituting their fuels. With the advent of more formal wildland fire protection, other reasons became important. The association of logging debris and conflagrations led to programs of hazard reduction; the desire for reforestation argued for protective networks of fuelbreaks; the need for fire control in difficult fire regimes led to searches for non-flammable cover or the eradication of highly flammable exotics for cheatgrass.

In most fire management programs, fuels management constitutes only one project among many, and then it is usually conducted for more objectives than solely fuel modification. This is especially true where prescribed fire is used. Multiple outcomes are inevitable. Burning for hazard abatement, for example, does more than simply reduce fuel, and prescribed burning for other objectives has the added result that it reduces or rearranges fuels. Historically, fuel activities were secondary to larger goals. But as fire control through prevention and rapid initial attack reached a point of diminishing returns, the strategy of fire control shifted to one that emphasized modification of the fire environment as a means of lessening fire intensity. For this strategy fuels management became fundamental. And as fire ecology research pointed to the value of fire in many ecosystems, broadcast fire became increasingly the prescription of choice. Most fire agencies added specialists in fuels management to their staffs, and the U.S. Forest Service undertook development of a National Fuel Appraisal System to assist in decisions about suitable technologies.

### *Fuel Reduction*

Fuel reduction is a generic term which decribes programs that seek to reduce fire hazard (fire intensity) through the diminution of available fuels. Fuel reduction programs may be broadcast or site-specific, irregular or periodic. To be effective, reduction must keep pace with accumulation. The specifics of a fuel reduction program will depend on the mechanisms of accumulation. If the fuels result from human activity—logging or road construction, for example—a program of systematic reduction can be designed to follow upon it. If the fuels are the product of natural accumulation—as in the case of the southern rough—periodic broadcast fire can keep fuels within acceptable limits. In both these cases the buildup of fuels is predictable. But fuels accumulate episodically, too, and in great quantities—the result of a blowdown, for example, or a large fire. Suddenly phytomass has become fuel. Hazards like these, moreover, become kindling for fires that can spread to sites far beyond the point of origin, creating new fuels through a process of positive feedback.

A fuels program cannot develop in isolation. A successful program must be closely integrated with prevention and suppression functions that can exclude wildfire until fuels are processed. Its range of treatments may be circumscribed by ecological considerations, which look upon wildland residue as something other than fuel; by air quality boards, which may frown on open burning as a means of disposal; by technological capabilities and market economies that may transform waste residue into a marketable commodity; by historical circumstances, backlogs of untreated or partially treated fuels which must be handled for site preparation, hazard reduction, or preprescription objectives prior to the restoration of natural fire. What appears simple as a strategy may become hideously complex in practice.

A range of techniques exists (Pierovich et al., 1975; Cramer, 1974). For broadcast fuels that accumulate as a result of natural processes broadcast treatment is needed. Practically speaking, this means fire. For activity fuels, limited to specific sites and one-time processing, fuel managers may select between prescribed open burning, intensive utilization of residue, yarding of unmerchantable materials (YUM), conversion of the debris into chips, portable incinerators with the prospect of smoke-free burning, and even burial of debris (Figure 8-6). Where large amounts of fuel have resulted from episodic catastrophes such as blowdowns or fires a primary technique

*Figure 8-6. Fuel reduction methods. Slash burning (upper photo); mechanical destruction by bulldozer team and anchor chain (middle); portable burner (lower). Disposal is necessary after activities that do not completely utilize or remove their residue. Local considerations will determine what method of hazard reduction is most appropriate. Courtesy U.S. Forest Service.*

is salvage logging. This removes potentially merchantable standing timber. The remaining fuels may be treated like activity fuels.

Somewhat different in its approach are mechanical treatments of debris and increased fire protection measures. Mechanical treatment may be a prelude to prescribed fire, or it may, like increased suppression capabilities for the site, act as an alternative to it. The idea is that slash does not maintain its hazard forever; after a few years, its potential flammability is greatly reduced by changes in the composition and bulk density of the fuel complex. Less fuel is available, and less available fuel is in a form that encourages intense fires. Crushing slash accelerates this natural process of physical decomposition. Paradoxically, keeping fire out of such sites by temporarily increasing control forces may be cheaper in some cases than the direct processing of the fuel.

Slash burning is, in a sense, an intermediate process. It reduces more fuel than does the broadcast burning used for extensive land management, but less fuel than the complete utilization of debris characteristic of intensive management. It offers certain advantages which attend the process of burning, but it also transfers responsibility for debris treatment from the activity, like logging, which generated it to the fire agency which must dispose of it. To assist in the selection process the U.S. Forest Service developed the Activity Fuel Appraisal Process. The program converts fuel information into forecast fire behavior information for each alternative, and weighs the alternatives against the likelihood of such a wildfire (Hirsch et al, 1981).

### *Fuel Conversion*

The concept of fire management through fuel (type) conversion is a tested, venerable one. Its greatest expression came through the process of settlement and reclamation. Domesticated species replaced wild species, farms replaced prairies and forests, pastures replaced grasslands. Each such substitution, of course, changed the fire regime. In most instances, fire frequency and fire intensity diminished. But in some instances they increased: brush invaded timber areas, cheatgrass spread throughout the Great Basin, slash coated forest floors.

The most treacherous period occurred during the transition when the conversion process was incomplete or when conditions allowed a mixture of types to remain in quasi-permanent association. Precisely this situation characterized the intial periods of agricultural settlement, and it accounts for the holocausts that swept the Lake States during the late nineteenth century. Eventually agricultural reclamation ceased its penetration of new wildlands, and even began ebbing away from lands it had once converted. The fire situation improved. The recent incursion of suburban or recreational dwellings into wildlands threatens, however, to recreate analogous conditions in a more modern form.

As fuel management techniques the preceding examples were incidental to larger processes for reclamation and settlement. But the deliberate conversion of fuel types is a legitimate mechanism of fire management. For wildlands destined for permanent wildland status, conscious programs of fuel management are essential. They may substitute for the creation of a physical plant for fire control that is otherwise prohibited by an agency's general charter. By and large, however, broadcast conversion solely on the grounds of fire control is rarely justified. Large conversion projects demand a wider purpose, like the reclamation of rangeland or forest from brush. Conversion in the name of fire management alone is generally restricted to strips or spots.

The problems with type conversion, and its prospects, affect all categories of fuel. Where fire has been excluded, the conversion of grasslands of wild species to pastures of introduced species such as Kentucky bluegrass has successfully reduced flammability. Where fire has persisted, however, the process has sometimes resulted in even greater flammability—cheatgrass being an example. In this instance, fire control and range management techniques must combine to deliberately replace cheatgrass with a more palatable and less fire-prone species. Almost everywhere in which brush thrives extensively, there are efforts to replace it either with timber (in areas in which timber can be grown) or with grass (where the climate prohibits trees). When the conversion is to timber, fire exclusion is mandatory for at least the transitional period. When the conversion is to grass, a medley of mechanical and chemical techniques, in addition to prescribed fire, is typical (Green, 1977).

Even when the cover type remains forested, the conversion from one dominant species to another is common—hardwood to pine in the South, commercially valueless conifers to Douglas-fir in the Northwest, aspen to pine in the Lake States. In most cases the result is an increase, at least temporarily, in fire hazard. Where reforestation or afforestation are practiced, the increase in flammability may be large. The same holds true in instances of deforestation—logging, for example, or the removal of pinyon–juniper forests to enhance the western range. And there are even instances involving ornamentals in which fire considerations are significant. The replacement of native species with eucalypts brought a terrific increase in hazard to portions of California, including cities such as Berkeley. The reconversion of eucalypts to the more malleable Monterey pine promises to abate somewhat that hazard.

In most instances of deliberate conversion, transformation by type rather than by specific species was a sufficient outcome. It was enough that grass replaced brush, or timber, grass. Often, however, the objectives call for specific species. A variation on this theme is the search for relatively nonflammable plants. Most of this exercise in biotic engineering has concentrated on Southern California. Here watershed cover is imperative, fire control in chaparral difficult, and the opportunities for prescribed fire limited. Fire management wanted a plant that could stabilize hillslopes, grow in a Mediterranean climate, and burn with less intensity than chaparral. The best candidates were certain varieties of desert vegetation. Drought-resistant, they could thrive under semiarid conditions; high in salt content, they pyrolyzed poorly. No organism has satisfied all the requirements, however, to say nothing of justifying the costs of planting (Green, 1977). But the concept is valid, and in the more restrictive form of fuelbreaks, it enjoys success. An interesting variant involves the use of fire retardants as a presuppression measure. Here the inhibiting chemicals are not intrinsic to the plant, but added to it.

### *Fuel Isolation*

The essence of the fuel isolation concept is to segregate high hazard fuels from high value resources, or to break up the continuity of high hazard fuels. In fuel reduction, the objective is to prevent the site from becoming a source for fire; in fuel isolation, to prevent fire from entering or leaving the site. If the object is to make fuel complexes more manageable on a larger scale, then fuels can be isolated into predetermined administrative (preattack) blocks through selective modification along strips—fuelbreaks or greenbelts. This technique differs from type conversion only in scale. It applies best to areas where fuel complexes are mixed—wildlands and suburbs or wildlands and plantation forests, for example. These are areas that can neither be fully converted one to the other, nor abandoned by organized fire protection. Fuelbreaks and greenbelts offer a compromise, though one that brings most of the problems of type conversion with only some of the benefits.

*Fuelbreaks.* Fuelbreaks exist for all kinds of fuels: grass, brush, and timber (Figure 8-7; Green, 1977). They may be nothing more than a plowed furrow, or they may take the form of elaborate landscaping through wooded mountain crests. They may be temporary, designed to assist in type conversion or hazard reduction; or they may be quasi-permanent, entrenched fortifications to support fire control in areas in which wildfire is endemic. Commonly, fuelbreaks accompany other land management activities, such as timber management. They work best when, as with pine plantations, they are built into the initial design for the activity. And they serve a variety of fire

*Figure 8-7.* *Fuelbreaks. Sample fuelbreaks in grass and grass-timber (above), brush and brush-timber (facing page). The grass fuelbreaks illustrate the system put in during afforestation on the Nebraska National Forest. The brush and timber photos show construction of the Ponderosa Way. Courtesy U.S. Forest Service.*

control purposes—improving access to remote areas, creating lines for indirect control according to preattack blocks, and fashioning safety zones for firefighters.

Historically, fuelbreaks of some variety have appeared whenever fuels were experiencing wholesale modification. As quasi-permanent features of fire management plans, however, they have several limitations. The fundamental paradox is that they are most justified for the control of large fires, for which indirect attack methods are appropriate. But large fires require wide fuelbreaks, increasing installation and maintenance costs, and large fires do not propagate solely along the surface. No fuelbreak can be wide enough to halt the long-range spotting that will almost certainly accompany high-intensity fires. An optimum size for a fuelbreak, thus, is difficult to determine, even considered solely from the perspective of fire control (Omi, 1977). Fuelbreaks take many forms, however, and judgment about their efficiency must be empirical and site-specific.

Fuelbreak design will vary with fuel type. Where grasses provide a ready fuse from roads and railroad right-of-ways to wildlands, small firebreaks in the form of furrows or graded roads are common. For brushfields, broad fuelbreaks often frame the natural burning blocks along ridgetops and wide valley bottoms. Within the fuelbreak proper, brush is converted to a perennial grass. In timberlands, many fuelbreak designs are possible. They may form conflagration barriers, combining selective logging and grazing with fire control; pasture fuelbreaks, swaths of grass through timber which are heavily grazed to keep flammability low; shaded fuelbreaks, in which mature trees are retained but the understory is replaced with grass. On heavily burned timberland, fuelbreaks in the form of roads and corridors of felled snags may supplement salvage logging. During the 1950s and 1960s, when conflagration control was a strategic concept, the U.S. Forest Service conducted experiments in fuelbreak design for both timber and chaparral-dominated landscapes in California (Figure 8-8; White and Green, 1967; Green and Schimke, 1971). It was this program that firmly distinguished the concept of the fuelbreak from that of a fireline.

Whatever the criteria for deciding on fuelbreaks as a fire control strategy, and whatever particular design might be selected, there still remains the problem of installation and the often greater problem of maintenance. Both activities may involve considerable costs. Depending on design, installation requires wholesale or selective removal of existing vegetation and its replacement by other vegetative cover—that is, type conversion—along a strip 100–300 feet wide (29–92 m). Fire can be effective in disposing of the original cover and in maintaining its replacement; but almost never can fire do either alone. Mechanical and chemical treatments must precede the initial burn, and they must accompany, in some diminished form, subsequent maintenance burning. In brush this means preliminary crushing and dessication of fuels, with at least spot application of herbicides to retard resprouting.

The cost of annual maintenance can be high, particularly as mounting environmental concerns restrict the use of machinery and chemicals. One old solution has been to use cheap labor, like the CCC or prison inmates. It may be argued, in fact, that the availability of such labor has often been the principal stimulant to ambitious programs of fuelbreak and firebreak construction. A more recent approach, under trial in California, involves the use of Spanish goats to keep resprouting under control (Green and Newell, 1982). In other regions, where fuelbreaks are constructed as a part of reforestation or plantations, annual burning requires little preparation beyond the reclearing of control lines.

The best fuelbreaks are built into the design of a fuel complex—whether it be a postburn forest, a pine plantation, or a residential development in wildlands. The poorest are installed after the fact, when fire behavior considerations cannot inform the design. Like any control line, fuelbreaks perform best against surface fires. And like so much of the apparatus for fire suppression, fuelbreak

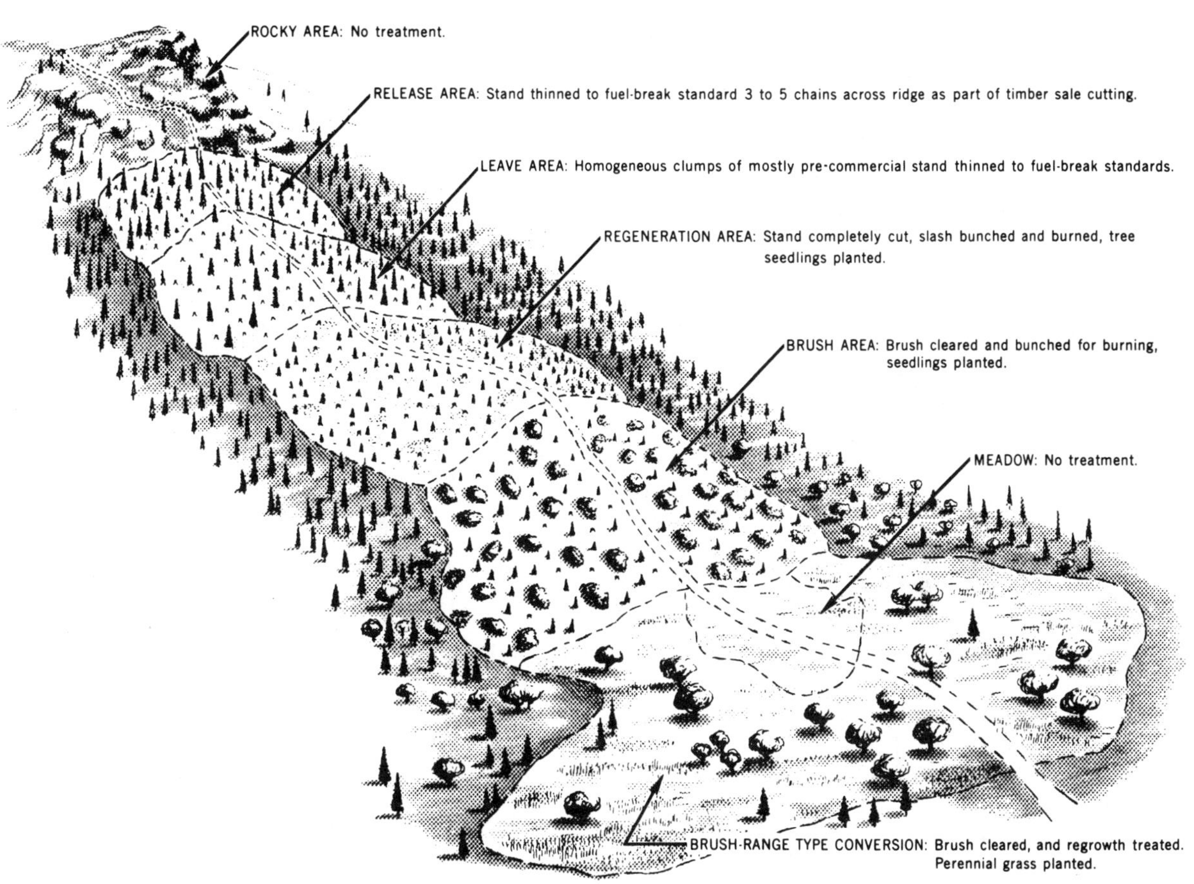


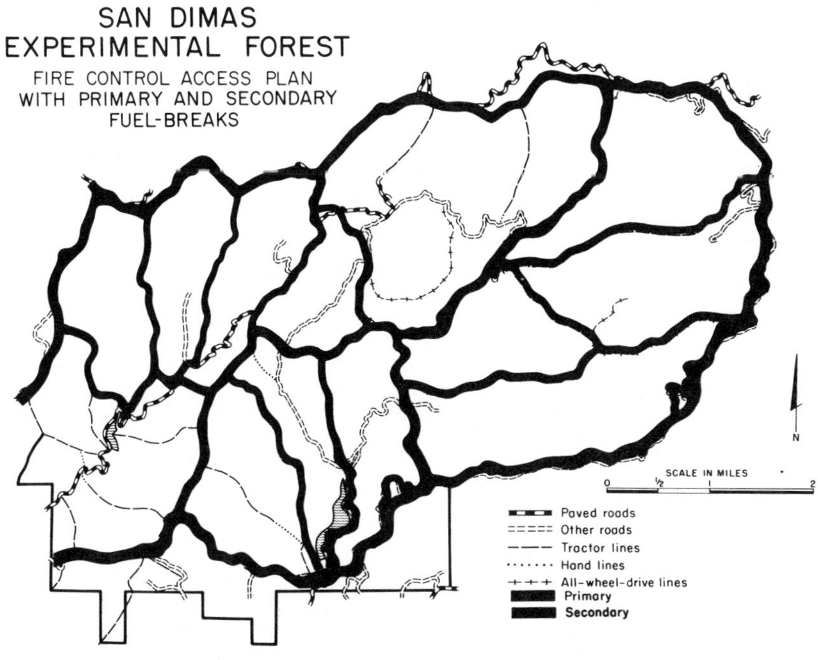


*Figure 8-8. Conflagration barriers. Design plans for an elaborate fuelbreak through mixed conifer in northern California (above), and for a network of primary and secondary fuelbreaks for the San Dimas Experimental Forest in Southern California (below). Both designs were created during the Fuelbreak Program sponsored by the U.S. Forest Service and California Department of Forestry. From Green and Schimke (1971) and Bentley and White (1961).*

systems may find their greatest use in the application of prescribed fire—not merely fire used for the maintenance of the fuelbreak system, but for broadcast burning within the blocks defined by the network. In this case they function in support of broadcast fuel reduction. The value of fuelbreaks is repeatedly questioned, but their utilization is continually affirmed, not only in the United States but throughout the world. Everywhere reforestation and afforestation projects rely on fuelbreaks, especially where a suppression system cannot supply concentrated force in very short periods of time. Initial attack and fuelbreaks together can make for effective fire control: the combination of two principles of defense, fixed fortifications and rapid counterrattack.

*Greenbelts.* A slightly different version of the fuelbreak is the greenbelt, a form of selective type conversion along designated perimeters. Although some forms of greenbelts exist in wildlands (the use of aspen groves, for example), the greenbelt concept is especially relevant for the boundary between wildland and suburb. Consequently, it belongs with urban as much as with wildland planning. Orchards, parks, and golf courses are all examples of greenbelts that can combine open space, recreational use, or agricultural production with fire protection. The problem that restricts the use of greenbelts is the same that plagues the urban–wildland fire scene generally. Zoning pressures and real estate values will try to push developments through and beyond a proposed greenbelt. Where developments are well confined, however, the greenbelt concept—long advocated by urban planners for a variety of reasons—may be pertinent.

The greenbelt concept accents a somewhat larger problem with structural fire protection in those areas where dwellings are intermingled with or surround wildlands. Not only do wildlands and urban sites abut each other with ever-increasing frequency, but a major demographic migration to rural landscapes for recreational and residential living has placed large numbers of dwellings and even whole communities in hazard. Few of these developments incorporate into bona fide towns, few plan for any degree of fire services, and what fire protection does exist comes, as often as not, through the state forester. In the old days, such towns appeared within a context of agricultural settlement; the surrounding lands were converted from wildlands to farms, and the prospects for wildfire diminished. But the newer settlements have deliberately located in wildlands, and generally resist extensive modification of the vegetative cover.

In these circumstances, however, fuels management is the only reasonable foundation for fire protection. If the development is a "planned community," then the possibility exists that fire protection may be designed into the site from the beginning. Building codes can specify nonflammable roofs, fire codes can promulgate standards for clearing debris from around structures, and fuelbreaks or greenbelts can form a protective perimeter around the area. But few such opportunities exist, or are taken advantage of. In nearly all cases agencies—wildland fire agencies for the most part, with help from volunteer brigades—must provide protection after the fact. This greatly complicates the prospects for fuel management, but only through substantial fuels modification programs can fire protection come. Otherwise such developments face not only intermittent fires but outright conflagrations. That the source of the fire need not arise from within the community only increases the precariousness of the situation. The number of people using the surrounding lands for recreation increases the possibility of wildfire ignition without improving the prospects for wildfire control.

### *Fuel Management: Selected Examples*

Most fuel management programs now in existence consist of many small projects that have developed over decades in response to local needs. The enthusiasm during the 1970s for fuels management as a strategy of fire management, however, has moved fuels projects from the periphery of fire planning to the center. Part of the reason was a general interest in prescribed fire, for which fuels management offered brilliant opportunities. Yet fuels management remains only one component among many within the fire management mission. Part of its function belongs with planning, part with suppression, and part with prescribed fire. Only in special areas and for special periods of time have fire programs as a whole developed principally out of fuels management. The case of broadcast burning in the South and slash disposal in the Northwest are examples of fuels programs that are both more less continuous and fundamental to fire management in their respective regions. Because of their use of fire, however, they will be examined within the context of prescribed burning.

Consider instead a few historic examples, spectacular instances of type conversion and fuelbreaks, in which massive fuels management programs underscored the entire fire program. These particular examples could be multiplied many times. The salvage after the Tillamook fire had its counterpart in the preventive cleanup program that followed the 1938 New England hurricane. The Sand Hills fuelbreaks apply to most plantation forests, especially those associated with reforestation. Similar programs developed in the South and the Lake States, though never on so grand a scale. The Ponderosa Way can serve as a model of a quasi-permanent fuelbreak system, one that has been extended to certain remote settlements (such as the Canadian Yukon) and those metropolitan areas that border wildlands (such as Los Angeles).

It should be emphasized, however, that all these historic examples were used to break a cycle of fire and fuel. They assisted in the transition from one regime to another. When that task was accomplished, the programs vanished. All were heavily dependent on cheap, abundant labor. When that source of labor disappeared, the project was abandoned, the technique fell into disuse, and the fuelbreaks lapsed into disrepair. The value of such projects, in brief, will be specific as to time and place, and it will probably depend on the successful substitution of technology for hand labor. Heavy machinery for crushing and helitorchs for prescribed burning are possible examples of technology that can make installation feasible and maintenance possible.

*Tillamook Burn, Oregon.* In August, 1933 the original Tillamook fire burned over 270,000 acres (109,000 hectares) of dense Douglas-fir forest in the Coast Range of Oregon. The devastation was nearly total. Equally disturbing, the burn held the prospects for a initiating a cycle of fire that would progressively radiate outward from the scene of the original site into unburned forests. There were many reasons for removing the dead timber —the recovery of a charred but still usable resource among them. But whatever the technique, fire protection demanded that the fuels be modified.

The outcome took the form primarily of salvage logging on a prodigious scale. Time and again a consortium of logging companies went back to the Burn, successively culling out usable timber. Access to the Burn required roads, which functioned as firebreaks and as points of entry for suppression forces. To break up the expanses of standing fuel, corridors of snags were felled into a network of fuelbreaks. The presence of the CCC made such labor-intensive programs possible, and the presence of the Burn encouraged the availability of CCC camps. Prescribed burning for hazard reduction was not used, but reburns from wildfire removed more fuel in 1939 and 1945. Brush appeared in natural succession, another fuel hazard, and had to be removed over many areas.

Nor was it sufficient simply to reduce the fuels left from the cycle of fire. Ultimately, for both fire protection and land use, another cover type had to replace the existing fuels. After an amendment to the state constitution in 1949, replanting of Douglas-fir began. Site preparation demanded more snag removal, and survival of seedlings required vigorous fire protection during the early years (Figure 8-9). The outcome, however, was a new forest, a new fuel complex, and a new fire regime from that left after the original Burn. Without deliberate fuel modification, the process

*Figure 8-9. Tillamook Burn. Construction of snag-free corridors (left), and graph of removal and replanting efforts beginning in 1949 (below). Much of the dead fuels had already been removed by salvage logging, reburns, and fuels treatment prior to this time. Courtesy Oregon Department of Forestry.*

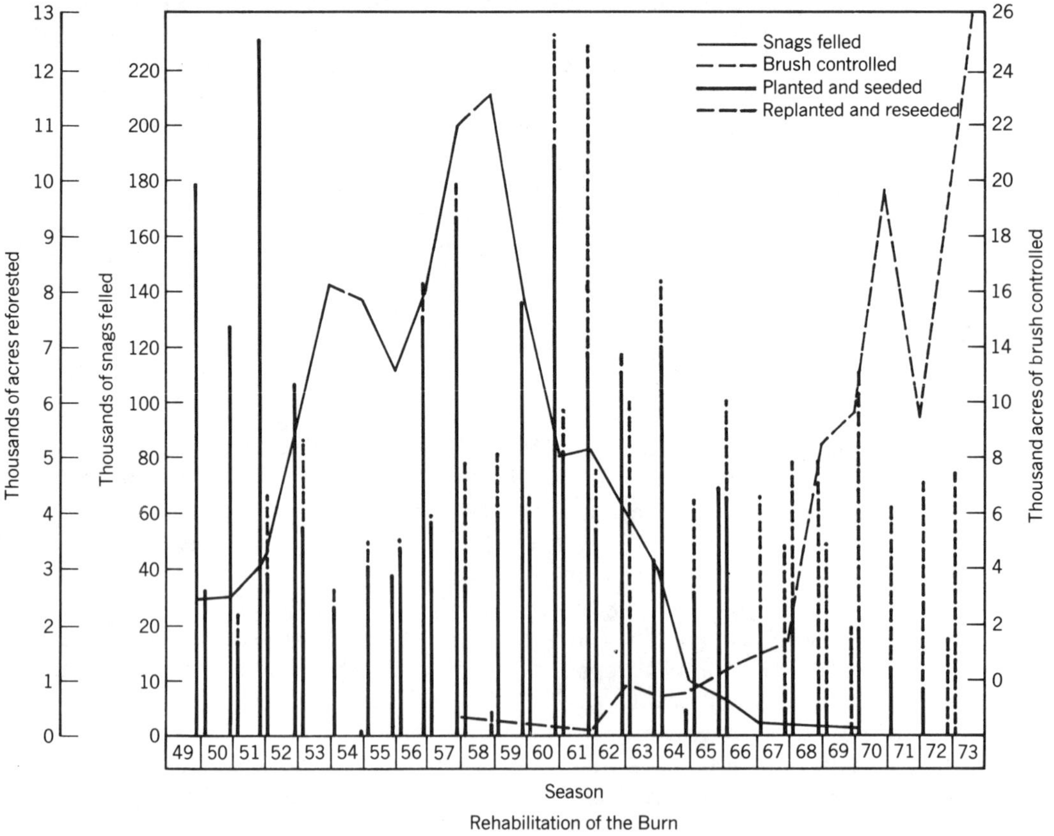


Rehabilitation of the Burn

might have taken many more decades—or even centuries—and extended over far more acres (Forestry Department, n.d.; Cronemiller, 1934).

*Sand Hills, Nebraska National Forest.* The desire to afforest portions of western Nebraska was a dominant reason for the establishment of a national forest there, but early attempts to transform prairie into plantation pine were repeatedly wiped out by fire. Problems with site and climate complicated the undertaking. Until the prospect for controlling the wildfires that swelled out of the surrounding grasslands improved, however, all other successes were tenuous. The speed of grass fires roaring through the unbroken landscape made reliance on a ground-based suppression program impossible. The only solution was fuel modification—temporarily, until the conversion of grass to pine could be accomplished; then continuously, to modulate the fuel accumulation that would develop within the plantation.

The result was an ingenious system of fuelbreaks, intimately integrated with the design of the plantation as a whole (Figure 8-7; Davis, 1951). There were exterior fuelbreaks to segregate the plantation from the prairie and interior fuelbreaks to break up a homogeneous monoculture of trees into a mosaic of fuels and to provide access for control forces. The interior fuelbreaks could, in time, also define burning blocks for fuel reduction through prescribed fire. Afforestation succeeded. Gradually, too, the lands surrounding the plantations saw their fuels change—cropped by cattle, divided by roads, submerged under reservoirs or towns. With more intensive use of the Sand Hills, heavy grazing has replaced broadcast fire as a means of controlling the grassy cover that supports fire.

*The Ponderosa Way and Truck Trail, California.* In northern California and along the west slope of the Sierra Nevada, early foresters confronted a serious fuel complex of brush and forest—a case of brush encroachment, or from another perspective a problem of forest reclamation. To reforest brushfields and to protect existing forest from further brush invasion, fire control was essential. Not incidentally it was from this region that the concepts of systematic fire protection were developed. The Sierras were especially vulnerable to wildfire. Chaparral-sustained burns rushing up into the timber gradually reduced forest to savannah and then to brush. The relationships were complex, but any long-term solution demanded the segregation of forest from brush, or at least of potential forest land from endemic brushfields.

As a solution S.B. Show proposed the Ponderosa Way and Truck Trail, a 650 mile-long (1046 km) fuelbreak and road that spanned the length of the Sierra Nevada (Price, 1934). With the CCC to furnish labor, most of the Ponderosa Way was eventually completed—a fuelbreak of varying width with road access along most of its length. The system was integrated with the hour control plans promulgated at the same time (Figure 8-7). With removal of the CCC, much of the Ponderosa Way fell into disrepair. But its example inspired a renewed program of research and development (the Fuelbreak Project) by the U.S. Forest Service between 1958 and 1972. Meanwhile the California Department of Forestry, outfitted with inmate labor (the Conservation Corps), undertook selective renovation of the Ponderosa Way and auxiliary fuelbreaks. The new guidelines, in turn, underwrote the proliferation of fuelbreaks on the Southern California national forests after the disastrous fires of 1970 made available additional funds for fuel management (Green, 1977).

Unlike the Nebraska National Forest example, however, this new system of fuelbreaks did not grow up with the resources they were designed to protect. They came after the fact and could not expect that, over time, there would occur type conversion to a more manageable fuel. On the contrary, the proportion of available fuel would only increase with each year of successful protection. In an absolute sense, the value of the fuelbreaks was questionable. But in a relative sense, there were few alternatives. Demand for watershed protection, air quality standards, and difficulties with fire control limited the range for broadcast fire; real estate pressures and the forest reservation system prevented complete type conversion to either wildland or city; encroachment by urban structures introduced fuels over which fire agencies had little control. Even within the wildlands proper fuel accumulation was extensive rather than spotty, the result of natural growth rather than human activity.

Fuelbreaks were a half-measure, improving the capabilities of suppression forces but not substituting for them, converting fuels but not sufficiently to eliminate the prospect for conflagrations. By the 1970s there were perhaps 500 miles (805 km) of fuelbreaks throughout Southern California, many established and maintained by the state. Particularly successful was the International Fuelbreak which paralleled the border with Mexico for over 40 miles (64 km). Here the intention was defensive. Fires from the south were not controlled, but were prevented from entering the United States by backfiring from the fuelbreak.

## 8.4 FIRE RESEARCH

The problems faced by organized fire protection on wildlands are as much intellectual as practical. The creation and administration of a fire protection system demanded considerable knowledge, above and beyond practical know-how. Because they declared their profession to be scientific, foresters could not rely on folk knowledge. But because wildland fire in America was not a subject included within academic forestry, neither could they look to European precedents. Urban fire, with its association to insurance ratings, developed in ways that made it unsuitable as an exemplar for a wildland fire program grounded in land management. Scientific research scrutinized heat machinery—regulated fires fed by coal and gas, not open, free-burning fires responding to general environmental conditions. If wildland fire protection was to succeed, it had to create a sound base of information. Once emplaced, if it wished to progress, it had to sustain that production of knowledge. From the beginnings of organized fire protection, then, fire research was fundamental to administration. All aspects of fire protection were open to investigation.

### *Historical Synopsis*

From its inception the U.S. Forest Service has dominated fire research as it has fire policy and fire administration. The present structure of wildland fire research either derives from the programs of the Forest Service or depends on them, and until the 1960s the history of fire research is largely the record of research within the Service. The evolution of that research program parallels the development of Service fire policy, previously outlined.

*Systematic Fire Protection.* During the period in which systematic fire protection was developed for frontcountry lands and then extended into the backcountry, fire research was envisioned as an administrative process. Its purpose was to preserve policy from political attack, to put protection systems on a firm administrative base, and to assess the economic investment appropriate to adequate fire control. Researchers and administrators exchanged positions freely. The methods of fire research were, by and large, those of silviculture and forestry economics, adapted to new topics. Fire research was conceived of as research into fire protection, an operational problem whose management required systematization more than experimental discovery. The great products of this era were manuals of occupation, working plans for the establishment of a fire protection system. Not fire itself, but fire control was its subject.

The breakdown of the fire organization during 1910 showed the necessity for fundamental analysis, and the period that followed was no less a time of experimentation for research than it was for administration. In both cases the most important trials came from California, where they were under the direction of one man, Coert duBois. As regional forester, duBois initiated a host of investigations that included the origin of individual fire case studies, the origin of statistical methods in fire research, and the origin of fire behavior and fire effects research. The search for a scientific methodology that underlay these efforts was thus complementary to the search for administrative methods that motivated concomittant experiments in light burning, letburning, and so forth. Both inquiries were amalgamated into duBois' treatise, *Systematic Fire Protection in the California Forests* (1914). *Systematic Fire Protection* revolutionized the methodology of fire research by introducing other models, such as those of industrial engineering; by establishing the individual fire report as the fundamental unit of information; and by demonstrating how, through careful statistical milling of such information, fire planning could attain rigor.

The techniques were ideal for translating administrative objectives into field programs. About proper objectives there was much debate, but not about the procedures for converting those goals into practical fire control. *Systematic Fire Protection,* moreover, was not merely innovative but comprehensive. Most of the great names in fire research for the next 30 years came under the personal influence of duBois or his successor, S.B. Show. The LCPL, economic efficiency, and least damage concepts of fire protection all emerged from the group centered around duBois. In 1916 the Forest Service created a Branch of Research, which included plans for fire research but did not outline specific projects. DuBois' group did.

*Hour Control.* By 1930 the concepts of systematic fire protection had evolved into the precepts of hour control. Fire research followed suit. Fire behavior meant the calculation of rates of spread for the determination of hour control standards, not the physics and meteorology of free-burning fire. Fire effects meant the study of fire damages, not fire ecology. The primary source of data was to be the individual fire report, not laboratory or field experiments; the chief methodology was the statistical analysis of thousands of such reports. Fire research was asked to comment on the appropriateness of fire protection goals such as the relative merits of the LCPL and least damage theories. It was promoted, too, as a "tool for perfecting performance." It did more than merely standardize: it furnished standards by which fire problems in different forests could be evaluated and individual competency judged. Both administration and research agreed that their shared concern was to establish, direct, and evaluate fire protection systems.

By 1930, too, the institutional framework for fire research had been strengthened. The McSweeney–McNary Act (1928) authorized the Forest Service to engage in extensive forest research, including fire research. All federal research on forest fire, in fact, was to come through the Service. Not until 1974 were other federal agencies specifically empowered to investigate wildland fire. The Shasta Experimental Forest dedicated some of its land to fire research, a national demonstration of the principles of organized fire control. The success of a "made-to-order" invention for fireline construction, the bulldozer, encouraged interest in formal equipment development. Although some research remained in administrative hands, most gradually transferred to regional experiment stations directed by the Branch of Research.

Under the hour control program, researchers concentrated on refining, and then adapting to new regions, its most important variables—fuel types, rate of spread, and fire danger rating. Out of these investigations came Lloyd Hornby's concept of fuel types, defined by rate of spread and difficulty of control; H.G. Gisborne's fire danger meter, a guide to presuppression activities; and, tentatively at first, field and even laboratory experiments in fire behavior, for better determination of the rate of spread component of planning. The adoption of the 10 AM Policy eliminated, for the time being, the search for a relationship between resource values and an appropriate level of protection. The pressures built, too, for a national division of fire research. By the mid-1940s only nine researchers worked on fire for the Forest Service full-time, though many others labored intermittently and not a little research passed under an administrative label. In late 1948, however, the

Service formally established a Division of Forest Fire Research within the Branch of Research.

*Mass Fire.* The era of mass fire brought great reforms to fire research. There were changes in funding, with new revenue coming from military and civil defense sources; changes in methodology, with fire analyzed as a physical, chemical, meteorological subject suitable for laboratory experimentation; and changes in institutional composition, with a National Academy of Sciences-National Research Council Committee on Fire Research providing advice on national programs. Fire research left the confines of academic forestry for investigation of fire as a physical phenomenon, and it broadened the prescriptions of the hour control program to include the tenets of conflagration control.

All this opened new fields for study and redefined, in physical terms, most of the old ones. The problem of mass fire behavior and control reoriented all aspects of fire research. The Service established three laboratories: the Northern Forest Fire Lab (NFFL) in Missoula, Montana; the Southern Forest Fire Lab (SFFL) in Macon, Georgia; the Riverside Forest Fire Lab (RFFL) in Riverside, California. By the mid-1960s most fire research was located within the labs. Fire physics provided a new common denominator for concepts about fire—one founded on dynamic processes, not on statistics from fire reports. When a national fire danger rating system was proposed in 1958, the physical processes of water and heat exchange underwrote the research program.

Meanwhile, Forest Service fire research expanded into new geographic territories such as the northeastern United States, the Lake States, and Alaska. It achieved the same kind of hegemony enjoyed by the fire protection program of the Service. Complementing the labs, two equipment development centers were established to assist in the mechanization of fire control, in the transfer of military surplus to the Forest Service and, through it, to state cooperators. In 1970 the program title was changed to Forest Fire and Atmospheric Sciences Research. From being an obscure topic relevant only to the remote backcountry, wildland fire research had acquired a sound scientific footing and national backing. It had, in brief, escalated into Big Science.

*Wilderness Fire.* By the late 1960s, however, mass fire as a program of research began to lose ground before the advance of new needs: the administrative need to manage fire in legally designated wilderness areas, the perceived need to use prescribed fire, and the intellectual need to incorporate the biology of fire within a research agenda. The institutional hegemony over wildland fire research created by the Forest Service broke up. A congressional resolution forbade research into the military applications of fire, the U.S. Fire Administration promised to sever Forest Service connections to urban fire, and the programs under civil defense guidance shriveled up, dissolving the NAS-NRC Committee on Fire Research. Other agencies, like the BLM, invested in equipment development. For the investigation of wildland fire the Forest Service faced challengers like the Tall Timbers Research Station, the National Park Service, the National Science Foundation (under its Forest Biome Program), and international bodies like the FAO. Private consulting firms within the United States offered a host of research services, and other nations developed independent research programs.

The result was a major reorientation of Service fire research around the question of fire biology, much on the model of the reformation that previously centered on mass fire. Investigation into fire ecology, into prescribed fire as a means of ecological manipulation, and into the biology of fire as a mechanism of integrating fire management programs with land management plans underscored many of the new research projects. Meanwhile, new land management statutes, Office of Management and Budget directives, and policy revisions complemented the fire ecology emphasis with a revival of interest in fire economics.

The new topics supplemented rather than replaced the old ones. Descriptions of fire as a physical phenomenon, in particular, were essential to the new orientation: fire behavior parameters provided a common denominator for nearly all investigations into fire ecology, fire economics, and fire management. Programs persisted, with modifications, for fire prevention, fuels analysis, planning, fire danger rating, and fire control technology. To them were added programs for smoke management, fire ecology, fire economics, and fire and land management through the concept of prescribed fire. As in other dimensions of fire management, fire research acquired a degree of interagency integration—although by being a participant in virtually all programs, the Forest Service assured its overall dominance. The major data banks of fire research, for example, are due to Service investments; the primary laboratories remain in its hands; the trade journals like *Fire Management Notes* and *Forest Fire News* come under its editorial guidance; and most of the major symposia have had their proceedings published through Forest Service assistance.

### *Structure of Fire Research in the United States*

*Federal Research.* A survey of contemporary fire research may justly begin with the programs of the U.S. Forest Service. Not only did the Service create a science of wildland fire, but it continues to support fire science to a degree unparalleled in the United States. Institutionally, that research belongs under the purview of the Forest Fire and Atmospheric Sciences Research program (FFASR). FFASR distributes its projects through the Forest and Range Experiment Stations maintained by the Service's Branch of Research.

Not all Stations have an equal responsibility for fire research; most of the projects are concentrated within the three forest fire labs administered through the Intermountain Station (Northern Lab), Pacific Southwest Station (Riverside Lab), and Southeast Station (Southern Lab). Originally, each lab was a regional duplicate of the others, but they now share a division of labor—the NFFL handling fire behavior and fire control technology, the SFFL, smoke management and prescribed fire, and the RFFL, systems analysis for projects such as Firescope and Focus. Some fire research is also conducted through the Pacific Northwest Station (mostly on slash-related problems), the North Central Station (on fire problems of the Lake States), the Rocky Mountain Station (meteorology and fire danger rating), the Southern Station (fire prevention), and the Institute of Northern Forestry at Fairbanks, Alaska. The general research strategy combines fundamental investigations with multi-disciplinary, mission-oriented programs that can concentrate on special problem areas and then transfer the results to field users. Most of the research is in-house, but some is conducted with universities through the medium of cooperative agreements and the rest is contracted out to private firms.

Its projects are not limited to the needs of the national forest system. FFASR undertakes a range of research, often functioning as a service research organization for Forest Service cooperators and other government agencies. Much of the work on smoke management in the South and slash disposal in the Northwest, for example, has more relevance for state forestry departments and even for private industry than for national forest management. Similarly, many of the contracts undertaken by FFASR for civil defense and the Department of Defense, the international symposia published under its auspices, and its close ties to other federal land management agencies all show it to be a fire research organization of broadly conceived proportions. Few programs of long-term relevance, even in regions of the country (such as the interior of Alaska) where the Service has little land management responsibilities, lack connections to FFASR. And if the range of inputs into FFASR is broad, no less varied are its outputs. The dissemination of information comes through abundant publications, NWCG training manuals and courses, symposia and conferences, and the exchange of personnel.

This is not to say that all Forest Service fire research is limited to FFASR. Some proceeds through the Forest Products Laboratory. Some remains under the direct control of the national forest system. The equipment centers at Missoula (MEDC) and San Dimas (SDEDC), for example, are administered through the regional offices of the national forest system and directed by the engineering staff of the national forest system (Washington Office). As particular needs for information are felt, moreover, regions and even individual forests have created staff positions in such areas as fire ecology, contracted outside the FFASR system for special research projects, and undertaken cooperative agreements with local academic foresters or consulting firms.

Other federal agencies have, from time to time, contributed to fire research. During its tenure, the NAS-NRS Committee on Fire Research coordinated a broad spectrum of research conducted by the Bureau of Standards, civil defense, and the Department of Defense. Needless to add, most of this work concentrated on urban fire; virtually all of the wildland fire research done through the Committee, regardless of funding source, was conducted through the Forest Service. The emergence of different fire and land management policies during the late 1960s, however, led to a demand on the part of other federal agencies to supplement Forest Service research with programs of their own. This meant primarily the National Park Service, and the research that resulted focused on local fire history, local fire ecology, and the means by which natural fire could be reintroduced into wilderness areas. The Park Service established a small in-house research division headed primarily by wildlife biologists stationed at important parks; other work was contracted out, or done under cooperative agreements with universities. The Bureau of Land Management, too, contracted for research, orienting their program more to equipment development than to fire behavior or ecology. In 1972 the National Science Foundation created its Coniferous Forest Biome program, which gave high priority to the quantitative modeling of fire ecology; through NSF, academic researchers and private research organizations came into play. After 1974 the National Bureau of Standards maintained a Center for Fire Research, with authority to investigate wildland fire but little intention of doing so.

*State and Private Research.* Their need for information had prompted several states to establish formal mechanisms for research early in the history of their fire programs. In 1929 Michigan created its Forest Fire Experiment Station, today synonymous with the Roscommon Equipment Center funded by a consortium of eastern states. Beginning in the 1950s California developed a research program to examine aspects of prevention, fuelbreaks, control technology, and equipment; often the State contributed funds to a joint investigation with the Forest Service. Other States have contracted for research on important topics, such as North Carolina's work with organic soils, or persuaded the Forest Service to direct its own research programs in a direction favorable to state needs. Other work—notably on rangelands—has come through agricultural stations at state universities.

The private sector has made its contributions, too. The Tall Timbers Research Station in Florida contributed a seminal program with its Fire Ecology Conferences (1962–1975), a forum for research conducted outside the Forest Service and an early consolidation of prescribed fire information. Of longer duration have been the fire councils under the administrative aegis of the Western Forestry and Conservation Association, including the California–Nevada, Southwest, Intermountain, Pacific Northwest, Alaska, and Rocky Mountain.

Curiously, academic interest in fire research has never acquired much strength. Most forestry schools have concentrated on topics other than fire, another legacy of academic forestry's origins in Europe. When fire was a topic of interest, academics either deferred to the Forest Service or worked with the Service under cooperative agreements. In more recent years a number of private consulting firms have sprung up to offer services, though most are staffed with retirees from the Forest Service.

*International Research.* That science recognizes few national boundaries is something of a truism. As they have industrialized, other countries have established formal research institutions for the investigation of wildland fire. Canada, Australia, the Soviet Union, and to a lesser extent Spain, France, and Great Britain have all undertaken research on fire behavior, ecology, and control. There are researchers in other countries, mostly biologists, who have investigated fire effects in Israel, Sweden, Finland, Germany, India, Greece, Chile, and South Africa. As industrial forestry spreads, so will fire management and fire research.

Much of the information exchange is in the form of technology transfer—the establishment, often for the first time, of an organized form of fire protection. Typically, such programs accompany reforestation efforts. Often the exchange of information comes from one nation to another through advisors such as those the United States has sent under its Agency for International Development (AID) program or the Peace Corps. Occasionally, an international research program is created, like the study of mass fire done by the United States, Britain, Australia, and Canada under the Technical Cooperation Program during the 1960s. The Food and Agriculture Organization of the United Nations has promoted study tours, and under its aegis the North American Forestry Commission has included a Fire Management Study Group for the exchange of information among the United States, Canada, and Mexico. UNESCO has sponsored symposia and research that deal with fire under its Man and the Biosphere Program (MAB). Among non-governmental bodies, the ICSU Scientific Committee for the Problems of the Environment, the East–West Environment and Policy Institute, and the Tall Timbers Research Station have sponsored international symposia on fire.

With more extensive utilization of wildlands in the tropics and the boreal forest, the prospect for even further international programs of fire research appear good. Virtually all such programs involve the U.S. Forest Service as the American participant, and on behalf of the North American Forestry Commission, FFASR publishes *Forest Fire News* as a kind of international newsletter on fire mangement.

## 8.5 PLANNING FOR FIRE MANAGEMENT

All aspects of fire management include plans. The plans may be strategic, concerned with large areas and long time periods, or they may be operational, tied to actual practice. There are plans for fire prevention, for detection, for presuppression manning and fuel treatments, for prescribed fires of both natural and anthropogenic origin, and for suppression. These specific plans for fire management must be integrated with, or at least made compatible with, other plans for resource use or protection. The number of plans required for all the various activities that make up fire management can number in the dozens, and the difficulty of integrating them can increase exponentially with each additional plan (Figure 8-10). Moreover, so complicated has the planning process become that large agencies even have plans by which to conduct their planning process. But all plans must somehow reconcile stated objectives with the character of the administering institution and with the properties of the fire regime (Barney, 1977). All require information, supplied by special staff or by research institutions. And all show a history.

### *Fire Planning: A Synoptic History*

The conditions which plans confront continually change, and the history of planning is a part of that change. Like theories of fire economics, theories of planning continually evolve. The problem was never simply to devise an ideal fire protection scheme, but to create and maintain one in the context of specific environmental and historical circumstances. As those circumstances have changed, so have strategies of fire protection, techniques of fire planning, and planning

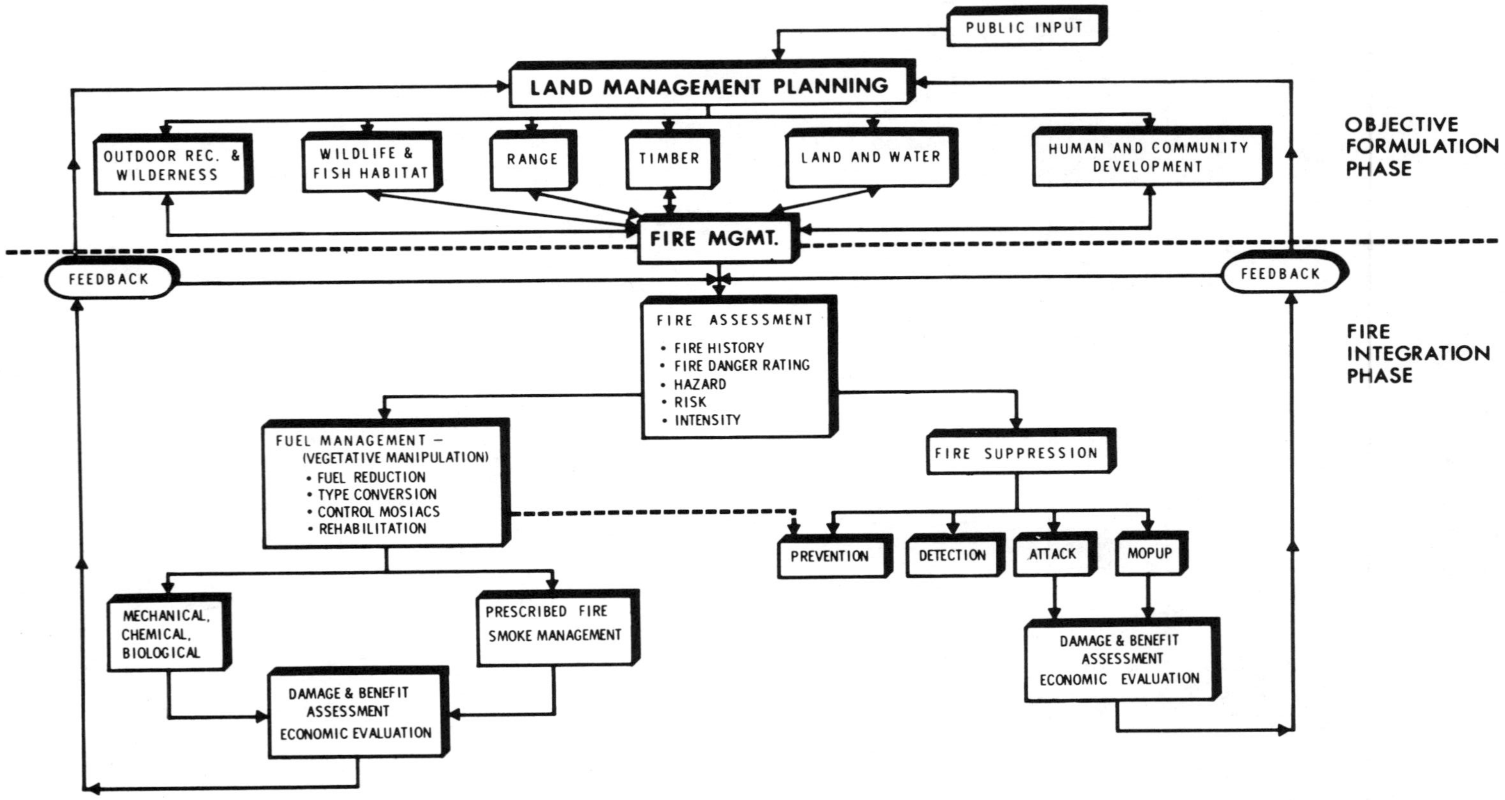


*Figure 8-10. Fire management and land management planning relationships. From Barney (1977b).*

documents. The plans for fire management currently in operation reflect not only large ensembles of contemporary plans but a historical composite—each set of plans evolving out of planning documents and techniques made in earlier times.

Once again, because of its research efforts and its capacity to distribute its programs, consider the history and process of planning in the U.S. Forest Service. Few fire agencies will have planning efforts of equal complexity, but virtually all of the plans in use derive from, or in some way imitate, Forest Service plans or the Forest Service planning process.

*Systematic Fire Protection.* Prior to 1910 improvement in fire protection for the national forests came about through local, pragmatic adjustments in fire programs—a few more miles of trail or phone line, some more fire guards when fire season approached, a better routine for patrols in the backcountry. After 1910 more formal planning documents appeared. The most important, and most comprehensive, was *Systematic Fire Protection in the California Forests* (1914) written by regional forester Coert duBois. Applying the systems analysis of his day, duBois left no aspect of fire protection unexamined and he gave his scheme impressive rigor by quantifying large portions of it. Assuming as an objective to keep fires under 10 acres (4 hectares) for high-value timber land, duBois could harmonize values at risk for different cover types with typical rates of fire spread for those cover types and, hence, with those time intervals within which it was necessary to deliver suppression crews to the fire in order to hold burned acres to an acceptable value. This last value, known as *speed of control* or *elapsed time,* measured the time from the discovery of a fire (or the known start of a fire) until control forces arrived on the scene. The concept of elapsed time was thus integral: it measured the total efficiency of the fire organization. It was, moreover, easily calculated and routinely recorded on fire reports.

Obviously for a given cover type the system needed numbers for resource values, for rate of spread, for crew travel times, and for the rate of line construction. DuBois supplied them by recourse to fire reports, by certain field experiments, and by guess. More important than his specific figures, which were bound to change with improvements in fire control technology, was his method for their analysis. Fire report forms reflected the duBois orientation, giving the pertinent information, and refinements in systematic fire protection came about by the statistical processing of thousands of more reports. The Mather Field Conference (1921) sustained the duBois approach, and by 1928 systematic fire protection as a planning concept and a set of planning techniques for northern California was embodied in an experimental fire district on the Shasta National Forest.

Yet the very comprehensiveness of systematic fire protection was also a limitation. The program only really solved one side of fire management, the question of fire suppression; and it solved that dimension to a degree perhaps greater than the problem demanded. It required such large quantities of information and represented, in most cases, such a high level of investment in transportation and communication networks that it was limited to certain frontcountry lands having large timber resources. The concept was thus well-suited to the lands of northern California, which had considerable mature timber, and to the historical circumstances under which the document was created—the problem of large wildfires (like those of 1910), and the question of light burning. The program was ideal for breaking the previous fire cycle. But it had the unintended effect of driving off alternative concepts of fire planning, like those from the South, that sought to use controlled burning as a method of fire protection, and of selecting against researchers not connected with the duBois circle. In this it resembled the impact of the LCPL concept on fire economics.

Gradually, the concept of systematic fire protection was refined by further research into the concept of hour control, and the program posited by systematic fire protection was extended into lands other than timber and regions other than California. When the LCPL theory appeared, its goals and those of elapsed time were implicitly accepted as compatible. However the objections of fire protection might be stated, systematic fire protection offered a methodology for plans that would meet them.

*Hour Control.* By 1930 the concept had metamorphosed into a refinement called hour control. Under the direction of S.B. Show and E.I. Kotok, both of California, the hour control program gave a generalized solution to fire protection, at once theoretical and programmatic. Hour control was not limited to duBois' test area with its intrinsic high timber values. Instead it outlined the kind of physical plant that would be needed to extend fire control into any region, the chaparral watershed of Southern California among them, and it stated what kind of capital improvements such as roads and lookouts were needed to bring about an acceptable elapsed time, which it expressed in increments of hours.

Total elapsed time from fire start to initial attack actually consisted of four distinct parts: discovery, the time between start and sighting; report, the time between sighting and the report to a dispatcher; getaway, the time between report and actual dispatch of suppression forces; travel, the time between dispatch and arrival. Discovery thus measured the efficiency of the detection function. The report time tested the communication network. The getaway time tested the dispatching and presuppression functions, and the travel time, the initial attack organization.

Hour control sought, in particular, to refine the travel category. Hour control zones ranged from 2 to 4 hours, according to fuel type, and they had as an objective to prevent more than 15% of the total number of fires from reaching 10 acres (Class C). A grass fire thus required a quicker response than a timber fire, for it would reach 10 acres (4 hectares) more rapidly. Such calculations suggested where suppression forces ought to be located and what type of transportation network ought to exist.

The abundance of manpower and funding unleashed by the Roosevelt Administration during the 1930s helped the hour control program to become widely disseminated. Further refinements came with the development of a fire danger rating meter, which promised numerical guidance in short-term presuppression investments and rate of spread calculations; with the evolution of cover types into the more general concept of fuel types, which differentiated the suppression problem between those components intrinsic to the fire and those associated with control efforts; and with the adoption by 1937 of refined hour control methods as a national standard for planning, which further enhanced the need for general solutions to fire planning. All of these refinements furnished better measurements on speed of control within certain time periods, or hour control zones. In theory, policy determined what hour control zone should apply to what lands, but in practice hour control favored a minimum damage approach such that strong initial attack was universally prescribed.

Despite its complexity, the concept suffered from the number of parameters it had to assume as constant—the fuel types and the weather, which set rate of spread and resistance to control, and the travel times and line construction rates, which set the response and control times. Like systematic fire protection, hour control as a concept of planning and a program of fire management addressed the problem of establishing a fire protection system where, for the most part, none previously existed. For its source of data it remained heavily dependent on the fire report form, and for statistical analysis, on the methodology of silviculture. By its very nature, however, it emphasized those aspects of fire management which aided the control of wildfire. Of prescribed fire it said little. Its own concerns—continental in scale—were large enough.

It was an extensive program as well as an expansive one, liberally funded for presuppression and suppression activities. It was intimately connected with the construction of new roads, trails, lookouts, and guard stations, with bringing fire protection into the backcountry regardless of existing resource values. For such purposes, hour control was ideal. Though many areas could not realize its full program, all recognized the power of its methodology and accepted its concepts as a planning ideal. As a result fire protection in the national forest system became at least a first order approximation of the hour control program as revised by Hornby (1936).

*Conflagration Control.* During the 1950s the hour control program was amended according to the concept of conflagration control, and it addressed new problem fires, new technologies, and new policies. The program was peculiarly adapted to the Southern California fire scene, much as systematic fire protection had been to northern California and hour control to the Northern Rockies. Conceptually, the program argued for an elaborate system of fuelbreaks, preestablished lines of control for ease of construction and burning out; for the description of fires by intensity, not merely by rate of spread; and for means by which to incorporate the added muscle brought about by mechanization, especially by aircraft. The competition between fire and suppression forces was not limited to the geometry of the fire perimeter—to the relative rates of growth between fire and fireline—but to their relative intensities. The concept of resistance to control eventually referred to fireline intensity, not simply the difficulty of line construction. Strength was required by a suppression organization as well as speed if fires were to be prevented from making the transition to conflagrations.

To existing hour control facilities and zones, planners thus superimposed the added punch of aerial attack by helicopter, air tanker, and smokejumper; disciplined fire crews ready for dispatch to any portion of the national forest system threatened by large fires; and improved initial attack crews whose primary duty was fire protection, not some other assignment for which fire control was secondary. Hour control planning had been almost wholly circumscribed by ground crews, ground-based detection, roads and trails. The presence of a helitack crew or an air tanker changed the old standards for initial attack, and they made possible the achievement of hour control standards in areas whose resource values otherwise made such goals inappropriate. But this very capability also brought fire protection to a point of diminishing returns, made manifest by the planning efforts associated with the 10 Acre Policy adopted in 1971.

*Fire By Prescription.* By the early 1970s it became apparent that further improvement in fire control would have to come through other means—through modification of the fire environment, through other strategies of fire management, and through shared resources among fire agencies. Shifting attention from fire control to fire management—from breaking an old fire cycle to the reconstitution of a new one—coincided with a slough of legislation mandating large-scale planning by the Forest Service. Long-range plans were to be based on principles of land management, loosely coinciding with concepts of an ecosystem and subject to review by the public and other agencies (including the Environmental Protection Agency). New techniques for planning became available, and new strategies for planning were promulgated—devised first by the U.S. Forest Service, then adapted by other federal agencies.

The consequences for fire management were, in abstract terms, a somewhat different set of goals and, on practical grounds, a diverse set of options available for response to fire. The replanning began in a serious way after the Forest Service revised its 10 AM Policy in 1971. A fire planning handbook was issued, providing general national guidelines but leaving to individual forests the construction of the actual plans. In general, the exercise sought to strengthen presuppression activity (in accordance with the 10 Acre Policy) and to liberalize opportunities for prescribed fire (especially in wilderness areas). Agencies in the Interior Department adapted the methodology for use in what was termed "normal fire-year planning."

The replanning process was soon revised, however, as a result of the Forest and Rangeland Renewable Resources Planning Act (1974) and amendments contained in the National Forest

Management Act (1976). This legislation mandated long-term planning on a massive scale, directed by land management objectives and resource values. The resulting Forest Plan for each unit of the national forest system would contain a fire management plan, but the fire plan would have its goals set by the Forest Plan. In brief, fire plans would again vary according to the value of the resources under protection or the larger land use purposes, such as wilderness, to which they were put. The adoption in 1978 of a new fire policy by the Forest Service, a de facto policy of fire by prescription, agreed with these provisions. Again, fire management undertook another national planning program. The plans for this new planning process, embodied in an additional manual, were in themselves a major enterprise.

The new policy and plans promoted a pluralism of options with regard to fire. Plans specified different responses for different areas, or for the same area as conditions changed. Prescribed fire was encouraged, both from scheduled (deliberate) and unscheduled (natural) ignition sources. Similarly, the response to a reported wildfire could take a variety of forms. A wildfire could be *confined* to a preestablished area, *contained* along its active perimeter, or *controlled* by extinguishment. If a prescribed fire exceeded its prescriptions or if a wildfire escaped initial attack, then an *escaped fire analysis* was required. Based on its conclusions, the fire might again be confined, contained, or controlled through such tactics as *aggressive attack, modified attack,* and *delayed attack.*

This range of options introduced an extraordinary flexibility into fire management, but it also dramatically multiplied the number of decisions and of plans for decisionmaking that were needed for routine fire management operations. Wherever possible, the agency sought to regulate this decision process through the approval of preplanned decision charts (Figure 8-12). The volume of data and number of steps required, moreover, argued for automation, and a variety of computer programs have been devised to assist with both planning and operations.

The new plans, however, built upon the old. The tendency was to evolve from the agency's strengths. This meant the suppression organization, and planners could assume the existence of a first-order fire control system upon which to erect new programs. In some cases, this meant introducing complementary plans to promote prescribed fire or fuels management. In other cases, it meant placing restraints on suppression—on its choice of equipment, on its potential geography, on its relative merit in comparison with other management options. Gradually, suppression programs acquired not a single plan, but a matrix of plans, one for each aspect of fire detection, prevention, presuppression, and control. To these were amalgamated other plans. The development of suitable plans required new techniques and guidelines for planning, and the simultaneous accommodation of all these various plans demanded, in turn, large commitments of agency resources and time.

### *Contemporary Planning: A Sampling*

*General Strategies.* The new strategy directed that all fires be considered as either *wildfire* or *prescribed fire,* that administrative units be organized around geographic zones known as fire management areas and that an array of options be available for every fire management activity. These were embodied in new strategic plans. Behind them was the belief that natural fire belonged in wilderness areas, and that some prescribed fire would be useful in the management of most wildlands. The new plans sought means by which to inject more fire under controlled conditions. For the states, however, law continued to mandate fire control. A few states (like Florida and California), outfitted with special legislation, practiced prescribed burning on public and private lands, but their primary mission remained fire control, not land management.

To the old geography of hour control zones that had previously guided suppression planning, there appeared fire management areas, which provided a first-order classification of fires and appropriate responses. A fire management area (FMA) defined the area subject to planning. In turn, it could be subdivided into fire management zones or fire management units, or both. Fire management zones (FMZ) included land within the FMA that shared some important characteristic—similar fuels, for example, or a similar management response. Fire management units (FMU), by contrast, broke FMAs into distinct burning blocks, defined wherever possible along natural features. Typically, detailed planning will be conducted by FMU, one by one (Figure 8-11). The units could be considered equally as preattack blocks or burning blocks, depending on whether the appropriate response was suppression or prescribed fire.

For every activity in fire management there appeared a range of options. A single area might include full suppression, modified suppression, natural fire, and prescribed burning. A single fire might, in the course of its history, be considered a prescribed natural fire to be monitored, then a prescribed fire exceeding its prescription (escaped fire) and requiring containment along one flank, and then a wildfire to be extinguished. A wildfire in one area might, in adjacent areas, be considered a prescribed fire. In general, a fire could be confined, contained, or controlled through a flexible scale of tactics ranging from all-out suppression to delayed or partial containment. Correspondingly, there had to be criteria and guidelines for distinguishing wild from prescribed fire, for specifying which option among many was desirable for any fire, and for selecting prescribed fire from among many alternative land or fuel treatments. Such criteria were expressed as fire management units and as prescriptions that forecast expected fire behavior. Appropriate response options were formatted as decision trees.

The amount of information available for decisions increased exponentially, and increasingly complex plans, with their multiple pathways for response, called for even further data. By relying more and more on quantitative measures of fire behavior, which already underscored fuel modeling and fire danger rating, it was possible to bring computer-assisted analysis to the planning process and to interrelate one plan with another. Some form of mechanization was necessary; once begun, automation at one level of planning meant, in the end, perhaps automation at nearly all levels.

*Selected Planning Techniques.* Consider, by way of illustration, the following examples of fire planning techniques. All of the programs were devised by the U.S. Forest Service to assist in its almost continual replanning projects of the 1970s. All have been incorporated, in some form, into standard planning manuals. Most are built out of, and presuppose the existence of, preexisting plans, just as prescribed fire assumes the infrastructure of fire suppression. With regard to technique the major advances have been to introduce decision trees as a means of incorporating preplanned variety into the response to fire and to reach for computer assistance for preplanned simulation of possible fire behavior and effects. As an example of decision trees, one can look to DESCON or to most prescribed natural fire plans. For examples of computer simulation or data processing, one can look to FOCUS, for suppression planning; to FORPLAN, FIRELAMP, FIREFAMILY, and the Activity Fuel Appraisal Process, for prescribed fire. With the exception of FOCUS, these techniques have sought ways to actively incorporate the more desirable consequences of fire into fire management plans.

The DESCON (DESignated CONtrol Burn) program was developed for the Francis Marion National Forest in South Carolina. Here deliberate prescribed burning was commonly practiced, but the decision tree was designed to accomodate any fire, regardless of ignition source, so long as it satisfied a predetermined prescription (Devet et al., 1974). The technique has disseminated to other areas, but its principal use has become to support prescribed natural fire programs. Here the decision tree can guide the response to an unscheduled (lightning) ignition according to many considerations—fire behavior forecast, fire load over the whole forest, smoke, resources at risk,

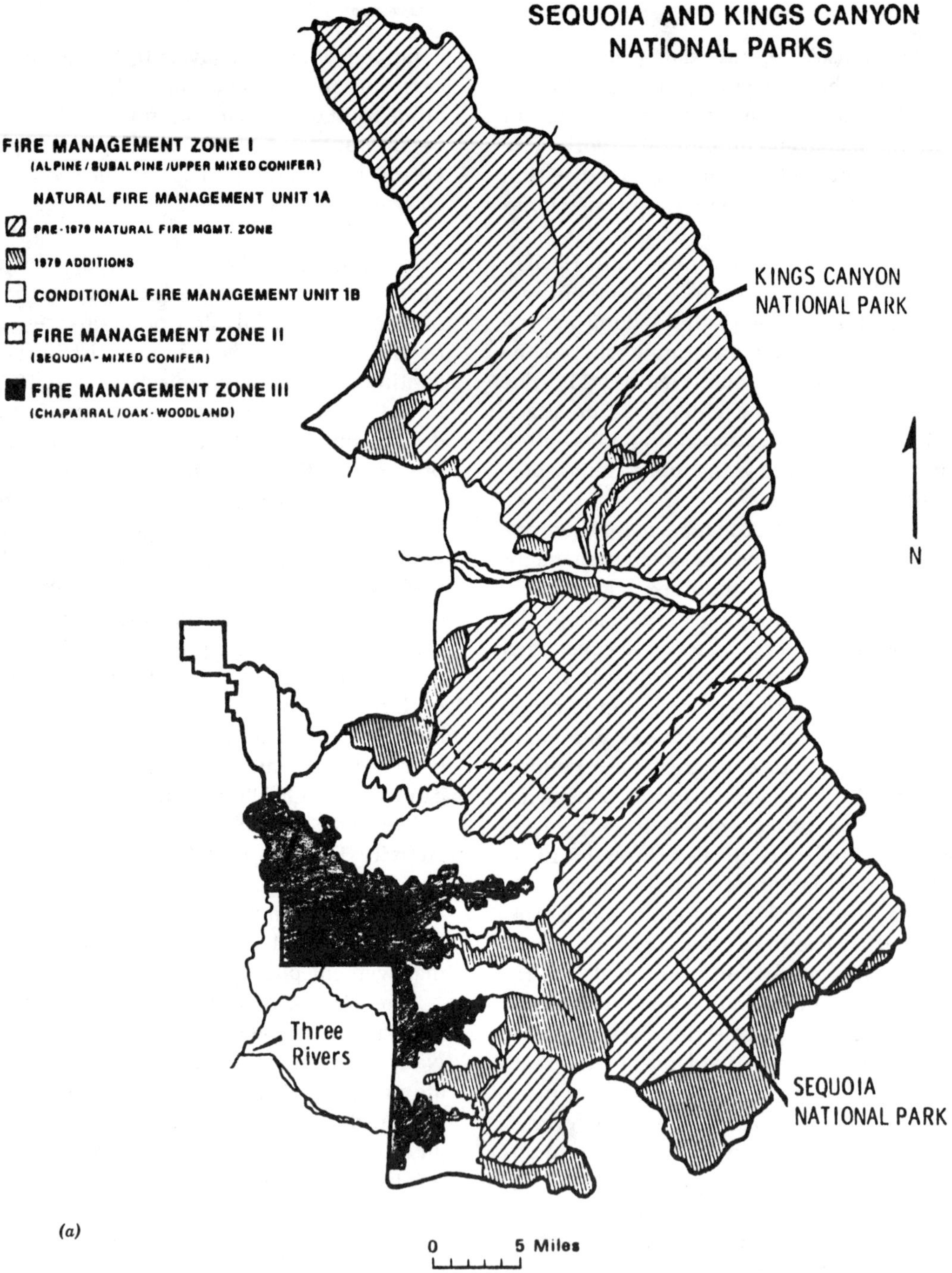


*(a)*

*Figure 8-11. Planning techniques. (Upper left) Map shows delineation of fire management zones for Sequoia–Kings Canyon National Park. The zones correspond broadly to elevation and fuel type, with natural fire designated for upper elevation conifers, prescribed fire and suppression for mid-elevation Sequoia groves and conifers, and suppression for lower elevation chaparral and grass. (Upper right) Map defines the actual units of planning, which in most cases correlate to natural burning blocks. Maps courtesy of Sequoia and King's Canyon National Park.*

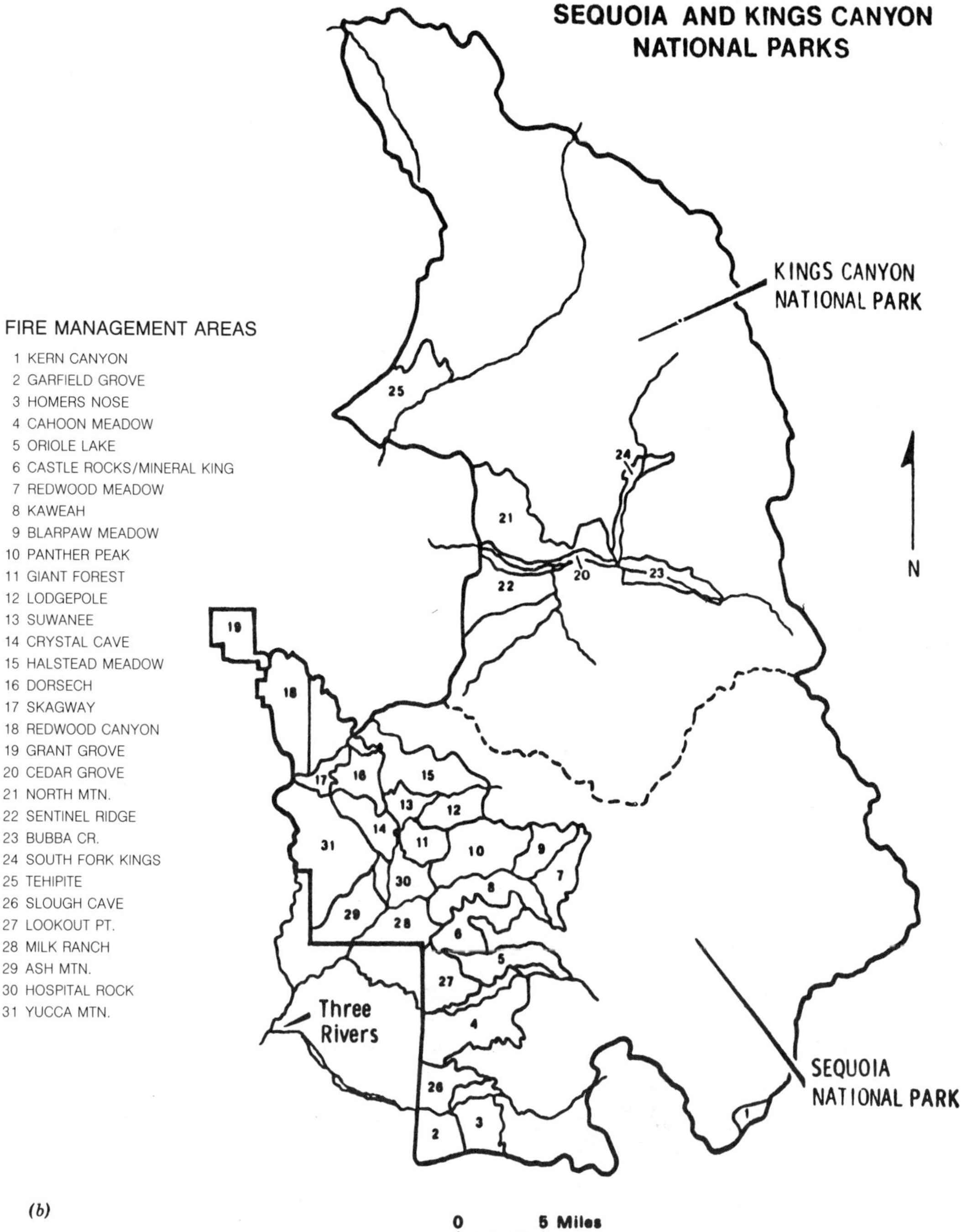
SEQUOIA AND KINGS CANYON
NATIONAL PARKS
KINGS CANYON
NATIONAL PARK
SEQUOIA
NATIONAL PARK
Three
Rivers
N
0
5 Miles
FIRE MANAGEMENT AREAS
1 KERN CANYON
2 GARFIELD GROVE
3 HOMERS NOSE
4 CAHOON MEADOW
5 ORIOLE LAKE
6 CASTLE ROCKS/MINERAL KING
7 REDWOOD MEADOW
8 KAWEAH
9 BLARPAW MEADOW
10 PANTHER PEAK
11 GIANT FOREST
12 LODGEPOLE
13 SUWANEE
14 CRYSTAL CAVE
15 HALSTEAD MEADOW
16 DORSECH
17 SKAGWAY
18 REDWOOD CANYON
19 GRANT GROVE
20 CEDAR GROVE
21 NORTH MTN.
22 SENTINEL RIDGE
23 BUBBA CR.
24 SOUTH FORK KINGS
25 TEHIPITE
26 SLOUGH CAVE
27 LOOKOUT PT.
28 MILK RANCH
29 ASH MTN.
30 HOSPITAL ROCK
31 YUCCA MTN.

*(b)*

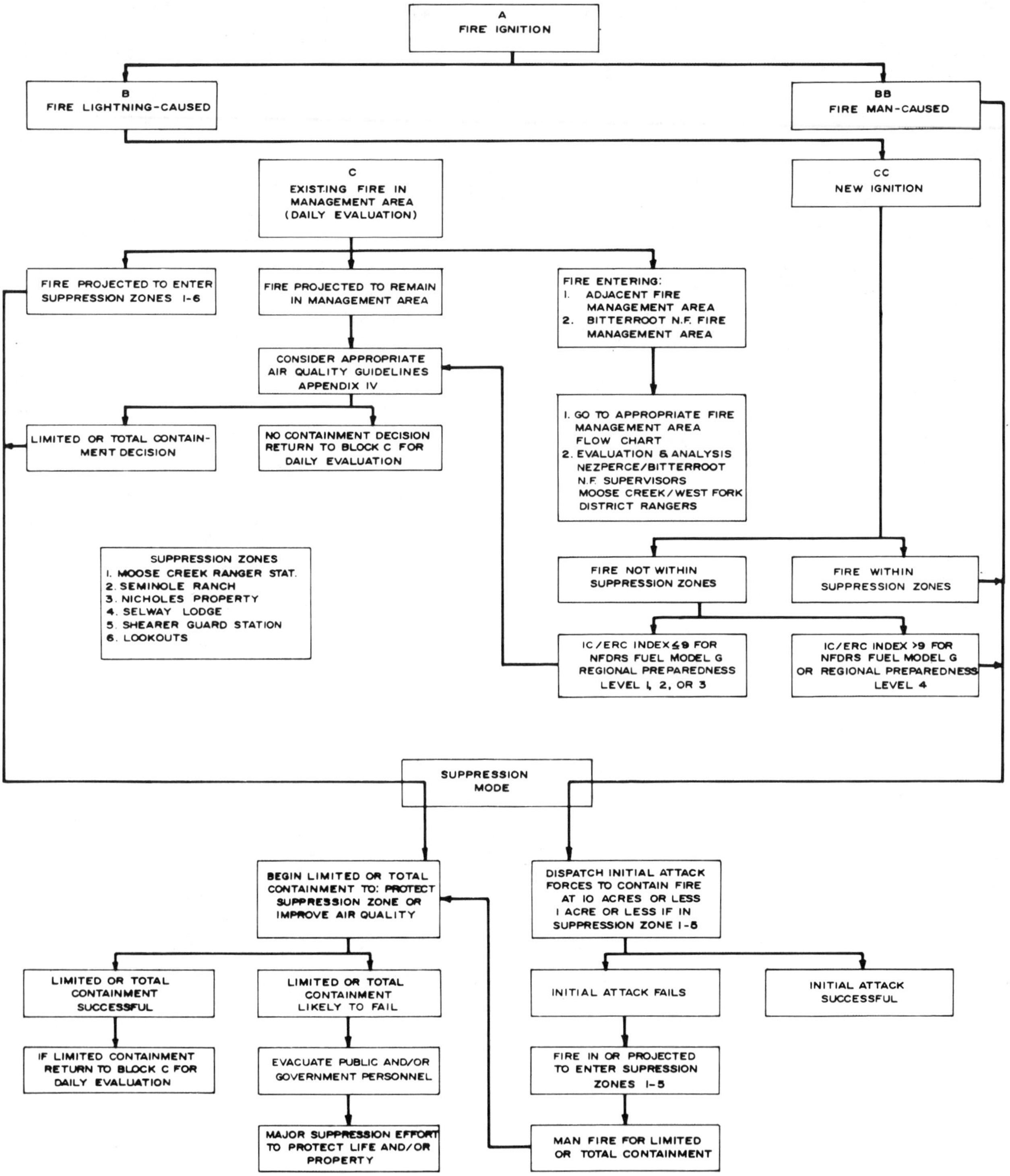


*Figure 8-12. Decision chart for prescribed natural fire. Developed for Moose Creek Wilderness Fire Management Plan, Selway-Bitterroot Wilderness, Idaho. From Thomas and Marshall (1980).*

and so on (see Figure 8-12). Flow charts such as these have also provided a basis for interagency coordination for fire management areas (such as that between Grand Teton National Park and the Teton National Forest) which allow natural fires to burn across jurisdictional lines.

FOCUS (Fire Operational Characteristics Using Simulation) developed as a means of supporting the national replanning effort of the 1970s (Bratten et al., 1981). A computer model, it represents, through gaming, the response of a wildland fire protection organization to simulated fire occurrences (Figure 8-13). The program can be used on any sized unit for a seasonal (or yearly) time scale. For fires that escape initial attack (over 100 acres) FOCUS requires gaming by fire experts rather than by computer. It assumes that suppression is the appropriate response to a fire report, that the kind of data developed for hour control is generally applicable, and that a suppression organization is already in place. One can select from among particular kinds of resources (e.g., helitack, groundtanker, air tanker) and among different locales for fires and suppression forces; by gaming, the program shows the probable consequences of selecting certain suppression resources over others. The FOCUS model can be used, in turn, as an initial attack module for other forms of analysis, such as the economic inquiries of FEES. As an operational analogue there are computer-assisted dispatching systems which rely on similar projections of fire behavior and suppression response. A prototype is the Firescope Information Management System (FIMS) (Kessell and Cattelino, 1978).

The essence of the new policies, however, was not to sharpen initial attack and suppression but to broaden the mission of fire management by assimilating it within land management planning and by incorporating prescribed fire. For assessing the role of fire programs within the context of multiple-use land management, fire planners appealed to FORPLAN (FORest PLANning; Potter et al., 1979). FORPLAN could simulate the impact of different activities, fire and fire management programs among them. For effects specifically related to fire, FIRELAMP could simulate fire effects on different components of an ecosystem. With the aid of such simulation, planning can incorporate state-of-the-art knowledge about fire

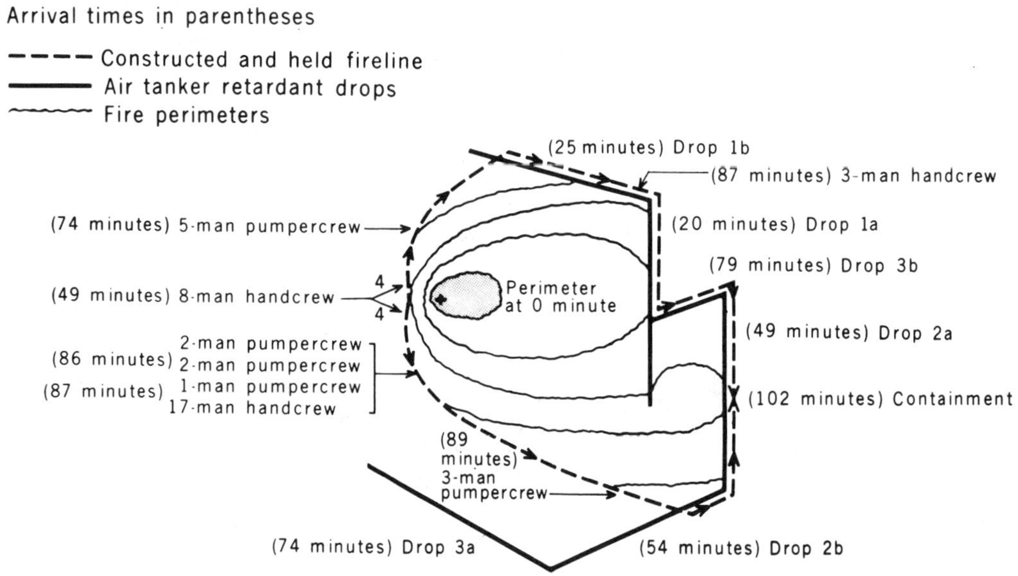


*Figure 8-13. FOCUS modeling of suppression. The fire perimeter (derived from fire behavior modeling) is contained through both retardant drops and handlines. Any fire that reaches 100 acres is considered, for planning purposes, an escaped fire—beyond initial attack. From Bratten et al. (1981).*

effects—a fundamental need for the design of prescribed fire projects. For coping with slash, the Activity Fuel Appraisal Process, which translates various fuel management techniques into probable fire behavior outcomes, is available (Hirsch et al., 1981). For the scheduling of prescribed fire, a host of computer programs collectively known as FIREFAMILY has been developed (Main et al., 1982).

All of these programs rely on archival climatological data stored in the National Fire Weather Data Library—data accessioned automatically for any administrative unit that participates in the Affirms network. General climatological summaries are available through a program known as Firedat (Main et al., 1982). Rxwthr summarizes climatological data relevant to prescribed fire. The information may be organized by individual weather element or by co-occurrence groupings, say, by windspeed, temperature, and relative humidity. Using the same database Rxburn converts this information into fire prescription parameters and summarizes the frequency distribution for those parameters. The output can take several forms, all of which describe the number of days in which conditions suitable for prescribed burning are likely to be met. If the prescription is written in terms both of preferable days and of acceptable days, then Rxburn will list the likelihood of preferable, acceptable, and unacceptable day occurrences for each 10-day period of the specified fire season. It does for prescribed fire planning what the "average worst" calculation or risk assessment does for fire control planning.

### *Fire Management Planning: The Alaska Example*

Perhaps the most spectacular instance of contemporary planning is occurring in Alaska. The reason is, first, the late arrival of organized fire protection within the interior, and then its domination by one agency, and, secondly, the absence of a period of agricultural reclamation. Fire protection would not need to assist in the transition from agriculture to industry, but from subsistence economies to wildland (and wilderness) management. An Alaska Fire Control Service was organized by the Interior Department in 1939, and then transferred to the Bureau of Land Management in 1946. Perhaps 1–2 million acres (405,000–810,000 hectares) a year burned. Serious fire control efforts did not get underway until after the great fires of 1957, and they eventually led to a program similar to that of the U.S. Forest Service. An organic act for the BLM strengthened its hand after 1976. The administrative situation in the interior was unique, with the BLM as practically the sole fire agency. In addition to its own lands it was the contract source for other federal agencies and the State of Alaska.

What makes the Alaska situation special, however, is the fundamental reconstitution of land ownership now under way, the timing of this transfer relative to the national history of fire management, and the particular options open to fire management as a result of the Alaskan environment. The transfer of land ownership was set into motion by the Alaska Statehood Act (1958), the land transfer provisions of which (from federal to state control) did not become effective until the 1980s; the Alaska Native Claims Settlement Act (1971), which granted some 44 million acres (18 million hectares) to Alaska native corporations; and the Alaska National Interest Lands Conservation Act (1980), which designated considerable portions of the interior for wildlife refuges, parks, and wilderness. These land acts effectively guaranteed that the human population would be small, the land areas large, and wildland fire fundamental to land management.

The timing of this transfer, moreover, coincided with the conclusion of more than a decade of intensive reevaluation of fire management—its purposes and its institutions. Much of this debate had centered on the question of wilderness, which was also the focus of the Alaska land acts. Coming when it did, the need for fire planning meant that wilderness fire would be an informing concern, prescribed fire a basic technique, and interagency coordination an administrative assumption. The transfer thus encapsulated many of the transitions that had transpired in fire management from the 1960s on.

Yet there was much that was different, too. Permafrost limited the use of certain suppression equipment such as heavy tracked vehicles and some techniques of line construction that might lead to serious postfire erosion. Even apart from the consideration of possible damage from ground suppression, the tyranny of distance in the interior meant that all aspects of fire management would be dependent on aircraft. Detection, initial attack, logistical support—all rely on aerial observers, air tankers, helitack crews and smokejumpers, and paracargo delivery. The success of the BLM fire program had resulted from its reliance on aircraft. It is likely that much prescribed burning will likewise be conducted by air. With further settlement in the interior, there will be some transfer to ground-based operations tied to roads. Most of these operations will fall under the control of the state, and a probable administrative model is California, where a mixture of structural and wildland fire protection has developed over the course of many decades (Alaska Department of Education, 1977).

Unlike the Transfer Act of 1905, the Alaska land acts did not consolidate control over many lands into one agency, but broke up lands for distribution among many agencies. Unlike the situation in Southern California, where interagency management had to integrate many well-established fire programs, the Alaska situation sought to design an interagency program from the beginning. Interagency planning and coordination would be an initial condition, not an end product. Fire plans would, more or less at the start, begin with the modern complex of fire management concepts. In a sense fire management would be recapitalized wholesale—its physical plant, its planning procedures, its institutions and its administrative concepts. Certainly, agencies like the NPS and FWS would bring with them organizations developed elsewhere, and it would be impossible to delete the history of BLM involvement any more then one could ignore the suppression infrastructure created by that agency.

But compared to the situation elsewhere in the United States, Alaska offered the opportunity for a remarkable degree of reconstruction. Among other considerations was that much of the land would be designated as wilderness, that large portions of the rest would continue to be subject to subsistence lifestyles, and that the U.S. Forest Service would, except for certain research functions, be absent from the scene. Prescribed fire would be built into the system as a precondition, not as an aftereffect.

The instrument for planning was the Fire Management Project Group within the Alaska Land Use Council, formed under provisions of the Alaska National Interest Lands Conservation Act. Among agencies represented were Doyon, Ltd. (Alaska Federation of Natives), Alaska Department of Fish and Game, Alaska Department of Natural Resources, NPS, FWS, BLM, BIA, U.S. Forest Service, and the Institute of Northern Forestry maintained by the Forest Service; public input was added through meetings. The Fire Management Project Group provided for an Alaska Fire Management Plan by designating planning areas, establishing priorities for the release of plans for each area, and working out an agreed upon set of objectives and procedures for producing each plan. The land transfers are still proceeding, and only a few plans have been published. As an illustration of how the system works, consider the Tanana-Minchumina Fire Plan (Alaska Land Use Council, 1982).

The Tanana–Minchumina Planning Area encompasses 31,000,000 acres (12,550,000 hectares) in the center of Alaska (see Figure 8–14). Within this block—roughly the size of the State of New York—17 separate fire management units have been designated. The units correspond to natural features, like watersheds, rather than to normal administrative boundaries. All participants agree to an operational policy for the entire area, all accept certain operational procedures (options) for the response to fire, and through the medium of cooperative agreements all share in common suppression resources. But final jurisdiction remains with the responsible agency on whose

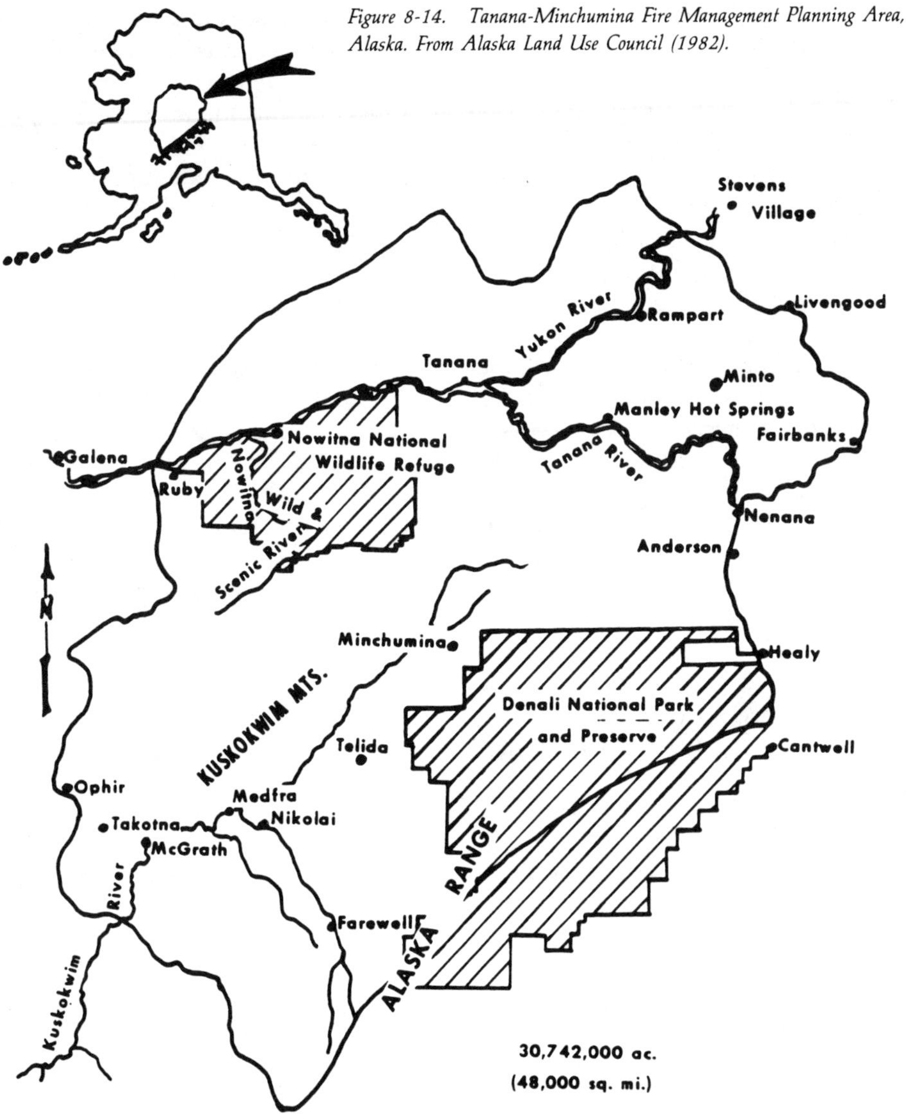


*Figure 8-14. Tanana-Minchumina Fire Management Planning Area, Alaska. From Alaska Land Use Council (1982).*

lands a fire occurs. By specifying its fire season, moreover, each agency defines the range of available options. What the plan provides for is interagency coordination before the fact—not merely to the fact of a particular fire, but to the full establishment of fire programs.

The fire management options are four. *Critical protection* applies to areas where wildfire could threaten human life and developments. The designation of such sites is left to the individual agency, but all participants agree to grant "unquestioned priority" to the response to fires that occur in such sites. *Full protection* grants high priority to natural areas and sites with cultural and historic significance. If invoked, the option requires aggressive initial attack and continued suppression effort until the fire is declared out. *Limited action,* by contrast, applies to those areas where a natural fire program is desirable or where the values at risk do not warrant intensive suppression. The emphasis of fire control should be to contain wildfires or prescribed fire within the zone or, where special sites might be involved, to protect those selected sites. Intermediate between the full action and limited action options is the *modified action* option. The intention here is to protect lands that require protection during critical burning periods, but only nominal protection otherwise.

The agency determines, on a 10-day review cycle, just when the critical burning periods occur. During critical periods, all fires will receive initial attack. Should the fire escape, an escaped fire analysis will be made. Outside the critical periods, the limited action option will apply. The use of calendar dates, adjusted for actual conditions, substitutes for a fire danger rating system, which has so far seen limited application in the interior because of sparse coverage by weather stations. Similarly, the use of calendar dates offers an alternative to predetermined prescriptions, which have also demonstrated limited usefulness under Alaskan conditions.

## *REFERENCES*

### *Fire Prevention*

Fire statistics are published annually by the U.S. Forest Service. AFM publishes statistics for the national forest system, and CFP for all lands protected by the Forest Service and its cooperators. Two special studies of lightning fire have been conducted: Barrows et al. (1977), "Lightning Fires in Northern Rocky Mountain Forests," and Barrows (1978), "Lightning Fires in Southwestern Forests." For the distribution of southern fires, see Doolittle (1977), "Forest Fire Occurrence in Southern Counties, 1966–1975"; and for their causes, consult Cole and Kaufman (1966), "Socio-Economic Factors and Forest Fires in Mississippi Counties." As a western analogue, see Folkman (1969), "Residents of Butte County, California." Lightning modification is summarized by Fuquay (1974), "Lightning Damage and Lightning Modification Caused by Cloud Seeding." An excellent series of practical handbooks in prevention has been written by Moore (1978–1981).

Barrows, Jack S., 1966, "Weather Modification and the Prevention of Lightning-Caused Forest Fires," pp. 169–182, in W.R. Derrick Sewell (ed.), *Human Dimensions of Weather Modification* (Chicago: University of Chicago Department of Geography).

———, 1978, "Lightning Fires in Southwestern Forests," Final report, Northern Forest Fire Lab, U.S. Forest Service.

Barrows, Jack S. et al., 1977, "Lightning Fires in Northern Rocky Mountain Forests," Final report, Northern Forest Fire Lab, U.S. Forest Service.

Bertrand, Alvin L. and Andrew W. Baird, 1975, "Incendiarism in Southern Forests: A Decade of Sociological Research," Social Science Research Center, Bulletin 838, Mississippi State University.

Brown, Arthur A. and Kenneth P. Davis, 1973 rev., *Forest Fire: Control and Use,* 2nd ed. (New York: McGraw-Hill).

Bureau of Land Management, Division of Fire Control, "Project MOD," in "Annual Reports, 1969–1973," (Anchorage).

California Department of Forestry, 1980, "Fire Safe Guides for Residential Development in California," (Sacramento: California Department of Forestry).

Christiansen, John R. and William S. Folkman, 1971, "Characteristics of People Who Start Fires...Some Preliminary Findings," U.S. Forest Service, Research Note PSW-251.

Cole, Lucy W. and Harold F. Kaufman, 1966, "Socio-Economic Factors and Forest Fires in Mississippi Counties," Social Science Research Center, Preliminary Report No. 14, Mississippi State University.

Doolittle, M.L., 1977, "Forest Fire Occurrence in Southern Counties, 1966–1975," U.S. Forest Service, Southern Experiment Station, SO-227.

Dunkelberger, J.E. and A.T. Altobellis, 1975, "Profiling the Woods-burner: An Analysis of Fire Trespass Violations in the South's National Forests," Agricultural Experiment Station, Bulletin 469, Birmingham, Alabama: Auburn University.

Folkman, William S, 1966a, "Modifying the Communicative Effectiveness of Fire Prevention Signs," U.S. Forest Service, Research Note PSW-105.

———, 1967, "Evaluation of Fire Hazard Inspection Procedures in Butte County, California," U.S. Forest Service, Research Note PSW-145.

———, 1966, "'Children-with-Matches' Fires in the Angeles National Forest Area," U.S. Forest Service, Research Note PSWS-109.

———, 1969, "Residents of Butte County, California—Their Knowledge and Attitudes Regarding Forest Fire Prevention," U.S. Forest Service, Research Paper PSW-25.

———, 1972, "Studying the People Who Cause Forest Fires,"

pp. 44–64, in *Social Behavior, Natural Resources, and the Environment* (New York: Harper & Row).

———, 1973, "Fire Prevention in Butte County, California..-.Evaluation of an Experimental Program," U.S. Forest Service, Research Paper PSW-98.

———, 1977, "High-Fire-Risk Behavior in Critical Fire Areas," U.S. Forest Service, Research Paper PSW-125.

Fuquay, Donald M., 1974, "Lightning Damage and Lightning Modification Caused by Cloud Seeding," pp. 604–612, in W.N. Hess (ed.), *Weather and Climate Modification* (New York: John Wiley).

Fuquay, Donald and Robert G. Baughman, 1969, "Project Skyfire Lightning Research: Final Report," Unpublished report, U.S. Forest Service.

Gaylor, Harry P., 1974, *Wildfires. Prevention and Control* (Bowie, Maryland: Robert J. Brady Company)

Haug Associates, Inc., 1968, "Public Image of and Attitudes Toward Smokey the Bear and Forest Fires," Unpublished report, U.S. Forest Service.

Moore, Howard E., 1980, "Industrial Operations Fire Prevention Field Guide" (Sacramento: California Department of Forestry).

———, 1978 rev., "Power Line Fire Prevention Field Guide" (Sacramento: California Department of Forestry).

———, 1978, "Railroad Fire Prevention Field Guide" (Sacramento: California Department of Forestry).

———, 1981, "Protecting Residences From Wildfires: A Guide for Homeowners, Lawmakers, and Planners," U.S. Forest Service, General Technical Report PSW-50.

Murphy, James L., 1975, "The National Interagency Wildfire Prevention Task Force," pp.30–33, in "Global Forestry and the Western Role," *Proceedings of Permanent Association Committees* (Portland: Western Forestry and Conservation Association).

National Wildfire Coordinating Group, n.d., "Prevention (P) Series Training Courses."

National Fire Protection Association, 1974, "Homes and Camps in Forest Areas," NFPA 224–1974.

Rayonier Incorporated v. United States, 352 U.S. 315 (1957).

Roussopoulos, Peter J. et al., 1980, "A National Fire Occurrence Data Library for Fire Management Planning," in American Meteorological Society and Society of American Foresters, *Sixth National Conference on Fire and Forest Meteorology* (Society of American Foresters).

Pyne, Stephen J., 1982, *Fire in America: A History of Wildland and Rural Fire* (Princeton, New Jersey: Princeton University Press).

U.S. Forest Service, annually, "Wildfire Statistics," Cooperative Fire Program.

———, annually, "National Forest Fire Report," Aviation and Fire Management.

U.S. Forest Service et al., 1975, "Wildfire Prevention Analysis: Problems and Progress" Unpublished report, U.S. Forest Service.

## *Detection and Communication*

Amidon, Elliot L. and E. Joyce Dye, 1981, "SCANIT: Centralized Digitizing of Forest Resource Maps or Photographs," U.S. Forest Service, General Technical Report PSW-53.

Fuquay, D.M. et al., 1979, "A Model for Predicting Lightning Fire Ignition in Wildland Fuels," U.S. Forest Service, Research Paper INT-217.

Gray, Gary, 1982, *Radio for the Fireline: A History of Electronic Communication in the Forest Service* (Washington, D.C.: U.S. Forest Service).

Innes, William, Jr., 1978, "Gathering Fire Danger Data by Use of Satellite," *Fireline* (March–April): pp. 2–5.

Latham, Don J., 1979, "Progress Toward Locating Lightning Fires," U.S. Forest Service, Research Note INT-269.

McBride, Fred, 1981, "The Fire Management Electronic Age," *Fire Management Notes* **42**(4): 3–5.

Mees, Romain M., 1976, "Computer Evaluation of Existing and Proposed Fire Lookouts," U.S. Forest Service, General Technical Report PSW-19.

Rothermel, Richard C., 1980, "Fire Behavior Systems for Fire Management," in American Meteorological Society and Society of American Foresters, *Sixth National Conference on Fire and Forest Meteorology* (Society of American Foresters).

Show, S.B. and E.I. Kotok, 1937, "Principles of Forest Fire Detection on the National Forests of Northern California," U.S. Department of Agriculture, Circular 574.

Show, S.B. et al., 1937, "Planning, Constructing, and Operating Forest-Fire Lookout Systems in California," U.S. Department of Agriculture, Circular 449.

Travis, Michael et al., 1975, "VIEWIT: Computation of Seen Areas, Slope, and Aspect for Land Use Planning," U.S. Forest Service, General Technical Report PSW-11.

Vance, Dale L., 1978, "Lightning Detection Systems for Fire Management," pp. 68–70, in Michael A. Fosberg (chairman), *5th National Conference on Fire and Forest Meteorology* (Boston: American Meteorological Society).

Wilson, Ralph A. et al., 1971, "Airborne Infrared Forest Fire Detection System: Final Report," U.S. Forest Service, Research Paper INT-93.

## *Fuels Management*

The literature on fuels management is large, but it blends easily with other topics. A good summary of residue problems and programs is given in Cramer (1974), "Environmental Effects of Forest Residues Management in the Pacific Northwest." A fundamental compendium on fuel treatments—especially good on fuelbreaks—is Green (1977), "Fuelbreaks and Other Fuel Modification for Wildland Fire Control." For fuel treatment around residential develoopments, see references under *Fire Prevention.* For fuel descriptions and the fuel appraisal process, with its criteria for management decisions on various fuel treatments, see references in Chapter 3. For the use of prescribed burning for fuels management, see Chapter 10.


Davis, Wilfred S., 1951, "Nebraska Firebreaks," *Fire Management Notes* **12**(1): 40–43.

Bentley, Jay R. and Verdie E. White, 1961, "The Fuel-Break System for the San Dimas Experimental Forest," U.S. Forest Service, Pacific Southwest Experiment Station Miscellaneous Paper No. 63.

Brown, Arthur A. and Kenneth P. Davis, 1973 rev., *Forest Fire: Control and Use,* 2nd ed. (New York: McGraw-Hill).

Clar, C. Raymond, 1969, *California Government and Forestry,* Vol. 2 (Sacramento: California Department of Forestry), pp. 252–258.

Cramer, Owen (ed.), 1974, "Environmental Effects of Forest Residues Management in the Pacific Northwest. A State-of-Knowledge Compendium," U.S. Forest Service, General Technical Report PNW-24.

Cronemiller, Lynn, 1934, "Oregon's Forest Fire Tragedy," *American Forests* **40**: 487–490, 531.

Davis, Lawrence S., 1965, "The Ecomonics of Wildfire Protection with Emphasis on Fuel Break Systems" (Sacramento: California Department of Forestry).

DeByle, Norbert V., 1981, "Clearcutting and Fire in the Larch-/Douglas-fir Forests of Western Montana—A Multifaceted Research Summary," U.S. Forest Service, General Technical Report INT-99.

Dell, John D. and Lisle R. Green, 1968, "Slash Treatment in the Douglas-fir Region—Trends in the Pacific Northwest," *Journal of Forestry* **66**(8): 610 614.

Forestry Department, State of Oregon, n.d., Statistical Information about Tillamook Burn and State Forest (Salem: Office of State Forester).

Green, Lisle R., 1977, "Fuelbreaks and Other Fuel Modification for Wildland Fire Control," U.S. Forest Service, Agriculture Handbook No. 499.

Green, Lisle R. and Harry E. Schimke, 1971, "Guides for Fuel-Breaks in the Sierra Nevada Mixed-Conifer Type," U.S. Forest Service, Pacific Southwest Experiment Station.

Green, Lisle R. and Leonard A. Newell, 1982, "Using Goats to Control Brush Regrowth on Fuelbreaks," U.S. Forest Service, General Technical Report PSW-59.

Hibbert, Alden R. et al., 1974, "Chaparral Conversion Potential in Arizona," U.S. Forest Service, Research Paper RM-126.

Hirsch, Stanley N. et al., 1981, "The Activity Fuel Appraisal Process: Instructions and Examples," U.S. Forest Service, General Technical ReportRM-83.

Hornby, Lloyd G., 1936, "Fire Control Planning in the Northern Rocky Mountain Region" (Missoula, Montana: U.S. Forest Service).

Moore, Howard E., 1981, "Protecting Residences From Wildfires: A Guide for Homeowners, Lawmakers, and Planners," U.S. Forest Service, General Technical Report PSW-50.

Omi, Philip, 1977, "Long-term Planning for Wildland Fuel Management Programs," (PhD thesis, University of California, Berkeley).

Pierovich, John M. and Richard C. Smith, 1973, "Choosing Forest Residues Management Alternatives," U.S. Forest Service, General Technical Report PNW-7.

Pierovich, John M. et al., 1975, "Forest Residues Management Guidelines for the Pacific Northwest," U.S. Forest Service, General Technical Report PNW-33.

Price, Jay H., 1934, "The Ponderosa Way," *Journal of Forestry* **40**: 387–390.

Radloff, David L., 1980, "Coping with Uncertainty in Fuel Management Decisions," in American Meteorological Society and Society of American Foresters, *Sixth National Conference on Fire and Forest Meteorology* (Society of American Foresters).

Roussopoulos, Peter J., 1980, "Improving Wildland Fuel Information for Wildfire Decisionmaking," pp. 138–142, in American Meteorological Society and Society of American Foresters, *Sixth National Conference on Fire and Forest Meteorology* (Society of American Foresters).

Shepherd, W.O et al., 1956, "Pasture Fuelbreaks. Construction and Species Trials on Pond Pine Sites in North Carolina,"

Agricultural Experiment Station, Bulletin 398 (Raleigh: North Carolina State University).

Smart, Charles N., 1973, "Decision Analysis of Fire Protection Strategy for the Santa Monica Mountains: An Initial Assessment" (Menlo Park, California: Stanford Research Institute).

White, Verdie E. and Lisle R. Green, 1967, "Fuel-Breaks in Southern California, 1958–1965," U.S. Forest Service, Pacific Southwest Experiment Station.

Wright, Henry A. and Arthur W. Bailey, 1982, *Fire Ecology. United States and Southern Canada* (New York: Wiley).


## *Fire Research*

No general summary of fire research in United States currently exists. Prior to its demise in 1977, the NAS-NRC Committee on Fire Research published a *Directory of Fire Research* every few years. Old issues are still useful. Pyne (1982), *Fire in America,* covers the history of research. FFASR issues an annual list of publications which ably surveys its current activities. The international scene is perhaps best described in *Forest Fire News,* a periodical published by FFASR on behalf of the North American Forestry Commission, the international symposia on fire that assemble every few years, and the World Forestry Congresses sponsored by FAO. Until their conclusion in 1975, the Tall Timbers Fire Ecology Conferences (1962–1975) conducted by the Tall Timbers Research Station offered an ongoing panorama of world research in fire.

### *Planning for Fire Management*

The best summary of current planning practices is contained in the planning handbooks issued by fire agencies. Planning has been an almost constant undertaking since 1970, however, and handbooks are under constant revision. Fire management programs—and by implication, fire planning processes—are taught at the National Advanced Resource Technology Center in Marana, Arizona. Workbooks are used, but not published.

Alaska Department of Education, 1977, "Applicability of Wildlands Firefighting Techniques for Structural Fires," (Anchorage: Alaska Department of Education).

Alaska Land Managers Cooperative Task Force, 1979, "The Fortymile Interim Fire Management Plan."

Alaska Land Use Council, 1982, "Alaska Interagency Fire Management Plan. Tanana/Minchumina Planning Area."

Barney, Richard J., 1976, "Land Use Planning—Fire Management Relationships and Needs in the U.S. Forest Service," (PhD thesis, Michigan State University).

——— (compiler), 1977a, "Proceedings of the Fire Working Group. Society of American Foresters," U.S. Forest Service, General Technical Report INT-49.

———, 1977b, "How to Integrate Fire with Land Use Planning and Management Activities—A Process," pp. 85–95, in Connie M. Bourassa and Arthur P. Brackebusch (eds.), *Proceedings of the 1977 Rangeland Management and Fire Symposium* (Missoula: Montana Forest and Conservation Experiment Station).

Barney, Richard J. and David F. Aldrich, 1980, "Land Management—Fire Management. Policies, Directives, and Guides in the National Forest System: A Review and Commentary," U.S. Forest Service, General Technical Report INT-76.

Barney, Richard J. and Toni Rudolph, 1981, "PATTERN—A System for Land Management Planning," U.S. Forest Service, Research Note INT-318.

Bratten, Frederick W. et al., 1981, "FOCUS: A Fire Management Planning System—Final Report," U.S. Forest Service, General Technical Report PSW-49.

Chapman, John F., 1977, "The Teton Wilderness Fire Plan," *Western Wildlands* (Summer): 11–19.

Chase, Richard A., 1980, "FIRESCOPE: A New Concept in Multiagency Fire Suppression Coordination," U.S. Forest Service, General Technical Report PSW-40.

Daniels, Orville L., 1980, "Land Management Planning: Where Fire Management and Resources Meet," pp.90–98, in Richard J. Barney (ed.), "Fire Control in the 80s," (Missoula, Montana: Intermountain Fire Council).

Dell, John, 1972, "Region Six—Preattack Handbook," U.S. Forest Service.

Devet, D.D. et al., 1974, "DESCON Plan," Francis Marion National Forest, South Carolina.

Dubois, Coert, 1914, *Systematic Fire Protection in the California Forests* (Washington, D.C.: U.S. Forest Service).

Egging, Louis T. et al., 1980, "A Conceptual Framework for Integrating Fire Considerations in Wildland Planning," U.S. Forest Service, Research Note INT-278.

Fischer, William C., 1978, "Planning and Evaluating Prescribed Fires—A Standard Procedure," U.S. Forest Service, General Technical Report INT-43.

———, in press, "Wilderness Fire Management Planning Guide," U.S. Forest Service, General Technical Report INT-xxx.

Hornby, Lloyd G., "Fire Control Planning in the Northern Rocky Mountain Region" (Missoula, Montana: U.S. Forest Service).

Johnson, Von J. and William A. Main, 1978, "A Climatology of Prescribed Burning," pp. 59–62, in Michael A. Fosberg (chairman), *5th National Conference on Fire and Forest Meteorology* (Boston: American Meteorological Society).

Kessell, Stephen R. and Peter J. Cattelino, 1978, "Evaluation of a Fire Behavior Information Integration System for Southern California Chaparal Wildlands," *Environmental Management* **2**(2): 135–159.

Main, William A. et al., 1982, "FIREFAMILY: Fire Planning with Historic Weather Data," U.S. Forest Service, General Technical Report NC-73.

Nash, Roderick, 1982 rev., *Wilderness and the American Mind,* 3rd ed. (New Haven, Connecticut, Yale University Press).

National Park Service, 1981, "FIREPRO. Normal Fire Year Planning Handbook," (Washington, D.C.: National Park Service).

Potter, Meredith W. et al., 1979, "FORPLAN: A FORest Planning LANguage and Simulator," *Environmental Management* **3**(1): 59–72.

Pyne, Stephen J., 1982, *Fire in America: A Cultural History of Wildland and Rural Fire* (Princeton, New Jersey: Princeton University Press).

Smith, Freeman et al., 1979, "Final Report FIRELAMP," (Unpublished Report, Northern Forest Fire Laboratory).

Tall Timbers Research Station and Intermountain Fire Research Council, 1976, *Proceedings of the Tall Timbers Fire Ecology Conference and Fire and Land Management Symposium,* Vol. 14 (Tallahassee, Florida: Tall Timbers Research Station).

Thomas, David A. and Sandra J. Marshall, 1980, "1979—Test Year for Prescribed Fires in the Northern Region," *Fire Management Notes* **41**(4): 3–6.

U.S. Forest Service, 1980, "Preattack Handbook," FSH 5109.-15.

———, 1978, "Fire Management and Planning Handbook," FS Handbook 5109.5 (Washington, D.C.: U.S. Forest Service).

*Chapter Nine*

# *Fire Suppression*

For fire management the ultimate fact is that humans can, within certain limits, start and stop fires. Upon this capability rest all administrative assumptions, plans, and operations. The capacity to initiate and terminate a powerful natural process is what most differentiates fire management from other forms of resource management. Starting and stopping fires, moreover, are complementary activities. Without the ability to start fires, humans would never have achieved their dominance upon the Earth. Without the ability to stop fires, however, the power to start them would have resulted in only a kind of opportunism or vandalism. The standards for control have always varied, relative to needs, but even today without the ability to control fire the whole edifice of fire management would collapse.

Historically, fire control meant more than the ability to halt individual fires. It was the means by which old fire practices were broken down and whole fire regimes reconstituted. Fire control remains fundamental to the contemporary array of options that has replaced those old fire practices with a new set. Without fire control a single wildfire or fire complex could undo decades of previous management efforts in prevention, presuppression, fuels treatment, and control. Nor can suppression be entirely replaced by other functions, either alone or in combination. Even in theory not all ignitions can be prevented, not all management fires can be kept under control, and not all wildfires can be replaced with prescribed fires. Weather alone will ensure that at least some wildfires will occur, and that a few, an irreducible quantum, will escalate into large fires. Fuels cannot be eliminated nor wildlands made fireproof; even to significantly reduce the flammability of a biota may, in many areas, conflict with the resource management objectives for which the land has been reserved. Paradoxically, both prevention and prescribed fire programs may actually increase the probability of wildfire—prevention, by increasing fuels, and prescribed fire, by increasing ignition. No fire management program that includes only the suppression function can be entirely successful, but no program can succeed without it.

The suppression function is neither self-defining nor unchanging. The purposes and conduct of suppression have continually evolved to meet new problem fires, to satisfy agency mandates, and to harmonize with the other components of a fire program. Reform in the methods of fire suppression will almost certainly mean changes in other fire programs, while alterations in other programs will likely bring about reforms in the conduct of suppression. Suppression goals and techniques have changed with redefinitions in the concept of control, with the availability of new resources for control, with new strategies and tactics for applying those resources.

A few factors remain constant, nonetheless. One is that all fires are most easily controlled while they are small. Another is that everything must be done quickly, and in the case of a large fire, with urgency. Consequently, resources are prepackaged, attack strategies preplanned, and dispatching procedures preordained. Another constant is that, although no fire is identical to another, all fires require some action. There is no neutral position possible. Some response is demanded; not to act will have consequences as profound as any chosen activity. Each fire demands at least an assessment, nearly all will require an initial attack, and a few will make the transition to large fires. Against that handful a remarkable institution, the large fire organization, will be mobilized. In the United States the apparatus for fire control has attained great sophistication—conceptual, technical, and administrative.

## 9.1 SUPPRESSION STRATEGIES

In contemporary thinking there are three categories of fire: wildfire, prescribed fire, and escaped fire. All require control, but "control" may mean different things for each. In general control can take three forms. It may consist of direct control, in which the source (the fire) or its fuel is contained; perimeter control, in which the flaming front rather than the entire fire is contained; prescription control, in which the properties of the fire environment contain fire spread and intensity such that the fire does not threaten resources and may, indeed, enhance them. A single fire may show different types of control around its perimeter or at different times in its history. The tactics needed to make control a reality will vary with the type of fire, the type of fuel, and the type of suppression resources on hand.

### *Concepts of Control*

Fire suppression describes the process by which control is achieved over a fire. It is not identical with fire extinguishment. Obviously, all fires end in extinction, but a fire need not immediately expire in order for it to be considered controlled. Other forms of control are possible, and these forms have changed over time. For systematic fire protection, control was achieved with the arrival of suppression units within certain predetermined intervals of time. For hour control it meant control over the fire perimeter, not merely arrival at the scene. For conflagration control it indicated control over fire spread and intensity, such that a fire could not make the transition to large fire. Perimeter control was a part of this process, but its area alone did not describe a fire and control did not result simply from containing fire size. For the policy of fire by prescription, control expanded to include other meanings. A fire behaving under prescribed conditions could be considered controlled even if its perimeter was unconfined or if its intensity allowed for occasional runs. Consequently, fire control does not describe one state but many, and it does not indicate one response but an array of potential responses as circumstances dictate.

*Varieties of Control.* The simplest form of control is *direct control.* Suppression in this case leads to immediate, complete extinguishment. The advantage of this approach is that a fire, once controlled, will not rekindle again into wildfire, that the suppression function shows a direct evolution with a well-defined end. Its disadvantage is that direct control can only apply to relatively small fires or to fires with brief residence times such as those in grass and brush. Large area fires and fires that burn in heavy fuels with long residence times (usually characterized by smoldering combustion) demand heavy investments in mopup in order to bring them to extinction. Direct control is usually restricted to initiating fires, to steady-state fires that have not reached large sizes, and to selected portions of large fires. A variant on the concept is the converse practice of *exposure protection* in which critical resources such as houses are shielded from a fire. As with direct control, such activity is specific and local, restricted to small sites. In this case it is not the fire but the fuel that is the object of attention, not the source but the resource.

More common are forms of *perimeter control,* a strategy that seeks to confine the flaming front. To surround a fire with a fireline (a complete break in fuels) is to *contain* that fire. If the fireline is strengthened such that flareups within the fire cannot break through the line, then the fire is said to be *controlled.* Most fires are suppressed through a strategy of perimeter control. But most fires, too, are subjected to a medley of suppression efforts. Some portions of the perimeter—the most threatening—may be attacked selectively through direct control, a process known as *hot spotting.* Similarly, some portions of the interior of the burn may be selectively guarded through exposure protection. The most obvious illustration involves houses along an urban-wildland interface. Here perimeter control alone may contain the fire but lose the resource.

The emphasis in perimeter control is not on immediate suppression of the entire burn, but on halting and diminishing over the active zone responsible for fire spread. As a strategy the concept does not embrace a single process, but a continuum of responses that may range from partial containment to total containment, from partial to total control, and from partial to total mopup. *Mopup,* the actual extinguishment of all fire, will proceed after control over the fire is established and it will progress from the fireline inward. For relatively small fires or fires with short residence times, mopup may extend over the entire burned area. For larger fires, mopup will proceed inward to the extent necessary to protect the line (e.g., 2 chains).

The concept of *prescription control* considers fire to be controlled as long as it burns within specified geographic boundaries and predetermined burning properties. These parameters are contained within a written *prescription.* The prescription allows for those fires seen as advancing management goals to be considered controlled. The ignition source may be scheduled or unscheduled. If the prescription is violated—if the fire escapes its geographic bounds or assumes different burning properties—then the fire becomes a wildfire. What mode of control is then adopted will vary, just as it does for a fire that is initially reported as a wildfire.

Between these two categories of fire, however, comes a third, a transitional form known as an *escaped fire.* All wildfires are subject to initial attack. Should initial attack fail, the fire is considered an escaped fire, subject to an escaped fire analysis. Based on this analysis, the fire may be classified as either a wildfire, for which some form of suppression is suitable, or a prescribed fire, for which control takes the form of a predetermined prescription. Similarly, a prescribed fire that leaves its prescription also becomes an escaped fire. Large fires have the effect of reducing options and of accelerating the rate of suppression buildup. The escaped fire concept has the opposite effect of introducing more options and of providing a momentary pause in the evolution of suppression, an opportunity to ponder which of the many forms of control and the many suppression options is best suited to the management of this particular fire.

### *Model Suppression Evolution*

The suppression function consists of an evolution of activities: dispatching, initial attack, buildup to control (perhaps up to a large fire organization), mopup, demobilization, and postsuppression. All fires contain all these stages, but only for large fires are these phases recognizable as distinct activities. Consider the following analysis as a model of how suppression develops.

Suppression begins when a wildfire is reported. Who is responsible for sending control forces to a fire depends on the nature of the fire agency. For small organizations or rural scenes, it may be a warden or a crew boss, responding to a smoke report. For most organizations, however, there exists a distinct position, a dispatcher, who initiates responses to reported fires and who orchestrates the whole suppression apparatus when multiple fires occur.

Dispatching may be organized in various ways. It may take the form of a *threat-determined response* (TDR), in which the response follows from the nature of the reported fire, or it may proceed according to precepts of *preplanned area dispatching* (PPAD), in which the appropriate response to a fire is preset according to its location and the current fire danger rating. Most dispatching for initial attack occurs at a forest or park level. Integrated dispatching systems, such as those in Firescope or the Initial Attack Management System developed by the BLM, are bringing computer assistance to the dispatcher. It is possible to make fire behavior predictions, even to simulate the consequences of initial attack with various suppression forces, in advance of actual dispatch. If initial attack fails, or if the fire load escalates beyond the capabilities of that administrative unit, then calls for assistance go out to zone dispatchers, regional dispatchers, or BIFC dispatchers. In areas of mutual jurisdiction, joint interagency dispatching may be practiced. At all levels dispatchers mobilize suppression resources through a system of requisitions called *fire orders.*

*Initial attack* begins when suppression units reach the fire scene. But getting there may be half the fun. Firefighters may proceed by foot, boat, ground tanker, helicopter, or parachute. In the case of small fires in large backcountries, the problems in locating a single burning tree may be formidable, requiring a set of skills commonly known as *smokechasing.* Once at the fire, initial attack forces confront one of two possibilities: either they can control the fire with the resources on hand or the fire is beyond their control. It may be apparent that the fire, even at this stage, is beyond the control of the whole administrative unit. Of all the phases in the evolution of suppression, initial attack is the most critical. All wildfires experience an initial attack, but only a fraction make the transition that requires a large fire organization.

Moreover, fires are most different in their initial attack stages. Once a fire has transformed into a large fire there is a certain uniformity about its behavior which compels similar strategies for control and similar organizational structures for its control. With initial attack there is room for maneuver, for tactics; with large fires, strategy and logistics tend to control fireline activities. The fire boss of initial attack forces is, in many respects, the pivotal figure in the suppression organization, around whom the entire apparatus swings.

Often a few timely reinforcements can rescue an initial attack. Another squad of firefighters rushed to the fireline, another ground tanker or engine, an aerial retardant drop—all can salvage an initial attack that might otherwise fail if left to the first units on the scene. But if such emergency reinforcements are unavailable or if the fire still exceeds control, then an escaped fire analysis must be made. If the fire remains a wildfire, then the initial attack phase gives way to the *project (*or *campaign) fire* orchestrated under the precepts of the *large fire organization* (LFO).

Here resources of all sorts are built up to match the buildup of the fire. More crews, more supplies, more equipment, and more overhead (fireline supervisors) are rushed to the scene. Most of these resources come in prepackaged units—organized crews, catering services, kits for radios, chain saws, and tool sharpeners, to name a few. Specialists in the control of large fires, *overhead teams,* or *project fire teams,* assume direction over the fire from local fire officers until control—however it is defined—is attained.

Typically, the LFO will remain in effect through the *mopup* stage. The standards for mopup will vary according to fuels, fire behavior, the resources under protection, and the suppression resources available for extinguishment.

With control in sight, the process of buildup reverses into a process of *demobilization.* It is almost axiomatic that the buildup of suppression forces is out of synchronization with the buildup of the fire, that there will be too few resources on hand initially and too many at the conclusion. To contain costs and release national suppression resources for use elsewhere, demobilization must proceed rapidly. For very large fires, or fire complexes, staging areas may be established to oversee this process and to exchange resources among various fires. With demobe complete, control over the fire—whatever remains of it—is returned to local administrators.

A host of *postsuppression* activities remains. Rehabilitation efforts, salvage operations, and the restoration of initial attack capabilities on the local unit all clamor for attention. Postsuppression may amount to nothing more than cleaning up the trash around a firecamp and sharpening tools back at the fire cache. Or it may involve broadcast seeding, replanting, fuels management projects, erosion control structures, and the repurchase of equipment or supplies used up or damaged during the fire. There will be reports to file, bills to pay, accounts to clear, and (for fires of any consequence) a review of performance. Occasionally, for very large or problematic fires, an institution known as a *fire analysis* or board of review may convene to critique the overall handling of the fire.

### *Methods of Control*

Although the goals for control can vary and the responses to a fire can be many, there do exist some fundamental principles common to all suppression efforts. Some principles relate to the means by which a fire can be extinguished, some to the process of sizing up a fire during initial attack, and still others to the methods by which prescriptions may be established. A suitable method of control will vary according to the stage of suppression evolution (initial attack or project fire), to the type of fire involved (ground, surface, or crown), and to the type of fuel (grass, brush, timber, or slash). But in all these situations direct control requires that fuels, oxygen, and heat be segregated; perimeter control, that a fireline be emplaced in a suitable location; prescription control, that fire behavior be predictable within acceptable limits.

*Fire Triangle Concept.* The traditional way to summarize fire control methods is through the concept of the *fire triangle.* Every fire requires oxygen, fuel, and heat. If any of these components is lacking in sufficient quantity, then the fire will go out. A fire requires, moreover, that these components be brought together in certain ways, and this process of mixing may control the effective quantities of these components available to the fire. To inhibit this chemical chain reaction is also to extinguish a fire or to change its burning characteristics in fundamental ways. Extinguishment is thus more a process of subtraction than of addition. To remove fuel, heat, or oxygen from a fire, or to interfere with the physical chemistry of pyrolysis and combustion is to put that fire on the way to extinction.

For free-burning fires the only practical techniques are to remove fuels, primarily in the form of a fireline; to reduce heat, through the application of water or dirt; and to inhibit combustion processes by the addition of chemical retardants. Only in very restricted circumstances can a fire be smothered. These techniques apply to any scale of fire, from a smoking branch to a conflagration of thousands of acres, from a shovel of dirt thrown on a burning snag to elaborate firelines punched out of dense forest. For most wildfires, perimeter control—based on the removal of fuels in front of the flaming front—is the only practical method of containment. This assumes, among other things, that the fire is a surface fire. With a deep ground fire, fuel removal is too laborious, and often too uncertain, a technique to be effective. Against a crown fire, with spotting, firelines are also impractical.

*Methods of Attack.* The process that matches available suppression resources with the character of the fire is called *sizeup.* The ability to evaluate the fire in whole and in its parts, the assessment of the entire control job, and matching of the fire against the suppression forces on hand or quickly available are skills that can be sharpened by training, but can only come through experience. Initial attack is the foundation of fire suppression, and sizeup, the essence of initial attack.

Sizeup may be likened to practice of triage in emergency medicine. Some points of the fire may be beyond control, others show little need of immediate attention, and still others, if untreated, threaten to escalate beyond control. Obviously, those areas that are controllable but have the potential to become uncontrollable command first attention. If the fire is small or if those portions of the fire that threaten to escape control are small, they can be attacked directly—hotspotting. The snag may be cooled and felled, the head of the fire knocked down and contained by a fireline, and hot spots ready to flare into crown fires cooled and isolated. If the fire is too large for direct control, either in whole or in parts, then some form of perimeter control is necessary.

A fireline may be put in next to an active fire, with wildfire burning out to the line; it may be put in some distance from the active fire, with the intervening fuels removed by deliberate burning inward from the constructed line; or it may simply incorporate portions of the fire perimeter that have extinguished naturally. Commonly a fire will have different portions of its perimeter controlled by constructed lines placed at different distances from it. A line adjacent to the fire indicates a *direct attack.* A line removed a short distance from the fire perimeter, but whose shape conforms to that perimeter, indicates a *parallel attack.* A line removed some distance from the main fire —perhaps to incorporate roads or take advantage of natural firebreaks within its perimeter—belongs with an *indirect attack.* When portions of burned out fuels are incorporated into the control line, the process is known as *cold trailing.* In practice the perimeter of fires of any size is controlled through a mixture of these modes of attack. Sizeup may indicate a few potential flareups, handled by direct control. Perimeter control may follow with cold trailing at the rear, direct attack along portions of the flank, and a mixture of parallel and indirect attacks along the head (Figure 9-1).

Sizeup may be a simple task, no more complex than deciding in which direction to fell a single burning snag. Or it may be quite complex. For fires that require initial attack forces on the order of 10 firefighters or equivalent, a special position, the initial attack fire boss, is recognized. Like fire behavior officers, initial attack bosses are well acquainted with local conditions. During sizeup the important tactical questions are usually whether to begin control at the head of the fire and how to remove any fuels that might remain between

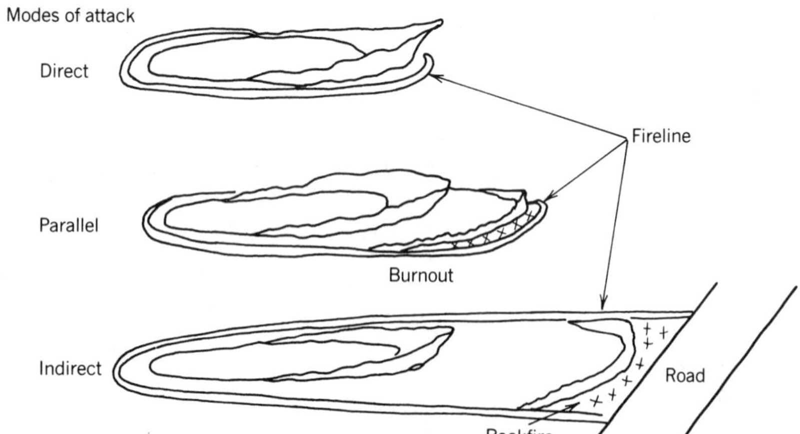


*Figure 9-1. Modes of attack. (Upper) Direct attack; (middle) parallel attack; (lower) indirect attack.*

the main fire and the fireline. Control of the head is desirable, but whether it is feasible will depend on fire behavior, fuel type and difficulty of line construction, and the suppression resources on hand. Whether fuels between the fire and fireline need to be treated, and by what means, will depend on what mode of attack is chosen. The choices are plural, not singular. A spreading fire does not propagate uniformly. Its flaming front will show different properties along different points of the fire perimeter and at different times. Methods of attack may vary according to location at any one time, or be phased to coincide with the burning period.

A more direct attack offers the advantage of immediate control, minimum size of burn, and an escape route—into the old burn—for crews. Against these considerations are the prospects for a long perimeter of fireline, the result of following the irregular contours of the fire; poor control over line location, which may be a factor in attempting to hold the line against flareups; and the requirement that crews work in heat and smoke, reducing efficiency and encouraging injuries.

Indirect methods of attack require less line construction, promote better line location, permit a buildup of suppression forces, and allow crews to work under less stressful circumstances. Many preattack plans anticipate indirect methods for any fire that escapes initial attack, and plan fuelbreak systems and points of control accordingly. Against the indirect mode, however, is an increase in area burned, problems attendant to the firing operations that are necessary to remove intervening fuels, and (with the amount of lead time required before control) the possibility of fire behavior changes that would render firing operations either unnecessary or hazardous.

For most fires initial attack will attempt as direct a mode of control as possible. Indirect modes will be saved for fires, or portions of fires, that are beyond the capability of initial attack forces. And always life safety—that of civilians and firefighters both—takes precedence. The choice of attack is an empirical art, not a theoretical science (Figure 9-2). Local circumstances will dictate the response for particular fires, and local personnel, trained in the principles of initial attack, will make the best sizeups.

*Firelines.* A fireline is a break in fuel continuity (Figure 9-3). The fireline may consist of a natural

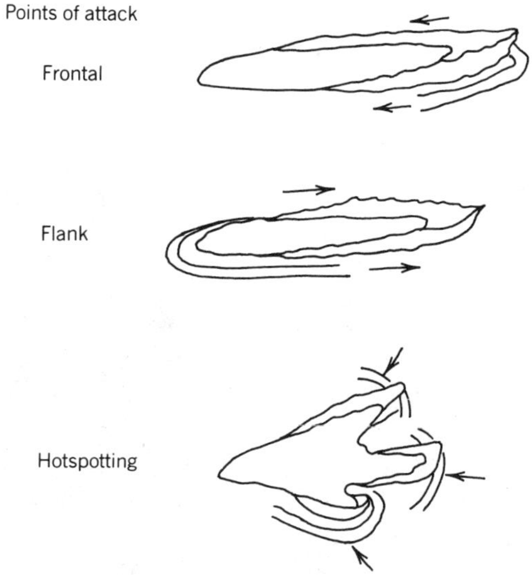


*Figure 9-2. Points of attack. (Upper) Frontal attack; (middle) flank attack; (lower) hotspotting attack. Sizeup will determine the appropriate mode or mixture of modes.*

*Figure 9-3. The fireline. (Below) Handline through timber. If ladder fuels or heavy brush are present, then a wider swath should be cleared and the fireline proper placed on the outside of the larger swath, farthest from the main fire. Courtesy U.S. Forest Service. (Left) Double tractor-plow lines through heavy surface fuels. The space between the lines will be burned, then a counterfire set from this improved blackline. Courtesy North Carolina Forest Service.*

barrier, like a river or a cliff. Most often it is deliberately constructed, and it must be built according to considerations of location and size. Fireline width will depend on fire behavior characteristics such as intensity; on fuel types, particularly the fuel load and distribution of aerial to ground fuels; on the availability of suppression forces, their total number and type; and on fireline location, which helps determine whether fire behavior will assist the line or resist it.

To put in a fireline ahead of a fire on a steep slope, for example, practically begs for trouble. Radiant heat, convective heat, and spotting will call for a very wide fireline at the head of the fire; meanwhile rolling debris may ignite fuels below or to the side of the original head, sending up still more heads. Careful placement of lines to account for predictable events like these will mean a more secure fireline with less work. Similarly an ideal location with regard to fire behavior may be inaccessible to mechanized line construction equipment, and direct attack may be impossible due to slow line construction by hand crews or to considerations of safety.

Line location will, in general, follow natural burning blocks, as defined by fuel arrangements and topography. Fireline width should vary with fire and fuel. A good measure of fire and fuel traits is flame height, and a rule of thumb for firelines holds that they should be 1.5 times the flame height. Line construction should proceed in such a way that crews will not be outflanked by the fire. An essential practice is to begin construction from a secure location, an *anchor point.* In order to protect the line, fuels between it and the main fire must be removed or rearranged. Heavy concentrations of fuel can be broken up, reproduction and brush trimmed, cut, or crushed, and surface fuels burned up.

Depending on their mode of construction, many kinds of fireline are identified. Fireline constructed in direct attack is often referred to as *hotline.* Fireline which consists of burned out portions of the fire is known as a *cold trail.* Cold trailing is only possible in fuels such as grass and brush that show short residence times and that burn out completely. *Handline* refers to a fireline put in by hand labor. A preliminary handline used for initial attack but later improved for control is known as a *scratch line. Catline, tractor line,* or *plow line* all refer to firelines constructed by tracked vehicles, primarily bulldozers and tractor-plows. A *wetline* refers to firelines made by putting down a strip of water in fine fuels. A variant is the *retardant line* made with chemical retardants. The line will not persist as long as one scraped to mineral soil, but it may hold long enough for firing operations to clean out the fuels between it and the main burn. This process of burning—which accompanies all forms of attack other than direct control—widens the constructed fireline measurably, with a result commonly referred to as a *blackline.* To be effective, all firelines should end in a blackline.

Firing operations that lead to a blackline may consist of little more than a firefighter with a fusee following behind a handline crew, or it may become a complex, relatively independent function in suppression. When firing accompanies a parallel method of attack, cleaning out pockets of fuel, the process is known as *burning out.* When firing is used in conjunction with indirect modes of attack, it is traditionally referred to as *backfiring.* The procedure is complicated, and potentially hazardous. Over the years backfiring has acquired such a pejorative connotation, especially in legal usage, that a variety of synonyms have appeared. *Counterfiring, suppression firing,* or simply *firing out* are all expressions that describe the process by which large amounts of fuel are burned away in advance of the main fire.

The same conditions that make for a large fire, and therefore require an indirect attack, may make control of a counterfire difficult. Careful placement of firelines with regard to fire behavior, shrewd selection of ignition techniques, and proper timing can make for a favorable counterfire—and a safer operation. In mountainous country this means that lines will be constructed along ridgetops or, where valleys are wide, along valley floors; that ignition will begin along the top and corners of the line before advancing into the main burn; that the fires will be set at times when weather and fuel conditions favor the counterfire against the wildfire—in the West, often at night. Fundamental to the process of putting in any fireline is the selection of anchor points, secure locales from which line construction and firing operations may begin without danger of being outflanked.

Firing techniques usually proceed by igniting strips—substituting a backing fire for a head fire, or putting down multiple small head fires (strip firing) for a single line of fire (Figure 9-4). By varying the timing and the width between strips, fire intensity can be modulated—dampening fires under conditions that favor intense burning, and increasing burning rates when conditions favor a fire less vigorous than the situation requires. Timing can use indrafts from the main fire to pull a counterfire quickly away from a fireline and reduce the potential for spotting.

Its execution may be difficult, but the need for a counterfire is usually apparent. In effect, a fire boss substitutes a counterfire for the wildfire, replacing a fire over which direct control is impossible with a fire for which some control is likely. Poor technique may eliminate this advantage, however. A badly timed firing operation may actually aggravate the main fire, accelerating it up a slope or generating a virtual firestorm, complete with long-range ember showers, when wildfire and counterfire meet. Conversely, firing may proceed too cautiously, with a poor burn of surface

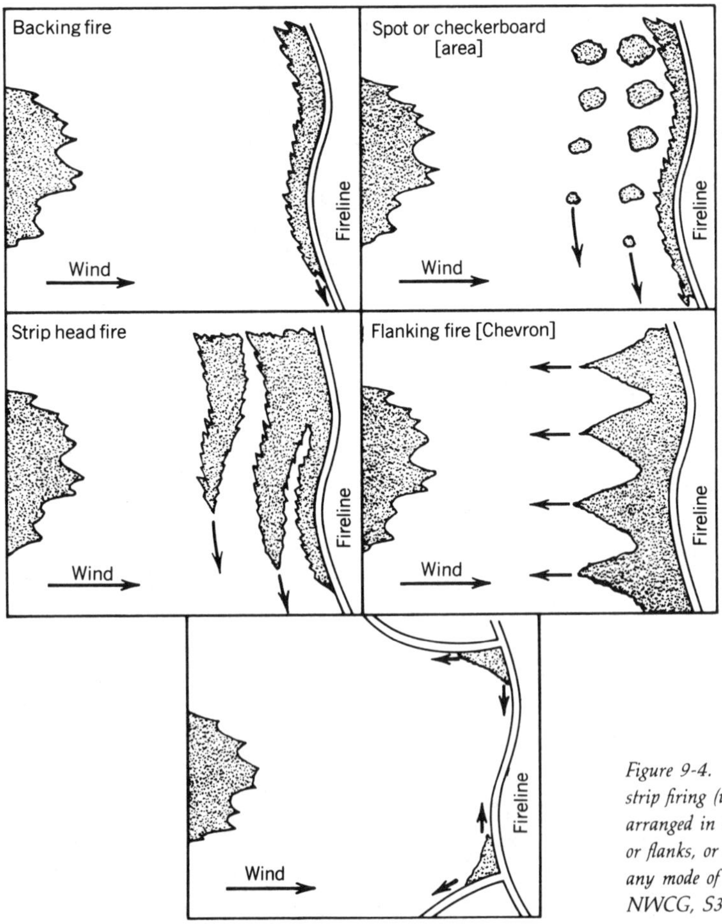


*Figure 9-4. Firing techniques. Fundamentally, there are two techniques: strip firing (upper left) and spot firing (upper right). These methods can be arranged in various ways, however. Strips may be placed across the head or flanks, or run parallel to the direction of fire spread (chevron). As with any mode of attack, a system of anchor points is essential (bottom). From NWCG, S390.*

fuels and a drying out of aerial fuels. At peak burning times the main fire may reburn the site. There are problems, too, with crew safety. A number of fatalities have occurred in association with firing operations. In areas of mixed jurisdiction and ownership, there may be serious legal questions surrounding the decision to counterfire and the caliber of its execution. And, of course, as with any prescribed fire, there exists the possibility that portions of the counterfire will escape control and proliferate as wildfire. To supervise these operations, a special line position, the firing boss, is recognized.

*Control in Special Zones.* The modes of control described so far pertain to normal wildlands. But in the past few decades two special categories of land have appeared that vastly complicate the strategy of control. One is the proliferation of suburban developments next to or within wildlands; the other, the designation of certain wildlands as wilderness. In a curious way the two zones form mirror images of each other. For wildland suburbs the control problem is to keep fire from entering the zone; for wilderness areas, it is to prevent fire from leaving the zone. The solution for wilderness areas has been to adopt a strategy of prescription control. As long as a fire burns within prescription, it is considered a prescribed fire with no action required beyond monitoring. The strategy is new, and its problems not fully apparent. Conversely, the problems of controlling fire within the wildland suburb are well known, but a solution is still uncertain.

Structural fire protection, like wildland fire control, works best through direct attack. This requires both the apparatus and the techniques of urban fire services. Adequate reserves of water, suitable protective clothing (including breathing apparatus), ventilation and rescue equipment, high volume engines, appropriate hoses and nozzles—only in a few select cases are wildland fire agencies prepared for these demands. Adequate protection in this regard can only come through integration with urban, or special rural, fire services.

The important issue here is not so much the suppression of an isolated structural fire, but the protection of structures within the context of a wildland fire. Against an approaching wildfire, wildland agencies can offer some degree of assistance. They can concentrate their attack on those portions of the fire perimeter that endanger developments, isolate those developments by firelines and counterfires, assist in exposure protection, and control spot fires just as they begin in individual houses. Spot ignitions on a roof or under eaves can be extinguished by ground tankers or dual-purpose engines.

Few urban fire services are equipped to meet conflagrations, just as few wildland fire agencies are prepared to control fires that no longer burn as a spreading perimeter but dissolve into dozens of spot fires which, in turn, propagate further spots. Where wildlands and suburbs intermingle, there may be a zone of transition between predominantly wildland fire, amenable to control by wildland fire agencies, and predominantly urban fire, manageable by urban fire services. In this case, the strategy of fire control will depend more on communications between agencies than on fireline tactics. The possibility of protecting small villages or clusters of dwellings against a wildland fire, however, is within the realm of most wildland fire agencies. Success will depend more on fuel management and presuppression programs than on suppression itself. Suppression should be the end of the protection process, not the beginning.

## 9.2 SUPPRESSION RESOURCES

Of all the resources available for fire control, manpower remains the most fundamental. Though more and more functions of fire control are being mechanized—even the decisionmaking process—the effect is to supplement manpower with machinery, not to replace it with some species of automation. Until there is a mechanical equivalent for the initial attack firefighter, able to meet a small fire, this condition will persist. The paradox, however, is that even for large fires where tracked equipment can put in line and power flamethrowers or aerial ignition devices can complete firing operations, there remains a tendency to build up large numbers of firefighters.

To support these numbers on the fireline, an elaborate logistical organization is necessary. The complexities of supporting a large fire can be staggering. The type and availability of resources will dictate, to some extent, the strategy of fire control. The development, deployment, and mobilization of resouces, especially in the early stages of a fire, will often determine the relative success or failure of the suppression effort. For project fires the support, or service, function will often drive the entire fire organization.

Again, it is the uncontrollable nature of wildfire that powers the service function. Prescribed fire, by definition, demands far less support apparatus, though the true costs of its overhead are often obscured by the fact that they are financed through the infrastructure established by suppression. The substitution of equipment for manpower is likely to proceed more rapidly in prescribed fire than in fire suppression programs. Indeed, the ability to use equipment has made prescribed fire feasible in many areas such as the South.

Most of the resources dedicated to suppression are primed for initial attack. Fire control apparatus thus tends to be widely dispersed, organized into initial attack units and administered by local officials. These are the bases for the national fire suppression system. But some fires will escape initial attack. To control them large numbers of crews and equipment, managers and service functionaries must be rounded up and brought to the fire.

For an administrative unit to routinely staff at a level adequate to meet the occasional fire is irresponsible, and unnecessary. Plans for mobilization, preestablishment of fire positions and organization, and arrangements for securing the necessary supplies constitute much of presuppression planning. But the suppression resources themselves will come from a larger, perhaps national pool dispatched under the direction of a larger, perhaps interagency authority. These large fire resources, moreover, cannot include the initial attack units of the affected unit. That one large fire is in progress does not mean that no other fires will start. Even in the face of a large fire or complex of fires, some initial attack capability must be retained. In effect each administrative unit sponsors its own initial attack force, and shares with other units the cost and control of those special resources needed for large fires.

This sharing of resources has been formalized under the concept of *total mobility.* Total mobility extends to crews, equipment, supplies, and overhead. To standardize its product it was necessary to devise a nationally recognized certification process, the *National Interagency Fire Qualifications System* (NIFQS). And, as an operational necessity, total mobility required a central dispatching center. This was created with the establishment of the Boise Interagency Fire Center (BIFC).

## *Manpower*

*Initial Attack Forces.* Manpower remains the fundamental resource for fire suppression. Especially for initial attack, the prospects for automation are dim. Initial attack forces generally consist of either handcrews or tanker crews. Hand crews will make an initial attack with hand tools, though they may brought to the fire by foot (smokechaser), truck (ground tanker), helicopter (helitack), or parachute (smokejumper). Initial attack systems put a premium on speed, and their individual size is small. A smokechaser team may consist of two firefighters; a ground tanker crew, of three to four firefighters; a helitack crew, of five to six. Dispatching is done by the local unit, a forest or park.

Where road access is good and fires start along roads, initial attack can be made with ground tankers, a combination of engine and hand tools. Here the analogy to urban fire control has some validity, and most fire agencies will rely on some form of ground tanker. Where wildland fire units are integrated with rural fire units, the tanker may be a dual-purpose engine. Where road access is poor, helitack crews or smokejumpers may be used, with smokejumping preferred for long-distance hauling (over 50 miles). In the United States smokejumping is concentrated in the Northern Rockies, Pacific Northwest, northern California, the Southwest, and Alaska. Most administrative units will rely predominantly on one initial attack system or another, but because such units do not show uniform conditions they will typically need a mixture of capabilities.

*Organized Crews.* If a fire escapes initial attack, the first response will be to add more initial attack units to meet it. But there will come a point at which this pattern of reinforcements will break down. The power of a suppression force comes from its organization, not simply its size. Patching together a string of initial attack units can be inefficient, and it strips a unit of its initial attack capability. Instead the fire boss, through the dispatcher, will request organized crews. Three categories of such crews are available, differentiated according to their abilities. *Category I crews* can be used on all aspects of fire control—handline construction, felling and firing operations, holding operations, and mopup. *Category II crews* are more restricted, less likely to be used with power equipment or hotline work, more often reserved for holding operations and mopup. *Category III crews* are generally reserved for patrol and mopup. Often, as requirements along a given sector of fireline change, so will its crew composition. Category I crews may build and burn out a line, then leave it for Category II or Category III crews to hold and mop up.

The premier Category I crew is the *interregional (IR) fire suppression crew,* of which over 40 exist nationwide. Each crew consists of 19 firefighters, organized into three squads, each with a squad boss and all under the direction of a crew boss. Highly trained and well-equipped, stationed on forests where they are likely to be used but responsive to dispatchers at BIFC, the IR (or hot shot) crew is the backbone of large fire suppression (Figure 9-5). Collectively, these crews comprise a kind of rapid deployment force. Such a crew may travel from one large fire to another, over a period of weeks, before returning to its primary duty station.

Category II crews include organized crews such as those from the Southwest Forest Fire Fighters (SWFFF) or Snake River Valley (SRV) programs, who are assembled in advance of the fire season, trained together, and shipped to large fires on demand. They may be stationed on a forest during periods of high fire danger as a presuppression measure, but once that danger has passed or the fire is concluded they are released and returned to their point of origin. Somewhat different are those special crews made up of regular employees in a unit who are trained for use on project fires. Depending on their training and experience, regular crews may be designated either Category II or Category III. Crews formed of casual, intermittent labor, or of regulars not trained as a fire crew make up the Category III crews. Appeal to labor centers, impressment of citizens, mobilization of national guard units—all sources of manpower in the past—have been steadily replaced in the name of safety and efficiency by organized crews. National guard units, when activated, usually assist with service functions, not line duty.

*Project Fire Teams.* Matching the organized crews are prepackaged cadres of large fire supervisors (overhead) known as *project fire teams*. Project fire teams will fill all the essential functions demanded for the management of a large fires, not only line assignments but the plans, service, and finance functions as well. Short teams fill those functions down to the first or second level; long teams, to the second or third level. The members of a team train together, and the team goes to a fire as a unit.

Three categories of teams are recognized. *Class II teams* are composed of overhead drawn from smaller administrative units (e.g., a forest), and are used on smaller project fires. *Class I teams* pool overhead from larger units, a region or an agency, and are generally used on larger fires. In special cases of very large fires or a complex of closely associated fires, a *GHQ team* may be mobilized to coordinate overall strategy and allocate limited

*Figure 9-5. Organized fire crews. Marching along a fireline (above), traveling by medium-sized helicopter (middle), and mopping up (below). Courtesy U.S. Forest Service.*

resources to different portions of the complex. A GHQ team fills only three managerial functions: command, plans, and service. A special overhead category is recognized, moreover, for initial attack: the *Class III fire boss.* Not part of a team, the Class III fire boss requires special training and exercises special responsibilities, not the least of which is the preparation for an overhead team should initial attack fail. In keeping with the total mobility concept, project fire teams can consist of members from different agencies, and teams composed of members from one agency may be used on fires within the jurisdiction of another agency. Forest or zone dispatchers control Class II teams; BIFC, the Class I and GHQ teams.

The duty of the project fire overhead team is to supervise the large fire organization. The team is reponsibile to the local agency on whose land the fire burns. A formal delegation of authority is made from the agency to the team; briefings will explain agency wishes with regard to the conduct of the fire suppression effort; the agency will normally assign one or several liaisons in an advisory capacity, to ensure that the control efforts will satisfy agency policy. A resource adviser, for example, may point out sensitive areas or special constraints—sites that should have a high priority for protection, or prohibitions against the use of tracked vehicles or chemical retardants. The project fire team is intended to make the suppression function more efficient, and to free local officials from an emergency so that they may return to their normal duties. But the suppression function is not autonomous from the larger land management obligation of the agency, and the project fire team is not independent from the authority of the responsible agency. Theirs is a support function.

Training for all personnel follows the courses developed by the NWCG and certified under the NIFQS. Three series of courses exist: fire prevention (P series), fire management (M series), and fire suppression (S series). Most states combine these courses with others, often to meet rural fire obligations. Some states, such as California, New York, and Pennsylvania, maintain academies with special fire training courses; others, such as North Carolina, engage in annual field exercises. A fire simulator has been developed for training overhead locally, a supplement to lectures and role playing.

At lower levels instruction takes the form of lectures, demonstrations, and practical exercises; at higher levels, of gaming and simulation. Certificates are issued as fire job qualification cards, popularly known as red cards. To qualify for any particular position requires both classroom training and job experience, and experience is usually the limiting factor. Most project fire teams will consequently carry a few trainees, in this way building up a supply of overhead for future teams. In theory, at least, staffing a fire follows according to fire job rating, not any other administrative rankings. A helitack crew boss may have a higher fire rating than a forest supervisor, and would assume command should the two work a fireline together.

### *Equipment*

Equipment development for fire control has shown great zest and imagination, and modern fire management without machinery is almost unthinkable. Some degree of mechanization has affected all levels of suppression, and prescribed burning is almost everywhere successful because of its connections to machinery. The administrative problems of manning a fire, however, have their counterpart in the problems of equipping that force. Like crews, fire equipment must be appropriate and reliable, and like crews there must be a degree of standardization. Without interchangeability the total mobility concept could not apply to equipment. The challenge is to both develop special equipment and then standardize that equipment.

One solution is the use of equipment development centers, of which the U.S. Forest Service maintains two: the Missoula Equipment Development Center (MEDC) at Missoula, Montana, and the San Dimas Equipment Development Center (SDEDC) at San Dimas, California. The centers design new equipment, adapt existing machinery, establish standards, and test purchased equipment to ensure that it meets specifications. At agency levels, examples of standardization are the ground tanker models developed by the Forest Service and the dual-purpose engines worked out by the California Department of Forestry. To make total mobility a nationwide reality, a National Fire Equipment System has been created. Presently the system includes hand tools, power tools, safety gear, firing devices, camp supplies, rations, and, through the National Radio Cache at BIFC, some communications.

*Hand Tools and Power Tools.* The basic hand tools for fireline construction include implements for cutting, digging, and scraping or raking. In the United States these take the form of shovels, axes, hoes, and assorted fire rakes. Hand tools for direct control, or flame suppression, include swatters and some sort of bucket for carrying water. In the United States these include fire flappers and backpack pumps for spraying water. Finally, there are hand tools for igniting counterfires and controlled burns. All such devices are known as torches. Historically, the tendency has been strong to combine several functions into one tool. Even the shovel is specially designed to scrape as well as to dig. A number of hand tools are now powered. Examples include chain saws, portable pumps, trenching devices for line construction, and power flame throwers. Explosive cord for line construction has proved successful in field trials, but is not widely used.

*Line Construction Vehicles.* Mechanized line construction vehicles are generally of two types: the bulldozer and the tractor-plow. Both are tracked vehicles, and both are formally referred to as *tractors* (Figure 9-6). The plow, of which there are many varieties, is restricted to level terrain. For many decades, however, the tractor-plow has been fundamental to fire control in the South and the Lake States. Hydraulic plows, moreover, have been attached to vehicles outfitted with slip-on pump units, making a dual-purpose vehicle (tanker-plow) that is especially well-suited for rural fire protection in much of the eastern United States. Interestingly, the bulldozer (a tractor with a movable front blade) was invented for fire control by a Forest Service employee in the late 1920s. Though barred from use in select areas such as wilderness sites, and restricted from practical use in other areas such as permafrost zones in Alaska, few suppression organizations lack a dozer or "cat." And, of course, there are special adaptations. In boggy areas such as coastal North Carolina, tracked vehicles are outfitted with low ground pressure (high flotation) treads. In Michigan a tracked flailing machine (the sand caster) constructs fireline through sandy soils somewhat in the manner of a snow-blower, showering debris to one side.

*Ground Tankers and Engines.* For vehicles that carry and pump water there are two general expressions. Most wildland fire agencies refer to such a vehicle as a *ground tanker* (as distinct from an air tanker), while urban and rural fire services call it an *engine.* In areas of overlapping use, engine is the preferred term, with ground tanker reserved for those vehicles that merely haul water.

Two varieties of the tanker-engine are recognized: the *slip-on unit* and the *integral unit.* A slip-on unit consists of a tank and pump assembly that can be removed from or added to a vehicle as a single unit. In part the device represents an adaptation to the seasonal nature of fire control. A pickup truck, for example, may be converted to a ground tanker during fire season and then returned to duty as a general purpose vehicle when fire season ends. In part, too, the slip-on unit reduces the costs of purchasing special fire vehicles. Trucks designed for other purposes can be converted easily by inserting a prepackaged assembly. The integral unit, by contrast, has the pump built into the engine system of the vehicle and the water tank constructed as part of the entire body. Most dual-purpose engines are of the integral variety. Both slip-on and integral unit engines can substitute chemical retardants or wetting agents for simple water. The U.S. Forest Service uses seven basic models of ground tanker (Figure 9-7; U.S. Forest Service, 1974).

*Figure 9-6.* Tractors. (Upper) Bulldozer constructing line in western forest, and (lower) tractor-plow cutting line in South. Note protective cage for operator. Courtesy U.S. Forest Service.

*Figure 9-7. Ground tankers. (Below) Dual-purpose engines, California Department of Forestry; and Dragon Wagon, BLM. (Facing page: upper) Slipon unit and portable tank, National Park Service; (middle) tanker-plow converted from military 6 × 6, Michigan Department of Natural Resources; (lower) integral tanker, a model 60, U.S. Forest Service. Photos courtesy of identified agency.*

There are, of course, adaptations of these standard forms. Surplus military 6 × 6 trucks have been converted to fire vehicles through the addition of a 1000 gallon slip-on unit and a hydraulic plow. For sagebrush fires in the Great Basin the BLM developed an 8-wheel drive twister vehicle, commonly known as the Dragon Wagon. The vehicle has the capability to lay down a swath of water or retardant along a fire edge, or to put down a wetline (or retardant line) on one side the vehicle while burning out the other side with power torches.

*Aircraft.* Aircraft are used to reconnoiter fires, to deliver firefighters, and to drop supplies, retardants, and ignition devices. In general two types of aircraft are recognized: *fixed wing* (airplane) and *rotary wing* (helicopter; Figure 9-8). Within each category aircraft are differentiated according to the payload they can carry. Fixed wing aircraft are used for heavy payloads carried over long distances, and for reconnaissance when long observation periods are required. Some light aircraft have been adapted for aerial ignition, and others for infrared mapping. Most air tankers began as surplus military aircraft, generally of World War II vintage. As the planes have aged, other aircraft of civilian origin, such as the DC-6, have been adapted to aerial attack. One plane, the CL-215 manufactured by Canadair, has been designed specifically for fire control. And one military plane, the C-130, can be outfitted with a Modular Airborne Fire Fighting System (MAFFS) that carries over 3000 gallons of retardant. Some planes, moreover, have special scooping devices that allow them to fill up their interior tanks with water from lakes while in flight. In the United States, however, most planes require a fixed retardant base at an airport.

The versatility of helicopters has made their presence ubiquitous. They can be used to reconnoiter, to haul supplies and firefighters, to drop retardant, lay fire hose, and deposit incendiaries in a variety of ways. Special night-vision glasses make evening flights possible. As a means of delivering retardant or water, the great virtue of the helitanker is its accuracy. Its limitation, a small payload, can be overcome by the establishment of portable retardant bases conveniently placed near the fire. For many areas helitack is a basic mode of initial attack. A few places have added the technique of rappeling from a hovering helicopter (helirappel) to more traditional helitack operations. The ability of medium-sized helicopters to quickly transport large numbers of firefighters has made them a common vehicle for moving IR crews on project fires. In special cases, helicopters have even transported heavy equipment (such as tractor-plows) to otherwise inaccessible firelines.

### *Support Services*

The logistics of mobilizing, outfitting, transporting, feeding, and otherwise managing the forces massed for a project fire can be awesome. Collectively, such operations comprise the *service function,* and on every overhead team there will be a service chief. Whenever suppression units remain on a fire, but not on a fireline, they are the responsibility of the service organization. Camp management, equipment management, heliport management, transportation, supplies—all are supervised by special officers who have met training and experience requirements laid down by the NIFQS.

The service organizaton oversees all fire orders, from shovels to crews, aviation gas to radios, infrared mapping planes to timekeepers. Its service duties extend beyond support for the line to include support for other staff positions such as plans and finance. For sources of supply, the service organization looks to local caches, then to regional and zone caches, and finally to national caches such as BIFC. BIFC is, in fact, a national service center for the support of fire suppression. Contracts made out well in advance of fire season bring in other equipment and services. This may mean nothing more than the rental of a bulldozer or water truck, or it may involve one of several private outfits who specialize in catering to large fires. Again, the emphasis is on preplanning and packaging. Emergency requisitions may be placed

*Figure 9-8. Air tankers. (Upper) Small aircraft used in groups, North Carolina; (middle), B26 laying retardant along a flaming perimeter, Arkansas; (lower), C-130 with MAFFS unit. Courtesy North Carolina Forest Service; U.S. Forest Service.*

through the General Services Administration (GSA), the federal government's procurement agency. And many goods may be simply purchased from local vendors. After control of the fire, the service organization must reverse the operation. Every unit has to be systematically released, returned to its point of departure, or otherwise be accounted for.

## 9.3 *ORGANIZING FOR FIRE SUPPRESSION*

Whether attacked by two smokechasers with hand tools or by dozens of organized crews with sophisticated equipment, every fire requires a degree of organization. Certain functions must be performed, and it is a truism that, in some form, all these functions must be done on all fires. Not only do fires require a division of labor, an organization by function; they need an integration of that division, an organization by complexity. In the United States two patterns exist for the management of big suppression efforts: the large fire organization (LFO), the traditional means of wildland fire control, and the incident command system (ICS), a method that generalizes the response from fire to any emergency, and from one agency to many.

In either form, the management of the suppression effort itself may become as difficult a problem as control over the fire. The amount of organization required does not increase in direct proportion to the size of the suppression forces, but geometrically. To increase fireline forces from two crews to four or from four crews to eight does not merely double the organizational complexity of suppression, but quadruples it. This increase in size demands specialists to fill intermediate roles. That more crews are on the line means many more support personnel will be needed at staging areas and firecamps. Moreover, the question of organization is never really resolved. Nearly all suppression efforts are either building up or building down, and the organization they exhibit is the product of these evolutions. Only on certain very small and very large fires is the fire organization stabilized for any significant length of time (Figure 9-9).

### *Organization by Function*

*Fire Level Organization.* The division of labor for fire suppression traditionally follows five functions: command, line (suppression), service, plans, and finance. The *command function* belongs to the fire boss, of which there is only one. The *line function* has responsibility for actual work on the fire; the *service function,* for logistical support; the *plans function,* for gathering intelligence and making plans; the *finance function,* for processing all fiscal obligations. To these functions some agencies have added a sixth: *safety,* usually in the form of a safety officer who advises on all aspects of the operation.

The line, service, plans, and finance functions, however, will consist of separate suborganizations; how large these suborganizations become will depend on the fire. For a fire manned by two smokechasers, one will exercise command, and both will share equally in all other duties. For a fire of large size, supervision will reside in an project fire team, specialists who will fill the key positions for each of the required functions. To integrate the team with local practices and administrative wishes, the agency on whose lands the fire burns may add several liaisons: a comptroller, to assist the finance chief; a resource advisor, to advise the plans chief on strategy; an agency liaison, to coordinate administrative procedures if different agencies are involved (Figure 9-10).

Most of the personnel, of course, will go to the line organization, under the direction of a line boss. How large the line organization will get depends on the size and complexity of the fire. Full-blown, a line organization will consist of two divisions, each headed by a division boss responsible to the line boss. Each division boss will direct two or three sectors, each with a sector boss. At the sector level are integrated the basic suppression forces—crews, tractors, tankers, aircraft. Each crew will have a crew boss; each comple-

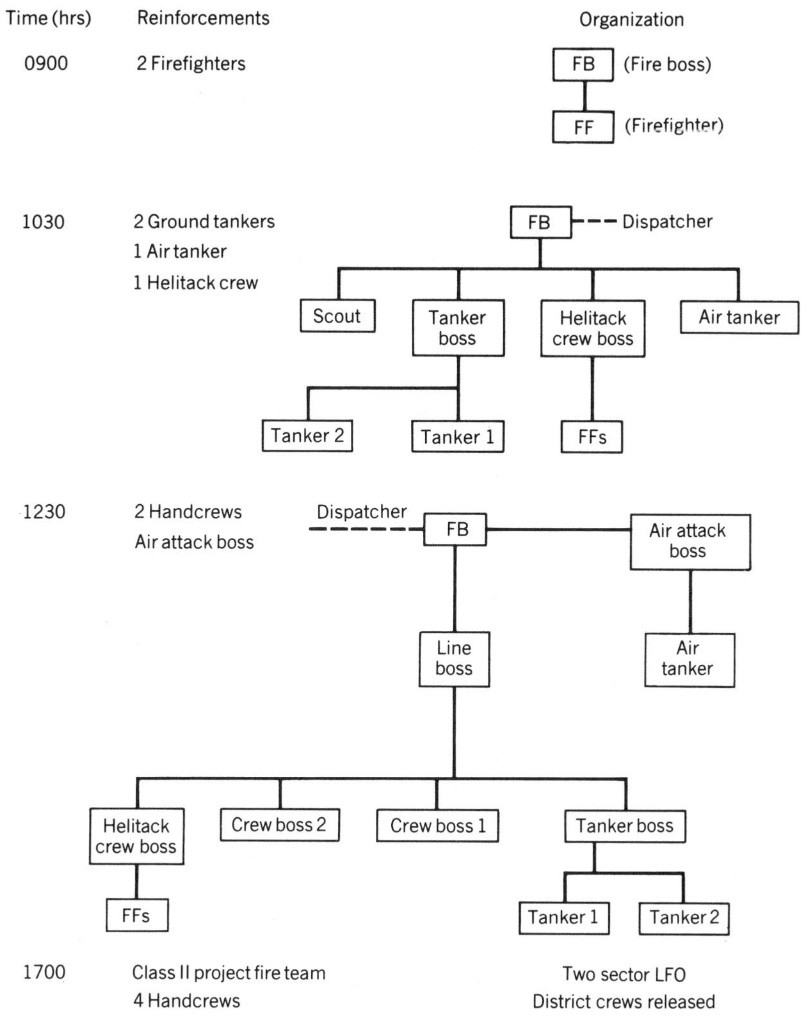


*Figure 9-9. Hypothetical evolution of a fire organization, from IA to arrival of a Class II project fire team.*

ment of tractors, a tractor boss; each set of ground tankers, a tanker boss. In large crews the crew boss will exercise authority through squad bosses. In smaller crews—a helitack crew, for example, or a felling crew—the crew boss will personally direct all operations. Because of their special requirements, air attack and firing operations fall under the supervision of an air attack boss and firing boss, respectively. Both report directly to the line boss.

The sector is thus the smallest unit of fireline organization, and the division the largest. Of course, not all fires will reach sector size, or be contained within two divisions. For fires smaller than one sector, positions are telescoped, with one person assuming many roles. All suppression forces report directly to a line or fire boss. For a single crew fire, the crew boss may exercise the roles of line boss and fire boss. For very large fires, the two-division fireline organization will be re-

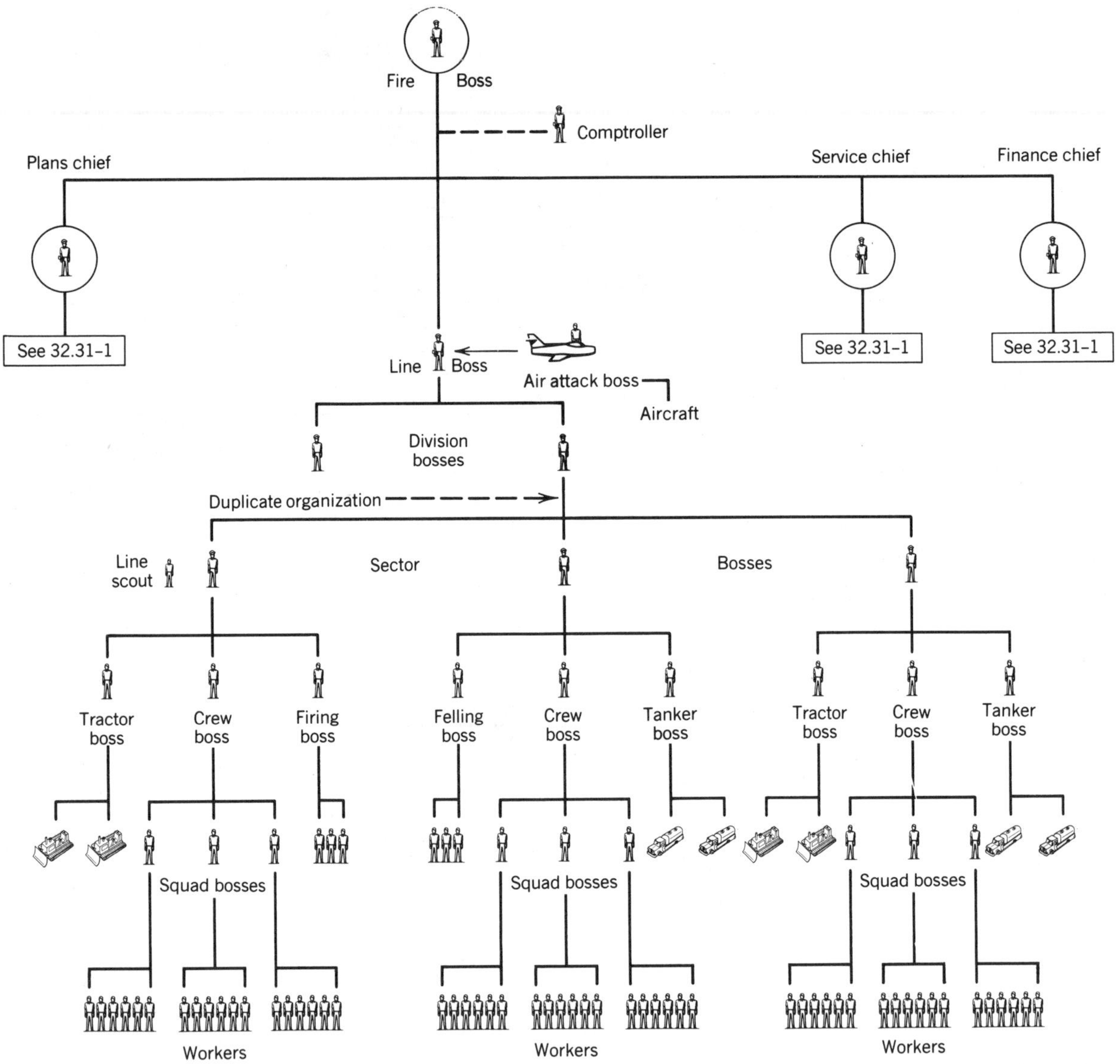


*Figure 9-10. Large Fire Organization. Detail emphasizes line functions, not the full complexity of support functions such as service, plans, and finance. From U.S. Forest Service,* Fireline Handbook.

peated around the perimeter until the fire is wholly contained within it. Such a fire is treated as multiple fires; each two-division perimeter is considered a zone, and it is assigned a fire boss and overhead team. A GHQ team will provide umbrella supervision to the zones.

For any unit on the fire, however, the entire organization is less critical than the immediate chain of command surrounding that unit—those to whom it reports and those whom it directs. For most fires, moreover, no fireline organization as such is set up. The fire is attacked more or less directly, or a line is established in short order and forces turned to mopup. Every function must be done, but not every position needs a special occupant. For small crews accustomed to initial attack, for example, experience leads to an implied organization that is understood by all but need not be formally stated in order to be effective. The important considerations are that suppression be systematically organized, that the fire organization (however constituted) have the potential to expand as a fire escalates, and that fire overhead recognize the need for reorganization as the complexity of the fire changes.

*Crew Level Organization.* Although there is a panoramic splendor in very large fireline organizations, successful suppression depends most heavily on organization at the tactical level: the organization of a crew for handline construction, firing operations, and mopup; the integration of a crew with tractors, of ground tankers with hand crews and water trucks, of air strikes with ground operations. Sample organizations are given in Figures 9-11 and 9-15.

Methods of organization for handline construction appeared with the development of organized crews in the 1930s, the one defining the need for the other. In contrast to a system of individual assignments, in which particular tasks are performed by designated individuals, or a system of squad assignments, in which each squad

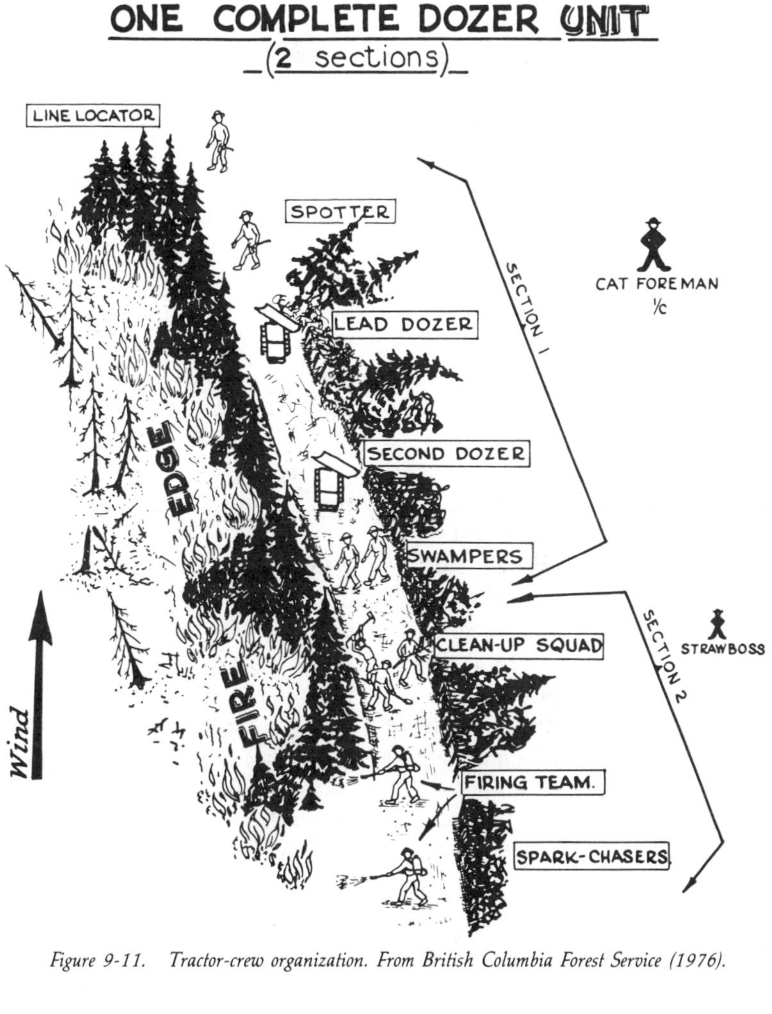


*Figure 9-11. Tractor-crew organization. From British Columbia Forest Service (1976).*

completed a task and then leapfrogged to a new position along the line, tacticians proposed *progressive methods* of line construction. The essence of the progressive method was that the crew advanced as a unit, with each member retaining the same position within the crew.

Two major variations developed. The *one-lick method* assigned a particular tool to each firefighter and arranged the order of tools according to fuels and line standards. Cutting tools preceded digging and scraping tools. Each crew member took one "lick," or several, then advanced along the line. By the time the last crew member contributed, a complete fireline would result. By contrast, the *step-up* or *bump* method required that each firefighter complete a small section of line. That job ended when the crewmember's line touched the section completed by the crewmember ahead. The first crewmember would then advance along the completed line until reaching the crewmember ahead, take his place ("bumping"); the bumped firefighter would advance to the crewman ahead of him, bumping him along. The one-lick method works well in all fuels; the bump method, in relatively homogeneous fuels where a single tool, like a McLeod (a rake), can construct complete line. Both methods apply to forest fuels. In grasslands, firelines take different forms, and where handcrews are used, local patterns of crew organization have been developed.

Similar principles apply to mopup. Mopup is intrinsically tedious and surprisingly hazardous. Fatigue, let-down after control of the fire, indifferent supervision—all contribute to frequent injuries. Mopup standards will vary according to the larger strategy of the fire. In general, for small fires or fires in light fuels mopup will be total. For large fires or fire in heavy fuels, mopup will concentrate along a strip paralleling the fireline. Within this zone, however, extinguishment will be total. This can only occur where techniques are good and organization efficient. Small groups work better than large, and specific assignments are superior to general exhortations.

This means assigning jobs by function or by sector. Saw teams may be designated to cut logs and drop snags, hose teams to spray water, and shovel teams to mix smoldering fuels with dirt. Alternatively, small teams may be assigned to particular sectors of fireline, with each team responsible for total mopup within its zone. The work is unglamorous, but it must be done thoroughly. Unless supervision is good and jobs are distributed systematically, the work will proceed poorly and interminably, with flareups possible and decaying morale a certainty.

### *Organization by Complexity*

A large area fire does not always require a large fire organization. For statistical purposes, fires are categorized by burned acreage. But for managerial purposes, fires are ranked according to their administrative complexity. A 1000-acre (405 hectare) grass fire may be controlled by a few pieces of equipment reporting more or less directly to a fire boss; a 200-acre (81 hectare) forest fire may require a small army of firefighters and machinery, supervised by a special project fire team. It is not enough to identify and fill special jobs; they must also be integrated into a whole. There has to be a unity of command to match the division of labor. A comparison of the two systems for ranking fires is given in Figure 9-12. For evaluating the fire load of an administrative unit, fire complexity is a more useful index than fires of a particular size, though for a given fire regime some correlation between size and complexity will be apparent.

Matching management needs to fire complexity is not simple. Any particular supervisor can only exercise a certain *span of control,* direct supervision of a small handful of people. This number typically varies from three to eight, with smaller spans of control at higher levels of the organization and longer spans at the lowest levels. The formal organization charts developed for the fireline reflect decades of empirical experience in finding the right spans of control for different positions. The same principle applies to project fire teams. The prepackaging of project fire teams ensures a degree of internal coordination, and a

| Fire Classification | |
|---|---|
| *Fire Size by Acreage* | |
| Class A | Under 0.25 acre |
| Class B | 0.25–9 acres |
| Class C | 10–99 acres |
| Class D | 100–299 acres |
| Class E | 300–999 acres |
| Class F | 1000–4999 acres |
| Class G | Over 5000 acres |
| *Fire Size by Management Complexity* | |
| Level 1 | Multi-zone fires or multiple-fire situations involving two or more project fire teams |
| Level 2 | Multi-division fires |
| Level 3 | Multi-sector fires |
| Level 4 | Multi-crew fires and initial attack activities involving 10 or more firefighters |
| Level 5 | Initial attack situations with fewer than 10 firefighters |

*Figure 9-12. Fire classes. (Upper) By acreage, and (lower) by management complexity.*

team can accomodate a certain amount of further adjustment. If fire complexity continues to grow, then another team of greater experience or size will replace it in toto. The mere presence of specialists, even in great numbers, does not guarantee organization. The practice of prepackaging and pretraining groups of supervisors does mean that, even after transitions from one set of supervisors to another, a minimal amount of coordination is always present. An organization does not have to be built from scratch. It is rare for too much organization to exist on a fire, but only too common to have too much overhead.

The project fire organization has several objectives. It should restore the initial attack capability of the administrative unit, free the unit to return to its routine duties, and meet—through appropriate fire strategies—the land management goals of the responsible agency. The actual organization of the suppression effort may take several forms. The pattern and terminology described so far pertain to the *large fire organization* (LFO). This is an organizational structure which has grown up with wildland fire control and it is the form most commonly used throughout the United States (Figure 9-10). New positions are constantly added to meet new needs, and new position descriptions are written to integrate these functions into the organization as a whole. For the U.S. Forest Service, the entire LFO is synopsized into a field reference work known as the *Fireline Handbook.* For the Department of the Interior, there exists an analogous volume, the *Fire Control Notebook.* Several States with large fire organizations such as California and North Carolina have developed similar guides, and the BLM in Alaska has adapted a version to suit its own particular needs. Through the auspices of the NWCG and the necessities of total mobility, these slightly different forms of the LFO are being combined into one interagency pattern.

At the same time, the Firescope program in Southern California led to a new organizational schema known as the *incident command system* (ICS). The ICS rationalized the somewhat empirical structures that characterized the LFO, provided a common language by which to merge wildland and urban (or rural) fire services, and generalized the response such that it could apply to any emergency, and from one agency to many. A fire boss, for example, becomes an incident commander; handcrews, strike forces; ground tankers, engines. Initial attack is considered as one form of initial response; reinforced attack, as a variety of extended action; a burning period, one manifestation of an operational period. The ICS, in brief, offers a total systems approach to all risk incident management; it can be used as readily to respond to a flood as to a fire. The NWCG has promoted the ICS into a National Interagency Incident Management System (NIIMS) (Figure 9-13). Among federal agencies, the NPS, BLM, and FEMA have adopted NIIMS, and the Forest Service plans adoption in 1986. Accordingly, the NIFQS and its library of training courses are being revamped.

Of necessity the similarities between the LFO and the ICS are striking, yet the systems are in-

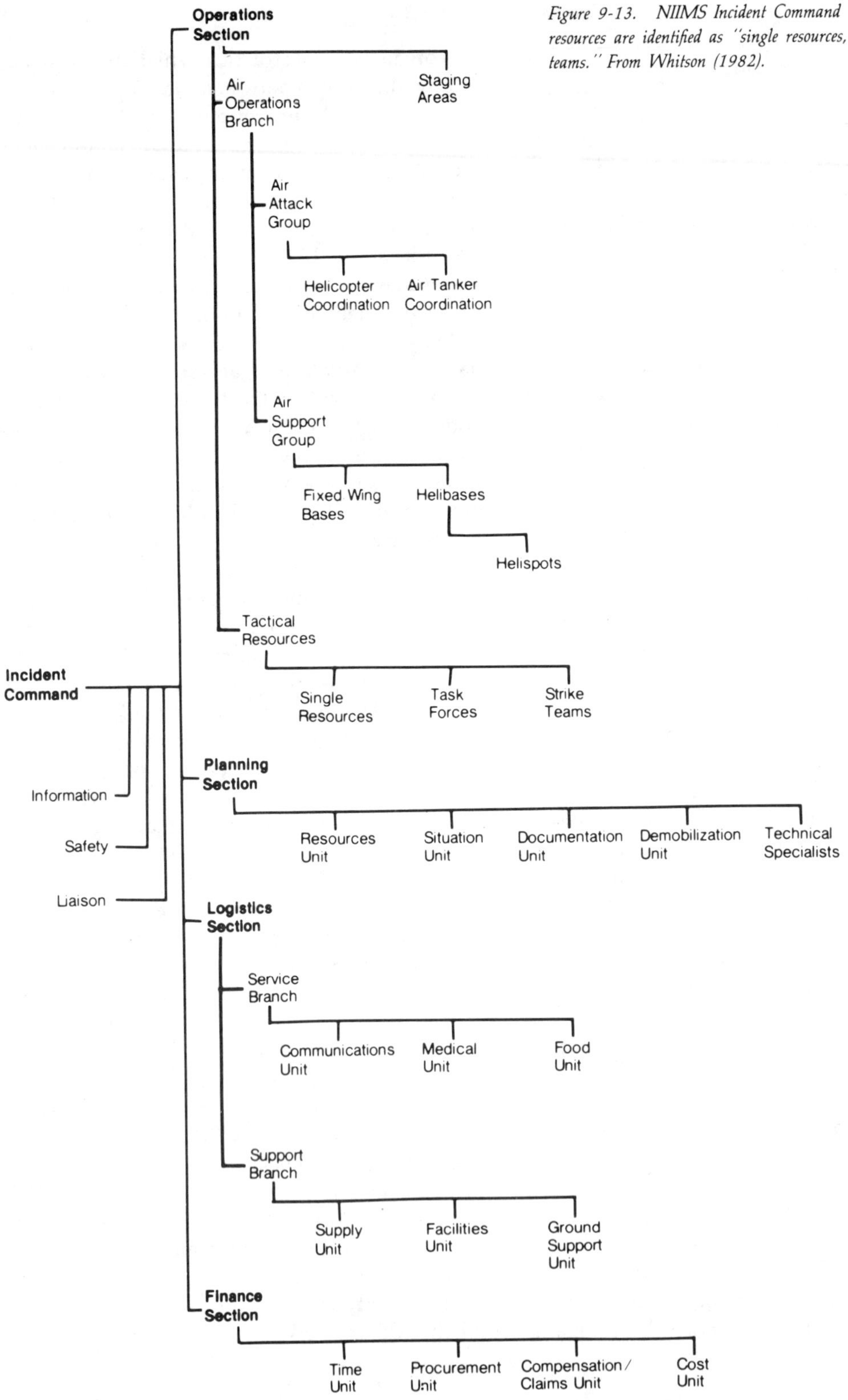


*Figure 9-13. NIIMS Incident Command Organization. Tactical (line) resources are identified as "single resources," "task forces," and "strike teams." From Whitson (1982).*

compatible in that, for any given situation, one system or the other must be used. Through the ICS, NIIMS can accomodate different functions, with fire as one incident among many for which an agency might organize. Through MACS, or its regional analogues, NIIMS can accomodate different agencies, not only in their response to an incident, but in the preplanning for that response. The ICS divides incident (suppression) resources into *strike forces,* or units of one type (such as handcrews or tractors), and into *task forces,* or a variety of different kinds of units that function together at one job or along one section of fireline. Similarly, ICS divides the organizational structure into *divisions,* which represent geographic units, and *groups,* which comprise functional units. Thus a sector in the LFO has its equivalent in an ICS division. An advantage of the ICS organization, of course, is that it can accomodate other "functions" or "task forces," such as emergency medicine, rescue, or law enforcement assignments, under the same umbrella structure that includes a large fire suppression effort.

Even more fundamentally, ICS replaces the concept of unity of command with the concept of a unified command. Within a unified command, representatives from those agencies that are involved in the incident, but have different jurisdictions, can agree on a set of objectives. These objectives—communicated to operational commanders—drive the entire organization. In this way many agencies, not only different land management agencies, but agencies with other missions, can coordinate their efforts on the emergency they all confront. More than a mechanism for sharing resources, the NIIMS program offers a means of sharing, or at least of reconciling, objectives.

## *9.4 SUPPRESSION TACTICS: SAMPLE WILDFIRES*

No policy of fire management or strategy of fire suppression can be effective unless its objectives are translated into actual operations. This is the realm of fireline tactics. The range of tactics available to a suppression force is vast, but not unlimited. Considered nationally, the deployment of crews and equipment shows a startling variability. Viewed locally, however, only a handful of procedures is generally needed for the management of a particular fire regime or administrative unit. Similar kinds of fires will reoccur, and similar tactics will be used against them. The greatest variety of tactical options, moreover, will occur during the initial attack stage. By the time a fire reaches project size, a distressing uniformity of methods results. A large fire, or one experiencing blowup, reduces drastically the range of options available for control. As examples of commonly used tactics, consider the following survey of fire situations.

### *Initial Attack: Snag Fire and Smokechasers*

A lightning-struck tree with ground fire around the base, more or less remote from roads, is a common wildfire situation in the western United States. If discovered quickly and if the lightning storm brought precipitation, fire spread will be limited, probably under a quarter-acre (0.1 hectare). It is easy for such a fire to smolder for days, however, undetected, only to erupt as a "sleeper" or "holdover" fire several days after ignition, when burning conditions become more suitable for rapid spread. Assume here that the fire was reached while still confined to the vicinity of the source tree.

The strategy will be one of direct control. Normally, two firefighters outfitted with handtools and possibly a chain saw and backback pump will make the initial attack. The firefighters may belong to a tanker crew, smokejumpers, or helitack crew, but whatever their primary mode of conveyance, they will conclude on foot. If burning conditions warrant and travel time will be long and arduous, an air tanker may drop retardant on the fire to quell it until the crew can arrive. For directions into the fire, the crew may follow geographic descriptions, proceed according to a compass bearing given by an aerial observer, or peri-

*Figure 9-14.* *Smokechasers. Felling a burning snag (upper left). Note the safety gear worn by the sawyer and the presence of a spotter. Travel to the fire may be by foot, ground tanker, helicopter (lower left), or parachute (lower right). Smokejumper photo courtesy U.S. Forest Service.*

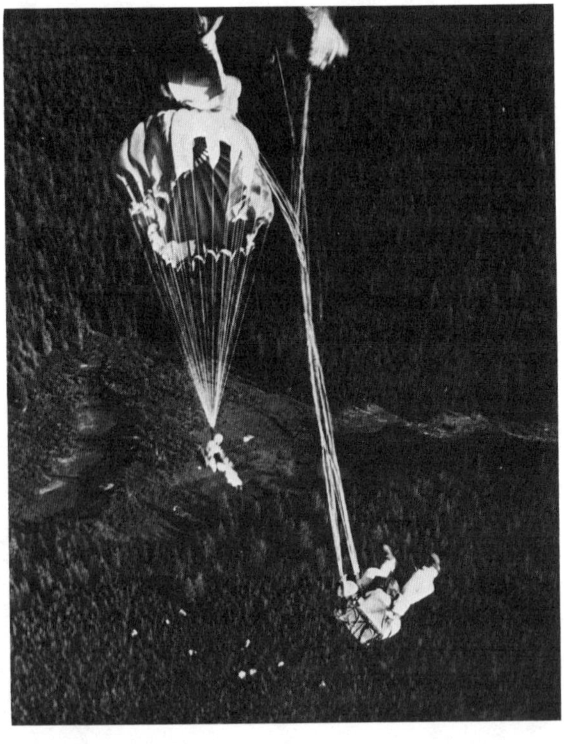

odically climb trees or find lookout posts. If the route is crosscountry, it will be marked by tying strips of plastic flagging tape to tree limbs. Upon arrival at the scene, sizeup will be promptly followed by hotspotting. Then attention will turn to the burning snag.

A source of embers, the snag should be brought to the ground as speedily as safety allows. If possible it should be felled. If it cannot be cut, then it may be burned through at the base by piling debris on one side. Felling requires preparation of a safe place to drop the tree, cooling the base of the bole so that a sawyer may work in close, and then cutting, with one crewmember, a lookout, watching for falling limbs ("widowmakers") and sparks (Figure 9-14). Once on the ground the log can be cooled with dirt and water. Often it must be bucked into manageable sections that can be individually treated.

Control merges indistinguishably into mopup. Hot material is segregated from cold, sorting the latter into a "boneyard" where it will not reignite. From large fuels to branchwood to needles and duff, mopup proceeds steadily until no smokes remain. If a complete fireline is not already in place, it should be put in, a protective measure against reignition and escape. Twenty-four hours after the last smoke, firefighters will reinspect the scene and remove any materials left behind, including the flagging tape.

### *Initial Attack: Surface Fire and Handcrews*

The flame in this case is not confined, as with a snag fire, but running. Initial attack forces consist of perhaps 10–20 firefighters equipped with handtools. Sizeup will dictate both strategy and tactics. Accordingly, initial attack may begin with hotspotting or an assault on the head of the fire. An air strike may assist in containing the fire at this stage. If the fire is too intense for direct control, then the crews will begin a flanking operation, building from a secure anchor point in the rear of the fire around the perimeter and pinching off the head. How close the crews work to the fire will depend on the strategy chosen. Usually there will be a combination of direct and parallel attacks—direct to the rear, and parallel around the flanks and head. A common technique is to divide a large body of firefighters into two groups, each to proceed simultaneously along the flanks.

From the beginning, that is, questions of crew organization become as important as the means of directly extinguishing the fire. Unacceptable from the standpoint of both efficiency and safety is the free-for-all approach—each firefighter racing to different portions of the fire in random hotspotting or desultory line construction. Instead the available manpower should be treated as a single unit, or broken into squads with each squad handled as though it were a separate crew. Even if initial attack begins with hotspotting, suppression will require perimeter control, which means fireline construction.

All line should be put in according to some systematic program. Line may be constructed and burned out by squads within a single large crew, or by squads operating on different flanks of the fire. For crew organization, there should be a line locator (possibly performed by the crew boss), a cutting squad (probably with chain saws), a digging and scraping squad, and a burnout squad. Of all these operations line construction will be the slowest; it will control the tempo of the entire effort, and its organization is essential.

With the fire contained, mopup will begin. Initially, mopup will concentrate along the fireline, further securing it; then it will spread inward over the entire burned area. Again, organization is mandatory. A saw team can buck up large logs and drop snags. Teams of two firefighters, one with a shovel and the other with a backpack pump, make good mopup units, and can be assigned to different sectors of the fire. Area by area, smoke by smoke, the fire is extinguished.

If the effort will require more than one burning period, then preparations for the evening must be made. Food and water must be brought in, sleeping bags and headlamps secured, perhaps reinforcements requested. The fire boss must decide whether to work late into (or through) the night, or whether to plan on mopup over the course of another day or two. All these decisions must be made with sufficient lead time. Considerations of crew welfare compete with those of crew organization, and gradually the fire boss segregates himself from the category of line worker.

*Initial Attack: Surface Fire and Ground Tanker*

Ground tankers are a common feature of suppression, useful at all stages of control. But only in some fuels such as grasses, light brush, and open forest, can tankers make an initial attack. In other areas, the tanker serves more as a vehicle for transporting firefighters, or for use during holding and mopup operations. For initial attack the tanker can be used in two ways. It can put down a wetline from which counterfiring may commence, and it may directly extinguish the flaming front. Either way the principles of line construction or direct control are identical to those involving handcrews or tractors. There are, of course, some adjustments peculiar to the tanker. Some tankers include plow attachments with which to construct fireline, some can drive and operate pumps at the same time, and some (like the Dragon Wagon) can simultaneously put down wetline and burn out. Tactics will also vary with the ability of the tanker to maneuver around the fire.

Sizeup will determine whether a direct or flanking attack is appropriate. If the fire is small, or slow moving, or concentrated into localized hot spots, direct control of the flaming zones is the tactic of choice. This may mean attacking the head of the fire, driving the tanker from hotspot to hotspot, or running out hoses from a fixed locale. The positioning of the tanker must also take into account the need to recharge the tank. It may be wise to operate hose lines from a fixed locale such as a road, which offers the possibility for refilling the tanker by other water trucks (nurse tankers). Alternatively, several tankers may work the fire, with some on the line and some in various stages of travel for recharging at any one moment.

With a fast-moving fire or a fire on the verge of blowup, there is always the danger of being outflanked. The proper strategy for tankers, as for handcrews and tractors, is to secure an anchor point, then proceed along the flanks until the head is eventually pinched off. A direct attack on the head of a rapidly moving fire, however, is an invitation to disaster. In some cases, the problem can be solved by attacking the flaming front from within the burned zone itself. This technique is common in flashy fuels like cheatgrass. Where such mobility is not possible, then hose lines must be laid down along the flanks—in effect, constructing a wetline (Figure 9-15).

In contrast to a *simple hose lay,* in which there is a single direct line from engine to nozzle, flanking action requires a *progressive hose lay,* in which each extension of the trunk line ends in both a nozzle and a wye. This allows for indefinite expansion of the system. The same principles, moreover, are used to put down hose lines for support of firing operations or mopup. The hose is placed within the fireline. Whatever its purpose, the hose lay evolution demands considerable tactical supervision. Where multiple tankers are working in concert, logistics as well as tactics must figure in the plans of the tanker boss.

For many areas the abundance of water may be the limiting factor in the success of a tanker attack. Most ground tanker models hold from 200 to 500 gallons (757–1893 liters) of water, with some rural equipment showing capacities of up to 1000 gallons (3785 liters). A ground tanker program is only as good as its road system and its water system. Where surface water is abundant, tanks may be recharged from streams or lakes. Where water is naturally scarce, artificial tanks or reservoirs may be created, with water-carrying vehicles held in reserve. Additional water may be the principal reinforcement for a ground tanker attack. Conservation of existing supplies will also help. Prudent practices include the on-again-off-again method, opening the nozzle only when the water contributes to control; chemical additives, reducing viscosity; low pressure pumping; reliance on fog streams from the nozzle rather than straight streams. An understanding of hydraulics will assist in maximizing the water on hand, and relay systems and temporary tanks may help to both store and move existing supplies.

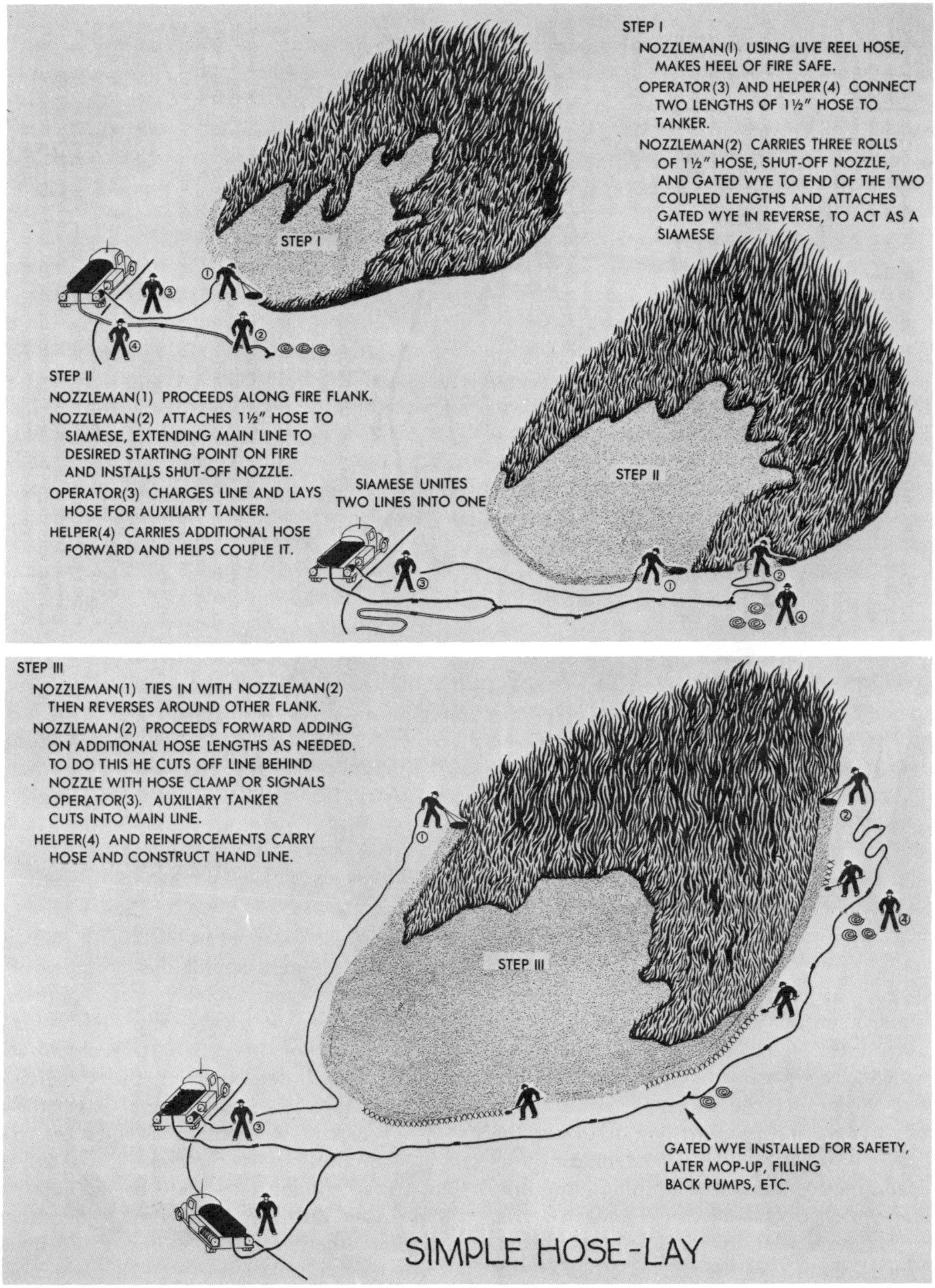


*Figure 9-15. Hose lays. (Above) Simple hose lay, and (next page) progressive hose lay. From U.S. Forest Service (1973 rev).*

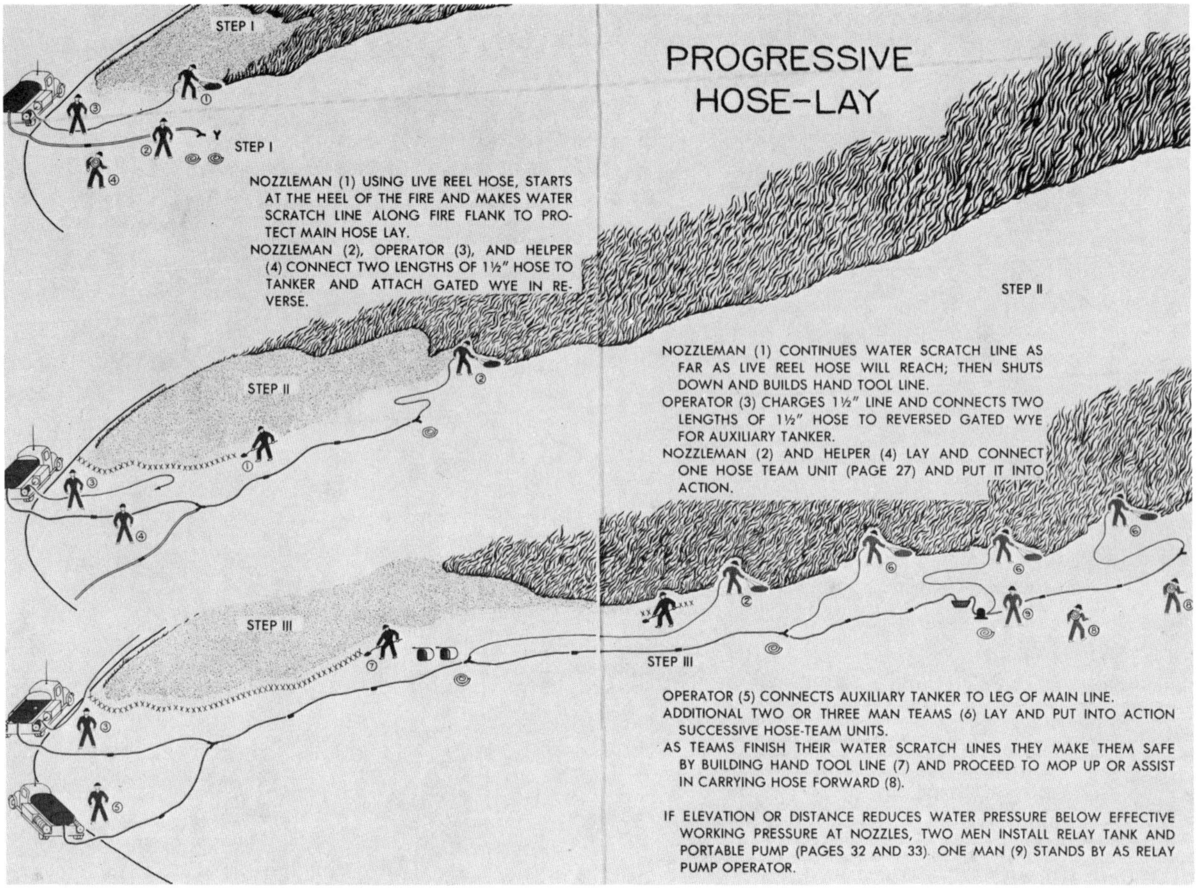


*Figure 9.15 (continued)* *Progressive hose lays.*

### *Initial Attack: Surface Fire and Tractors*

Initial attack by tractor always involves a strategy of perimeter control, and typically a mode of parallel or indirect attack. The tractor constructs a line which is then burned out and held by handcrews. The line may be single or multiple; the tractor may be a tractor-plow or a bulldozer; the plows may be used singly or in tandem; the burnout operation may consist of nothing more than a solitary torcher or of several organized crews. The same considerations of line construction apply in all cases. Yet the tractor offers special assets. Its strength means that it can put down line at great speed, even in difficult fuels. Compared with a crew, it simplifies problems of organization and logistics. For many parts of the United States the tractor-plow and tanker-plow are the basis of the initial attack organization.

Sizeup will determine whether the attack will begin at the head of the fire or along the flank. The fuel type will determine whether a tractor-plow or tanker-plow can make line unassisted, or whether a bulldozer should be used to clear an advance swath. Fire behavior considerations will decide whether a single line is adequate, or whether multiple lines (with their interior fuels burned out) will be needed to stop an advancing fire. Torches follow the tractors, burning out the remaining fuels. For a small fire a tractor and a torch may be adequate to stop a fire. The success of these tactics has meant for many regions the successful mechanization of fire control.

Yet the tractor has limitations, too. Although it constructs line rapidly, its absolute velocity is slow. If trapped it cannot easily maneuver for escape. It must be transported to fires on special trailers (low-boys). Plows are restricted to level, firm terrain. Even bulldozers are restricted in mountainous areas by steep slopes. Everywhere access to the fire can be a problem. Some areas, such as Florida, have partially overcome the problem of access through the use of heavy helicopters, capable of slinging tractors. Others, such as North Carolina, have invented special portable bridges with which tractors can span the drainage ditches that flank highways in the coastal plain and have outfitted tractors with low ground pressure treads for operation in boggy terrain. In the mountain West, except for areas of intensive use such as logging sites, bulldozers are too remote from fire sites to make initial attacks. Instead they are reserved for project fires.

Moreover, there are policy constraints that are no less restrictive. Only in special cases can tractors be used in national parks and wilderness areas. Their use in permafrost soils or on steep slopes can lead to soil erosion. More and more, fire control through tractors leads to postsuppression activities, mostly erosion control, aimed at containing the damages that can result from mechanized line construction. In part, such effects are inherent in the use of the machine. But much can be reduced by sensitive tactics. Such environmental considerations must form a part of initial attack planning.

### *Initial Attack: Crown Fire and Ground Fire*

Effective initial attack assumes that the fire is a surface fire. For crown fires, it is necessary to either bring the fire back to the surface or, more often, wait until it returns to the surface after its run. Until then control will concentrate on those portions of the fire perimeter that burn as a surface fire, and build up suppression forces in preparation for the time when the run has exhausted itself. To the extent that some action can be taken against a head, it will involve indirect attack from a preexisting break in the fuels. To a limited extent a few spot fires beyond a prepared fireline can be individually controlled, and lines can be widened to great distances through counterfiring. Eventually high-intensity fire behavior, which is necessarily transient, will subside, allowing for control along the surface. But the only effective strategy against a crown fire is prevention. Once in the crown only broad changes in the fire environment can influence fire behavior.

Similar considerations apply to ground fires—to fires burning in organic soils like muck or peat. When the fuel is wet, a surface fire entering into it will expire. But when the fuel is dry, the soil will burn. Its burning properties are diametrically opposite to those of a crown fire. Combustion is smoldering, not flaming; the rate of spread is slow, not rapid; intensity is mild, and fuel consumption almost total. But ground fires share with crown fires the property that control is almost impossible by other than indirect means. A ground fire burns down as well as out, so perimeter control is impossible. The best strategy of control is to prevent a surface fire from entering organic soils during droughts. With ground fires, however, it is possible to alter the fire environment in significant ways. The fire burns slowly enough that control forces can be marshalled.

With surface fires tactics vary according to the rate of spread and intensity of the fire. With ground fires, the controlling variable is the depth of burning. This depth is set by the water table. If only a shallow surface layer overlying sand is burning, then the fire may be treated more or less conventionally. The organic soil acts like a thick duff, complicating mopup, but not preventing perimeter control through plowing. But if the organic soil is deep, then the only way to control the depth of burning is to raise the water table. Sometimes harrowing the fire will turn over enough lower, moist soil to extinguish the burn. Most often only flooding with sprinkler systems or high-volume pumps will end the burn. The alternative is to control the perimeter of the organic fuels, surrendering direct control over the fire in favor of control over its spread into other, surface fuels.

### *Initial Attack: Aerial Attack*

In many regions aircraft comprise a primary vehicle of initial attack. Aerial reconnaissance may discover a fire, air tankers make the initial attack, smokejumpers, helitack, or helirappel crews provide ground reinforcements, and paracargo, the necessary supplies. The tactics of aerial fire control differ only in agency, not in principle, from those of fire control on the ground. Like any suppression resource, air attack is most effective in the early stages of a fire. Like most others, too, it can be used for a variety of purposes, both during initial attack and in the project fire phase. In a curious way, air tankers and ground tankers are mirror images in this regard. Air tankers are most effective during initial attack, and ineffective during mopup. Ground tankers are excellent for mopup, but access may limit their utility for initial attack. In light fuels both can make fireline, but more commonly both air and ground tankers are used in support of line construction and line-holding by other means.

Some tactical considerations are peculiar to air tankers. Tankers differ by size, as measured by the number of gallons they hold. Equally important, they differ according to the tank and gating system they contain. Each tank and slurry combination will produce a particular rate of flow from the tank and a special ground pattern, called a *footprint.* Maneuvering can alter the footprint slightly, but its shape remains a function of the tank-gating system. Some flexibility can be achieved by multiple tanks which can be opened in various combinations. If several tanks are released simultaneously, a *salvo* results. If tanks are released sequentially according to some predetermined delay time, a *trail* is produced (Figure 9-16; Swanson et al., 1976). (The MAFFS unit operates somewhat differently, with evacuation of the tanks under pressure through nozzles; but the C130 which carries it is not used for initial attack.)

Salvo drops are effective for direct control of small fires, spot fires, or hot spots, and for punching retardant through dense canopies. Trail drops are used to construct retardant lines or in support of ground lines. The desired density of retardant will vary according to fuel type, fire behavior, and certain characteristics of the drop itself. Ideally, when the retardant evacuates from the tanks, it atomizes and descends as a cloud. Good drop heights are from 150 to 300 ft (46–107 m). Below that height cloud momentum presents a safety hazard and it fails to give as wide a coverage as it ought to; above that range, the cloud will tend to dissipate and wind will carry it away from the drop zone. With winds greater than 30–35 mph (49–56 km/hr) control over the retardant cloud is difficult.

Ineffective drops can also result from simple misses. Errors result from poor range, in which the release comes too early or too late; from poor cross-range, in which alignment is incorrect or the cloud drifts; from inadequate momentum, in which drops are made too high, forward speed is too slow, or the wind and convective plume cause the cloud to break up early. To correct errors of range and cross-range it is customary to have a lead plane direct the air tanker to its target. For many initial attack situations, however, a lead plane is not possible and experienced air tanker pilots will make the run on their own. In such a situation the full load will be divided into several drops, such that each pass can learn from the mistakes of the previous ones. Accuracy is not so critical a problem where helitankers rather than air tankers are used, but the range and payload of the helitanker is less than that of the air tanker. Helitankers are more likely to be used for close support during project fires than as a vehicle for initial attack (Figure 9-17).

Air tankers can deliver retardant for direct control or perimeter control, direct attack or indirect attack. If retardant can penetrate through the can-

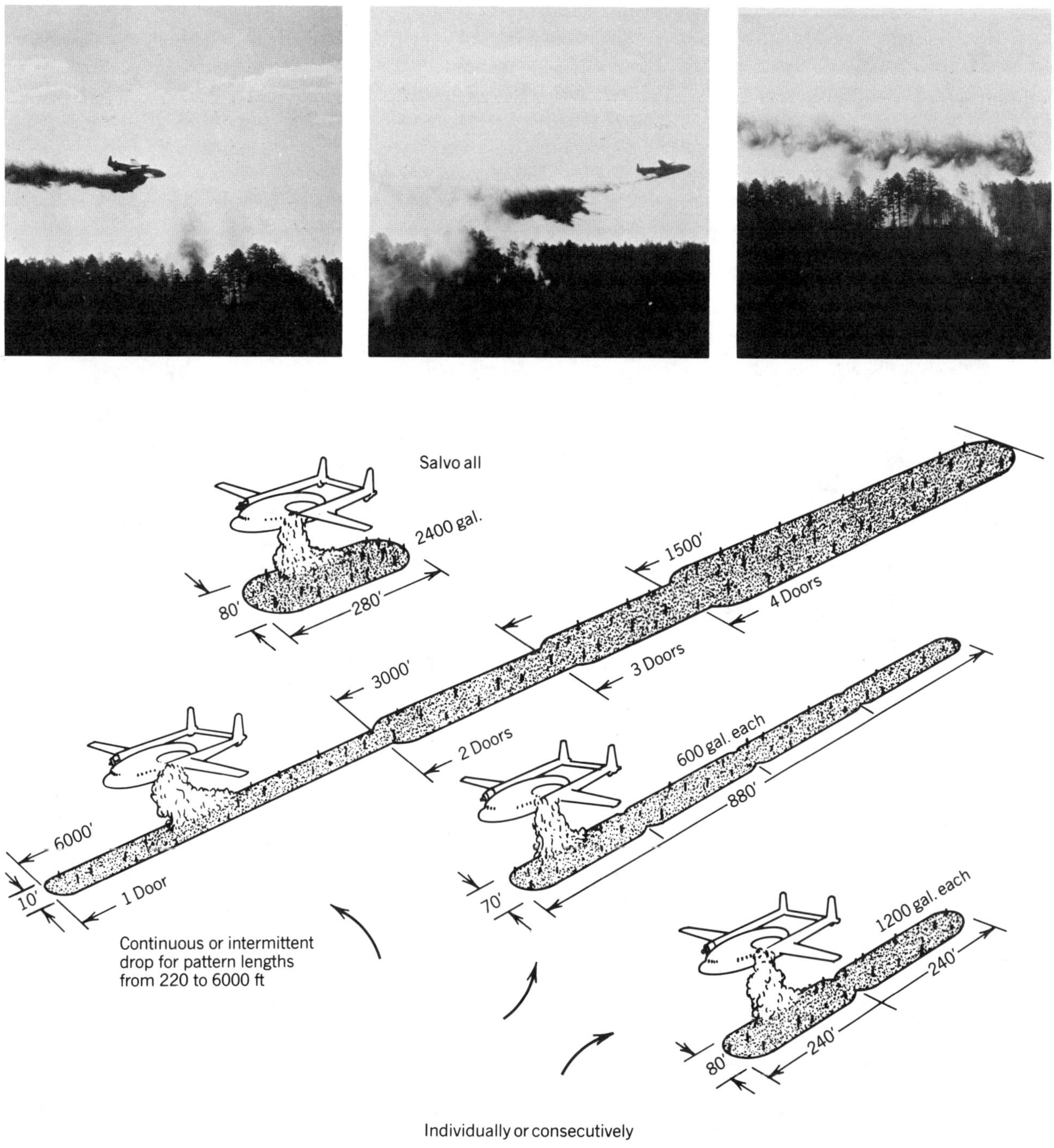


*Figure 9-16. Aerial attack. A C-119J dropping retardant on a small wildfire, Grand Canyon; and idealized drop system patterns for the C-119. The particular selection of a drop pattern will depend on fuels, fire behavior, and maneuverability. Photos courtesy Jim King; diagram courtesy Northern Forest Fire Lab.*

*Figure 9-17. Helitanker. Bell 205 picking up water from a lake. Courtesy U.S. Forest Service.*

opy and the convective plume, it can cool the flaming zone, coat surrounding fuels, and inhibit fire spread. Only in light fuels is direct control through aerial retardant possible, but in heavier fuels or dense brush a direct drop can slow fire growth and buy time for ground forces to arrive. Success, of course, depends on drop accuracy, but even a small fire may be hit more than once. For air tankers with large capacity, it is common to split the load into several drops. For smaller capacity air tankers, several planes may be used on a single fire, striking the fire in series, or a single plane may make repeated passes provided it can refill its tanks quickly. Examples include North Carolina, where single-engine air tankers operate in groups of three out of grassy fields, and Canada, where tankers can recharge their load by scooping water out of the innumerable lakes that are never far from a fire in the Canadian Shield. Drops are made in salvo.

For perimeter control trail drops are used. Control may begin with hotspotting, perhaps a direct attack on the head. Alternatively, a flanking action from a secure anchor point may pinch off the head. It is even possible to conduct indirect attack through aerial means. An air tanker may put down a retardant line, and a light plane or helicopter may begin counterfiring through aerial incendiaries. More typical is the use of these tactics in close coordination with ground control. Not only for efficiency, but for crew safety, good communications are imperative. On initial attack this may mean direct communication between the pilot and the fire boss, if any has yet arrived at the scene. On project fires, an air attack boss, reporting directly to the line boss, will direct all tactical air operations.

*Initial Attack: Structural Fire and Dual-Purpose Engines*

Control of structural fires differs from control of wildland fires in several respects. There is, first, the question of people—victims who may need medical attention, and onlookers who may require constraint. There is the matter of fire behavior. Compared to wildland fires, fuel loads are heavier, fuel moisture lower, residence time longer, and fire growth more rapid (the result of confined convective heat and radiation). Fuel–air mixing processes, however, are poorer, sometimes retarding fire spread, and usually allowing for a type of control over fire behavior, through ventilation, that is not available in wildland fire situations.

Then there is the problem of equipment, which determines the options open for control of the fire. Compared to ground tankers, dual-purpose engines generally have larger capacity tanks and higher volume pumping capacities, but they have less water on hand than a city with a hydrant network and lower pumping capabilities than urban fire engines. Nor do wildland agencies have the protective clothing, breathing apparatus, ventilation equipment, and so forth that makes direct control over a structural fire possible. Yet in some areas at least portions of this apparatus are available, some expertise and training are on hand, and responsibility for limited structural protection is present.

The best strategy is direct attack. This requires lots of water if the building is heavily involved with fire, or the ability to enter the structure, if it is less seriously involved. In rural areas water trucks will be dispatched along with engines. If the building is not heavily involved, control over the fire can come through a combination of two means: control over heat production and control over ventilation. For cooling, large diameter hoses will introduce water in the form of spray in quantity—water which will then absorb heat and leave the scene as steam. For ventilation, selective opening of the structure will allow heat and smoke to leave the building, making entry possible, and will bring some measure of control over the behavior of the fire. Improper ventilation, however, can worsen the situation, even leading to the explosive backdraft phenomenon.

An alternative form of direct control is to turn attention from control over the fire source to the protection of the resources threatened by the fire —a farm house near a barn fire, for example, or a summer home adjacent to a surface wildland fire. The structure may be protected against surface fires by surrounding it with a fireline; against radiant heat by dousing it with water; against spot ignition by closing points of entry and attacking spots as they appear. If multiple dwellings are threatened, however, the problem becomes one of coordination—of communication between the wildland fire control efforts and those protecting the structure, and of communication between wildland and urban fire services (NFPA, 1970).

*Project Fire: Tactics by Logistics*

Fireline tactics for a project fire do not differ in kind from those practiced on small fires. Control will mean perimeter control, and at least some portions of the perimeter will involve indirect attack. For tactical purposes, a large fire may be considered as a composite of many small fires. What occurs in one sector will not affect all other sectors, only those adjacent to it. There are no tactics required for a large fire that are not found on small fires. Where the question of scale becomes compelling is in the degree of organizational complexity demanded by a project fire and in the extent to which logistics controls tactics. For organizational and logistical considerations, unlike tactical ones, the question of scale makes for changes that are greater than the sum of their individual parts. Large fires are many times more complex than small fires, and this complexity, rather than tactics as such, often determines suppression success.

Between a strategy of control and its tactical execution, there exists a chain of logistical operations. In initial attack situations these operations are negligible. Ample equipment and supplies are carried to the fire directly with the firefighters, and resupply will only be necessary if reinforcements are needed. For project fires, however, the logistical chain intervenes into nearly every dimension of the suppression effort. The more complex the fire, the more pervasive is this intervention. It is, in fact, in good measure because of their logistical needs that large fires become complex. Normally the fire organization will plan on two shifts per calendar day: a day shift (0600–1800 hrs) and a night shift (1800–0600 hrs). This makes planning, feeding, sleeping, resupplying, and transporting a 24-hour operation. Because of the speed with which the organization must move, an error made at one time may be hard to correct later.

For project fires most tactical failures result from logistical failures—the inability to move crews rapidly on and off the line during shift changes; poor fire camp location, poor food and sanitation, or insufficient spike camps; inadequate suplies, improper equipment, or inappropriate crews and overhead; poor coordination between the servicing of aircraft and their tactical deployment. The opportunities for crippling interruptions are endless. Unlike a tactical breakdown on a fireline, whose effects tend to be local, breakdowns in the service function tend to multiply throughout the whole fire organization. Failures on the fireline often result from failures in the firecamp—not a product of a crew's inability to construct line, for example, but of assigning the wrong kind of crew to the job, of sending crews to the wrong sector, in delivering them too late in the shift to be useful, or in separating the crew from its handtools (which must be carried in a separate vehicle for safety reasons). There are, of course, fireline failures: blunders in counterfiring or poorly executed line construction. But even these, as often as not, result from breakdowns in communication, organization, or logistics rather than in actual tactics.

Accordingly, the service chief is a prominent figure in the project fire team; service officers must meet standards of experience and training in order to qualify for positions as camp officers, supply officers, and heliport managers. To the extent possible, supplies are prepackaged in kits—each kit containing all the necessary accessories as well as its principle item. A chain saw kit, for example, includes files, chaps, wedges, as well as a saw. A radio kit includes not one radio, but a set of portable radios with batteries, holding cases, relays, antennas, and base station. Standard lists suggest which kits to order for particular situations. In this way logistical needs, if not always complete, are at least a predictable quantity.

## 9.5 FIRELINE SAFETY

Accidents occur in all phases of wildland fire suppression, not merely during initial attack, and they result from all sources, not simply from fire. Vehicle accidents, snag falling incidents, aircraft crashes, handtool injuries, carbon monoxide exposure, contact with poisonous plants and animals, heart attacks, climbing accidents and rock falls—all contribute to a distressing roll call of fire-related injuries. National statistics consistently show firefighting to be among the most hazardous of all occupations. A firefighter is about 10 times as likely to suffer a fatal injury and 6 times as likely to endure a lost-time injury as the average worker.

Many of these hazards are not unique to fire situations, and they can be corrected by adopting commonly accepted procedures for good driving, felling, handtool use, and so forth. Fireline duty aggravates these situations with fatigue, smoke, and night-time assignments, but they are not special to fire suppression. Injuries related to the agency of wildland fire, however, are unique and they deserve special examination.

### *Safety Considerations*

The etiology of accidental injury or death is not so different from that of accidental fire. Origin does not lie with a single process of cause and effect but with the coincidence of several processes, the compounding of one event with others. Only a small number of ignitions result in large fires, and only a fraction of certain fireline procedures end in tragedy. That a particular practice does not inevitably end in an accident both mitigates and complicates a program of prevention. It means, on one hand, that there will be more near misses than accidents, and more accidents than fatalities. But it also means that experience will work to lessen the sensitivity of firefighters to danger. By repeating hazardous situations over and over again without disastrous consequences, the sense of danger disappears. Unsafe practices may become routine. Equally, normally safe, or marginally safe, practices may combine with other events in unexpected ways to make an accident far more likely. Low probability events will, on occasion, occur, and high probability events will fail to happen.

An accident prevention program thus confronts the same kind of problems as a fire prevention program. Accidents represent the compounding of source, place, and time. To a certain extent all can be modified. What fire suppression contributes in particular are a special source, the fire; a special place, the fireline; and a special timing, occurrence at a time and under circumstances which make control over fire behavior and human behavior exceptionally difficult. The haste of fire control becomes itself a hazard. Again, like fire prevention, accident prevention resorts to programs of education, engineering, and enforcement.

*Considerations of Source.* Source includes both human behavior and fire behavior. It is possible to identify hazard without knowing risk, and place without knowing time. Some categories of firefighters are especially prone to accidents, some fire environments lead to more fatalities than others, and some fireline procedures are more likely to result in casualties than are others. Among firefighters two groups are most susceptible to fireline accidents: those with less than 2 years of experience, and those with 10–15 years of experience (NWCG, 1981). For the young group, accidents result from the inability to recognize or properly respond to hazardous situations. For the experienced group, the explanation is somewhat more complicated. Members are more likely to be out of shape, perhaps removed from day-to-day fire operations, and overconfident—the result of having witnessed poor practices or normal fire behavior over and over without penalty or blowups.

High-risk fires are typically small fires, or small portions of large fires; they burn in fine fuels rather than heavy; they occur under sudden wind shifts rather than strong but steady winds, and concentrate on steep slopes or topographic chimneys. Many fatalities have occurred on fires considered under control, or fires that burned under routine circumstances until the fire suddenly flared up explosively. Fine fuels, variable winds, and steep topography make for fires that can respond quickly to environmental changes (Wilson, 1977). When a fire burns in conditions that have two or three of these elements, its risk can be exceptionally high. The identification of such situations forms a mandatory part of basic firefighter training, as is the recognition of those fireline procedures that bring with them the likelihood of putting firefighters into potentially explosive situations.

The hazards of the fire are greater than its flames, however. Even working next to an intense radiant heat source or within smoke—a common condition of direct attack methods—can lead to heat stress, dehydration, and in certain cases a degree of carbon monoxide poisoning (Countryman, 1971). Protective clothing, frequent work breaks for water, good line location, and sensible tactics will mitigate these hazards. It is important, too, that fatigue be recognized as nearly ubiquitous on the fireline, not only the exhaustion brought about by arduous labor, but that more insidious prostration which results from sleeplessness. Alertness drops, simple actions may become difficult, and normal precautions may be ignored.

*Techniques of Prevention.* One of the reasons that fires result in so many accidents is that only portions of the accident context can be controlled. By definition the principal source, the fire, is wild, uncontrollable. This wild source, in turn, dictates place and time. The opportunity to modify source, place, and time is vastly constricted. Obviously, low risk will only return with suppression of the fire. Until then prevention techniques focus on the most readily controllable aspect of the situation, its people. By methods of education, engineering, and enforcement it is hoped to reduce that portion of source which resides in the firefighter, and to lessen the likelihood that a firefighter will be jeopardized by being placed in high hazard areas at high risk times. Fire behavior forecasts can identify those places and those times that may contain an unacceptable degree of risk. Fireline tactics can proceed accordingly.

Educational techniques emphasize training—physical conditioning, instruction in fire behavior, and drills with tools and equipment. A certification program ensures that at least minimal standards of physical fitness and education have been met for even the lowest-ranking firefighter. Fundamental to classroom training is the ability to recognize certain inherently hazardous situations that can lead to rapid changes in fire behavior or fireline situations that signify a breakdown in organization. Essentially these are circumstances that put a firefighter in a position of ignorance—ignorance about fire location and behavior, uncertainty about orders or a chain of command, or confusion because of misinformation or poor communications. Written declarations must become tactical maneuvers if they are to be effective. Education must emphasize proper line location and construction, good counterfiring procedures, and suitable evasive actions in the event that a firefighter becomes trapped. Personnel must not only be outfitted with protective clothing, but must be trained in protective maneuvers.

In 1958 the U.S. Forest Service promulgated the 10 Standard Fire Fighting Orders as a kind of 10 commandments of safe firefighting (Figure 9-18). These were supplemented with "13 Situations That Shout Watch Out," which identified specific circumstances that increased the hazard to a firefighter. Both of these devices have been incorporated into basic fire training courses. To be effective, however, they must be translated into standard fireline procedures. A frontal assault with ground tankers, an undercut line on steep slopes, and burnout operations in topographic chimneys all ask for accidents. Every control plan should incorporate safety islands or escape routes into line operations, insist on good communications, and refuse to put personnel into high risk areas where the values under protection are far inferior to human life. Unlike urban fire services, few wildland fire fatalities have resulted from a situation where life was risked to save another life. Chance will ensure that, even under careful management, some accidents will occur. There is no excuse for raising those probabilities by poor planning or reckless tactics.

Engineering can protect firefighters by separating them from the fire to a limited extent. A hard hat, gloves, and sturdy boots will prevent common injuries to the head, hands, and feet. Special flame-resistant clothing offers protection against radiant heat and the likelihood of a flashover in clothes. Bright yellow in color, fire shirts improve the visibility of a firefighter at night; chemically treated, they force spot ignition to char rather than flame. Emergency fire shelters offer a further degree of protection, a last resort for a trapped firefighter. The shelter is packaged so that it can be worn on the waist. It can be assembled in 30 seconds, and it offers (if properly situated) a degree of emergency protection against radiant heat. The shelter is a blanket, however, not a bunker. It is most suitable for fast-moving fires (such as grass fires) where the flaming front will pass quickly. It will disintegrate on contact with hot flames (Figure 9-19).

Equipment can also provide a temporary shelter. Vehicles can protect against a rapidly moving flaming front; even glass windows will cut radiant heat in half. The air inside the cab may become foul because of smoldering debris, but the vehicle will not flame for many minutes—long enough for the fire front to have passed by—and the gas

| Ten Standard Firefighting Orders |
|---|
| 1. Keep informed of fire weather conditions and forecasts. |
| 2. Know what your fire is doing at all times. Observe personally, use scouts. |
| 3. Base all actions on current and expected behavior of fire. |
| 4. Have escape routes for everyone and make them known. |
| 5. Post a lookout when there is possible danger. |
| 6. Be alert, keep calm, think clearly, act decisively. |
| 7. Maintain prompt communication with your firefighters, your boss, and adjoining forces. |
| 8. Give clear instructions and be sure they are understood. |
| 9. Maintain control of your firefighters at all times. |
| 10. Fight fire aggressively, but provide for safety first. |

*Figure 9-18. Safety guidelines.*

tank will not explode unless it has been mechanically ruptured. Parking in a clearing or road further lessens heat stress. Less satisfactory are tractors. The cab is unshielded, but many operators have survived entrapment by clearing an area, placing the tractor within it, and taking refuge behind the blade. Again, it is only the flaming front against which one needs immediate protection. What is calculated to worsen the situation is simply to run. To run directly away from a fire is usually worse than to run through it, to run upslope worse than to traverse. The need for safety islands to provide a protected area and for lookouts to give ample time to get to such sites is obvious.

Protective clothing must be used in coordination with protective maneuvers. No fireline plan is complete until it has identified escape routes or safety islands or both, and has made them clearly known to all line workers. Lookouts should be

*Figure 9-19. Safety equipment. The fire shelter opened and in position. For a view of personal safety gear, see Figure 10-10. Photo courtesy U.S. Forest Service.*

used whenever fire behavior forecasts warrant, and when direct surveillance of the fire is not possible by line crews. In the event of entrapment, fire shelters can offer some protection; escape fires can be ignited, then entered; and, as the adage goes, one foot can be kept in the burn. Especially in fuels where burnout is rapid, the safest escape route is into the burned area. Another means of escape is to avoid places where flareups are likely to develop: chimneys, steep slopes with flashy fuels, ridgetops that can spawn firewhirls, the turbulent wake of low-flying aircraft.

Enforcement measures include the application of discipline and reward to ensure conformity with accepted procedures. All too often "safety" is a cosmetic, a mandated and barely tolerated veneer of declarations, memorandums, task force reports, safety officers, and exhortations that has little relevance to the conduct of practical affairs. Something is taught as a "safe" procedure rather than as the only procedure. Safety is something added to a program, not something integral to it. Instead, safety must be built into a system from the ground up, not dictated from the top down. It is learned by example, not by lecture; by experience, not by rote. Most safety programs fail at the bottom because they are not truly practiced at the top.

In an emergency, moreover, success will not depend solely on individual responses but on the ability of the organization to respond. A lack of coordination between line units, unclear instructions, poor control over personnel, failure to observe personally or to insist of proper procedures —upon the fireline supervisor rests much of the responsibility for fireline casualties. Good briefings, sound training, and careful observations can overcome of these difficulties. But the example of a crew boss or sector boss speaks louder than words. A supervisor's ability to command depends on two kinds of authority: on the legal authority vested in the position and on the moral authority apparent in the person. Appeals to legal authority—threats, orders, exhortations, paper shuffling through a chain of command—are necessary, but they carry far less effect than an appeal to moral authority.

Moral authority, by contrast, derives from confidence in leadership. Only through example can a supervisor, or an organization, ultimately inculcate personnel with the good work habits that reduce the likelihood of accident or that bring about during a time of emergency the response necessary to prevent casualties. The certification process of the NIFQS gives a certain legitimacy to an officer, but it does not grant the occupant *ex officio* moral authority. That authority is built into a crew or organization through long experience, not magically discovered during a moment of crisis. The problem is compounded by the practice, common to fire agencies, of having to staff fireline supervisors with personnel for whom fire duty is, at best, an infrequent exercise.

### *Fire Fatalities: Selected Case Studies*

Often the difference is only slight between a routine fire and a tragic fire, between a fatality and a near miss. By way of illustration consider the following case studies of wildfire fatalities. In all three episodes the conjunction of source, place, and time brought the fire and the fire organization into fatal association. In some, unsafe practices momentarily coincided with other unfavorable events; in others, normally secure practices combined with unusual or unexpected environmental circumstances. In all of the episodes the conjunction—for all of its lethal effects—was remarkably brief.

*Loop Fire, California, 1966.* The fire began from a faulty electrical distribution line on an Army missile base within the Angeles National Forest, November 1. Santa Ana winds pushed the fire rapidly, threatening not only the base but other facilities, including a hospital. Fuels were a chaparral mixture: sage, chamise, and sumac. The Los Angeles County Fire Department and the U.S. Forest Service jointly attacked the fire. County action concentrated on the lower reaches of the fire, primarily by means of catlines. Forest Service crews attempted to tie a handline from the ridgetops down to the catline. The fire was almost contained when the accident—the result of a local flareup—occurred. A photograph of the accident site is given in Figure 9–20. Except that it occurred during the day instead of at night and that it burned within a chimney rather than a bluff, the scene was identical to the Inaja fire which had burned a decade earlier and took the lives of 11 firefighters.

Initially a catline was to cross the gully (at C), while an IR crew cold-trailed and cut a handline down the topographic chimney (just to the right of the proposed line). A "diamond"-shaped clear area was located along the proposed line. Between this spot and the proposed tie-in was a distance of about 500 ft (152 m). For 300 ft (91 m) there existed a 3–10 ft (1–3 m) natural opening that could act as a line. The remaining 200 ft (61 m)

*Figure 9-20. Loop fire accident scene. Note the "diamond"-shaped clear area within the chute. Courtesy U.S. Forest Service.*

passed through light fuels. The fire simmered off to the side, the winds were favorable for completing the line, and the crew believed it would be able to cut a handline in about 10–15 minutes. At this time, however, the County crews realized that they would be unable to cross the gully (which was much too steep) with their tractors as they had planned. They began to scout out a line up the gully.

Meanwhile, the crew boss of the Forest Service crew elected to proceed straight down the ridge to the previously proposed catline. The handcrew was accordingly strung out within this topographic chute. At this same instant, a spot fire started at the base of the chimney. The spot was doused by a helitanker. But within five minutes a combination of circumstances caused the spot to explode up the chute, first in crippling waves of heat and then as flame. The fire raced through 22 men. Some found refuge in the "diamond" clearing, one rolled through the flames below, and the rest were simply overrun. All suffered burns, and eleven men died.

From the flaring of the spot until its attack by the helitanker about four to six minutes passed, and from water drop until until the surge that took flames through the crew, only five minutes. Its flashover through the chute took perhaps one minute. The topographic configuration, fuels, and burning conditions were fraught with hazard. Yet actual fire behavior at the time of the decision to proceed downward was favorable. Another 10 minutes and the crew would have arrived at the bottom, handline complete. Instead, it was precisely within this span of time that the flareup occurred.

Its specific cause is unknown. Obviously, the fire smoldered in unstable equilibrium. Perhaps the Santa Ana winds, then moderate, faltered, a pause sufficient to allow local upslope winds to redirect convective heat up the chute. Perhaps the sum of local conditions—local fuel concentrations, local winds, and local topography—became momentarily dominant. Perhaps downwash from the helitanker stimulated burning, an increase in intensity which coincided with the reassertion of local conditions. Once fire spread began virtually no evasive action was possible (U.S. Forest Service, 1966).

*Battlement Creek Fire, Colorado, 1976.* The fire started as part of a general lightning bust on July 11. Initially, it was located near an orchard and was attacked by the county rural fire department. Over the course of the next four days the fire rekindled twice, the second time under conditions that favored an explosive run. The fire entered cheatgrass and sage, developed a head, and eventually penetrated a large drainage filled with scattered timber and 6–12 foot high Gambel oak. A killing frost almost a month earlier made for unseasonably high quantities of dead, volatile, standing fuels. During the run of the 15th, the Bureau of Land Management assumed control over the fire, began air tanker drops, and put an LFO in place. The next day the fire made another run, burning out a drainage. A B26 air tanker crashed, killing the pilot.

Strategy now called for controlling the fire within the larger drainage by a combination of handlines and catlines, with firing operations moving from the upper ridgetops to the lower basin. On the 17th line improvement and firing continued, with one crew moving down the north side of the perimeter and a second moving down the south side (Figure 9-21). Both crews were Category I-rated IR crews. Burnout fared poorly due to sparse fuels and mediocre burning conditions. The south sector crew moved quickly; they soon reached the bottom of the spur ridge and continued firing within the base of the drainage. The north sector crew split into a line improvement squad and a burning squad.

Conditions that had favored high-intensity runs the previous two days encouraged yet another run as the afternoon developed. Burning that had not gone well earlier in the day, now picked up in intensity. In particular, the fire ignited along the lower basin began to spread up slope towards the ridgetop. In the course of 1.5 hours the fire would travel 3600 ft (1097 m). At the same time the burnout squad on the north sector

*Figure 9-21. Battlement Creek fire accident scene. In the photo, south is to the right, and north (where the accident occurred) to the left. Courtesy BLM and U.S. Forest Service.*

successfully ignited heavy pockets of fuel somewhat inward from the fireline and down slightly in a draw. They could not see the main fire.

Air tankers, helicopters, and a light plane all reported increased fire activity. The sector boss and the crew boss (north sector), who was working as part of the four-man burnout squad, agreed to speed up the firing operations and to send the line improvement squad to a preselected safety zone. A small spur ridge separated the line squad from the burnout squad, and when the line squad reported that it had reached the safety zone (just ahead of the fire), the sector boss misunderstood the report to mean that both squads were at the zone. Shortly afterwards, the burnout squad reported that its route to the safety zone was cut off by an arm of the main fire that had rushed up the opposite side of the spur ridge. Another arm—probably assisted by their successful burnout—was approaching up the side of the ridge in which they were located. An air tanker circled over the fire, but heavy smoke and turbulence prevented a retardant drop in the vicinity of the burnout crew, which now referred to itself as "trapped."

The four men scrambled up the draw to the fireline, wet down vests and clothes, lay prone in the shallow soil, and waited for the fire. The final surge of the fire covered 1800 ft (549 m) in about 15 minutes. As the flaming front struck, one man, then another, ran from the fireline; both died. Two remained, one of whom, despite serious burns, lived.

Fire organization, crew experience, support equipment, and communications—all were adequate by most standards, but each broke down in subtle ways. The timing of these lapses, individually minor, added up to tragedy. Following the episode, the U.S. Forest Service mandated that all of its employees engaged in fireline duty would carry fire shelters (Wilson et al., 1976).

*Ransom Road Fire, Florida, 1981.* The fire started from lightning on the Merritt Island National Wildlife Refuge in June. Rain postponed the appearance of the fire until the following day, and when the report of the smoke came, it was part of a complex of fires. Attack strategy was typical for the area: a tractor-plow would encircle the fire with a line, while intervening fuels were burned out. But fuels (palmetto and pine) were exceptionally dense and dry, other thunderstorms introduced serious problems with wind, and the suppression organization—for whom fire control was not a primary job—faced unusually heavy fire loads.

A tractor-plow completed control of one fire, then unloaded at the scene of another. Line was cut while counterfiring proceeded along other portions of the fire perimeter. Sudden downdrafts from thunderheads, however, caused an instantaneous acceleration in wind speed and a shift in wind direction from south to west. The tractor-plow working the east flank found itself imperiled. With a lookout riding on the machine, but the smoke almost impenetrable, the plow operator raised his bulldozer blade and began moving from the fire. The tractor caught on a raised stump, the men fled from the machine—one with his fire shelter and one without it—until the fire overtook them 60 ft (18 m) from a clearing. Both men died. The fire shelter had been pulled, but positioned amidst fuels, not within the plowed furrow; intense heat and flame contact disintegrated it.

Many normal safety precautions were lacking. There was no lookout, a role that an aerial observer could have filled; no real knowledge about the fire, its behavior, or the anticipated weather; no adequate control over the plow operation, such that escape routes could be designated. The firefighters had various levels of classroom training and experience, but for none, not even the supervisory personnel, was fire management a primary responsibility. Yet these identical considerations applied equally to the numerous other fires that burned in the vicinity, all of which were controlled safely.

What happened in the Ransom Road fire was the product of timing. The fire blew up and changed direction from winds spilling out of a thunderhead, and the fire virtually expired 20 minutes later from torrential rains brought by the same storm. All the things that had to happen occurred within that precise span of time—the redirection of the fire, the "stumping" of the tractor, the fateful footrace. The same practices that worked elsewhere did not coincide so perfectly with the arrival of the thunderhead (Fish and Wildlife Services, 1981).

## REFERENCES

The best summation of suppression is contained in the NWCG "S" series courses. Some of the courses are now published and available through the National AudioVisual Center; others are still in various stages of production. An overview of suppression equipment and strategy is given in Brown and Davis, *Forest Fire: Control and Use,* and a nicely illustrated, but less conceptually satisfying treatment is contained in Gaylor (1974), *Wildfires. Prevention and Control.* The pocket handbooks developed for fireline use give excellent summaries of fireline organization, tactics, and other miscellaneous information. The two most important guides are U.S. Forest Service, *Fireline Handbook* and Bureau of Land Management, *Fire Control Notebook.* Both are published by the Government Printing Office. The Florida Division of Forestry's *Fire Fighter's Guide* is recommended as a digest of southern fire control methods; especially good is its discussion of tractors. For a history of changing fire control practices, see Pyne (1982), *Fire in America.*


Barney, Richard J. (ed.), *Fire Control for the 80s* (Missoula, Montana: Intermountain Fire Council).

Brown, Arthur A. and Kenneth P. Davis, 1973 rev., *Forest Fire: Control and Use,* 2nd ed.(New York: McGraw-Hill).

Bureau of Land Management, *Fire Control Notebook,* updated periodically (Washington, D.C.: Department of the Interior).

Chase, Richard A., 1980, "FIRESCOPE: A New Concept in Multiagency Fire Suppression Coordination," U.S. Forest Service, General Technical Report PSW-40.

Countryman, Clive M., 1971, "Carbon Monoxide: A Firefighting Hazard," U.S. Forest Service, Pacific Southwest Experiment Station.

Davis, James B., 1979, "Building Professionalism Into Forest Fire Suppression," *Journal of Forestry* **77**(7): 423–426.

Fire Task Force, 1957, "Report of the Fire Task Force," Unpublished report, U.S. Forest Service.

Fish and Wildlife Service, 1981, "Ranson Road Fire Report," Unpublished report, U.S. Fish and Wildlife Service.

Florida Division of Forestry, 1975 rev., *Fire Fighter's Guide* (Tallahassee: Division of Forestry).

Gaylor, Harry P., 1974, *Wildfires. Prevention and Control* (Bowie, Maryland: Robert J. Brady Company).

Gemmer, Thomas V., 1980, "Fire Control Education in Forestry," pp. 207–213, in Barney, *Fire Control for the 80s* (Missoula, Montana: Intermountain Fire Council).

———, 1980, "Fire Management Curriculum Study: A Report," National Wildfire Coordinating Group and Society of American Foresters.

George, Charles W., 1981, "Determining Airtanker Delivery Performance Using a Simple Slide Chart-Retardant Coverage Computer," U.S. Forest Service, General Technical Report INT-113.

Hafterson, John, 1980, "Escaped Fire Situation Analysis," pp. 220–226, in Barney, *Fire Control for the 80s* (Missoula, Montana: Intermountain Fire Council).

Halsey, Donald G., 1980, "National Interagency Fire Qualification System (NIFQS)," pp. 192–206, in Barney *Fire Control for the 80s* (Missoula, Montana: Intermountain Fire Council).

Loop Fire Analysis Group, 1966, "The Loop Fire Disaster," Unpublished report, U.S. Forest Service.

McBride, Fred, 1981, "The Fire Management Electronic Age," *Fire Management Notes* **42**(4): 3–5.

National Wildfire Coordinating Group, "S (Suppression) Series Courses."

———, 1978, "Fire Management Curriculum Development Project. Part I. An Interagency Task Analysis for the Fire Management Specialist," National Wildfire Coordinating Group and Society of American Foresters.

———, 1981, "Preliminary Report of the Task Force on Fatal/Near Fatal Fire Accidents."

Newell, Marvin, 1981, "FIRETIP," *Fire Management Notes* **42**(3): 3–4.

Pyne, Stephen J., 1982, *Fire in America: A Cultural History of Wildland and Rural Fire* (Princeton, New Jersey: Princeton University Press).

Rothermel, Richard C., 1980, "Fire Behavior Systems for Fire Management," pp. 58–64, in Society of American Foresters, *Sixth National Conference on Fire and Forest Meteorology* (Washington D.C.; Society of American Foresters).

Swanson, D.H. et al., 1976, "Air Tanker Performance Guides: General Instruction Manual," U.S. Forest Service, General Technical Report INT-27.

Taylor, Alan R., 1977, "Transferring Fire-Related Information to Resource Managers and the Public: FIREBASE," in H.A. Mooney and C.E. Conrad (coordinators), "Proceedings of the Symposium on the Environmental Consequences of Fire and

Fuel Management in Mediterranean Climate Ecosystems," U.S. Forest Service, General Technical Report WO-3.

U.S. Forest Service, *Fire Management Notes* (formerly *Fire Control Notes)* (U.S. Forest Service, CFP).

———, *Fireline Handbook,* updated periodically (Washington, D.C.: Government Printing Office)

———, 1957, "Inaja Forest Fire Disaster," Unpublished report, U.S. Forest Service.

———, 1966, "Loop Fire Disaster," Unpublished Report, U.S. Forest Service.

———, 1969, "Intermediate Fire Behavior," U.S. Forest Service TT-81.

———, 1973 rev., "Water vs. Fire. Fighting Forest Fires with Water," U.S. Forest Service, Northeastern Area State and Private Forestry.

———, 1974 rev., "Water Handling Equipment Guide," U.S. Forest Service, AFM.

U.S. Forest Service and North American Forestry Commission, *Forest Fire News* (U.S. Forest Service, FFASR).

Weiner, Jack et al., 1975, "Forest Fires I: Equipment," Institute of Paper Chemistry, Bibliographic Series No. 261. National Technical Information Service (NTIS).

———, 1975b, "Forest Fires II: Methods, Meteorological Aspects, and Statistics," Institute of Paper Chemistry, Bibliographic Series No. 262, National Technical Information Service.

Wilson, Carl C., 1977, "Fatal and Near-Fatal Forest Fires. The Common Denominators," *International Fire Chief* **43**(9): 9–15.

Wilson, Jack et al., 1976, "Accident Report. Battlement Creek Fire Fatalities and Injury. July 17, 1976," Bureau of Land Management.

Whitson, Jim, 1982, "What is This Thing Called NITMS?" *Fire Management Notes"* **43**(1): 9–11.

Young, Edwin J. (ed.), "Fire Training. What? Why? How?" *Fire Management* **35**(4): entire issue.

## Chapter Ten

# Prescribed Fire

The concept of prescribed fire is integral to the concept of fire management. Simply put, a prescribed fire is any fire that achieves management objectives. Ignition for such a fire may be scheduled or unscheduled, but as long as the fire burns within a designated area and within a range of fire behavior properties, it will be promoted. Often it is not the physical or biological properties of the prescribed fire that differentiate it from a wildfire but its cultural context. A prescribed fire promotes the general land management goals of the responsible agency, a wildfire does not. The distinction between a wild and prescribed fire is a cultural, not a natural one. The boundary is often narrow, and it continually changes.

In some respects, prescribed burning is simply a new name for old practices. Slash burning and protective burning, for example, have been renamed prescribed fire. In other respects, prescribed burning means the adaptation of old practices to new ends. The agricultural uses of fire for landclearing and fertilization have been replaced by fire for site preparation within the context of industrial forestry. Fundamental to this transformation was the realization, derived from research in fire ecology, that fire could be biologically useful for those lands managed for extensive rather than intensive purposes. And in a few cases, new uses have been developed for new purposes—the promotion of prescribed natural fire, for example.

Between fire use and fire control there exists a reciprocity. The ability to use fire presupposes the ability to control fire, and the ability to control fire predisposes one to its use. Each new use created new needs for control, and each new technique of control made possible further uses. As concepts and techniques the use and control of fire show a remarkable symmetry. Some uses have made possible a new definition of fire control, control by prescription. Yet a prescription alone will not confine a fire to a particular area or to a predetermined span of time. It can only exploit those environmental circumstances which make up the fire environment. With the exception of fuels, it cannot modify them directly.

At the same time prescribed fire presents new opportunities for wildfire, new requirements for the control of fire effects (such as smoke), and new fireline hazards. Once escaped, a prescribed fire can only be controlled through the same methods as a wildfire. In techniques, too, prescribed burning and fire suppression are often mirror images. Similar methods of ignition, similar means of perimeter control, similar equipment, crews, and organization, similar preplanning devices—what differentiates prescribed from wildfire is that wildfire occurs under conditions that make control difficult and that lead to net resource damage, whereas prescribed fire can be set under conditions that favor control and that lead, on the whole, to the net enhancement of the resource.

## 10.1 PRESCRIBED FIRE STRATEGIES

Like strategies for fire suppression, those for prescribed fire hinge on concepts of fire control. Any fire that is out of control is, by definition, a wildfire. For prescribed burning control takes three forms: control over fire spread, though the establishment of firebreaks; control over fire intensity, by means of a burning prescription; control over frequency, through the ability to ignite favorable fires and to suppress unwanted fires. Through control over timing, place, and intensity a fire management agency can exercise control over fire effects. The goals of prescribed fire will follow from the land management plan of the administrative unit; the prescribed conditions, from a fire management plan; and the methods of control, from a burning plan. Just as with fire suppression, there is a predictable evolution of stages during the execution of a prescribed burn.

### *Objectives of Prescribed Fire*

A prescribed fire occurs within three sets of conditions. Considerations of fire behavior and fire ecology determine what the fire can and cannot do; considerations of fire culture, what the fire ought or ought not to do. Some things free-burning fire cannot do because it behaves in certain ways, yet some things are possible because those same physical properties allow the fire to be manipulated. Having wind or slope present will control a fire's rate of spread, intensity, and direction, but they also make control over a deliberately set fire easier. Backing fires, for example, may be substituted for heading fires with good prospects for control. Similarly, the biological potential of an ecosystem will determine fire effects. Yet the adaptations to fire present within the biota also make possible the manipulation of that biota through deliberate fire. Physically identical fires, moreover, may produce different ecological consequences. Biological considerations depend upon, but are not determined by, physical considerations.

To carry the issue one step further, to what ends fire may be used, or withheld, from ecosystems will depend on the civilization that controls it. Different cultures will distinguish wild from prescribed fires on the basis of different criteria, will apply and withhold fire for different purposes, and will exploit different techniques of control. In general, fire use and fire control will be paired. Those cultures that use fire widely will be prepared to control fire in some way, and those that exercise considerable control of fire will be disposed to use it in some forms.

Since most fire management in the United States is conducted by land management agencies or landholding companies, the goals of prescribed fire are expressed within the context of general land management plans. Contemporary practices include the use of fire for fuel management—by site or in broadcast treatment, in situ or after treatment. Where silvicultural purposes are primary, fire may be used in site preparation, thinning, control of species compositions, and fumigation against selected fungi, diseases, and pests. Type conversion remains a common fire use, either for restoration burning or maintenance burning. In wilderness areas, fire is preserved as one of many natural processes. For wildlife mangement through habitat manipulation, fire is often a superior instrument. Prescribed fire is most useful in areas where land is managed extensively rather than intensively, where management ambitions seek to harvest natural products or to preserve natural processes, and where at least some of the desirable effects stem from the nature of fire, a process for which no comparable surrogate may exist.

Instead of classifying prescribed fires by their purposes, as in the foregoing, it is also useful to consider them by reference to traits more intrinsic to fire. Prescribed fires may be distinguished by their size, intensity, and frequency. All of these properties influence fire effects, all can be manipulated by humans, and all relate directly to the complexity of a fire management program.

*Control Over Size.* With regard to fire size, the important variables are the nature of the burn (as defined by its fuels), the possibility for control over the fire, and economics. In most cases the purposes of the burn and prospects for control over the fire will determine the size of burning blocks. In burning clearcut slash, for example, the distribution of fuels sets the general size of the treatment area, and considerations of fire control and smoke management will set the size of discrete blocks within that area. With broadcast burning for habitat or underburning for fuel reduction the larger treatment area will be less well circumscribed. The actual size of the burning blocks will depend on where control lines are placed. This will vary according to considerations of fire behavior and whether firelines are deliberately constructed or natural barriers are incorporated.

Within a burning block, however defined, the object is to ignite the entire area. This may occur simultaneously (area ignition) in the case of slash burning, such that intense fires will develop that can combust larger diameter fuels and push smoke upward in a convective column. Or it may come through line fires, distributed in various ways. Leaving unburned fuels within the block not only voids some of the reason for the fire, but may present a control problem—just as unburned fuels do in wildfires. An exception is the prescribed natural fire. Here the burning block is defined by the boundaries of the fire management area; the fire needs to spread to all portions of the area or with similar burning properties; and the fire is allowed, within limits, to set its own size. In all cases, however, control is limited by the accuracy of fire behavior forecasts, and these will be restricted to the 48 hours or so over which weather forecasts are accurate.

Though fire size can be manipulated, how an economy of scale might apply to prescribed burning is not well understood. To the extent that control means perimeter control, large fires offer greater economies than do small fires simply because the ratio of perimeter to area is more favorable. Paradoxically, large area fires may be easier to control than small area fires, especially when natural barriers form part of the fireline. A qualifier, of course, is that large fires may require more control forces overall than small fires, even though the ratio of personnel to fireline may be better with a large fire. Similarly, the need to complete firing within certain spans of time may put limits on fire size. Fires that smolder through many burning periods and require attention will be more costly than fires that complete their task in shorter order.

For any particular burning block, the ideal size —other considerations apart—will be a maximum area for the control forces available. These forces should be far less than those needed to control a wildfire of identical size. Firelines can be emplaced well in advance of the burn, and mechanization can substitute for many of the costly tasks traditionally performed by hand labor. The ability to put in firelines by machine or to burn out by aircraft will affect the determination of ideal fire size.

*Control Over Intensity.* Control over fire intensity is achieved through a prescription, fuel preparations, and a choice of ignition patterns. Measures of fire intensity, like flame height, give a common denominator to both sides of the prescribed fire equation: fire control and fire effects. A prescription simply specifies those environmental parameters that will result in an appropriate intensity. By restricting the timing of ignition until those conditions are met, it is possible to control fire intensity.

Of the fire environment, only fuel characteristics and the presence of other fires can be manipulated to any degree. If the fire is likely to burn too intensely, fuels may be rearranged through crushing or chipping to lessen the quantity of available fuel or to increase its bulk density. If the fire may have difficulty reaching a high enough level of intensity to achieve its intended purposes, fuel preparation may take the form of crushing, dessicating, or otherwise increasing the ratio of dead to live fuels in the complex. It is also possible through various ignition techniques to manipulate some of the dynamic properties of a fire. Backing fires will show different intensities than heading fires, and line fires from area fires. Selecting a mode of ignition, in short, is not arbitrary. Residence time and intensity will vary with the mode of firing, and these will, in turn, select fire behavior properties and fire effects that will influence the entire strategy of the burn.

Fire intensity, of course, is not uniform except in special cases. A grassy site, even one large in area, may be fired and completely burned under conditions that are essentially uniform. But most sites will show differences in fuel distribution and will experience changes in weather that make fire intensity variable over the course of the burn. Most ecosystems consist of many species arranged in complex ways. Fire effects must be considered in statistical terms, not deterministic ones. To match an economy of scale there is what might be called an ecology of scale. It is difficult to predict a particular outcome for a particular site, with the difficulty increasing as the site becomes smaller and the outcome more specific.

Instead one can speak of ensembles of effects. To achieve these effects a certain size of fire is necessary. For some of the purposes to which prescribed fire is put, it is necessary to burn off the entire block under more or less uniform conditions. Consequently, firing of the large block proceeds by the firing of smaller internal blocks. But there are limits to this process, and eventually other forms of manipulation will be adopted. Fire, especially broadcast fire, is a technique for extensive management. As management becomes more intensive, free-burning fire is replaced by other means of manipulation.

*Control Over Frequency.* Prescribed burning programs can also vary the timing of fires. In one sense, this is a means to control fire intensity. A fire may be ignited at night, for example, to modulate its behavior, or during one day, when burning conditions are favorable, rather than at another time when they are not. But in a larger sense, fire frequency refers to the timing of fires according to biological and fuel histories. Wildfire obeys a logarithmic distribution through time, with many low-intensity and few high-intensity events. Prescribed burning programs tend to replace this history with a more cyclic pattern, or at least with one tied more directly to human activities, and to distribute fire intensities about a mean value.

In general, two patterns of prescribed fire may be identified as a function of frequency. *Restoration burning* refers to a single episode, or a single sequence, of prescribed fire. Landclearing, type conversion, and slash disposal are all examples of controlled fire applied one time. By contrast, *maintenance burning* refers to the periodic application of fire to sustain an existing environment. Examples include fuelbreak maintenance, broadcast fuel reduction, and habitat maintenance. For those programs that use broadcast fire for fuel modification, the cycle of burning will follow the pattern of fuels buildup; for those that seek to modify the biota, the historical patterns of succession.

Obviously, control over fire frequency will influence control over fire intensity and fire size. Where prescribed fire is routinely practiced, the tendency is to replace a stochastic process, wildfire, with a quasi-deterministic process, prescribed fire. To the extent that this exchange is successful, models of fire behavior, fire effects, and fire economics—all of which assume quasi-deterministic conditions—will also be successful. As prescribed fire becomes routinized, moreover, its true costs, including overhead, can be better appreciated.

### *Fire Use Plan*

A prescribed fire is executed under the provisions of a *fire use plan.* A written document, the plan will include, first, an identification of the treatment area and the objectives of burning; second, a fire prescription; and, third, a burning plan. The *prescription* specifies those environmental conditions and ignition patterns that will produce fire behavior which will yield the desired effects. The burning plan will tell how the fire will be executed in the field. The *fire use plan* is thus analogous to a preattack plan made as a part of presuppression; the prescription, to a fire behavior forecast around which plans for the control of wildfire are made; the burning plan, to the shift plan written to guide suppression (Figure 10-1).

The entire process must begin with a clear statement of objectives. Obviously, objectives are inherently limited by the properties of fire and the possibilities of fire ecology. But once the decision to use fire has been made and the more clearly the goals of the project are stated, the more likely a prescription and burning plan can be designed to satisfy them. If the objectives can be expressed in quantitative terms, so much the better. Where prescribed burning seeks to reduce fuels, quantitative statements on the amount of fuel to be eliminated in various categories can be made, and met. In some form the relevant information will include a statement of purpose, an identification of the treatment area, a summary of land management and fire use objectives, and a list of constraints and alternatives, if any. Where prescribed fire is used routinely, as with slash reduction or rough reduction, the goals will be assumed; the fire use plan can specify the particular objectives of the particular fire.

The prescription mediates between the fire environment and the fire. It must translate the properties of the fire environment into the physical properties of the fire, and the properties of the fire, into fire effects. The principal effects will include the biota, the fuel complex, smoke, and fire. The prescription must design a fire that will shape the biotic ensemble in favorable ways, reduce fuels of the proper type in the proper amount, ensure smoke dispersion, and lessen the likelihood of escape fires. Because fire size, intensity, and frequency can be manipulated, conceptual models that link fire effects with fire behavior can apply reasonably well. Precisely this sort of forecasting is ideally suited to computers, and the integrated systems being developed for wildfire suppression can be adapted to prescribed burns.

The prescription itself is usually given in one of two forms. It may be written as a matrix of environmental conditions, each of which shows a particular range. For example, winds might range between 5 and 10 mph (8–16 km/hr); fuel moisture (10-hr TL), between 8 and 12%; ambient temperature, between 65 and 80°F (18–27°C); relative humidity, 35 to 50%; and so on. Or it may be written in terms of important fire behavior parameters, such as fireline intensity or flame height, which integrate a variety of environmental conditions. In either case, some general limits or boundary conditions are specified, within which all prescriptions must function.

A prescription, however, is only a statement of conditions under which a fire will be set or the conditions under which an unplanned fire will be tolerated. To go from prescription to fire is the task of the burning plan. The plan outlines a scenario according to which the actual fire will evolve. It will specify any site preparations that may be needed—fuel treatments, firelines, weather stations, for example. On the model of a shift plan it will designate fire organization, firing operations, holding operations, patrol, and mopup (if any). It must examine contingencies in the event that the prescribed fire escapes, provide a means by which the prescribed fire organization can be quickly transformed into (or replaced by) a fire suppression organization. It should provide for any postburn treatments that may be necessary, for an evaluation of fire effectiveness, and for cost summaries, perhaps including an assessment of benefits. All of this information should be incorporated into a final report. The evolution parallels that of suppression (Fischer, 1978).

All this makes the production of a fire use plan seem more onerous than it need be in practice. Preprescription plans can be systematized just as preattack plans are. Large-scale data collection may be largely a one-time effort, with site-specific information needed only for those particular burning blocks under consideration. Prescriptions evolve out of previous prescriptions, they need not be computed from scratch for every prescribed fire. Reporting can follow standard

PRESCRIBED BURNING PLAN

| FOREST | DISTRICT | FY- |
|---|---|---|
| TRI COMPARTMENT (NAME & NUMBER) | GRID NUMBER | BURN UNIT |
| LOCATION: T______, R______, S______ | GROSS AREA | NET AREA |
| I&E CONTACTS | APPROPRIATION(S) | |
| | EAR PREPARED BY | DATE |

BURN AREA DESCRIPTION: DESCRIBE TREES & SHRUBS OVER 12 FEET TALL, SHRUBS & TREES UNDER 12 FEET, AND GRASSES AND FORBS.

ECOCLASS CODE(S)________ TOPOGRAPHY________ SLOPE________ ASPECT________

SOIL TYPE(S) & DESCRIPTION________

FUELS (NATURAL &/OR ACTIVITY)

0 -¼" SIZE CLASS________T/A SHRUBS________T/A TOTAL________T/A

¼"-1" SIZE CLASS________T/A HERBACEOUS________T/A

1"-3" SIZE CLASS________T/A DUFF DEPTH________IN FUEL CLASSIFICATION________

3" + SIZE CLASS________T/A SURFACE FUEL DEPTH________FT NFDR FUEL MODEL________

MANAGEMENT GOALS OF THIS BURN

| OBJECTIVES OF BURN | (CHECK) | (SPECIFICS) |
|---|---|---|
| HAZARD REDUCTION | ____ | ________ |
| SILVICULTURE | ____ | ________ |
| SITE PREPARATION | ____ | ________ |
| WILDLIFE HABITAT | ____ | ________ |
| RANGE MANAGEMENT | ____ | ________ |
| INSECT/DISEASE CONTROL | ____ | ________ |
| SPECIES MANIPULATION | ____ | ________ |
| OTHER | ____ | ________ |

BURNING PRESCRIPTION

SEASON________ TEMPERATURE (RANGE) ____to____ °F WIND DIRECTION:

TIME OF DAY________ RH (RANGE) ____to____ % PREFERRED________

DAYS SINCE RAIN________ WIND SPEED (RANGE) ____to____ MPH ACCEPTABLE________

FUEL MOISTURES: 1 HR TL______%, 10 HR TL______%, 100 HR TL______%,
1000 HR TL______%, DUFF______%, SHRUB______%, HERBACEOUS______%

FLAME LENGTH: MAXIMUM______, AVERAGE______ FLAME HEIGHT______

ALLOWABLE SCORCH HT (FT): CROWN____, BOLE____ NFDR DATA: BI____, ERC____, IC____, SC____

FIRING PATTERN________ IGNITION METHOD________

LOGISTICAL INFORMATION

CHAINS LINE TO CONSTRUCT: TRACTOR____ HAND____ OTHER (SPECIFY)____ TOTAL________

CHAINS LINE TO FIRE: EXTERIOR____ INTERIOR____ TOTAL____

MANPOWER NEEDS: UNIT PREPARATION________ BURNING________

HOLDING________ MOPUP________

EQUIPMENT NEEDS: UNIT PREPARATION________

BURNING________

HOLDING________

MOPUP________

BURN SUMMARY

DATE BURNED________ TIME OF DAY________ DAYS SINCE RAIN____ SEAS. PRECIP TO DATE____IN.

ACTUAL WEATHER: TEMP____ RH____ WIND SPEED & DIRECTION________ NFDR BI____

FUEL MOISTURES: 1 HR____ 10 HR____ 100 HR____ 1000 HR____ BRUSH____ HERBACEOUS____

FIRE BEHAVIOR: ROS______CH/HR, AVERAGE FLAME LENGTH______FT, HEIGHT______FT

SCORCH HEIGHT: BOLE______FT, CROWN______FT

BURN EVALUATION (If additional space is needed, use additional sheet)________

________

________

________

PLAN PREPARED BY:________ DATE________

PLAN REVIEWED BY:________ DATE________

PLAN APPROVED BY:________ DATE________

NOTE: ATTACH MAP OF BURN AREA

*Figure 10-1. Sample fire use plan. From Martin and Dell (1978).*

formats, encapsulated into suitable forms. The procedure can soon be routinized, much as an analogous procedure has been routinized by those states which regulate open burning through the issuance of permits. It may be argued, in fact, that a purpose of the fire use plan is to prevent the procedure from lapsing into habit by insisting that careful thought be given to each application. Fire suppression is also routinized, after a fashion, yet it requires constant, serious planning. A relatively comprehensive document, moreover, can protect an agency against charges of negligence should a fire escape and leave the agency open to litigation for damages.

### *Model Prescribed Fire Evolution*

Allowing for a few fundamental distinctions, a model evolution for prescribed fire shows an uncanny similarity to the evolution for wildfire suppression. In contrast to suppression, prescribed burning promotes rather than prevents fires. Its planning begins with fire effects and concludes with environmental conditions. And it exploits in the name of control those same fire behavior determinants that drive a wildfire. Ignition, rather than initial attack, marks the onset of line activity for fire use plans, however. For wildfire, control comes after ignition; for prescribed fire, control precedes it. If initial attack fails or initial control collapses, then the responses to wild and prescribed fires are indistinguishable. Its preconditions, not its techniques, distinguish the management of wild from prescribed fire. The procedures attending wild and prescribed fire are similar, but their arrangement varies.

For suppression, control begins with preattack plans and presuppression activities; for prescribed fire, with the fire use plan and any preparations it requires. Typical preparations include fuel modifications, the construction of firelines, a system of data-collection such as fuel samplings and weather readings, and the processing of any mandatory permits required by state law or local air quality boards. When the site borders other agencies or landowners, notification about the proposed burn is a normal precaution. If the public will encounter the fire or its smoke, public information sources should be notified. The possibility of overloading local airsheds with smoke at certain times of the year has led, in many regions, to interagency boards to oversee the amount of burning allowed by the various agencies. Prior to actual ignition, it is prudent to light a small-scale test fire.

For the actual burn, fireline procedures are identical to those used in suppression. In a sense, the organization proceeds as if the exercise were a counterfiring operation, with lessened opportunities for the fire to escape its designated perimeter. Should the fire escape control, however, or should its prescription be exceeded because of changes in weather and fire intensity, then the situation is identical to a failed initial attack. Suppression resources will build up, perhaps escalating to a large fire organizaton. With or without escape, some mopup is likely along the perimeter. Many prescribed fires in timber do not escape for several days after ignition; until then they may burn in cat-faces or snags which throw embers or fall across firelines some time after the flaming front has passed. Part of the problem with cost-accounting for prescribed fire is that escape fires are considered as wildfires, not as an extension of the prescribed burn.

Postburn activities will include an evaluation, a report, possible site rehabilitation or planned follow-up activities, and a review of the whole fire operation.

## *10.2 TECHNIQUES OF PRESCRIBED FIRE*

Most of the techniques needed for controlling a prescribed fire are identical to those used in the control of wildfires. That control over a prescribed burn is established prior to ignition, not after it, emphasizes some techniques over others and gives those activities a complexity that does not exist in fire suppression. Some of these special activities relate to preparations at the site, some to the actual firing of the site, and some to control over fire effects, notably smoke. Through modification of fuels and the selection of a suitable ignition pattern, it is possible to influence the fire and its environment such that it can be brought into prescription.

Part of the effects which the prescription seeks to control is smoke dispersal, a product of the fire and its surrounding air mass. The prescription may specify certain atmospheric conditions that must be satisfied prior to ignition, but the only form of active control possible is modification of the emissions source—the fuels and the behavior of the fire. Within limits, this can be manipulated independently of the air mass.

### *Site Preparation*

Site preparation includes monitoring the fire environment, establishing control lines, and practicing fuels management. All of these activities have analogues in suppression. Monitoring includes fuel inventories and weather collection, the same data-gathering that is needed for fire danger rating and fire behavior forecasting. Control lines require the same considerations used in constructing firelines around wildfires. The same fuelbreaks may define both preattack blocks and burning blocks.

Fuels management practices for preburning are roughly comparable in goals and technique to those used in presuppression. Preburn and presuppression activities are complementary, and in practice they are often synonymous. Fuel reduction through prescribed fire, after all, is often a means of presuppression fuels management. In some instances the object of fuels treatment is to reduce fire intensity. Slash, for example, is often crushed, chipped, or rearranged prior to ignition. In other cases, the purpose is to enhance flammability. This is frequently the case with brush. By dessicating or crushing swaths of brush, it is possible to fire those treated areas under conditions that make fire spread into the surrounding untreated brush unlikely. In the case of woody vegetation that have invaded grasslands, killing and toppling the brush and trees is a precondition to their successful burning.

### *Firing Techniques*

*Firing Equipment.* Most incendiary devices are intended for use with ground crews. Some, such as the fusee, use a solid fuel; others, such as the drip torch and flame thrower, a liquid fuel. A common mixture is 4:1 diesel fuel to gasoline, the gasoline providing flash ignition to the diesel, and the diesel a longer burning flame to wildland fuels. Propane gas is also popular. Incendiary bullets, Very pistols, thermite grenades—all allow for ignition at greater distances. Power equipment has been applied to flame throwers for some time. The choice of an ignition device is not arbitrary. Ignition costs money, some devices work better in particular fuels than do others, and the pattern of ignition may argue for a certain incendiary because of safety considerations. Power equipment is limited to ignition along the fireline, but handheld incendiary devices may penetrate well into the burn by being carried, thrown, or shot from the line.

Aerial ignition devices have appeared more slowly. In a sense this is surprising, given the dramatic demonstration of aerial incendiaries during World War II. Throughout the 1970s, however, as the need for broadcast burning on a large scale became apparent, the search quickened for better aerial incendiaries (Figure 10-2). Some were designed for light planes, most for helicopters. A number of delayed action ignition devices (DAIDs) became available, a popular version consisting of chemicals injected into pingpong balls. There were instruments for dropping fusees, for ejecting lightweight fuses, and for slinging tanks of jellied gasoline (helitorch). All of these devices were as useful for suppression as for prescribed fire. But the availability of new incendiaries, such as the techniques of aerial ignition, opened up new possibilities for strategy. Large areas, and center areas, could be safely ignited without the need for handcrews.

*Ignition Patterns.* Control over the flaming front of a prescribed fire does not come from eliminating the fire environment but by exploiting it. As long as ignition is manageable, the presence of light fuels, wind, and slope can bring added control

*Figure 10-2. Aerial ignition. (Upper) Helicopter ignition with DAIDs in Everglades National Park, and (lower) helitorch use in the Northern Rockies. Courtesy U.S. Forest Service.*

over the fire, not less. The direction of fire spread can be reversed, rate of spread can be lessened, and fire intensity regulated. This can be achieved by varying the pattern and timing of ignition.

Burning against the wind or slope substitutes a backing fire for a heading fire. By putting down fire in strips, it is possible to simulate the effects of a line fire but speed up its rate of spread. Putting down fire in spots, it is possible to simulate many growing steady-state fires and burn larger areas in a given time—again, effectively increasing the rate of spread. If the spots crowd together, they may interact to create an area fire with an increase in the rate of combustion and fire intensity. Encircling a site with fire reverses the normal perimeter growth of a fire, making the fire draw inward rather than outward and increasing its intensity at the center rather than along the advancing edge.

Similarly, by choice of timing it is possible to manipulate the history of a fire. Igniting a fire late in the burning period will produce a hot fire, but one that will abate during evening. Conversely, a fire set early in the burning period can be expected to increase in intensity as the day progresses. The combinations of timing and geome-

| Technique | Where used | How done | Advantages | Disadvantages |
|---|---|---|---|---|
| Head fire | Large areas, brush fields, clearcuts, under stands with light fuels | (1) Backfire downwind line until safe line created (2) Light head fire | Rapid, inexpensive, good smoke dispersal | High intensity, high spotting potential |
| Backfire | Under tree canopy, in heavy fuels near firelines | (1) Backfire from downwind line; may build additional lines and backfire from each line | Slow, low intensity, low scorch, low spotting potential | Expensive, smoke stays near ground, the long time required may allow wind shift |
| Strip head fire | Large areas, brush fields, clearcuts, partial cuts with light slash under tree canopies | (1) Backfire from downwind line until safe line created (2) Start head fire at given distance upwind (3) Continue with successive strips of width to give desired flames | Relatively rapid, intensity adjusted by strip widths, flexible, moderate cost | Need access within area; under stands having 3 or more strips burning at one time may cause high intensity fire interaction |
| Spot head fire | Large areas, brush fields, clearcuts, partial cuts with light slash, under tree canopies; fixed-wing aircraft or helicopter may be used | (1) Backfire from downwind line until safe line is created (2) Start spots at given distances upwind (3) Adjust spot to give desired flames | Relatively rapid, intensity adjusted by spot spacing, can get variable effects from head and flank fires, moderate cost | Need access within area if not done aerially |
| Flank fire | Clearcuts, brush fields, light fuels under canopy | (1) Backfire downwind line until safe line created (2) Several burners progress into wind and adjust their speed to give desired flame | Flame size between that of backfire and head fire, moderate cost, can modify from near backfire to flank fire | Susceptible to wind veering; need good coordination among crew |
| Center or ring fire | Clearcuts, brush fields | (1) For center firing, center is lighted first (2) Ring is then lighted to draw to center; often done electrically or aerially | Very rapid, best smoke dispersal, very high intensity, fire drawn to center away from surrounding vegetation and fuels | May develop dangerous convection currents; may develop long distance spotting; may require large crew |

*Figure 10-3. Ignition patterns and fire behavior. From Martin and Dell (1978).*

try are legion (Figure 10-3). Their power to modify fire behavior is not unlimited, but they can significantly modify the behavior that a prescription alone might anticipate.

Most ignition patterns use spot or line firing. *Spot firing* produces a multitude of point ignitions, which then spread and interact. Whether the spots grow individually, form into a line, or coalesce into an area will depend on the initial pattern and the properties of the fires that result, especially their time–temperature growth curve. *Strip firing* produces a line of ignition, which behaves like a flaming front. The resulting line may be a heading fire or a backing fire. In most cases the backing fire—with lower fireline intensities, slower rate of spread, and more complete combustion—is the preferred mode. The strips may be arranged in various patterns for ease of control or to stimulate certain effects.

Frequently a series of parallel strips are laid down. Each line of fire will only burn to the next line. If the strips are sufficiently wide, there will be little interaction between the separate lines of fire, leaving the burning zone as a set of independently burning lines. If the strips are crowded together and the residence time of the flames is long, the outcome may be an area of simultaneous burning. The expression *strip fire,* as such, is usually reserved for fires set parallel to the fireline and perpendicular to the wind, slope, or advancing head (in the case of a wildfire). When strip firing occurs along the sides of the burning block or advancing head, it is called a *flank fire.* If it occurs parallel to the direction of spread, or perpendicular to the fireline, such that it forms a series of widening triangles, it is termed a *chevron fire.* The chevron pattern is especially useful when a lot of fire must be put down against wind and slope effects. The pattern neatly reverses the normal head development of a fire so that all portions of the fire are continually driven into the old burn. Since all these fires are backing fires, putting down multiple strips can effectively increase the rate of spread of the entire burn. The spacing between strips can be varied to prevent or enhance their interaction.

Sometimes area ignition is preferable to line ignition. Heavy fuels such as slash may require higher intensities and longer residence times in order to combust sufficient quantities of fuel, and they may need a strong convective column to assist in smoke dispersal. *Area ignition* generally follows one of two forms. The entire area may be simultaneously ignited through electronic devices or aerial incendiaries, or a strong center fire may be generated which is then widened by surrounding the zone with concentric lines of fire known as *ring fires.* Each of these rings is then drawn inward to the center, further stimulating the rate of combustion. A strong convective column develops and the indrafting winds help to control embers that might otherwise lead to spot fires.

These same firing techniques can also be used for counterfiring, with two important qualifiers. The overall ignition pattern for counterfires is dictated by the wildfire, not the main fire by the ignition pattern. And the counterfire will, at some point, interact with the main fire. In many cases, a strong indraft from the main fire can assist the spread of a counterfire away from control lines. But the interaction cannot be avoided, as it can with prescribed burning. Again, as with prescribed fires, a variety of patterns is possible. The important consideration is not their nomenclature but the kind of fire that will result from a particular geometry of ignition and from the interaction of that fire with the fires around it.

### *Smoke Management*

Unlike the smoke from wildfire, which can only be controlled by the suppression of the fire, the smoke from a prescribed fire can be managed to a certain degree. In some instances smoke may even be considered an asset, a desired ecological agent. In any event its quantity, physical chemistry, and dispersal can, within limits, be regulated. In part, control results from sensible scheduling of prescribed burning during those periods that favor maximum mixing and dispersal through the atmosphere. In part, it comes from manipulation of fire behavior and the way in which a fire can interact with its surrounding air mass. Primarily, this means modulating fire intensity such that flaming rather than smoldering combustion is favored and that a strong convective column rather than diffuse smoke plumes develop. With control, however, comes responsibility. No other aspect of a fire carries its effects so far from the site, no other is subject to such regulation by other agencies, and no other compromises programs of routine prescribed fire or prescribed natural fire in the West. Much like fuels management, smoke management is rapidly becoming an important specialty within a general program of fire management.

Smoke management involves control over three phenomena: the fire, the atmosphere, and the regulatory environment surrounding the production of smoke. Each places certain constraints on the management of smoke, each set interacts with the others, and the object of smoke management is to reconcile all of them at one time. The point of interaction is the smoke plume. Fire intensity and fuels determine the character of the emitted matter—the type and amount of gases and particles—as well as the structure of the plume that transmits them to the atmosphere. The properties of the air mass set limits on the dispersion process. The airshed can only absorb so many combustion products, or within a certain rate, and it controls the means by which that absorption takes place. Regulatory agencies, established by law, set the cultural standards for the amount of discharge that will be accepted and control the means by which that discharge occurs. Where much prescribed fire is conducted, or where prescribed fires must coincide with industrial or agricultural burning, some system for airshed allocation will be necessary. For a national summary of prescribed fire and smoke production, see Figure 10–4.

Smoke management programs manipulate all three processes. Fire intensity can be varied through suitable fuel treatment, prescriptions, and ignition patterns. Dispersion can be encouraged by scheduling prescribed fires such that they are set under favorable atmospheric conditions. Regulation can be accomodated through participation in air quality boards, the use of permits, and conformity to approved guidelines. A fire use plan will calculate the type, amount, and rate of discharge for combustion byproducts; determine what the structure of the smoke plume will be, and how the trajectory of that plume will affect target areas; and, in conjunction with regulatory bodies, will decide whether the temporary discharge of smoke, when compounded with other sources of pollutants, will push air quality beyond acceptable limits (Figure 10-5).

*Fire and Fuel.* Fire and fuel interact to generate byproducts. The rate of discharge, or *emission rate,* may be defined as the weight of suspended particulate matter produced per unit length of fireline per unit of time. For a given fuel, the proportion of fuel load (available fuel) that is emitted tends to be constant, a quantity known as the *emission factor.* The emission rate is the product of this factor and fire behavior characteristics. In qualitative terms, combustion is distinguished as either flaming or smoldering, fire spread as either heading or backing, and the smoke plume as either convective or nonconvective. Obviously, a fire may show different characteristics at different times in its history. A heading fire may flame and show good convective lift during its active period, then smolder with little convective lift at a later time. For intense fires, convection may be organized into a column, carrying particulates and gases to great height before releasing them to the action of atmospheric processes. For less intense fires, the particulates will be dispersed by surface winds.

These properties can be converted into quantitative statements in various ways. Multivariate regression equations can describe empirically derived emissions rates for flaming and smoldering combustion. Where stylized fuel models are used, it is possible to assign constant emission factors for each state of burning. Fuel moisture then becomes the principal fuel variable which influences emission. For a given fuel model, then,

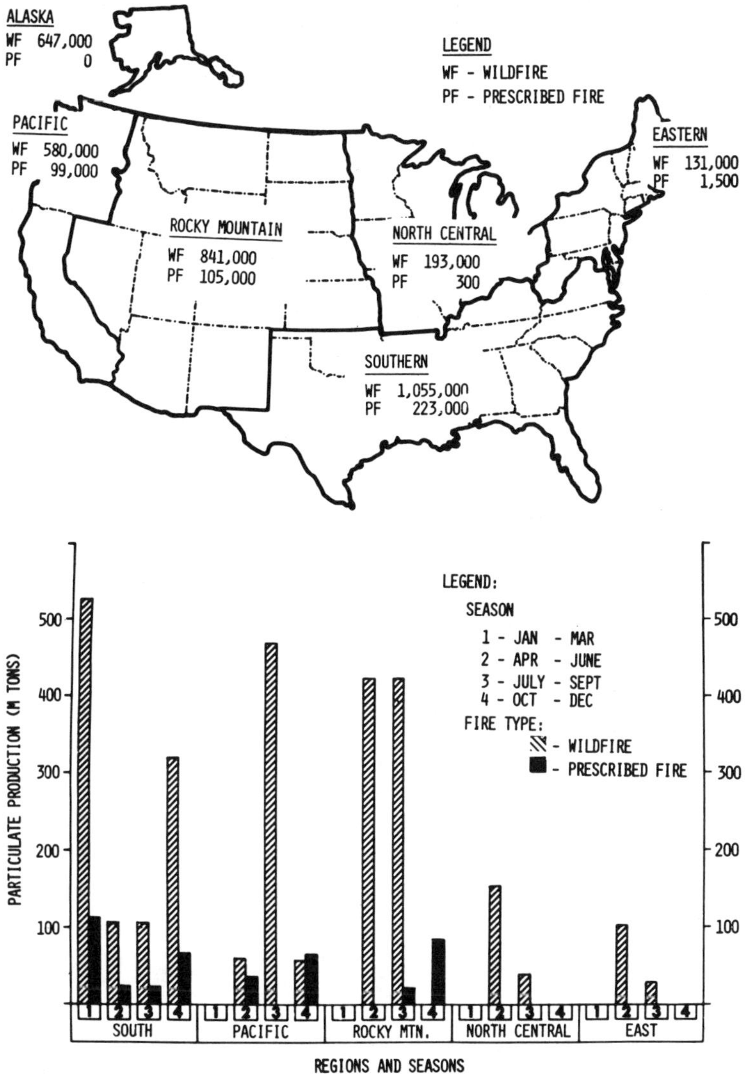


*Figure 10-4. Prescribed fire and air pollution: national and regional statistics. From Sandberg et al. (1979)*

it is possible to determine constant emission factors according to the mode of combustion, mode of spread, and mode of convection (Southern Forest Fire Laboratory, 1976). With further computer modeling, such charts will be replaced by precise quantitative estimates for each fire.

*Fire and Air.* The atmospheric variables that govern dispersion are mixing height, the speed and direction of transport winds, and the presence of local wind systems (Figure 10-6). The vegetative canopy, moreover, intervenes in several ways. It influences the surface winds, thus affecting the fire. It defines the roughness of the fuel–air boundary, introducing turbulence and defining the effective lower boundary of the mixing layer. *Mixing height* describes that portion of the air mass that is unstable, those strata that will support

| Parameters | Variables |
|---|---|
| *Fuel* | |
| Total litter layer FMC | Previous litter layer FMC, age of rough, previous day's duration of precipitation |
| Fuel loading | Fuel type, age of rough, stand basal area, understory height (palmetto only), average dbh (logging residue only), cords cut (logging residue only). |
| Available fuel | Fuel type, fuel loading, total litter layer FMC |
| Emission factor | Fuel type, combustion stage, age of rough, burning method |
| Fine fuel moisture | Temperature, relative humidity, sky condition (grass only) |
| Stand characteristics | Preburn inventory or preharvest inventory |
| *Fire* | |
| Fire phase | Heat release rate |
| Rate of spread | Fine fuel moisture, windspeed at midflame height, fuel type |
| Combustion stage | Fire behavior |
| Length of fired line | Prescription, plot geometry (for ring fires only) |
| Heat release rate | Available fuel, rate of spread, length of fired line |
| Particulate matter emission rate | Available fuel, rate of spread, emission factor |
| *Air Mass* | |
| Mixing height | Observed and forecast weather |
| Transport windspeed and direction | Observed and forecast weather |
| Stability class | If not forecast by NWS: solar angle (shadow length), cloud cover and height, 10-m windspeed |
| Target-area background concentrations | Effects of other emissions sources |

*Figure 10-5. Smoke prediction. Summary of parameters needed to forecast downwind concentration of smoke from prescribed fire. From Southern Forest Fire Laboratory Personnel (1976).*

vertical movement. The height of the mixing layer will be set by the temperature structure of the air mass and its lower boundary by the vegetative canopy. An inversion, if present, will cap the layer. Within this zone vertical movement will follow according to lapse rates. The mixing layer, in brief, defines that portion of the whole air mass which is available for interaction with the smoke plume.

Transport mechanisms include atmospheric turbulence and the speed and direction of the ambient winds. Collectively, they describe how well a smoke plume will be broken up, and how far and in what concentrations particulates will be carried. Turbulence describes the degree of mixing possible, transport windspeed the distance the diluted smoke will travel, and transport wind direction where those emissions will go. Recall that the atmosphere is stratified, that wind speed and direction will vary with height. The plume may travel in directions and at rates quite different from those apparent at the fire.

Local wind systems, including those winds created by the fire, can modify these processes. Of special concern is the character of the plume. If it organizes into a convective column, smoke can

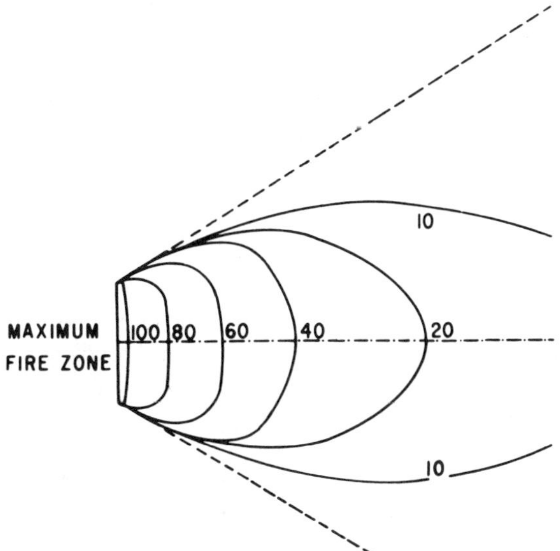


*Figure 10-6. Smoke dispersal. (Upper) Photo shows convection upward, horizontal transport, and rapid dispersion. (Lower) Idealized graph of downwind dilution. From Southern Forest Fire Laboratory Personnel (1976).*

resist transport at lower, surface levels, break through weak or surface inversions, and effectively extend the mixing height. In general, the greater the height of the plume the better the prospects for dispersal. The plume height, moreover, is determined by fire intensity, not by the depth of the mixing layer. By varying fire intensity it is possible to "manipulate" some of the properties of the air mass.

Most mathematical models for smoke dispersion follow formulas developed for point source emission by industrial combustion. The initial pollutants are concentrated both in chemistry and in place of emission. There are problems with extending such models to prescribed wildland fires. Some difficulties are inherent in the model, simplifying assumptions that limit nearly all models for meteorological phenomena. Some are peculiar to the question of particulate dispersal, assumptions about how homogeneous the mixing process is. To counter this difficulty smoke models tend to view the atmosphere not as a tank, within which smoke is distributed uniformly, but as a box, for which dispersion occurs along an axis determined by the transport winds.

Other problems pertain to the mechanics of open burning. A fire is not a point source, and only small, intense fires with convective columns approximate an industrial smokestack. Only a portion of all the smoke generated by combustion is entrained into a plume. With low-intensity fires, with strong surface winds, or with large portions of the burning zone given over to smoldering combustion, only a fraction of the smoke is concentrated. For precribed fires in the South, the proportion of entrained to nonentrained smoke is 60:40. With higher intensity fires, more smoke is entrained and ejected away from the surface. Most models assume that the mode of combustion is flame, that the mode of convection is a developed plume, and that the effective range of dispersion is under 65 miles (105 km). The models work best for slash and grass fuels for which the fire is intense and the burnout rapid. Broadcast underburning, however, may smolder in duff and heavy fuels for weeks, and most of the serious air pollution episodes associated with prescribed fire have occurred with prolonged residual burning (Southern Forest Fire Laboratory, 1976).

*Regulatory Context.* Excepting only the question of liability for escaped fires, no aspect of prescribed fire experiences more legal constraints than does the production of smoke. Beginning in 1955, congressional legislation has mandated ever more stringent air quality standards. The Environmental Protection Agency oversees the provisions of the Clean Air Act, largely through the establishment of primary and secondary National Ambient Air Quality Standards and approval of mandatory state air quality plans. Administration comes through the establishment of Air Quality Control Regions (AQCR) and, where deterioration of existing airsheds is expected to occur, through Air Quality Maintenance Areas (AQMA). Prescribed wildland fire is considered as open agricultural burning and, for most regions, wildland agencies can obtain variances for burning.

In its favor, prescribed burning is an episodic rather than a continuous source of pollutants, its primary constituents (particulates) are more of a nuisance than a hazard to public health, and prescribed burning is not extensively practiced near populated areas. Against it, however, is the likelihood that prescribed burning will become more widely and routinely practiced, that it can contribute significantly to overcharged airsheds during pollution alerts, that it compromises many of the scenic values for which wildlands are managed, and that future legislation may further confine its use, especially when alternative methods exist for fuels management. What is considered good weather for the control and predictability of prescribed fire is often bad for its dispersal. In some parts of the country, such as the Central Valley of California and the Willamette Valley of Oregon, open burning for any purpose is strictly regulated. In other locales—from Phoenix, Arizona to Jackson, Wyoming—smoke from prescribed burning has resulted at times in a significant deterioration of visibility, with drops in tourism and even public health problems.

Virtually everywhere prescribed fire, whether conducted by private parties or by public agencies, requires permits. In the old days, the chief purpose of the permit system was to regulate fire. Increasingly, it is to regulate smoke, and control over the permits by state forestry departments has passed to, or is shared with, air quality boards. Since the number of days suitable for prescribed burning is limited, most agencies do their burning at the same time. To cope with this overload, interagency burning boards are becoming common as a means of coordinating programs.

*Techniques of Smoke Management.* Smoke can be managed by selecting favorable atmospheric conditions and fires with suitable burning charcteristics. These are expressed as a prescription and incorporated into a fire use plan. For the most part, atmospheric conditions are opportunistic. Long-term climatological analysis can estimate how many appropriate days exist for prescribed burning and how, over the course of a season, these days are distributed. More control exists over the fire, through intervention into its fuel complex; by manipulating the fire, some modifications can be gained over the air mass. In part, too, these choices belong with the fire agency, and, in part, with the larger regulatory bodies responsible for air quality.

Favorable atmospheric conditions include a deep mixing layer, reliable transport winds blowing in a desirable direction, and local wind systems that will assist rather than retard dispersion. As the engineering adage goes, the solution to pollution is dilution. Lots of available airshed, good ejection through that airshed, turbulence to assist mixing, and strong winds to carry particulates all help disperse particulates. Gases are swamped quickly.

The most serious difficulties are associated with stagnant high pressure cells and inversions, both of which have the effect of compressing the available airshed. The two phenomena may even be related, with an upper-level inversion caused by subsidence associated with high pressure. But inversions may be local, too, a product of certain topographic configurations such as mountain valleys that are susceptible to strong evening inversions. In the first case, dispersion is limited by the depressed height of the mixing layer; in the second, by the topographic confinements that restrict broadscale circulation. The tendency in valleys to form strong local inversions often restricts underburning or its agricultural analogue, field burning. Instead, smoke-free incinerators may be needed, or intense burning promoted, as with slash, such that a convective column can punch through the shallow ceiling of the inversion.

More manipulation is possible with the fire. Its intensity can be predicted, and its behavior modified through an appropriate prescription, fuel treatments, and ignition strategy. Good smoke management is generally promoted by flaming fires that exhibit convective lifing and show little residual smoldering. Such criteria select for a backing fire over a heading fire, fine fuels over large, and low fuel moisture over high. Ideally, this means firing off a burning block during a single burning period. Perhaps the most important tradeoff is the choice between a heading and a backing fire. Heading fires burn hotter, induce greater convection, and spread faster. But they also generate more particulate matter, lead to scorching, and are harder to control. Where grass is the primary fine fuel, burning plans often call for a combination of fires—backing fires around the perimeter, followed by heading fires through the interior. Where slash is the principal fuel, area ignition is often preferred—not only for better smoke discharge, but for the higher intensities that will promote the consumption of larger-diameter fuels.

## *10.3 PRESCRIBED FIRE TACTICS: SAMPLE FIRE USE PLANS*

Prescribed fire is nothing if not varied. In the purposes to which it is applied and the biota in which it occurs, its realm is vast, and its range and techniques promise to expand even further. As with wildfires, local circumstances may be critical in determining fire behavior and fire effects. Yet similarities are evident, too. All fires require fuel, and hence selectively consume and shape the vegetative cover through which they burn. All show combustion properties common to free-burning fires. For a given regime, similar kinds of prescribed fires will reoccur just as similar kinds of wildfire do. Compared to wildfires, however, prescribed fires show greater predictability. Their behavior is more uniform, their effects more consistent, and their distribution in time more regular.

By way of illustration, consider the following sampling of typical fire uses drawn from different regions of the United States. The survey is neither intensive nor exhaustive. Not included, for example, are those fires set for purposes of fire research, a common form of prescribed burning. The synopses should convey, nonetheless, a sense of the possibilities and the adaptability of prescribed fire for land management, and of the varied means by which prescriptions are expressed.

### *Hazard Reduction: Northern Rockies Slash*

Prescribed fire in larch–Douglas-fir slash has two purposes: it helps prepare sites for replanting, and it reduces fuel loads for better fire protection. Fire management is intimately related to timber management. Cutting blocks will become burning blocks, preparations for debris disposal and fire protection must coincide with plans for replanting and harvesting, and the fire cycle depends on the timber cycle.

Site preparations include a fuel inventory, weather readings, and control lines (Figure 10-7). Occasionally, fuels may be rearranged to ensure more even burning and prevent undue ground scorching. Burning blocks range in size from 100 to 400 acres (40–162 hectares), with fireline width on the order of 12–20 ft (4–6 m). Logging and skid roads will provide the basic grid of control lines, and bulldozers will put in the secondary lines. The burning of large diameter fuels is unpredictable, usually occurring only with droughts, so the fire prescription will focus primarily on fine fuels and duff and secondarily on fuel particles under 4 in. (13 cm). Removal of these fuel size classes will eliminate most of the fire hazard. Excluding duff, fuel loading may vary from 60 to 105 tons/acre (22–39 metric tons/hectare); duff will range from 15 to 29 tons/acre (5.5–11 metric tons/hectare). Paradoxically, although the fine fuels present the greatest hazard, they are also essential to carry the fire through the coarse fuels. Most sites require one to two summer months of drying, and most burns are scheduled for spring or fall within a year after logging.

Generally accepted prescriptions call for relative humidity of 20–50%; wind speed of 0–10 mph (0–16 km/hr); and a 10-hr TL fuel moisture of 6–15%. Considering small diameter fuels as a whole, moisture content between 10 and 17% gives the best burning. Within this range the fire will consume approximately 78% of small diameter fuels (DeByle, 1981; Wright and Bailey, 1982).

Firing operations seek to burn off the site with flaming combustion and a single, central convective column. Typically, ignition begins towards the end of the daily burning period, during the lull in the diurnal cycle of local winds. Winds are light, humidity low, and temperature at its peak. Ignition patterns vary. A common mode is to burn out a wide swath on the uphill or downwind side of the block. This may be done by strip firing or by an electronic ignition system. The flanks are fired next, and finally the downhill or upwind side. An intense, rapidly moving heading fire cleans out the interior of the block, dispersing particulates high up into the atmosphere by means of a well developed convective column. Since the residence time is long, the result is an area fire. Curiously, the amount of duff consumed remains constant despite the ignition pattern.

For average-sized burning blocks fired under unexceptional conditions, control forces can consist of 10–12 firefighters, a bulldozer, and several ground tankers. The chief control problem is usually spotting. The amount of residual burning—and of holdover fires—will depend on the fuel moisture of the larger fires and the extent to which fireline construction mixed logs with dirt,

*Figure 10-7.* *Hazard reduction. Burning larch/Douglas fir clearcuts in the Northern Rockies. (Upper) Burning block and preparations; (lower) high-intensity fire with strongly developed convective column, ignited just prior to sunset. Courtesy U.S. Forest Service.*

thereby replacing flaming combustion with smoldering combustion along the perimeter.

*Fuelbreaks and Type Conversion: Southern California Chaparral*

Fuelbreaks may equally define preattack blocks or burning blocks. Depending on their width they may represent strip or area type conversion. Compared to other forms of prescribed fire, site preparations in the form of fuel treatment are unusually complex. Prescriptions will be sensitive to the available fuel, to the proportion of live to dead fuel, and to the fuel moisture of live and dead fuels.

Generally speaking, the ratio of dead-to-live fuel increases with age, and prescriptions are adjusted according to the age of the fuel complex. Chaparral (chamise) under 20 years of age will not burn without crushing or dessication; chaparral over 50 years old may burn fiercely without any preparations. Most prescribed burns in natural fuels occur in the 35–50 year age group, a fuel complex that is 30–45% dead. Within this range, the respective fuel moistures for live and dead fuels will determine burning properties. Outside this range, special preparations are needed to either enhance flammability or diminish it (Green, 1981; Wright and Bailey, 1982).

Control lines are put in by bulldozer or rely on natural barriers, with a standard width of 10 ft (3 m). If the burning block is a fuelbreak, the material within it is commonly crushed or chemically pretreated. If the block is much larger, then the uphill, downwind side of the block is typically reinforced with an enlarged control line and the interior fuels treated in various ways. The widened fireline may take several forms. It may be a fuelbreak, in which case the burning of the fuelbreak will precede the firing of the entire block. Or it may be a double fireline with an interior of treated brush. The width of such a line will be about 500 ft (152 m), and the interior will be burned prior to using the line as part of a larger burning block.

If the chaparral within the larger block contains less than 30% dead fuels or if it is desirable to extend the timing of the fire outside the normal burning season, some of the interior chaparral should be crushed, too. This increases the proportion of dead to live fuels in the complex, improving its flammability. The entire fuel complex may be crushed or chemically deadened, or 10-ft (3 m) wide strips may be spaced at 66-ft (20 m) intervals, effectively increasing the flammability of the complex and ensuring a reasonably even distribution of dead fuels. Whatever the preparations, fuel treatments should be completed about two months prior to burning. In a typical scenario, this means June or July, with firing underway in August or September.

Planning charts exist with which to match fuels, fuel reduction objectives, and prescriptions (Figure 10-8). A fire of medium intensity would result from the following conditions: dead fuel moisture, 6–8%; live fuel moisture, 60–75%; relative humidity, 18–25%; air temperature, 60–80°F (6–27°C); windspeed, 5–8 mph (8–13 km/hr). Site-specific fuel inventories and fire behavior modeling can refine considerably these generic prescriptions.

Firing operations will vary according to local circumstances. An idealized pattern would begin by burning out the reinforced control lines. In large blocks, with wide lines, this burning may be conducted some time in advance of the main fire. For the interior block, strip firing will start a backing fire along the reinforced line, then proceed to flank firing, and conclude with firing along the downhill, upwind side. The interior will be swept by a heading fire. The afternoon, when burning conditions have stabilized, is a good time to burn. To fire a block in a single burning period, the size of the project will be limited to 40–150 acres (16–61 hectares) for handcrews. With aerial ignition, notably the helitorch, the prospects exist for significantly enlarging this unit size. Postburn activities may include reseeding. For the fuelbreak, the fire will occur within a cycle of maintenance burning.

| Age of stand (years) | Proportion of dead fuel (percent) | Prescription number, by fuel reduction (percent) | | |
|---|---|---|---|---|
| | | 75 or greater | 50 to 75 | Less than 50 |
| 7 to 20 | 5 to 20 | 1 | 1 | 1 |
| 20 to 35 | 20 to 30 | 2 | 3 | 4 |
| 35 or more | 30 to 45 | 5 | 6 | 7 |
| 50 or more | 45 to 65 | 8 | 9 | 10 |

| | Prescription | | | | | |
|---|---|---|---|---|---|---|
| | 2 | 4 | 5 | 7 | 8 | 10 |
| Windspeed, mi/h | 10 | 6 | 8 | 3 | 6 | 0 |
| Relative humidity, percent | 17 | 28 | 25 | 40 | 35 | 70 |
| Air temperature, °F | 90 | 65 | 85 | 60 | 75 | 40 |
| Fuel stick moisture, percent | 5 | 8 | 7 | 10 | 9 | 16 |
| Live fuel moisture, percent | 60 | 65 | 65 | 75 | 65 | 85 |

Note: Prescription 1 should include crushing, or spraying with herbicide. This brush is too green for successful prescribed burning. It may also be too flexible for successful crushing. Prescriptions 2 and 3 will also need treatment if the dead fuel is in the lower part of the range.

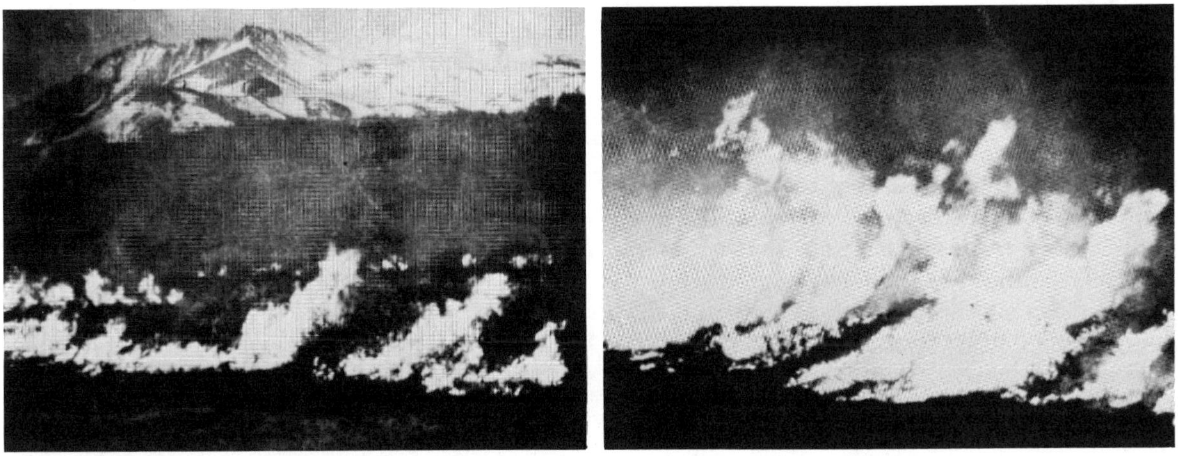

*Figure 10-8. Chaparral burning. (Upper) Method for adjusting prescription to fuel reduction and age of stand; (lower), center firing with expanding perimeter of ring fires. Note the excellent indrafts. From Green (1981).*

### *Underburning: South Carolina Pine*

Broadcast burning in southern pine plantations has for its purpose silvicultural improvement. The concern is not, as with slash disposal, for site preparation but, as with fuelbreaks, for the perpetuation of existing conditions—with maintenance burning rather than with restoration burning. In longleaf pine, fire at the proper stage can control brownspot disease. But in all the southern pines, fire retards those successional tendencies that would replace pine with hardwood, and it reduces natural fuel accumulations: the duff, grasses, and hardwood–shrubby understory known as the rough. Even in managed stands fine fuels build up rapidly: 10–14 tons/acre (3.7–5 metric tons/hectare) over a 10–15-year period. In dense forests, the increase is even greater. Moreover, most of this snowballing fuel load is available fuel, so that fire intensities increase proportionately. To be effective, prescribed underburning must proceed at a rate equal to the rate of fuel accumulation; a 3–5-year or 5–8-year cycle is common, depending on the particular species involved. The cycle will begin once the saplings have achieved a suitable level of protective maturity (Wright and Bailey, 1982).

Few preparations are necessary beyond normal fuel and weather monitoring and the clearing of control lines by tractor-plow. Where the trees in this case have been deliberately planted, burning blocks can be built into the plantation design. Firelines are generally spaced 65–130 ft (20–40 m) apart. Prescriptions call for a relative humidity of 30–50%; temperature between 20 and 50°F (7–10°C); winds from 2 to 10 mph (3–16 km/hr); recent rain, 24–48 hours previous to firing. Since backing fires are normally used, winds limit the amount of scorching that can occur to standing trees. The requirement for recent rains satisfies two conditions. Since burning is done during the winter, the rains will occur with the passage of cold fronts. The rain dampens somewhat the volatility of the rough, although fine fuels will have dried out sufficiently to carry the fire. And the passage of the front ensures reasonably stable weather for several days. If the burn were started prior to frontal passage, the fire might burn uncontrollably under the impress of prefrontal winds, and then expire with the arrival of the front and rain. For smoke dispersal considerations, the prescriptions of the *Southern Forestry Smoke Management Guidebook* published by the U.S. Forest Service may be followed.

Firing operations commence with a strip fire along the control line of the downwind flank (Figure 10-9). Similar strips are ignited along the interior lines within the block. Firing coincides with the burning period. Backing fires are the norm; and if the rough is more than four years old or the winds light, backing fires are necessary to prevent severe scorching. But where the prescription is marginal (cool), where stronger convective development is desired (for smoke dispersal), or where time for firing is short, heading fires may be used. For a heading fire, ignition begins with strip firing along the downwind side, then proceeds around the flanks, and finally concludes with strip burning along the upwind side. With backing fires, 1000 acres (405 hectares) constitutes a day's burn. Where heading fires are used, the burning block may be larger, or more blocks ignited by a single crew. These procedures have become routine; nationwide, more acres are broadcast burned under

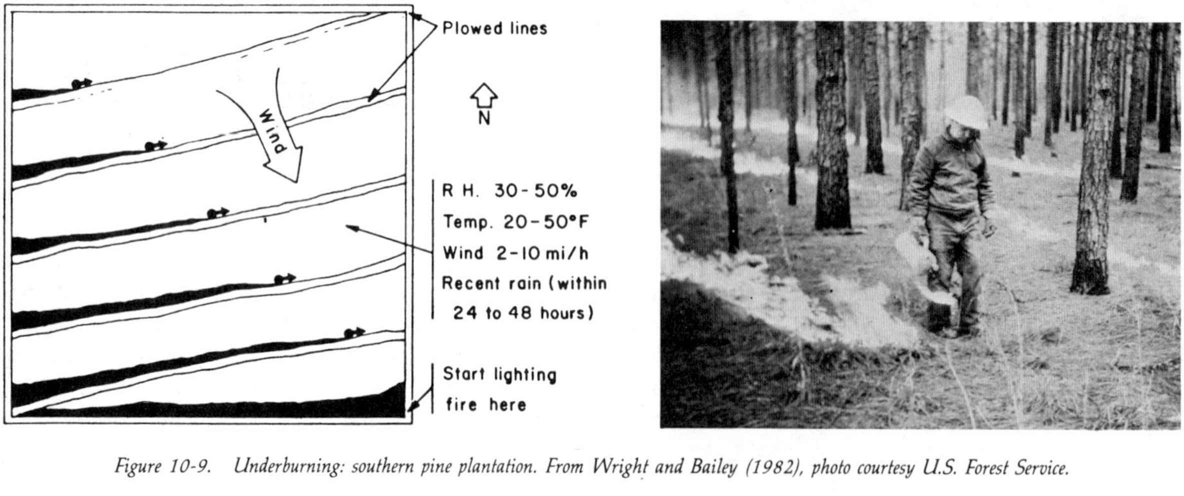


*Figure 10-9. Underburning: southern pine plantation. From Wright and Bailey (1982), photo courtesy U.S. Forest Service.*

this formula than by any other means, for any other region, or for any other purpose.

### *Underburning: Arizona Pine*

The techniques and purposes for broadcast burning in ponderosa pine closely resemble those for underburning in southern pine plantations. The chief differences for the Southwest are that the understory is not equivalent to the rough, that the opportunities for intensive fuel management will be less because the stand develops under a natural regime rather than by planting, and that the state of weather, rather than the stage of the rough, will guide prescriptions. In the South, fire intensity is largely controlled by fuels; in the Southwest, by weather. Burning blocks will conform more to topography, using natural barriers rather than interior control lines. Burning prescriptions will be more sensitive to weather, with less opportunity to modulate burning properties by fuel modifications. The object of fire is to shape the natural biota—to thin thickets of saplings, to reduce natural accumulations of fuel, to prepare sites for natural regeneration—not to replace it with semidomesticated species. Consequently, fewer fire and fuel variables can be directly manipulated.

Two general forms of prescribed fire are practiced. To remove heavy fuels or isolated clusters of debris, these concentrations may be separately ignited, a practice known as *jackpot burning.* To remove continuous, surface fuels over larger areas, *broadcast burning* is the norm. Often the two styles are used in cooperation, though the prescriptions for each varies. Jackpot burning requires that heavy fuels be dry and fine fuels moist; broadcast fire, that the fine fuels upon which the fire depends for continuous propagation be dry. For most broadcast burns, backing fires are used, with or without interior control lines. Where grass is the principal fine fuel, a sequence of backing fire, flanking fire, and heading fires around the perimeter may be followed depending on the desired fire intensity or the need to burn larger acreages in shorter periods of time.

Where fuels are moderate, ponderosa pine tolerates a relatively wide range of prescriptions (Figure 10-10; SWIFCO, 1968; Wright and Bailey, 1982). Perhaps the most important fuel criteria is whether the primary surface fuel is grass or duff (needles). For the Southwest, a reasonable prescription might specify a relative humidity between 20 and 40%; temperature of 40–60°F (4.4–16°C); 10-hr TL fuel moisture of 8–12%; wind of 5–15 mph (8–24 km/hr) for open areas and 2–7 mph (3–11 km/hr) for closed stands. Most prescribed burning occurs in the fall, when the weather has stabilized after the summer fire season. Unfortunately, this is also the time when high pressure cells stagnate over the region, leading to unfavorable circumstances for smoke dispersal. That the burning occurs in the mountains, and the urban populations concentrate in the valleys, means that inversions can seriously compromise smoke management plans.

### *Habitat Maintenance: South Florida Wetlands*

Prescribed fires are used in Everglades National Park, Big Cypress National Preserve, and conservation areas administered by the Florida Forest Service to perpetuate, as far as possible, natural conditions. The region has been severely disturbed by anthropogenic activities. Urban development, extensive drainage for agriculture and housing, incendiarism, the introduction of exotics (including fire weeds like maleleuca and Australian pine) all have upset natural patterns of fire, and constrain the kinds of prescribed fire that may be substituted for it. Fire alone did not upset the natural state, and the reintroduction of fire, natural or prescribed, will not by itself reinstate that prior condition. Yet as long as the land is to be kept as wildlands, some form of fire will be essential. Different fires are used for different components of the biota; but for purposes of illustration, consider that special biota which gave the Everglades its name, the sedge known as sawgrass.

The fire management plan encourages prescribed natural fires and tolerates accidental igni-

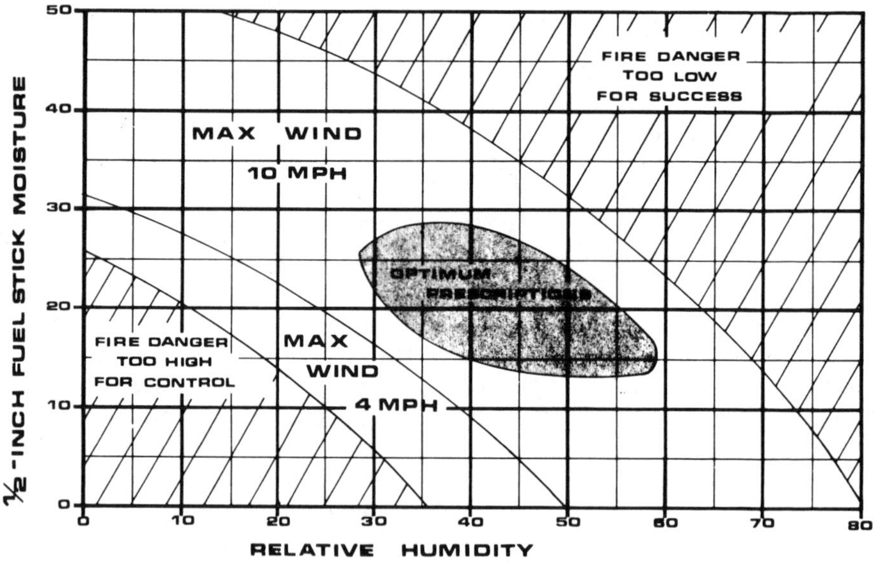


*Figure 10-10. Underburning: southwestern ponderosa pine. Prescription graph from Southwest Interagency Fire Council (1968).*

tions from humans if they burn within prescription. But considerable burning is also conducted through a program of planned ignition. Natural barriers furnish firelines, so preparations concentrate on monitoring those environmental conditions that are fundamental in determining fire effects—weather and moisture levels. Weather data will relate to questions of control and, with Miami nearby, problems of smoke dispersal. Moisture refers to soil moisture, both short-term and drought. By and large the biota of the glades reflects the dual influences of water and fire. For proper fire effects, the two must interact within certain ranges. Too much flooding reduces the potential for fire, too little water allows for deep ground burns in organic soils or the destruction of those organic islands, rich in exotic organisms, known as hammocks. The moisture question is compounded by the extensive drainage that has occurred around the preserves. During the dry season, the drought index, soil moisture, and water table are determined every 10 days at various sites. Remote sensing for these calculations has been attempted with Landsat imagery.

For Everglades National Park, the firing of the glades occurs in two forms. One is to burn along the park boundary, a general kind of fuelbreak, on a cycle of one to two years. The interior sections are then fired on a two to five-year cycle, a frequency that will prevent decadence of the sawgrass community. Lightning fires occur during the summer storm season, and prescribed anthropogenic burning follows during the winter dry season.

Control lines in the form of normal firelines are not used. Burning along the boundary provides a general fuelbreak. For interior control lines an airboat, by cruising back and forth across a strip of grass, can depress and soak the grass into a kind of wetline from which firing may commence. Otherwise differentials in biotic communities and water levels provide natural burning blocks. Further variability may be introduced through spot ignition patterns, and aerial ignition is widely used. Prescriptions call for relative humidity below 60%; winds of 10–15 mph (16–24 km/hr), with transport away from urban concentrations; a live to dead fuel ratio of 2:1; and a sufficiently high water table to prevent ground fires. This requires a soil moisture index above 67% for hammocks, a drought index below 600 for other organic soils, and a water table above 1.8 ft (55 cm). With high moisture levels, prescriptions may be broadened without loss of control or adverse effects; with low moisture, they must be narrowed (Bancroft, 1976; Wade et al., 1980).

### *Prescribed Natural Fire: Bitterroot-Selway Wilderness Area, Idaho*

What most differentiates fire use plans for prescribed natural fires (PNF) from other forms of prescribed burning is the question of ignition. Most plans carefully specify ignition timing and patterns. For prescribed natural fires, however, ignition is "unscheduled," "unplanned," opportunistic. This is not to say that it is unwelcome or unanticipated. On the contrary, the purpose of PNF plans is to promote natural fire for wilderness areas. But control over the important fire variables is minimal: timing is a probabilistic event, ignition takes the form of a point source rather than a pattern, and fuel pretreatment is minimal. This means that prescriptions must be broadly conceived, that areas designated for PNFs must be generously large, and that reliance on stochastic events for ignition must be acceptable. All of these conditions must be contained in an approved plan. There is no need to fine-tune a burning prescription: the variety of the burn, with its multiple effects, is a part of the desired outcome.

The common form for such a prescription is a decision flow chart (Figure 8-11). For each event, there is a go or no-go choice. The chief determinants are the source of ignition, which usually must be lightning; the location of the initiating fire, which must be within a fire management area; the burning properties, which include the ignition and energy release rate (intensity) components of the NFDRS fuel model applicable to the site; and the regional fire load or preparedness level, which assesses a unit's suppression capabilities should conditions later require suppression rather than observation. If the fire loads are high, an initiating fire may be suppressed immediately. Otherwise, the fire and site variables will be monitored. If the fire threatens to leave its zone, some level of suppression may be taken to keep it within the fire management area. Obviously, natural fire zones work best when they coincide with natural barriers. For a given fire management area and fire season, there will be all manner of responses. So, too, for a given PNF.

During 1979 in the Bitterroot wilderness areas, out of 20 fire starts, 11 were suppressed. Seven of these fires had intensities that put them outside of prescription; two occurred in proximity to the zone boundary; two were shut down for reasons of "public safety." From August 10 to 27, the regional forester ordered the suppression of all new starts because of heavy fire loads. No starts occurred. Of the eight fires left to burn, two eventually merged into a single fire and accounted for most of the activity. For 106 days the merged fire burned over 14,900 acres (6,030 hectares). Limited suppression to the extent of 4.5 miles (7.2 km) of handline was taken to protect scattered structures and prevent the fire from entering concentrations of heavy fuels that, it was projected, would generate unacceptable amounts of smoke (Thomas and Marshall, 1980).

### *Type Conversion: West Texas Rangelands*

Prescribed fire in West Texas rangelands takes two forms. It is applied for restoration, in which fire assists type conversion to grass, and for maintenance, in which fire perpetuates a grassy habitat. The second usually follows the first. For restoration burns site preparation requires killing and rearranging the dominant woody species. In Texas this means mesquite and juniper; further west, it may include pinyon pine as well. Through overgrazing and fire protection such woody species have taken over vast amounts of the western range, but through fire—if coordinated with other techniques—much of this range can be reclaimed.

The preparation of the fuel complex is basic to success. The techniques resemble those used for chaparral fires. Using bulldozers, brush and trees are piled, chained in situ, or pushed into windrows. This has two outcomes. It transforms live, standing fuel into dead, surface fuel, and it allows grasses to reestablish themselves. After three to five years there may be sufficient grassy cover to sustain broadcast fire between the woody fuels. Without such cover, piles and windrows are nothing more than slash deposits, each of which requires separate ignition. The purpose of the burn, however, remains the disposal of these woody fuels. The width of control lines will vary with fire intensity. For the downwind side, reinforced lines on the order of 500 ft (152 m), with burned out interiors, are common. Within this line the heavy fuels, perhaps piled, can be burned in May or early June when the grass is green with spring flushing; the fire will not spread. Prescriptions call for relative humidity between 40 and 60%; temperature, 40–60°F (4.4–16°C); and winds under 10 mph (16 km/hr).

The following January or February, when the grass has cured, the remainder of the firelines can be burned (Figure 10-11). A couple of days after the passage of a cold front, the residue of the burning block—its interior—can be fired. The procedure repeats that used for burning off the reinforced fireline. The piles are ignited first, then the perimeter. Where the grass is spotty, heading fires driven by stiff winds may be necessary for the fire to spread satisfactorily. The fire lurches rather than flows from fuel clump to clump. Where chaining rather than piling was used, the entire fuel complex can be treated as a single unit, subject to a single firing rather than a sequential firing. Prescriptions for burning out the interior call for relative humidities of 25–40%, temperatures of 70–80°F (21–27°C), and winds from 8 to 15 mph (13–24 km/hr; Wright and Bailey, 1982).

In pinyon–juniper complexes, where fine surface fuels can be sparse, interior strips may be chained, much as with chaparral. Along these chained swaths, separate strip firing can produce small heading fires of sufficient intensity to spread through the complex. For a prescription, a simple index has been developed. Sum up the numerical values for maximum windspeed (mph), shrub and tree cover (%), and air temperature (°F). If the sum is less than 110, the fire will burn but not spread. If greater than 130, the intensity and spotting potential are too great for control. Between these limits, however, the fire will spread successfully, perhaps assisted by occasional retorching (Bruner and Klebenow, 1979).

### *Preservation Burning: Iowa Prairie*

Historically, the tallgrass prairie relied on fire for its perpetuation. Where prairies are managed to recreate their appearance at the time of European discovery, prescribed fire is an essential tool. It favors native prairie grasses over exotic grasses,

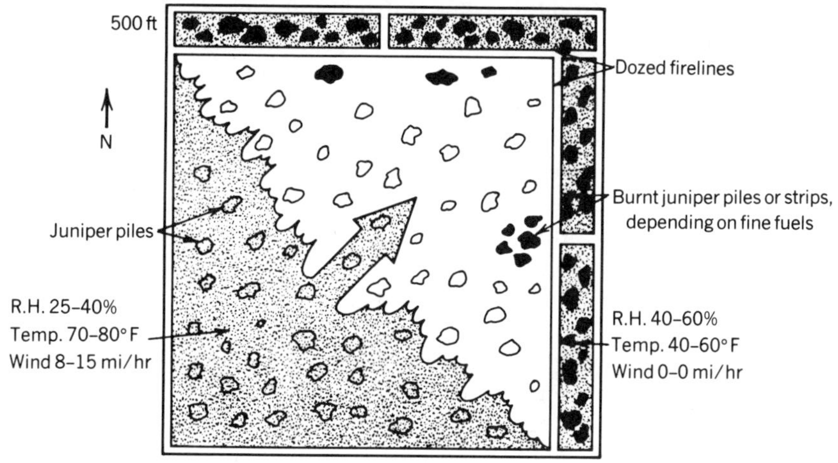


*Figure 10-11. Type conversion: woody vegetation to grass. From Wright and Bailey (1982); photo courtesy of Texas Tech University.*

and it keeps woody invaders in check. Once a prairie has been successfully established, the same regimen of maintenance burning can promote both goals. But to reestablish a prairie that has been taken over by exotics and trees demands slightly different approaches. The control of woody species requires somewhat hotter fires; control over exotic grasses and forbs requires fires of greater frequency. The timing of the fires, on a seasonal and annual basis, is critical (Christiansen, 1981). The object of prescribed fire is to recreate an historic scene, and the set of practices that it involves may be termed *preservation burning.*

Preburn preparations include sampling of vegetation to determine the nature of exotics, if any, and to set a suitable schedule for burning. This schedule will be based on a biological calender. Nearly all European exotics are cool-weather species that sprout early and persist late in the growing season, whereas native species flourish within a shorter summer season. Early spring burning will inhibit the cool-weather species and improve the site for native species; the increased vigor of native plants will help to keep the exotics under control. Accordingly, most prescribed fires occur between late March and late April, with the site burned annually or two years out of every three. If woody vegetation is to be treated, then some mechanical or chemical pretreatments will be necessary to kill the vegetation and put it on the ground.

The prescription will be more sensitive to biological timing (relative to annual growth cycles) than to physical properties (relative to fire behavior). Traditionally, prairie managers have not calculated normal fire behavior characteristics, simply burning when the biological timing was good and the grass would flame. Fire behavior forecasting, however, is comparatively simple. Intensity is determined by rate of spread, and rate of spread is directly related to fuel load and windspeed. As wind or fuel increase, so does fire intensity. Nearly all fuel is available fuel, and nearly all wind (up to about 35 mph; 56 km/hr) acts directly on the flame.

Properly ignited, control may be a relatively simple matter. Firelines will vary according to the fuel involved. If woody vegetation is prominent, then the treatment will mimic that described for mesquite–juniper type conversion. If only tallgrass prairie is involved, then firelines may be minimal—roads, trails, cowpaths, mowed swaths, or wetlines. Firing proceeds with a backing fire along the downwind perimeter, then flank fires, and finally a heading fire to burn out the interior. The speed of the burn will be rapid; a 100-acre (40-hectare) site may be fired in less than half an hour. Wind shifts account for most difficulties in controlling the flaming fire; smoldering dung or glowing woody fuels that may be blown across control lines are responsible for most holdover fires (Wright and Bailey, 1982).

## *10.4 PRESCRIPTION FAILURES: SELECTED CASE STUDIES*

Prescribed burning is no panacea for land management. It is one tool among many. There are times and places where it is inappropriate, and there are occasions where, as in any operation, breakdowns occur in the planning and execution of a burn. The failures of prescribed fire are not so well documented, perhaps, as the successes. The value, and for some managers the necessity, for prescribed burning has naturally led to an emphasis on its possibilities rather than its problems. Of one common failure—poor ignition and the inability to burn well—there may be no record at all. Of others, such as escaped fires, the full impact may be masked by not recognizing the fires as something other than the prescribed fire that initiated the situation or by reclassifying them as wildfires.

Breakdowns may occur at any stage in the evolution of a prescribed fire—planning and choice of prescription, site preparations, firing operations, and fire effects. Undesirable outcomes include escape fires, smoke leading to a deterioration of air quality, and fires of improper intensity. In a sense, being unplanned for, these effects are accidental, and like other forms of accidental behavior, they represent not a single cause but a compounding of many causes. A few select case studies illustrate the kinds of problems that can result.

### *Seney Fires, Michigan, 1976*

In the summer of 1976 two prescribed fires began on the Seney National Wildlife Refuge in the upper peninsula of Michigan. Refuge personnel ignited the Pine Creek fire as an experiment in habitat maintenance. Originally designed as a 40-acre (16 hectare) burn, actual ignition was kept to 1 acre (.4 hectare) when it was pointed out by state fire officers that the primary fuels included peat and that drought was deepening. The fire became uncontrollable, gnawing inexorably through organic soil. A few days later (July 30) lightning started the Walsh Ditch fire within a wilderness area of the refuge. The lightning fire was placed under aerial surveillance. All attempts to reconnoiter the fire on the ground were frustrated by dense growth and high water. The fire grew steadily but not explosively in size. It was expected that the fire would expire on its own. Consequently it was treated as, in effect, a prescribed natural fire.

For neither prescribed fire, however, was there a written prescription, a statement of objectives, nor contingency plans in the event of an escape. The refuge did not participate in the NFDRS or calculate moisture indices; personnel—none of whom were career fire managers—were not familiar with the kinds of fires that can, on occasion, occur in the region. Unable to penetrate by foot to the fire scene, uncertain about funding for suppression (logistical costs would be high), preoccupied with the Pine Creek burn, removed from normal fire danger rating calculations, believing that heavy tracked equipment would do more damage to the refuge than the fire, and somewhat isolated from the national fire organization, refuge officials allowed the Walsh Ditch fire to spread. Within three weeks both the Pine Creek and Walsh Ditch fires had grown well beyond local capabilities. Michigan Department of Natural Resources units began suppression, nearby national forests sent some assistance, and BIFC sent fire specialists to assist with management. The Walsh Ditch fire was 1,800 acres (782 hectares) in size; within another week, it had grown to over 20,000 acres (8094 hectares), crossed refuge boundaries onto State and private lands, and inspired a major mobilization of national fire resources.

Circumstances were not favorable for controlled fire. Humidity stayed well below 30%, fine fuel moisture remained below 10%, winds were frequently stiff, and the Keetch–Byram drought index continued to soar, exceeding 4.00 during the first major run of the Walsh Ditch fire. In retrospect it is is obvious that both fires should have been suppressed at origin. Without the apparatus with which to monitor fuels and weather, without experienced personnel, without written guidelines or contingency plans for escape fires, it was only a question of time until the proper combination of circumstances created a major incident. Those circumstances arrived on August 22. The fire raced 6–8 miles (10–13 km) along a well-developed head, left the refuge for state lands, and a LFO was mobilized under the direction of a project fire team dispatched from BIFC.

Rather than invade the wilderness area of the refuge and attempt the suppression of ground fire, the remainder of the wilderness area was included within a proposed burnout—adding 35,000 acres (14,145 hectares) and nearly doubling the size of the Walsh Ditch fire. Meanwhile, the Pine Creek fire was extinguished, after a complicated flooding process, at 200 acres (81 hectares). At 64,000 acres (25,900 hectares) on September 7 the Walsh Ditch was declared out and demobilization began, though the fire still glowed deep in the peat like a slow match. Five days later conditions worsened, winds rose, the fire reburned old areas (common in organic soils, where each pass of the fire dries out the soil beneath it), jumped control lines, and made a 7-mile (11 km) run on a mile-wide front. Mobilization began again, the fire was contained at 72,500 acres (29,341 hectares) along a perimeter of 88 miles (142 km), and arson fires sprang up. Prior to the Walsh Ditch fire, the local communities had been experiencing a small economic depression which the flood of money associated with fire control partially reversed; as long as more wildfires continued, there remained an incentive for incendiarism. Eventually, snowfall arrived in mid-October, the drought index dropped, and the fires went out. Suppression costs soared to $9 million (Fish and Wildlife Service, 1976).

*Pocket Fire, Georgia, 1979*

In January, 1979 prescribed burning was conducted along sections of the Okefenokee National Wildlife Refuge. The fire was ignited along a road, then allowed to back through heavy rough towards the interior of the refuge where it gradually expired among bogs. The fire was planned as the first stage—the establishment of a wide fuelbreak—for a larger burn. The remainder of the area was burned on February 14. By now the wind had shifted so that the previously burned zone would not be subjected to heading fires. The new burn approached the old, but considerable unburned fuels remained between the two when winds shifted suddenly on February 15. Flame height multiplied spectacularly as the surface fire, under the impress of the wind, roared into the dried but unconsumed crowns of the gallberry stand that comprised the old burn. The fire leaped across the road, then returned to the status of a surface fire.

A tractor-plow and tractor-harrow unit responded. The plow reworked an old firebreak along the flank of the escaped fire, and the harrow followed, widening the line. Again, the wind shifted. Flame heights on the order of 3–5 ft (1–1.5 m) swelled to 40 ft (12 m); rates of spread of several feet per minute accelerated so rapidly that the fire covered 150 yards (137 m) in seconds, burning ¾ of a mile (1.2 km) in 15 minutes; previously unavailable fuel (the upper rough) became involved. The tractor operators, working alone, found themselves along the new flaming front. The flames swept over the tractor-plow like a surf. The harrow operator fled from his machine and dashed back to the road, suffering minor abrasions and burns. The plow operator took refuge under his tractor until the flaming front had passed. When word of the catastrophe became known, state and industry cooperators rushed suppression units to the scene. The entire episode from flareup to control lasted less than two hours. Final size of the escaped fire was 125 acres (51 hectares). The tractor-plow operator died.

The fatality occurred on a wildfire, not a prescribed burn, but the prescribed fire was the source of the wildfire and the contributing factors in the failure of initial attack were, perhaps, the result of the fire's being attacked as a prescribed burn rather than as a wildfire. Had the escape fire not evolved from a prescribed burn, it might have been handled differently. But the belief that this was only a slopover from the prescribed fire perhaps suggested that it was not, in itself, a dangerous fire. The fire agency, in brief, failed to make the transition from prescribed fire organization to wildfire organization (Fish and Wildlife Service, 1979). The episode thus illustrates the problem with holdover fires, with the presence of unburned fuels that can encourage unacceptable fire behavior as environmental conditions change, and with the fireline hazards possible in prescribed burning. More and more fire-related fatalities are the product of prescribed fire operations, or of the transition between prescribed and wildfire operations.

In the case of the Pocket fire, that transition was delayed; the prescribed fire did not immediately change into a large wildfire. This is the typical situation with holdover fires. Such a fire may be handled like the first report of a wildfire, subject to normal initial attack procedures. More common as a safety problem, however, is the situation in which the transition comes abruptly. (Recall the Mack Lake fire described in Chapter 2.) Abandoning the original prescribed fire to attack the escaped fire may mean that still more fires escape, that organization may be inadequate for the escape fire, that resources designed for a controlled fire may attempt to control a wildfire—often a fire making the transition to large fire status. Fire agencies have, over the years, learned how to size up wildfires and to make the transition from initial attack to large fire organization. It remains for them to learn the analogous skills needed for sizing up escape fires and to execute the evolution from prescribed fire organization to wildfire organization.

### *Gallagher Peak Fire, Idaho, 1980*

The fire started from an early July lightning storm over the Targhee National Forest. Smoke rose out of a designated fire management area. The High Country FMA was defined by its elevation; above 8000 ft (2438 m), natural fires were allowed to follow their own course. The strategy was a common one: high elevation basins, it seemed, carved up the land into natural burning blocks. Nor was the Gallagher Peak fire an isolated event. A second PNF burned on another FMA in the Targhee, and over the course of the next month numerous others began throughout Idaho. None presented problems. Burning in light, sparse fuels, the Gallagher Peak fire doubled in size almost daily. Aerial surveillance was conducted, escaped fire analyses written up, and the decision made to allow the fire to burn, regardless of size, unless it traveled below the 8000 ft (2438 m) elevation that defined the FMA. There was some ground observation; the light grassy fuels, however, argued for letting the fire continue. On July 21–22 almost 0.40 in. (1 cm) of rain fell over the region, the burning index plummeted to zero, and no smokes could be seen within the fire. On the 26th three small smokes reappeared, wholly contained within the fire perimeter. The fire smoldered but did not expire.

Then traditional August fire weather set in. The region had already endured a stubborn drought. Now, after the ephemeral shower, the air warmed and dried, fine and intermediate-sized fuels were drained of moisture, and winds rose. Upper-air troughs passed over the region, and strong winds rushed out of the southwest. A wildfire on the Challis National Forest (the Mortar Creek fire) blew up. The Gallagher Peak fire built up heat, spread from 300 to 500 acres (121–202 hectares) in one afternoon, and the next day, August 4, made the transition to large fire. Private lands were threatened, and suppression was ordered. With temperatures in the 90s (30°C), relative humidity below 10%, and winds gusting to 40 mph (64 km/hr), the fire raged over 50,000 acres (20,235 hectares)—though, because of spotting, only 37,000 acres (14,974 hectares) were actually burned. The fire left the High Country FMA. An LFO was activated. The fire was contained on August 9, although mopup continued until August 17, when rains assisted in extinguishment. Elsewhere in the Northern Rockies the decision was made to suppress all new fire starts in FMAs until conditions abated. No new starts were reported. Meanwhile, the other PNF burning on the Targhee puffed up to 0.5 acres (0.2 hectares), then, unassisted, went out.

The Gallagher Peak fire was a prescribed natural fire that exceeded its prescription and escaped its zone. Following the fire, and in conformity with new policy guidelines promulgated in 1978, the fire management plan for the forest was rewritten. The revisions included a more precise prescription: in particular, it recognized the 1000-hr TL fuels. New methods for calculating the fuel moisture of 1000-hr fuels had been published, also in 1978, as part of revisions of the NFDRS. The 1000-hr fuels were integral to the history of the blowup. Prior to ignition they had plummeted to 8%. They held fire tenaciously when the fine fuels, dampened by rains, would not burn. The BI was zero after the July rainstorm, but the 1000-hr fuel moisture content only increased to 10%. The initiating fire did not spread, but it would not go out. Glowing in the heavy fuels, it persisted until, with a change in weather, the fine fuels again dried out and carried the fire. The revised fire plan put any fire that burned with a 1000-hr fuel moisture content under 10% out of prescription.

The fire served as a reminder, too, that hold-over fires are a serious problem—whether they are held within cat-faces, in a snag, organic soils, or slash piles; whether they occur along the fire-line perimeter of a wildfire or deep in the interior. The fire is a reminder, too, that coarse fuels can act as a heat source as well as a heat sink. In this case their role was, first, to act as a slow match and, second, to increase the residence time of the flaming zone. Whether such a fire rekindles is often the result of a competition between the fine fuels, which must carry the fire, and the coarse fuels, which hold the pilot flame.

### *Vista Fire, Arizona, 1980*

There were several purposes to the burn, ignited in late September on the North Rim of Grand Canyon. It was conceived as an initial phase in the preparation of a large forested peninsula for prescribed natural fire. The preliminary fire would reduce fuels, widening a road into a fuelbreak adequate to isolate the PNF zone. It would test prescriptions. And it would convey experience to the organization, one not especially well-practiced in prescribed fire.

Normal preparations were taken. Fuels were thinned along the fire roads used for firelines, snags near the line were dropped, weather and fuel data were collected. The size of the burning block was set by an economy of scale. Without natural barriers (other than roads), without mechanized line construction equipment, and without large handcrews, it would be easier to hold a big burning block defined by a road system than a smaller area block which would have a proportionately higher ratio of perimeter to construct and hold. The fire would be confined between a paved road on the east and dirt fireroads on the west. The burning plan called for the perimeter to be fired by hand, with aerial ignition of the interior. No special calculations were made for smoke management. It was felt that the burning would proceed quickly and that there would be little residual fire. Smoke would be temporarily dense, but soon vanish.

The general plan was approved. But the aerial ignition project was subsequently vetoed, and it was urged that the burning (originally planned for October) should be moved up to September because seasonal labor, upon which the fire program was dependent, would be terminated early. When the burning commenced, the firing along the paved road went smoothly. Forest Service ground tankers assisted in the holding operation. Tourist cars were convoyed through the active area by park rangers. Changing winds, however, made firing along the dirt road more troublesome, and ignition was temporarily shut down to wait out the remainder of the burning period.

At this point the organization began to lose its nerve. No further firing was allowed. That night, as an evening inversion developed, smoke poured over the rim into the Canyon. Much of it streamed into the developed trans-Canyon corridor. There were complaints about smoke from Park employees stationed within the Canyon, complaints from visitors troubled over the large smoke plume, inquiries from Congressmen (at the request of constituents then visiting the Canyon), and questions posed by various Park officials not affiliated with the fire program. Especially in view of national debates which then raged over some proposed coal-fired power plants in the region (with the prospect of industrial air pollution), it was considered impolitic to persist with the Vista burn.

Obviously, better smoke calculations should have been made prior to ignition. But had the original fire use plan been followed, smoke would have been a short-term irritant rather than a long-term nuisance. Once the Park decided that the smoke was, in fact, intolerable, it needed either to complete the burn as rapidly as possible or order an all-out suppression. Instead, with different portions of the Park administration seemingly at cross-purposes with one another and with inexperienced fire overhead making the critical decisions, the worst option was pursued. A large fire with lots of fuel was left to smolder. The burn flamed during the day and smoked at night, as it crept over the considerable unlit portions of the site. Eventually, the fire spotted across some meager firelines in the south of the burning block that had never been blacklined. Park officials ordered the fire to be extinguished. One firefighter, the solitary remainder of the seasonal crew, drove a ground tanker to the scene daily to mopup a 2000+ acre (800+ hectare) fire.

There have been other smoke-related episodes involving the NPS—around Grand Tetons, Yosemite, and Saguaro. And there have been serious air pollution alerts over the Southwest when, with a high pressure cell locked in the Four Corners area, tens of thousands of acres burned across the national forests and Indian reservations of the Mogollon Rim. All of these smoke-related problems were met and solved—the Park incidents without eliminating a prescribed natural fire program, and the burning along the Rim without the abolition of broadcast fire as a management tool. The aftermath of the Vista fire, however, left the fire management program at Grand Canyon in uncertainty. It was decided that future burning would take place only in small burning blocks by low-intensity fires that could be suppressed at the end of each burning period, with daily mopup if necessary. This would make large-scale broadcast burning impossible. It equally jeopardized the PNF program of which the Vista fire was a part. With each natural fire, as for the Vista fire, evening inversions would spill smoke into the Canyon. Such fires, moreover, might burn for weeks.

Whatever the final outcome, the fire illustrates the problems that smoke can make for a prescribed burn program, the complications that other parts of a land management agency can bring to the fire program, and the seriousness of events that appear to be purely local. A fire need not command national attention to become a compelling problem. The Vista fire suggests the sort of local dilemma that can profoundly affect a local program. Ultimately, the success of prescribed fire, as with suppression, will be determined by the cumulative successes and failures of many such small projects repeated by many agencies over many areas.

## *REFERENCES*

The literature on prescribed fire, like that on fire ecology, is vast and fragmented. A major encyclopedia of prescribed burning has, however, been assembled in Wright and Bailey (1982), *Fire Ecology.* The book contains prescriptions and burning plans for nearly all the environments in which prescribed fire is used. A good introduction to prescribed burning techniques which exceeds the scope of its title is Martin and Dell (1978), "Planning for Prescribed Burning in the Inland Northwest." Brown and Davis, 1973, *Forest Fire: Control and Use,* gives some treatment of prescribed fire, especially its objectives. Fischer (1978), "Planning and Evaluating Prescribed Fires—A Standard Procedure," describes the evolution of a burning plan. For a general history of prescribed fire use and policy, consult Pyne (1982), *Fire in America.*

Several symposia contain significant amounts of prescribed fire information. See, for example, U.S. Forest Service (1971), "Prescribed Burning Symposium. Proceedings"; U.S. Forest Service (1976), "Proceedings. Fire by Prescription Symposium"; Utah State University and Utah Division of Forestry and Fire Control (1976), "Use of Prescribed Burning in Western Woodland and Range Ecosystems"; Wood (1981), "Prescribed Fire and Wildlife in Southern Forests." The Tall Timbers Fire Ecology Conference Proceedings are rich in prescribed fire references; for a guide, see Fischer (1980), "Index to the Proceedings." For specific, regional references to prescribed fire, see the citations given within the text.

Smoke management is fast becoming a distinctive specialty within fire management. Two outstanding summaries are available. Sandberg et al. (1979), "Effects of Fire on Air. A State-of-Knowledge Review," provides an overview on combustion products and their treatment; the publication contains an excellent selection of references. For more detail on actual techniques, consult the Southern Forest Fire Laboratory (1976), "Southern Forestry Smoke Management Guidebook."

Bancroft, Larry, 1976, "Natural Fire in the Everglades," pp. 47–60, in U.S. Forest Service, "Proceedings. Fire by Prescription Symposium," U.S. Forest Service, Southern Region and Cooperative Fire Protection.

Beaufait, William R. et al., 1977, "Broadcast Burning in Larch–Fir Clearcuts: The Miller Creek–Newman Ridge Study," U.S. Forest Service, Research Paper INT-175, rev.

Brown, Arthur A. and Kenneth P. Davis, 1973 rev., *Forest Fire: Control and Use,* 2nd ed. (New York: McGraw-Hill).

Bruner, Allen D. and Donald A. Klebenow, 1979, "Predicting Success of Prescribed Fires in Pinyon–Juniper Woodland in Nevada," U.S. Forest Service; Research Paper INT-219.

Christiansen, Paul, 1981, "Prairie Management at Herbert Hoover Memorial Site," Unpublished report, National Park Service.

Committee on Fire Research, 1976, "Air Quality and Smoke from Urban and Forest Fires. Proceedings of an International Symposium," (Washington, D.C.: National Academy of Sciences-National Research Council).

DeByle, Norbert V., 1981, "Clearcutting and Fire in the Larch-/Douglas-fir Forests of Western Montana—A Multifaceted Research Summary," U.S. Forest Service, General Technical Report INT-99.

Dixon, Merlin J., 1965, "A Guide to Fire by Prescription," U.S. Forest Service, Southern Region.

Fischer, William C., 1978, "Planning and Evaluating Prescribed Fires—A Standard Procedure," U.S. Forest Service, General Technical Report INT-43.

———, 1980, "Index to the Proceedings of the Tall Timbers Fire Ecology Conferences: Numbers 1–15, 1962–1976," U.S. Forest Service, General Technical Report INT-87.

Fish and Wildlife Service, 1976, Seney Fire reports, assorted.

———, 1979, Pocket Fire reports, assorted.

Furman, R. William, 1979, "Using Fire Weather Data in Prescribed Fire Planning: Two Computer Programs," U.S. Forest Service, General Technical Report RM-63.

Green, Lisle R., 1977, "Fuelbreaks and Other Fuel Modification for Wildland Fire Control," U.S. Forest Service, Agriculture Handbook No. 499.

———, 1981, "Burning by Prescription in Chaparral," U.S. Forest Service, General Technical Report PSW-51.

Heinselman, Miron, 1978, "Fire in Wilderness Ecosystems," pp. 249–280, in John C. Hendee et al., *Wilderness Management*, U.S. Forest Service, Miscellaneous Publication No. 1365.

Hirsch, Stanley N. et al., 1979, "Choosing an Activity Fuel Treatment for Southwest Ponderosa Pine," U.S. Forest Service, General Technical Report RM-67.

———, 1981, "The Activity Fuel Appraisal Process: Instructions and Examples," U.S. Forest Service, General Technical Report RM-83.

Johnson, Von J. and William A. Main, 1978, "A Climatology of Prescribed Burning," pp. 59–62, in Michael Fosberg (chairman), *Fifth National Conference on Fire and Forest Meteorology* (Boston: American Meteorological Society).

Main, William A. et al., 1982, "FIREFAMILY: Fire Planning with Historic Weather Data," U.S. Forest Service, General Technical Report NC-73.

Martin, Robert E. and John D. Dell, 1978, "Planning for Prescribed Burning in the Inland Northwest," U.S. Forest Service, General Technical Report PNW-76.

Miller, Roswell K., 1978, "The Keetch-Byram Drought Index and Three Fires in Upper Michigan, 1976," pp. 63–67, in Michael Fosberg (chairman), *Fifth National Conference on Fire and Forest Meteorology* (Boston: American Meteorological Society).

Pase, Charles P. and Carl Eric Granfelt (coordinators), 1977, "The Use of Fire on Arizona Rangelands," Arizona Interagency Range committee, Publication No. 4.

Pyne, Stephen J., 1982, *Fire in America: A Cultural History of Wildland and Rural Fire* (Princeton, New Jersey: Princeton University Press).

Sandberg, D.V. et al., 1979, "Effects of Fire on Air. A State-of-Knowledge Review," U.S. Forest Service, General Technical Report WO-9.

Sandberg, D.V. and S.G. Pickford, 1976, "An Approach to Predicting Slash Fire Smoke," pp. 239–248, in Tall Timbers Research Station, *Tall Timbers Fire Ecology Conference*, Vol. 15 (Tallahassee, Florida: Tall Timbers Research Station).

Society of American Foresters, Task Force on Clean Air Act Regulations, 1980, "Wildland Fires, Air Quality, and Smoke Management," *Journal of Forestry* **78**(11): insert.

Southern Forest Fire Laboratory Personnel, 1976, "Southern Forestry Smoke Management Guidebook," U.S. Forest Service, General Technical Report SE-10.

Southwest Interagency Fire Council, 1968, "Guide to Prescribed Fire in the Southwest."

Tall Timbers Research Station, 1962–1976, "Tall Timbers Fire Ecology Conferences," Vols. 1–15 (Tallahassee, Florida: Tall Timbers Research Station).

Targhee National Forest, 1980, Gallagher Peak Fire reports, assorted.

Thomas, David A. and Sandra J. Marshall, 1980, "1979—Test Year for Prescribed Fires in the Northern Region," *Fire Management Notes* **41**(4): 3–6.

U.S. Forest Service, n.d., "Fuel Management Planning and Treatment Guide. Prescribed Burning Guide," U.S. Forest Service, Northern Region.

———, 1971, "Prescribed Burning Symposium Proceedings," U.S. Forest Service, Southeastern Forest Experiment Station.

———, 1972, "A Guide for Prescribed Fire in Southern Forests," U.S. Forest Service, Southern Region.

———, 1976, "Proceedings. Fire by Prescription Symposium," U.S. Forest Service, Southern Region and Cooperative Fire Program.

Utah State University and Utah Division of Forestry and Fire Control, 1976, "Use of Prescribed Burning in Western Woodland and Range Ecosystems. A Symposium," (Logan, Utah: Agricultural Experiment Station).

Wade, Dale et al., 1980, "Fire in South Florida Ecosystems," U.S. Forest Service, General Technical Report SE-17.

Wood, Gene W. (ed.), 1981, "Prescribed Fire and Wildlife in Southern Forests," (Georgetown, South Carolina: Belle W. Baruch Forest Science Institute of Clemson University).

Wright, Henry A., 1980, "The Role and Use of Fire in the Semidesert Grass–Shrub Type," U.S. Forest Service, General Technical Report INT-85.

Wright, H.A. and A.W. Bailey, 1980, "Fire Ecology and Prescribed Burning in the Great Plains—A Research Review," U.S. Forest Service, General Technical Report INT-77.

———, 1982, *Fire Ecology. United States and Southern Canada* (New York: Wiley).

Wright, Henry A. et al., 1979, "The Role and Use of Fire in Sagebrush–Grass and Pinyon–Juniper Plant Communities. A State-of-the-Art Review," U.S. Forest Service, General Technical Report INT-58.

# Index